# Organic Reactions

There was an *Oxford Cleric* too, a student,
Long given to Logic, longer than was prudent;
The horse he had was leaner than a rake,
And he was not too fat, I undertake,
But had a hollow look, a sober air;
The thread upon his overcoat was bare.
He had found no preferment in the church
And he was too unworldly to make search.
He thought far more of having by his bed
His twenty books all bound in black and red,
Of Aristotle and philosophy
Than of gay music, fiddles or finery.
Though a philosopher, as I have told,
He had not found the stone for making gold.
Whatever money from his friends he took
He spent on learning or another book
And prayed for them most earnestly, returning
Thanks to them thus paying for his learning.
His only care was study, and indeed
He never spoke a word more than was need,
Formal at that, respectful in the extreme,
Short, to the point, and lofty in his theme.
The thought of moral virtue filled his speech
And he would gladly learn, and gladly teach.

Chaucer, *The Canterbury Tales*.

*Pierre Laszlo* is Professor of Chemistry at both the École polytechnique in Paris, and at the University of Liège, in Belgium. His work, in the area of the methodology of organic synthesis, has led to a number of clay-based reagents and catalysts. He is the author of the books *Preparative Chemistry Using Supported Reagents*, 1987 and, with Maria Balogh, *Organic Chemistry Using Clays*, 1993. He has also published books on the language of chemistry and on science communication for the general public.

# Organic Reactions Simplicity and Logic

Pierre Laszlo
*École polytechnique, Palaiseau, France*
*University of Liège, Belgium*

Based on a course originally taught at the
École polytechnique, Palaiseau, France

*Translated by*
Hans-Rudolf Meier, *CIBA Ltd, Marly, Switzerland*

JOHN WILEY & SONS
CHICHESTER·NEW YORK·BRISBANE·TORONTO·SINGAPORE

*Other Wiley Editorial Offices*

John Wiley & Sons, Inc., 605 Third Avenue,
New York, NY 10158-0012, USA

Jacaranda Wiley Ltd, 33 Park Road, Milton,
Queensland 4064, Australia

John Wiley & Sons (Canada) Ltd, 22 Worcester Road,
Rexdale, Ontario M9W 1L1, Canada

John Wiley & Sons (SEA) Pte Ltd, 37 Jalan Pemimpin #05-04,
Block B, Union Industrial Building, Singapore 2057

*Library of Congress Cataloging-in-Publication Data*

Laszlo, Pierre.
    [Logique de la synthèse organique. English]
    Organic reactions : simplicity and logic / Pierre Laszlo.
      p.   cm.
    Includes bibliographical references and index.
    ISBN 0-471-93933-1 — ISBN 0-471-95278-8 (paper) :
      1. Chemistry, Organic.   2. Chemical reactions.   I. Title.
    QD253.L3813   1994
    547′.2—dc20                                    94-30633
                                                   CIP

*British Library Cataloguing in Publication Data*

A catalogue record for this book is available from the British Library

ISBN 0 471 93933 1 (cloth)
ISBN 0 471 95278 8 (paper)

Typeset in 11/13 Times by Dobbie Typesetting Ltd
Printed and bound in Great Britain by Bookcraft (Bath) Ltd
This book is printed on acid-free paper responsibly manufactured from sustainable
forestation, for which at least two trees are planted for each one used for paper production.

To the virtuosos of the pithy word and of the terse achievement:

E. J. Corey
Emily Dickinson
Richard P. Feynman
Glenn Gould
Seamus Heaney
the two Joes (Montana and Namath)
Vladimir Nabokov


to whom this book owes so much

# MODES OF REPRESENTATION

If you look in old chemistry books
you see
all those line cuts
of laboratory experiments
in cross-section.
The sign for water
is a containing line, the meniscus
(which rarely curls up the walls of the beaker),
and below it
a sea
of straight horizontal dashes
carefully unaligned vertically.
Every cork or rubber stopper
is cutaway.
You can see inside
every vessel
without reflections, without getting wet,
and explore every kink
in a copper condenser.
Flames are outlined cypresses
or a tulip at dawn,
and some Klee arrows
help to move gases and liquids the right way.
Sometimes a disembodied hand
holds up a flask.
Sometimes there is an unblinking observer's eye.
Around 1920
photoengraving
became economically feasible
and took over.
Seven-story distillation columns
(polished up for the occasion),
like giant clarinets,
rose in every text, along
with heaps of chemicals, eventually in color.
Suddenly
water and glass, all reflection
became difficult.
One had to worry about light,
about the sex
and length of dress or cut of suit
of the person sitting at the controls of
this impressive instrument.
Car models and hairstyles
dated the books more
than the chemistry in them.
Around that time
teachers noted a deterioration
in the students' ability to follow
a simple experimental procedure.

Roald Hoffmann
*The Journal*, **11**, 45 (1987)

# Contents

# Preface

This is the translation of my lecture notes from École polytechnique. This uniquely French, two century-old institution, prides itself upon giving its carefully selected and small student body a sound training in the sciences, centered on mathematics and on abstract, deductive reasoning. The trainees from École polytechnique, which Napoleon made into a military school—at present, something like a cross between West Point and MIT—go on to serve in the top technical and decision-making positions in French public administration and business.

Thus the challenge was to provide them with an introduction to modern organic chemistry that would be intellectually rigorous, not betray its experimental and inductive status, and sieve down the proliferation of reactions, reagents, and explanations to a small number of algorithms. It was made easier by another feature, also original to École polytechnique, the existence of an earlier course, taught by Professor Nguyen Trong Anh, initiating students to molecular orbital theory and to the Woodward–Hoffmann rules for electrocyclic reactions.

Back in 1988, the first French edition was entitled *Logique de la synthèse organique*, to emphasize reduction to a small set of mutually consistent and highly logical arguments. Two years later appeared, in a book with the identical title, the detailed presentation of Professor E. J. Corey's retrosynthetic analysis. In order to avoid confusion, we chose to entitle the present translation *Organic Reactions— Simplicity and Logic*.

Another feature distinguishes the present English translation from the French version. Whereas the original scheme was in two volumes, one with the core lectures and the other for the illustrative examples, we have conjoined the subject matter here into a single, unified story. It retains, though, the original two-track format in which the chemical formulas on one page face their commentary on the opposite page. Also the original writing was extremely terse, a house style that students at École polytechnique are familiar with, and which helps them to deal with an encyclopedic body of knowledge.

# Author's Note

A book can be read from page one. It can also be dipped into at random. Because we recognize this duality, we have designed this textbook both for studying and browsing. Each chapter is in two parts. The first part, in large type, presents the core material; what the teacher may want to feature predominantly in a formal lecture. The second part, in smaller type and with flags, presents the general culture material; what any of us would like to answer to a non-chemist friend with questions about the impact of chemistry in everyday life. We hope that this hybrid format will serve the purpose of what used to be termed in French 'instruire en amusant', i.e. knowledge can be (and should be) fun!

I am extremely grateful to Ms. Vivian Torrence for letting me use her collage on the cover. It conveys, together with the central importance to chemistry of highly symmetrical geometrical shapes such as the Platonic solids, the delicate charm, the exquisite refinement of this science: modern organic chemistry has given itself the capability of synthesizing almost any molecular object, under mild conditions, by highly rational methods. It offers a feast to the intellect and to the senses both.

I can do no better than dedicate this book to Professor Jerrold E. Meinwald, of Cornell University: together with his colleague Thomas E. Eisner, he has written an exciting new chapter of organic chemistry, that of chemical communication in nature. Thus, his example was influential in my selection of examples, many of which are taken from biology.

Finally, I wish to thank Professor Roald Hoffmann, also of Cornell University, for allowing me to reproduce two handsome pieces of his writing; Martin J. Röthlisberger and Dr Hans-Rudolf Meier who respectively launched and performed this translation: Valerie, who did the first dummy for the book cover, and who has had to cope at home with an absentee husband who spent weekends typing instead of hiking.

Pierre Laszlo
Summer 1993

# Introduction

Most organic reactions are placed under either charge control or orbital control. This book opens with this fundamental distinction. A relevant notion is that of the softness or hardness of an acid or a base. Indeed, interactions of a hard acid and a hard base are charge-controlled; those of a soft acid with a soft base are orbital-controlled. This introduction will end with some thermochemical data, a powerful tool. Adding together increments for each structural fragment provides an extremely simple evaluation of the enthalpy of formation of a molecule. This exemplifies transferability, a concept basic to chemistry: an atom, or a group of atoms retain their essential properties on passing from one molecule to another.

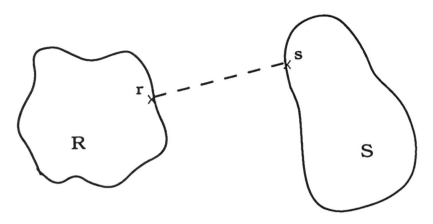

# 1 Orbital Control and Charge Control

Let us consider the supermolecule $[R+S]$ formed by combining two molecules R and S with the atoms r and s facing each other so that the predominant interaction is between these two atoms. There are two limiting cases:

(1) If the frontier orbitals of the donor (let us assume that R is the donor and S the acceptor) have energies of the same order, the predominant interaction is given by the simple perturbation equation

$$E = 2c_r^m \cdot c_s^n \cdot \beta$$

where $c_r^m$ and $c_s^n$ are the coefficients of the atoms r and s in the molecular orbitals m and n and $\beta$ is the resonance integral corresponding to formation of the r—s covalent bond. This is termed *orbital control*. The majority of the reactions to be considered in this text are under orbital control.

(2) If on the other hand, the frontier orbitals of donor and acceptor differ considerably in energy (criterion: $E_n - E_m \gg 4\beta$), electron transfer becomes negligible. In fact, the perturbation equation becomes

$$2(c_r^m)^2(c_s^n)^2\beta^2/(E_n - E_m)$$

If the denominator $(E_n - E_m)$ becomes very large, this covalent interaction term becomes very small. The R,S interaction does not completely vanish. Due to their proximity in space, the atoms r and s, with charges $q_r$ and $q_s$, can form an ion pair. In this second case, the r,s Coulomb interaction is accompanied by a partial desolvation $\Delta_{solv}$, and the variation in energy resulting from the perturbation R + S is of the type:

$$\Delta E = -q_r \cdot q_s \frac{\Gamma}{\epsilon} + \Delta_{solv}$$

where $\Gamma$ is the (r,s) Coulomb repulsion term and $\epsilon$ is the 'effective' dielectric constant (representing the permittivity of the local environment to atoms r and s). Such a situation is termed charge control. Many reactions to be discussed are under charge control. In general, orbital control and charge control coexist in chemical reactions.

Reference: G. Klopman, *J. Am. Chem. Soc.*, **90**, 223–234 (1968).

**Table 1**  Some examples of soft and hard acids and bases

|      | Acids | Bases |
| ---- | ----- | ----- |
| Hard | $H^+$ <br> $Li^+$, $Na^+$, $K^+$ <br> $Mg^{2+}$ <br> $Al^{3+}$ <br> $Ce^{3+}$ | $HO^-$, $RO^-$ <br> $F^-$, $Cl^-$ <br><br> $H_2O$, $NH_3$ |
| Soft | $Cu^+$, $Ag^+$, $Au^+$ $Tl^+$, $Hg^+$ <br> $Br^+$, $I^+$ <br> $Br_2$, $I_2$ | $H^-$, $R^-$, $I^-$, $CN^-$, $SCN^-$ <br><br> $R_2S$, $RSH$, $RS^-$ |

## 2  Hard and Soft Acids and Bases

Let us recall the distinction between charge control and orbital control. The former is observed when a donor (of electrons) has an energetically low frontier orbital (the highest occupied orbital or HOMO) compared to the frontier orbital (the lowest unoccupied or LUMO) of the acceptor. In other words, charge control involves an interaction between a poorly ionizable donor and a reluctant acceptor. Its contribution is increased when the reaction partners (charged, small atoms with low polarizability) are strongly solvated. Small values of $\beta$, the resonance integral between the interacting atoms, also favor charge control. To summarize: charge control involves small polarizable Lewis bases ($=$ donors) and Lewis acids ($=$ acceptors) which are small, charged, and of low polarizability. These are called hard acids and bases.

The opposite properties, large atoms with little or no charge, almost unsolvated and readily polarizable, are the characteristics of so-called soft Lewis acids and bases. The Pearson principle concerning the strength of Lewis bases and acids is therefore a chemical modification of 'similia similibus'.

---

*Note*: Pearson's principle is much older: 'Thus the soft searches for the soft, the bitter leaps towards the bitter, the acidic goes to the acidic and the hot is attracted to the hot', Empedokles, fragment 90.

**Table 2**  Hardness of some cations [Parr and Pearson]

| Cation | Hardness $\eta$, eV |
|--------|--------------------|
| $Li^+$ | 35.1 |
| $Na^+$ | 21.1 |
| $K^+$ | 13.6 |
| $Mg^{2+}$ | 32.5 |
| $Cu^+$ | 6.9 |
| $Ag^+$ | 6.9 |
| $Au^+$ | 5.7 |
| $Tl^+$ | 7.2 |
| $Hg^+$ | 4.2 |

**Table 3**  Hardness of some anions [Parr and Pearson]

| Anion | Hardness $\eta$, eV |
|-------|--------------------|
| $H^-$ | 6.8 |
| $F^-$ | 7.0 |
| $Cl^-$ | 4.7 |
| $Br^-$ | 4.2 |
| $I^-$ | 3.7 |

Hard (soft) acids react preferably with hard (soft) bases. If their hardness and softness match one another, their reactivity is increased. The reactivity is, however, weak whenever a hard (soft) acid is combined with a soft (hard) base. Charge control prevails for 'hard–hard' interactions, whereas orbital control is the rule for 'soft–soft' interactions.

One can define the hardness $\eta$ of a group, at least in the isolated state. Parr and Pearson, in order to unite the concepts of hardness and of electronegativity, have proposed the two complementary definitions:

Electronegativity $\varkappa = (I + A)/2$
Hardness $\eta \qquad = (I - A)/2$

where the electronegativity is half the sum of the first ionization energy $I$ and the electron affinity $A$, whereas the hardness is half the difference of $I$ and $A$. This definition has the merit of being very close to the distinction made by Klopman between charge and orbital control: in fact $I$ and $A$ are related to the monoelectronic energy of the frontier orbitals of the donor and of the acceptor:

$$I = E_{HOMO}$$

$$A = E_{LUMO}$$

Hence the difference $(I - A)$ is simply the gap $(E_n - E_m)$ between the frontier orbitals of acceptor and donor.

References: R. G. Pearson, *J. Am. Chem. Soc.*, **85**, 3533 (1963).
G. Klopman, *J. Am. Chem. Soc.*, **90**, 223–234 (1968).
L. Komorowski, *Chem. Phys. Lett.*, **103**, 201–204 (1983).
R. G. Parr and R. G. Pearson, *J. Am. Chem. Soc.* **105**, 7512–7516 (1983).
R. G. Pearson, *J. Am. Chem. Soc.*, **107**, 6801–6806.
R. G. Pearson, *J. Am. Chem. Soc.*, **108**, 6108–6114 (1988).
L. Komorowski, *Chem. Phys. Lett.*, **137**, 536–540 (1987).
L. Komorowski, *Chem. Phys.* **114**, 55–71 (1987).

**Table 4**  Hardness of neutral atoms (eV)

| | |
|---|---|
| Li | 2.38 |
| B | 4.01 |
| C | 5.00 |
| N | 7.27 |
| O | 6.08 |
| F | 7.01 |
| Na | 2.30 |
| Al | 2.77 |
| Si | 3.38 |
| P | 4.86 |
| S | 4.12 |
| Cl | 4.70 |
| K | 1.92 |
| Cr | 3.05 |
| Ni | 3.24 |
| Cu | 3.25 |
| As | 4.07 |
| Se | 3.86 |
| Br | 4.25 |
| Rb | 1.85 |
| Ag | 3.14 |
| Sr | 3.05 |
| Sb | 3.79 |
| Te | 3.52 |
| I | 3.70 |

M. K. Harbola, R. G. Parr and C. Lee, *J. Chem. Phys.*, **94**, 6055–6056 (1991).

**Table 5** Hardness of atoms in neutral molecules

| Atom | Molecule | Hardness $\eta$ (eV) |
|---|---|---|
| O | $R_2O$ | 16.2 |
| | $R_2CO$ | 18.8 |
| S | RSH | 9.94 |
| | $R_2S$ | 9.87 |
| N | $RNH_2$ | 14.7 |
| | $R_2NH$ | 14.3 |
| | $R_3N$ | 14.0 |
| | $ArNH_2$ | 12.1 |
| | pyridine | 15.0 |
| | RCN | 15.9 |
| P | $R_3P$ | 9.42 |
| F | RF | 45.3 |
| Cl | RCl | 10.9 |
| Br | RBr | 9.54 |
| I | RI | 8.14 |

L. Komorowski, *Chem. Phys.*, **114**, 55–71 (1987).

**Table 6**  Single bond energies (kcal mol$^{-1}$)

|     | H   | C   | N   | O   | F   | Cl  | Br  | I   |
| --- | --- | --- | --- | --- | --- | --- | --- | --- |
| H   | 104 | 99  | 93  | 10  | 135 | 103 | 88  | 71  |
| C   |     | 83  | 83  | 86  | 110 | 79  | 66  | 52  |
| N   |     |     | 38  | 53  |     |     |     |     |
| O   |     |     |     |     | 34  |     |     |     |

**Table 7**  Multiple bond energies (kcal mol$^{-1}$)

| C=C | 143 | N=N | 100 |
| --- | --- | --- | --- |
| C≡C | 194 | N≡N | 226 |
| C=N | 147 |     |     |
| C≡N | 213 |     |     |
| C=O | 178 |     |     |

**Table 8**  Group increments for hydrocarbons (approximate) (kcal mol$^{-1}$)

| | |
| --- | --- |
| $CH_3$—$C_{tet}$ | $-10$ |
| $CH_2$—$C_{tet}$ | $-5$ |
| CH—$C_{tet}$ | $-2$ |
| —$\overset{\mid}{\underset{\mid}{C}}$—$C_{tet}$ | $0$ |
| $CH_2$—$C_{tri}$ | $+6$ |
| CH—$C_{tri}$ | $+8.5$ |

# 3 Thermochemical Data and their Use

Table 6 shows the bond energies of some single bonds in (kcal mol$^{-1}$). These are average energy values for the (homolytic) dissociation of a molecule A—B into atoms. The heterolytic dissociation energies are much higher. To give an example, approximately 70 kcal mol$^{-1}$ is required to dissociate a carbon–bromine bond; the heterolytic dissociation (into $R_3C^{(+)} Br^{(-)}$) requires 220 kcal mol$^{-1}$ because a separation of the charges of opposite sign is necessary.

Table 7 displays the bond energies of some multiple bonds. The bond energies are very susceptible to the coordination of the atom: The carbon–hydrogen bond at a tetrahedral carbon has a bond energy of 92–98 kcal mol$^{-1}$, whereas that of a more electron-attracting trigonal carbon has a C—H bond energy of 108–110 kcal mol$^{-1}$.

References: S. W. Benson, *Thermochemical Kinetics*, Wiley–Interscience, New York, 2nd edn, 1976.
      R. Fuchs, *J. Chem. Educ.*, **61**, 133–135 (1984).
      D. E. Vitale, *J. Chem. Educ.*, **63**, 304–306 (1986).

Table 8 summarizes a certain number of thermodynamic increments characteristic of hydrocarbon fragments. Let us give some examples of its use:

(i) heat of formation $\Delta H^0$ of *n*-butane:

$$
\begin{array}{ll}
H_3C-C_{tet} \times 2 & -20 \\
H_2C-C_{tet} \times 2 & \underline{-10} \\
& -30 \ (-30.36 \pm 0.16 \ \text{exp.})
\end{array}
$$

(ii) heat of formation $\Delta H^0$ of neopentane:

$$
\begin{array}{ll}
H_3C-C_{tet} \times 4 & -40 \\
-C-C_{tet} \times 1 & \underline{\phantom{-}0} \\
& -40 \ (-40.27 \pm 0.25 \ \text{exp.})
\end{array}
$$

(iii) heat of formation $\Delta H^0$ of ethylene:

$$
H_2C-C_{tri} \times 2 \qquad +12
$$

$$
(+12.45 \pm 0.10 \ \text{exp.})
$$

**★★★**

$\Delta H^0 = +39.4 \, \text{kcal mol}^{-1} \, \text{exp.}$

$\Delta H^0 = -55 \, \text{kcal mol}^{-1} \, \text{exp.}$

Such thermochemical data are useful, for example, to predict the direction of a reaction. Therefore, the retro Diels–Alder reaction is a priori disfavored. It will be very slow, except if the initial state is destabilized (e.g. due to Baeyer ring strain) or if the final state is stabilized (e.g. because of the expulsion of small, neutral, and very stable molecules). An example of the latter process, a favorable (=fast) fragmentation is due to the high bond energy of the expelled diatomic nitrogen molecule $N_2$ (adjacent scheme).

**Table 9** Ring strain in cycloalkanes

| $N$ | $H^0$ calc. | $H^0$ obs. | $\Delta$ |
|---|---|---|---|
| 3 | − 15 | + 12.73 | + 28 |
| 4 | − 20 | + 6.78 | + 27 |
| 5 | − 25 | − 18.44 | + 6.5 |
| 6 | − 30 | − 29.50 | 0 |
| 7 | − 35 | − 28.21 | + 7 |

One can apply this additive formalism to the evaluation of (Baeyer) ring strain in cycloalkanes.

Table 9 shows:
  (i) the absence of ring strain in cyclohexane;
 (ii) the high ring strain in small 3- and 4-membered rings (about $30\,\text{kcal}\,\text{mol}^{-1}$);
(iii) non-negligible ring strain ($6-7\,\text{kcal}\,\text{mol}^{-1}$) in cyclopentane and cycloheptane.

**Table 10**   Ring strains of olefin containing rings

| Cyclic olefin | $\Delta$ |
| --- | --- |
| Cyclopropene | 54.5 |
| Methylenecyclopropane | 41.7 |
| Cyclobutene | 30.6 |
| Methylenecyclobutane | 28.8 |
| Cyclopentene | 6.8 |
| Cyclohexene | 2.5 |
| Methylenecyclohexane | 1.9 |
| *cis*-cycloheptene | 6.7 |
| *trans*-cycloheptene | 27.7 |
| Norbornene | 25.0 |

One can consider likewise the cycloalkenes, the Baeyer ring strains for which are listed in Table 10 (in $kcal\,mol^{-1}$).

Let us comment now on the standard formation enthalpies of some isomeric unsaturated hydrocarbons. The stability sequence of these olefins is: tetrasubstituted > trisubstituted > disubstituted > monosubstituted. This sequence stems from hyperconjugation of the alkyl substituents. It is to be noted that an additional alkyl substituent lowers the energy of the system by about $1.5\,kcal\,mol^{-1}$ (as an order of magnitude).

**Table 11** Baeyer ring strain in polycyclic compounds

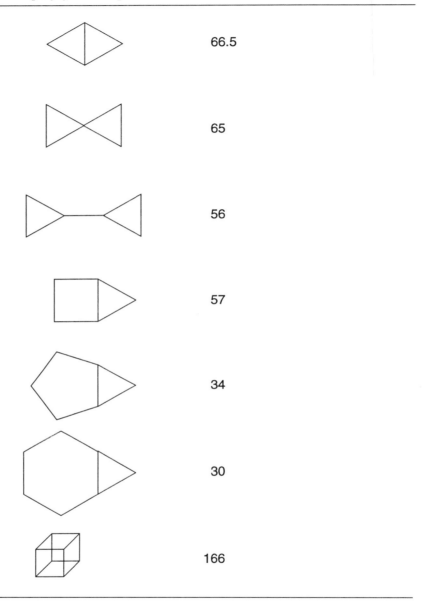

Another set of Baeyer ring strain values concerns polycyclic hydrocarbons including small strained rings (Table 11). For illustration let us quote the systems listed below, the names of which are:

bicyclobutane
spiropentane
bicyclopropyl
bicyclo [2.1.0] pentane
bicyclo [3.1.0] hexane
bicyclo [4.1.0] heptane
cubane

Cubane has a ring strain largely superior to the energy of a carbon–carbon bond; but this strain is distributed over all bonds in the molecule.

There follows a portrait of Berthelot: not only does thermochemistry owe a lot to him, but he was also a lucid, lyrical announcer of organic synthesis at its beginnings.

### Berthelot (1827–1907) (Pierre Etienne Marcelin)

The clever man! Incarnating the republican *savant*, he rushes up the steps of the social ladder: professor at the Collège de France (1865), Senator for life (1881), Minister of Public Instruction (1886–1887), then Minister of Foreign Affairs (1895–1896). Although he magnified his contribution out of all proportion, sometimes crediting himself for the discoveries by others (e.g. for the alcohol synthesis), his scientific work is nevertheless considerable. It extends from calorimetric measurements to studies of enzymes, and includes inorganic chemistry.

Among this large body of work, shine most vividly his first syntheses of organic molecules, during the two decades 1850–1870. These were often very simple constructs: methane; acetylene; hydroaromatic derivatives, glycerides and fats. Berthelot made himself the champion of chemical synthesis. From the hindsight of our historical perspective, we can see that he had the insight of announcing an organic chemistry based on synthesis, after the initial phase of identification and analyses of the natural products. But his narrow-minded positivism, his incurable hostility to atomic theory led him to an erroneous sectarian view of the structures of molecules. Due to these blinkers, he has failed to exert a long-lasting and beneficial influence. Furthermore, Berthelot abused his ministerial power and his authority as a national luminary to block the diffusion of the atomists' ideas in French Universities. Only few chemists dared to disregard his edicts: Wurtz; Barbier; Grimaux (at the École polytechnique). There is little doubt of Berthelot's responsibility for the decline of French chemistry between the two wars of 1870 and 1914. His historical studies, his rather unscholarly edition of Greek and Alexandrian treatises of alchemy, are not to be neglected. Berthelot wrote an elegant and spiced prose which was a major reason for his success:

'What is essential to determine is the fatal succession of changes that matter undergoes, the precise origin of the transformed substances, the influence of the medium, and the circumstances under which the metamorphoses occur'.

# 1.I
# Carbonyl Group:
# Additions and Addition–
# Elimination Reactions

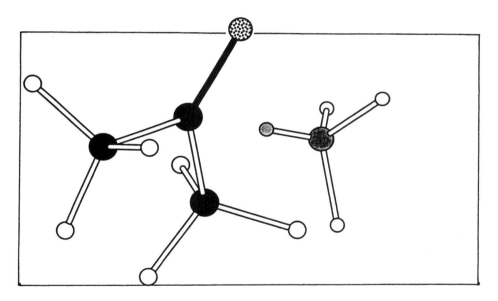

The reactivity of the carbonyl group is at the core of this course. Electrophiles add at the oxygen end of the carbonyl group, whereas nucleophiles, e.g. carbanions, attack the carbon end. This affinity to nucleophiles is used for protection of a carbonyl group.

A fact of primary importance is the destabilization of the tetrahedral intermediate resulting from the interaction of a non-bonding electron pair of a heteroatom with the (vacant) antibonding ($\sigma^*$) orbital of the carbon–oxygen single bond. The hereby defined class of addition–elimination reactions provides a first general synthetic method for the formation of a new carbon–carbon bond.

---

The chapter also comprises useful definitions for the following notions: $R/S$ configuration; enantiotopic and diastereotopic faces; activated methylene group.

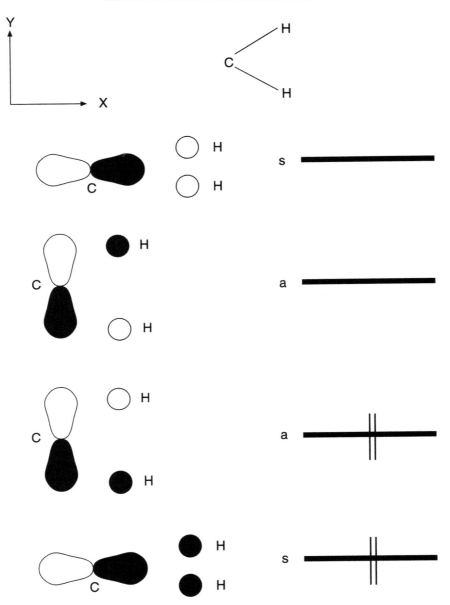

# 1 Carbonyl Group: Simple Molecular Orbital Description

The carbonyl group of aldehydes RCHO and ketones $R^1R^2C=O$ is built from a trigonal fragment RCH or $R^1R^2C$ and an oxygen atom.

This is illustrated for formaldehyde. Let us start with the molecular orbitals of the methylene fragment HCH. To build these, we use the atomic orbitals $2p_x$ and $2p_y$ on carbon and the 1s orbitals for the two hydrogen atoms. The $x$-axis is the HCH bisector and the $y$-axis is perpendicular to it, both axes are in the HCHO plane.

These *four* atomic orbitals give rise to *four* molecular orbitals either symmetric (s) or antisymmetric (a) with respect to the $x$-axis. Their energy increases with the increasing number of nodal planes.

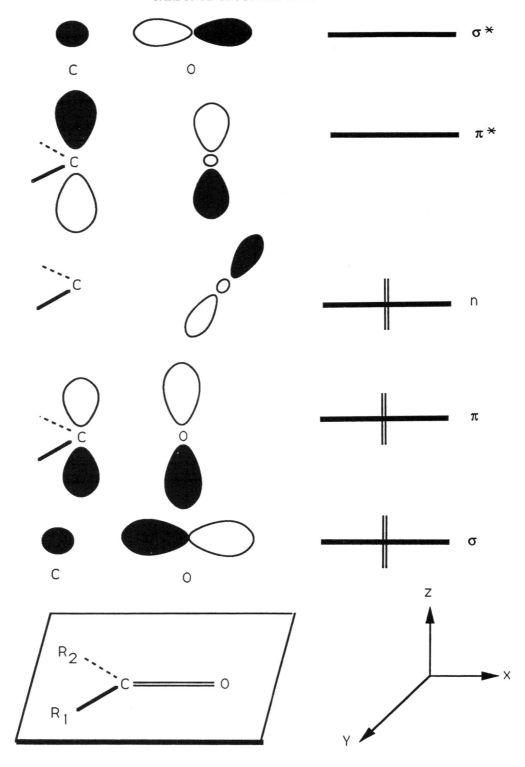

σ*

π*

n

π

σ

It now remains to attach oxygen by means of the 2s and $2p_x$ atomic orbitals on carbon atom and the $2p_x$ and $2p_y$ orbital on oxygen. The combination $2p_x$ (O) $\pm$ 2s (C) gives rise to two new levels $\sigma$ and $\sigma^*$. The second pair of molecular orbitals is obtained by the combination $2p_z$ (O) and $\pm 2p_z$ (C). Since oxygen is more electronegative than carbon, the $2p_z$ (O) level is lower in energy than the $2p_z$ (C). Hence the $\pi$ molecular orbital is nearer in energy to the $2p_z$ (O) level: it has a large coefficient at oxygen, a small coefficient at carbon; the opposite is true for the $\pi^*$ MO which resembles more the $2p_z$ (C) level. Finally, the n level, to first approximation, consists of the pure $2p_y$ (O) atomic orbital. The other two valence electrons remain in the 2s (O) atomic orbital.

This description entails two differing types of non-bonding electron pairs at oxygen: 2s and 2p (n). The HOMO is the non-bonding orbital n with a zero coefficient at carbon. The LUMO is the antibonding $\pi^*$ with a large coefficient on carbon and a small coefficient on oxygen (the molecular orbitals of the fragment $CH_2$ are in the nodal plane $xy$ with a zero interaction with the $\pi$ and $\pi^*$ orbitals of the fragment C=O).

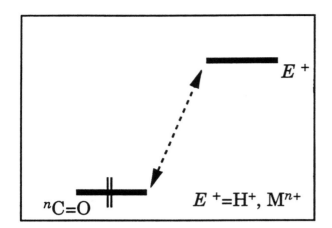

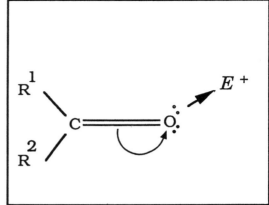

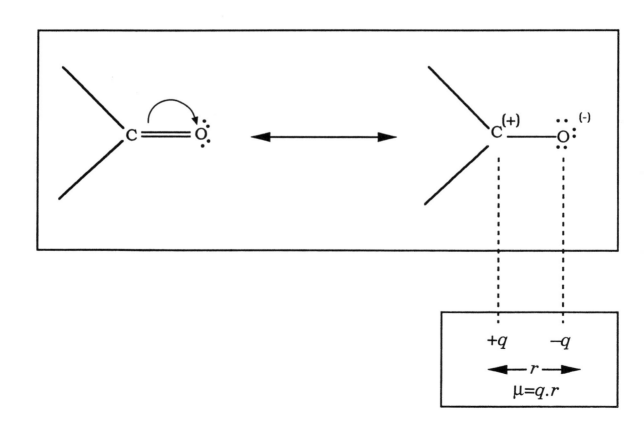

# 2 Reactivity of the Carbonyl Group: Addition of Electrophiles

Since the highest occupied orbital (HOMO) identifies with the n level, the predominant interaction with an electrophile will be with this orbital. Thus, the electrophile tends to attach itself to the oxygen atom since the n level is practically a 2p orbital on oxygen. Hence acidic particles such as protons and metallic cations should add to the oxygen atom. This indeed occurs. Furthermore, addition of an electrophile $E^+$ to the carbonyl enhances polarization of the C=O double bond. Even in the absence of an electrophile, the electron density is greater on oxygen: this accumulation of charge stems from the two non-bonding electron pairs localized at oxygen (2s and n), together with larger coefficients of both $\sigma$ and $\pi$ orbitals at oxygen. One should also keep in mind polarization of the $C-R^{1(2)}$ bonds and the fact that the lone pairs shift the center of charges towards oxygen.

It is common to express this polarization of the carbonyl group in the language of resonance. There is resonance between two forms, an ylene form without charge separation and an ylide form with a positive charge at carbon and a negative charge at oxygen. This charge separation creates an electric dipole. The dipole moment $\mu = q \cdot r$ is defined as the product of the absolute value of the charges $\pm q$ and the distance that separates them. Dipole moments $\mu$ measured for aldehydes and ketones are in the range 1.7–3.7 (Debye, CGS units); this translates to values between 0.4 and $0.9 \, e \cdot Å$, to use a more telling unit. The C=O bond length is about 1.20 Å. This means that the charge at both ends, carbon and oxygen, has a value between 1/3 and 3/4 of an electronic charge, which is considerable.

The resulting dipole–dipole interactions, overall attractive, between carbonyl compounds in the condensed phases are expressed as an increase of the melting and boiling points relative to hydrocarbons with the same number of carbon atoms. Hence these compounds are typically liquids at normal temperatures, and they dissolve other polar molecules and even certain salts: for instance acetone $(H_3C)_2C=O$ is one of the most common organic solvents.

| 1 | $EtO^{(-)}$ | $n\text{-BuLi}$ | $PhCOCH_2^{(-)}$ |

| 2 | $F^{(-)}$ | $NH_3$ | $PhS^{(-)}$ |

| 3 | $t\text{-BuO}^{(-)}K^{(+)}$ |

| 4 | $(i\text{-Pr})_2N^{(-)}Li^{(+)}$ |

# 3 Basicity and Nucleophilicity

A useful distinction is that between *bases*, their role being restricted to removal of protons, and *nucleophiles* (also known as Lewis bases), adding to atoms that offer them vacant orbitals. Given this distinction between bases and nucleophiles, how can we predict whether a reagent functions as a base by removing protons, or as a nucleophile by attacking electrophilic carbons? A first criterion is basicity: basicity and nucleophilicity are related properties, they go hand in hand. As to nucleophilicity, strong bases rank first, e.g. ethoxide anion, *n*-butyl lithium, the conjugated base of acetophenone. Weak bases such as fluoride anion, thiophenoxide and ammonia are, consequently, weaker nucleophiles; they rank second in nucleophilicity.

A second criterion is steric hindrance about the atom bearing the non-bonding electron pair(s) responsible for the basicity as well as for the nucleophilicity. When such hindrance is large, access to a carbon atom is much more impaired than that to the much smaller proton. Hence basicity is essentially not affected, whereas nucleophilicity drops. A sterically hindered base such as potassium tertiary butoxide ranks third in nucleophilicity. Bases with very strong steric hindrance, e.g. LDA (lithium diisopropylamide) rank fourth in nucleophilicity. To first approximation, they behave exclusively as bases, and not as nucleophiles.

---

Compare Chapter 8.II, complementary indications about nucleophiles.

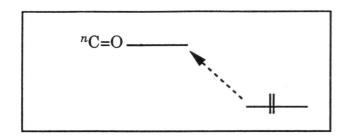

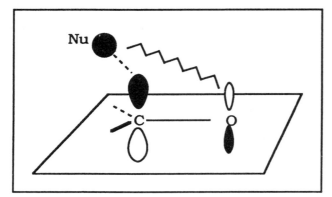

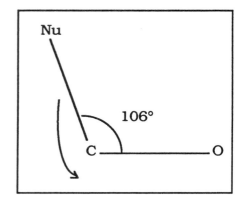

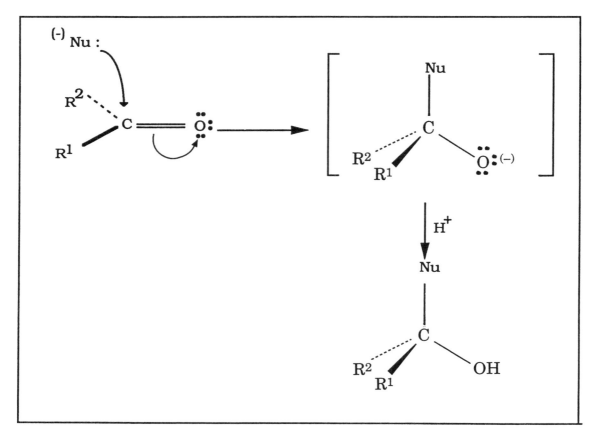

# 4  Reactivity of the Carbonyl Group: Addition of Nucleophiles

Since the energetically lowest unoccupied molecular orbital (LUMO) is the $\pi^*$ level which has a large coefficient at carbon, the predominant interaction with a nucleophile occurs with this orbital and involves mainly the carbon atom. The nucleophile approaches the carbonyl group: as soon as it enters into bonding overlap with the $2p_z$ atomic orbital at the carbon atom, this is automatically accompanied by an antibonding interaction with the $2p_z$ atomic orbital at the oxygen atom (remember that these two orbitals have opposite phases in the $\pi^*$). To minimize the antibonding interaction (nucleophile–oxygen) while at the same time maximizing the bonding interaction (nucleophile–carbon), the angle of attack O—C—Nu has to be obtuse: actually, it is about 106°, from both experimental and theoretical determinations. We shall refer to it as the Bürgi-Dunitz angle.

In the resonance formalism, addition of a nucleophile to the carbonyl group transfers one of the lone pairs of the nucleophile onto the carbonyl oxygen: the lone pair of the nucleophile serves to form a new $\sigma$ bond (Nu—C) and the $\pi$ electron pair migrates from carbon to oxygen, in such a way as to maintain the maximum of eight electrons around the carbon atom. In other words, the carbonyl carbon atom, initially trigonal, becomes tetrahedral. In the case of a negatively charged nucleophile, reprotonation of this resulting anionic tetrahedral intermediate affords an alcohol molecule as product of this addition. The metallic counter ion $M^+$ of the nucleophile $Nu^-$ is exchanged with a proton during product isolation.

$$H \overset{\cdot\cdot}{\underset{\cdot\cdot}{O}} \colon ^{(-)}$$

$$R \overset{\cdot\cdot}{\underset{\cdot\cdot}{O}} \colon ^{(-)}$$

$$\colon N \equiv C \colon ^{(-)}$$

$$R_3 C \colon ^{(-)}$$

$$H^{(-)}$$

$$H_2 \overset{\cdot\cdot}{O} \colon$$

$$R \overset{\cdot\cdot}{\underset{\cdot\cdot}{O}} H$$

$$\overset{\cdot\cdot}{N} H_3$$

$$R \overset{\cdot\cdot}{N} H_2$$

In this way, a great variety of nucleophiles add to the carbonyl group; to such an extent that it can be thought of as a 'nucleophile trap'. Nucleophiles belong to the two classes, anionic and neutral. Examples of anionic nucleophiles include hydroxides, such as potassium hydroxide in alcoholic solution, and alkoxides $RO^{(-)}$ obtained by dissolving an alkali metal in an alcohol. In the adjacent scheme we give two examples of carbon nucleophiles, the cyanide anion (KCN) and the carbanions (conjugated bases of several hydrocarbons).

Neutral nucleophiles consist of molecules having at least one heteroatom, i.e. one supplying a (non-bonding) lone pair. We show here two examples of oxygen and nitrogen nucleophiles, respectively water, alcohols ROH, ammonia and the amines $RNH_2$, $R_2NH$ and $R_3N$. Let us not forget that an anionic nucleophile is always accompanied by a counter ion, typically a metal cation $M^+$. This electrophile $M^+$ binds to the carbonyl oxygen; the resulting increased polarization of the $C=O$ assists the attack of the nucleophile.

$$\Delta H/\text{kcal} \cdot \text{mol}^{-1}$$

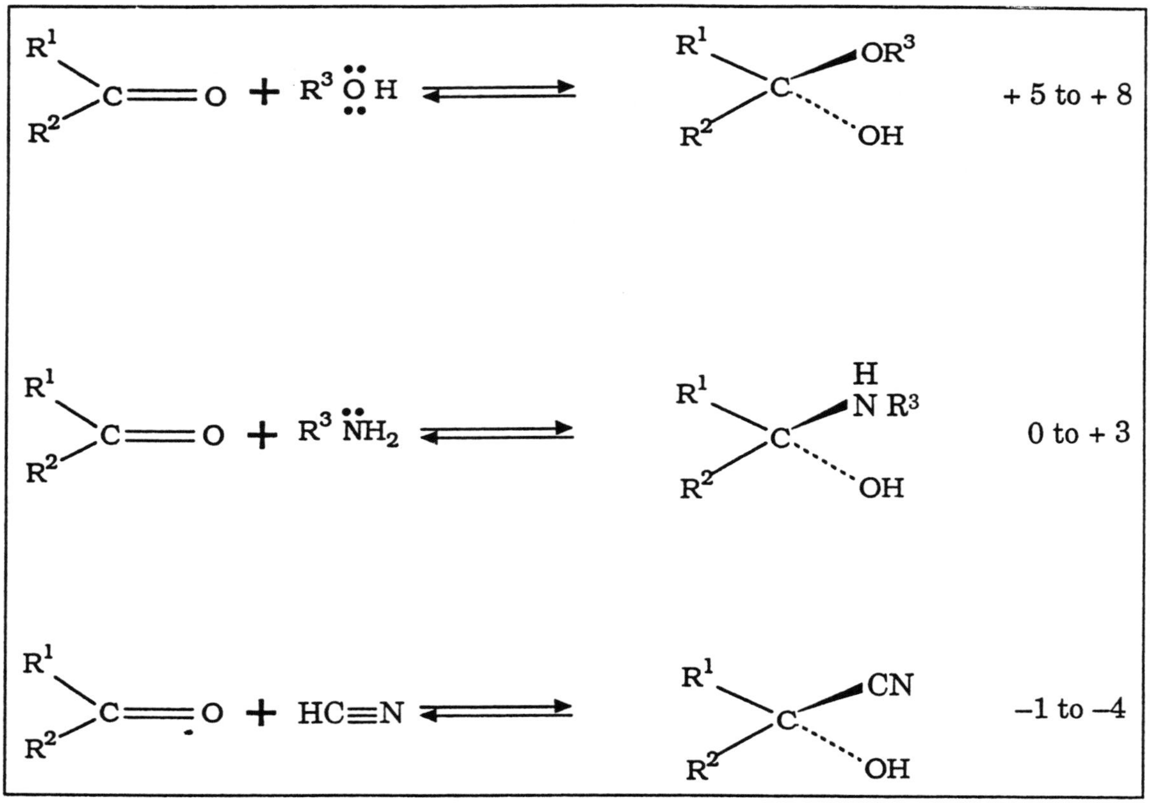

$+\,5$ to $+\,8$

$0$ to $+\,3$

$-1$ to $-4$

It is not possible to provide a universal scale of nucleophilicity. This depends on the electrophilic partner. We shall merely indicate a general tendency, in comparing the reaction enthalpies of the addition of alcohols ROH, primary amines $RNH_2$ and hydrogen cyanide HCN to ketones. The qualitative interpretation of these results is that the more readily available the lone pair that becomes the bonding pair of the new $\sigma$-bond, the more the equilibrium is shifted to the side of the adduct: carbon is a better nucleophile than nitrogen, and nitrogen is in turn a better nucleophile than oxygen. We shall return later to the stability of the tetrahedral charged intermediate. Let us keep in mind at this point the *equilibrium* between the reactants and this intermediate. This banal observation is of great practical importance. It is applied in the protection of aldehydes and ketones, in the form of their acetals. The final regeneration step of the parent aldehyde or ketone is done simply by acidic hydrolysis.

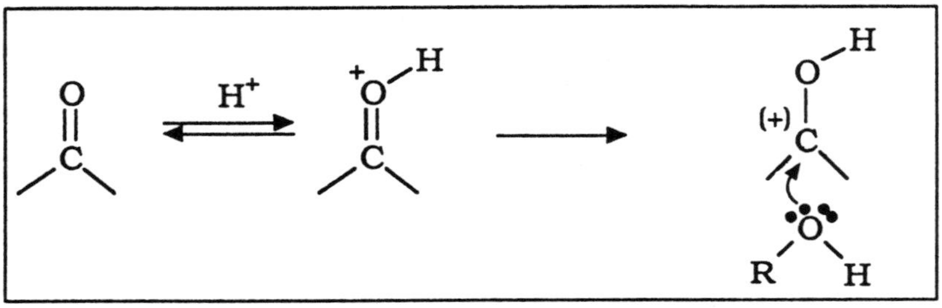
H+
(+)

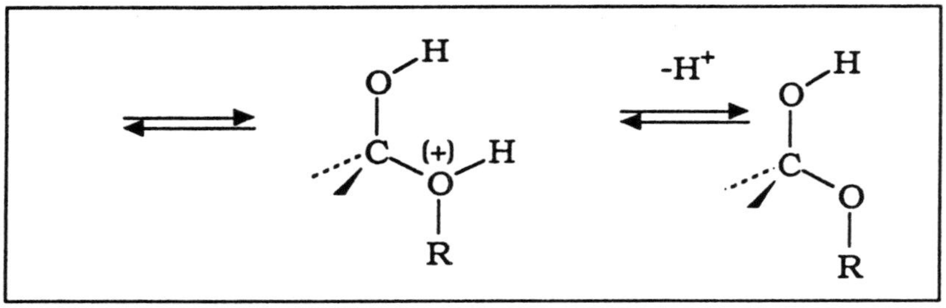
(+)
-H+

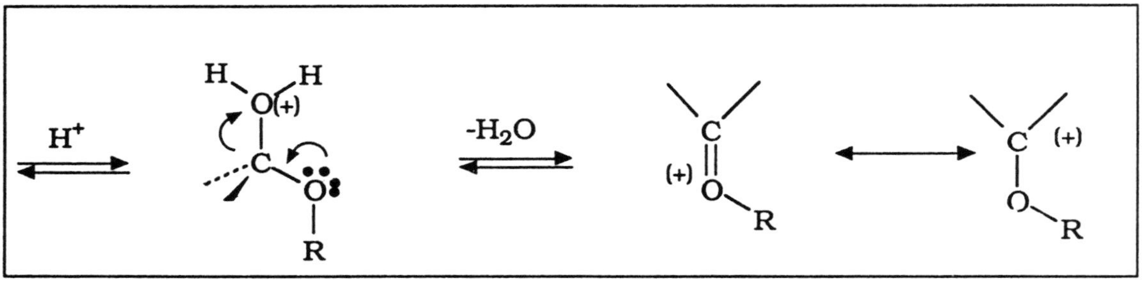
H+
(+)
-H2O
(+)
(+)

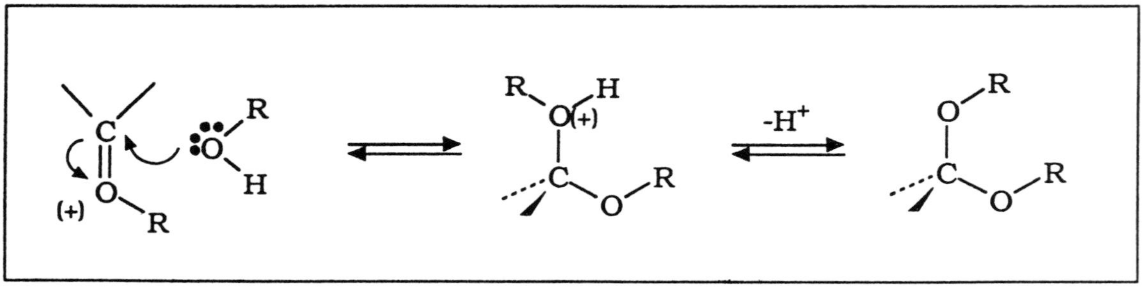
(+)
(+)
-H+

# 5 Protection/Regeneration

Addition of two molecules of an alcohol ROH transforms a carbonyl group into an acetal (cf. adjacent scheme). The first addition of the alcohol molecule is catalyzed by acids: the enhanced polarization of the carbonyl group favors the nucleophilic attack of the alcohol oxygen. In like manner, the process is repeated when a second alcohol molecule adds to the oxonium cation formed by elimination of a water molecule from the hemiacetal.

Overall, acetal formation is catalyzed by acids and implies the splitting off of a water molecule: hence dehydrating conditions favor the transformation of the carbonyl compound (aldehyde or ketone) to the corresponding acetal.

The overall process is a sequence of equilibrium steps. Therefore the reaction is reversed to the initial carbonyl compound by hydrolysis under acidic conditions.

1. $RH + B^{(-)} \rightleftharpoons R^{(-)} + BH$

2. $RX + R'M \rightleftharpoons RM + R'X$

3. $RX + M \rightleftharpoons RMX$

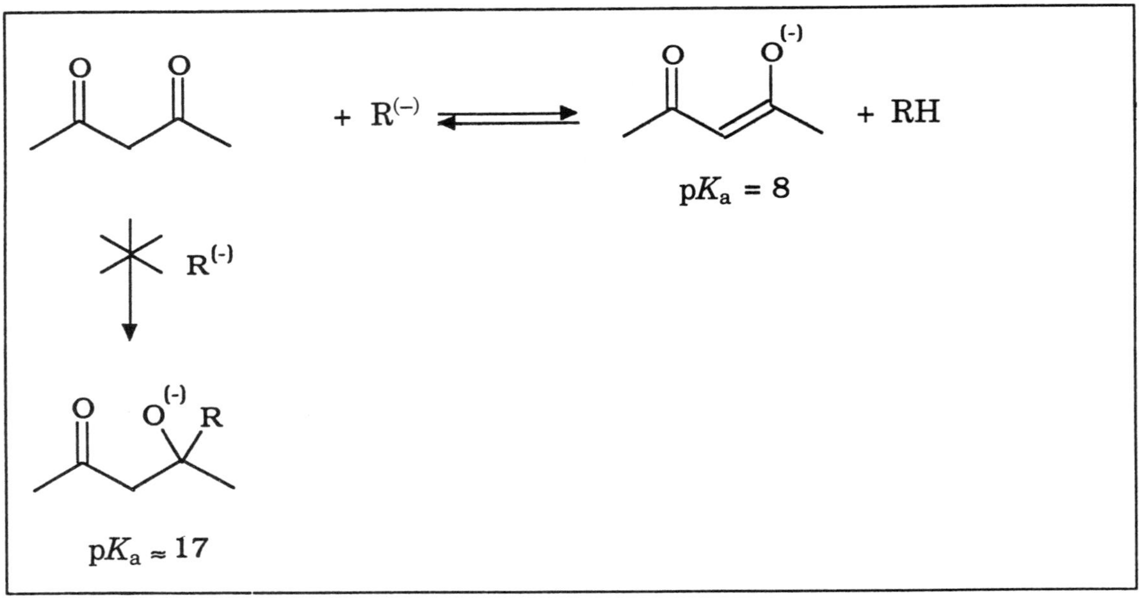

$+ R^{(-)} \rightleftharpoons$

$pK_a = 8$

$+ RH$

$R^{(-)}$

$pK_a \approx 17$

# 6  Preparation of Organometallic Reagents

The case of carbanions ($R^{(-)}$) used as nucleophiles will now be envisaged. They are obtained by three major methods:

(1) Deprotonation of a hydrocarbon RH by a base. This proton transfer is fast from acetylenic carbons, but slow from trigonal (unsaturated hydrocarbons) or from tetrahedral carbons (saturated hydrocarbons).

(2) An exchange reaction between an halogen X and a monovalent metal M, starting from an halide RX.

(3) Insertion of a divalent metal, such as magnesium, into the R–X bond of a halide (as in Grignard reagents).

Nucleophilic addition is as a general rule fast relative to a slow proton transfer, but slow relative to a fast proton transfer. Nevertheless, what determines which of the two processes (nucleophilic addition or proton transfer) occurs is the $pK_a$ of the conjugated acid of the resulting bases in the two processes: formation of the weaker base is favored. Let us expose, for example, acetylacetone $(CH_3 \cdot CO)_2CH_2$ to a reagent $R^{(-)}$. If it reacts as a nucleophile, the addition reaction would yield an alkoxide. If it reacts as a base, proton transfer would lead to formation of the conjugated base weaker by far than the carbanion. Therefore, the latter is expected to be the prevailing process, which is in fact observed.

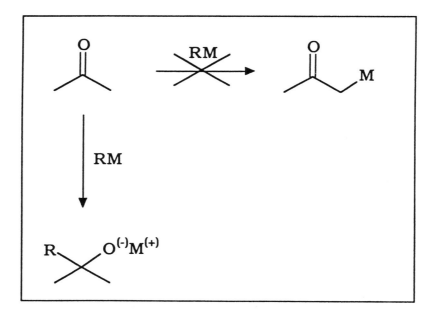

Proton transfer forming O—H or X—H bonds is also an extremely fast reaction. If the ketoalcohol in the adjacent scheme is allowed to react with an organometallic reagent RM, proton transfer is the prevailing process rather than the slow nucleophilic addition. On the other hand, with acetone as the substrate in the same reaction, nucleophilic addition is faster than proton transfer and the former is the observed reaction.

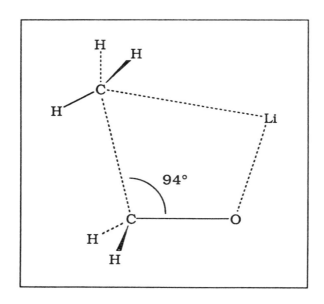

MR$^3$ : R Mg Br ; (RLi)n ; (R$_2$CuLi)$_p$ ; etc.

The second carbanion synthesis by halogen–metal exchange is a reaction with a rate of the same order of magnitude as nucleophilic addition. This synthesis is mainly used for preparation of organolithium compounds from chlorides (and never from fluorides). Organolithium compounds RLi exist as oligomers $(RLi)_n$ ($n = 4$ or 6 in most cases).

Organocuprates of the type $R_2CuLi$ also exist in oligomeric form $(R_2CuLi)_p$.

If an addition reaction takes place between one of these organometallic reagents MR and a carbonyl compound, the metal cation $M^{(+)}$ (or $M^{(n+)}$) binds to oxygen whereas the carbanion $R^{(-)}$ is bound at carbon. A charged (anionic) tetrahedral intermediate results. During isolation by acidic hydrolysis, the metal cation $M^{(+)}$ is replaced by a proton, and the resulting alkoxide is transformed into the final product, the corresponding alcohol (one enantiomer is represented only for both species). Hence, addition of organometallic reagents RM to aldehydes and ketones yields secondary and tertiary alcohols, respectively.

The calculated transition state for the addition (formaldehyde + methyl lithium) is shown in schematic form: note the double interaction of the lithium cation with the two negatively charged centers, the carbonyl oxygen and the carbon of the methyl group. Therefore there is a reduction of the Bürgi–Dunitz angle of attack ($106° \rightarrow 94°$) which remains nevertheless obtuse. The initially trigonal carbonyl carbon becomes tetrahedral and the C—H bonds on both ends of the new carbon–carbon bond adopt a staggered arrangement.

| RH | $pK_a$ |
| --- | --- |
| $CH_3$ | 50 |
| $CO\ CH_3$ | 20 |
| $CO\ OC_2H_5$ | 25 |
| $CN$ | 25 |
| $(COOC_2H_5)_2$ | 13 |
| $(CN)_2$ | 11 |

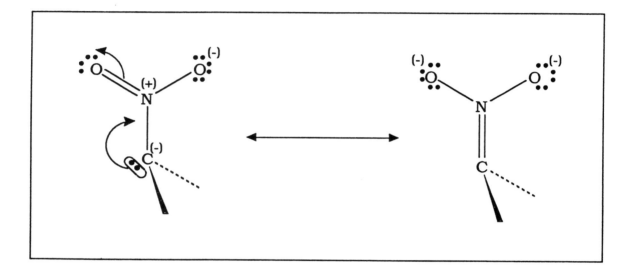

# 7 Activated Methylene Groups

Our first access to carbanions, which can then add to carbonyl groups, is then through deprotonation: a base removes a proton from the conjugated acid of the carbanion. The molecule RH has to be sufficiently acidic for the equilibrium to be shifted to the right (i.e. to carbanion formation). In other words, its conjugate base, the $R^{(-)}$ carbanion has to be stabilized.

The presence of certain attracting substituents denoted as A brings about such a stabilization. The adjacent Table demonstrates the increase of acidity of C—H bonds when the carbon bears one, or more dramatically, two A substituents such as acetyl $COCH_3$, an ester group $-CO \cdot OR'$ or a nitrile function $-C \equiv N$.

Halogens and the nitro group $-NO_2$ are also attracting substituents. We have indicated in the resonance formalism some of the limiting forms that describe the stabilization of a carbanion by one of these A substituents: the mesomeric effect adds indeed to the inductive effect of a nitro substituent.

Note: The curved arrows, in the canonical form on the left show the motion of electron pairs required for the transformation to the resonance form on the right. One should keep in mind that the molecule is a *simultaneous hybrid* of all these canonical (limiting) forms. Strictly speaking, displacements of electrons schematized by curved arrows occur physically only during the polarization of the molecule by a reactant.

Since molecules of the class $H_2CA^1A^2$ serve as efficient sources of carbanions $^{(-)}CHA^1A^2$, they are very often used for this purpose; these are termed activated methylene groups. For instance, ethyl malonate $H_2C(CO \cdot OC_2H_5)_2$ is a compound with a methylene group activated by the presence of the two ester groups $CO \cdot OC_2H_5$. In the following chapter we will examine in more detail the reasons for the activation of a C—H bond by an adjacent carbonyl group ($A = CO \cdot CH_3$ or $CO \cdot OC_2H_5$).

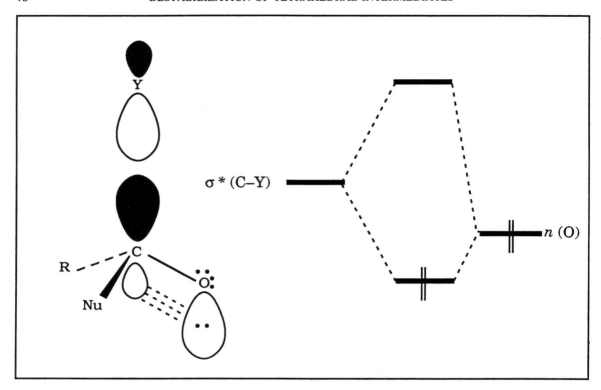

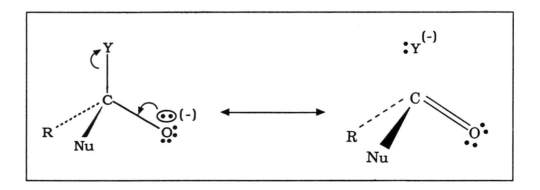

# 8 Destabilization of Tetrahedral Intermediates in Nucleophilic Addition

Let us return to the tetrahedral intermediates, negatively charged or neutral, formed by addition of a nucleophile Nu: to a carbonyl compound $R \cdot CO \cdot Y$. In these intermediates, a lone pair from the nucleophile has been transferred to the oxygen atom which now bears three non-bonding electron pairs. The Coulomb repulsion between these lone pairs makes their energy levels go up. Provided that there is an antiparallel orientation of the $C-Y$ bond and the axis of the lone pair n(O), an interaction of the n(O) orbital and a vacant $\sigma^*$ (C—Y) orbital is set up. It occurs through $\pi$ type lateral overlap of the n(O) orbital with the rear lobe of the $\sigma^*$ (C—Y) orbital.

In other words, since this interaction stabilizes (lowers) the n(O) level by the overlap with the antibonding $\sigma^*$ (C—Y) orbital, then this shortens the C—O bond and lengthens the C—Y bond: cleavage of this bond is made easier. The limiting forms depicted in the adjacent scheme indicate these structural changes in an equivalent way, in the resonance formalism.

Evidently, the greater the relative weight of the resonance form with the broken C—Y bond, in the latter formalism, the more apt will be the atom Y to bear the negative charge. If, on the other hand, the C—Y bond is already long (e.g. Y = Br or I), breaking of the bond is facilitated.

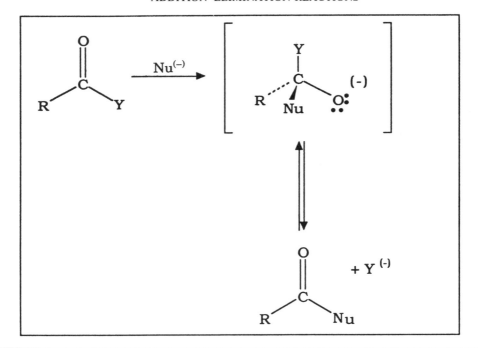

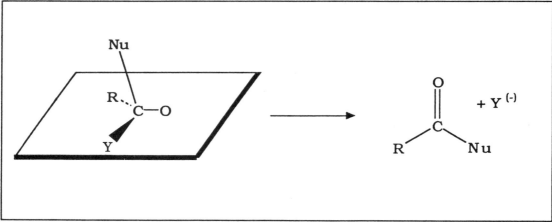

# 9 Addition–Elimination Reactions

Addition of a nucleophile to a carbonyl compound $R \cdot CO \cdot Y$ can go together with or be followed by cleavage of the $C—Y$ bond. The addition requires good nucleophiles, such as those already mentioned. The elimination—whether concerted with the addition or sequential, occurring in a second step—requires a good leaving group. Examples of good leaving groups $Y^{(-)}$, resulting from the $C—Y$ bond breaking are alkoxides $RO^{(-)}$ and halides $X^{(-)}$, especially iodides and bromides. Thus, esters $R^1 \cdot CO \cdot OR^2$ and acid halides $R^1 \cdot CO \cdot X$ are substrates set-up for such addition–elimination reactions.

An addition–elimination reaction involving as reaction partners a carbonyl compound and an activated methylene compound (in presence of a base to remove the proton) is called the Claisen condensation. This reaction forms a new carbon–carbon bond with breaking of the $C—Y$ bond (with an appropriate leaving group $Y^{(-)}$).

For example, the attack of the cyanide anion (= nucleophile) on ethyl acetate affords $H_3C \cdot CO \cdot CN$ and the ethoxide anion (= leaving group).

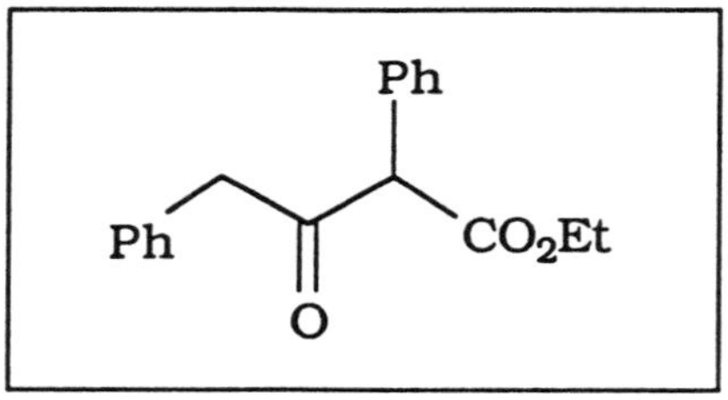
Ph
Ph
CO2Et
O

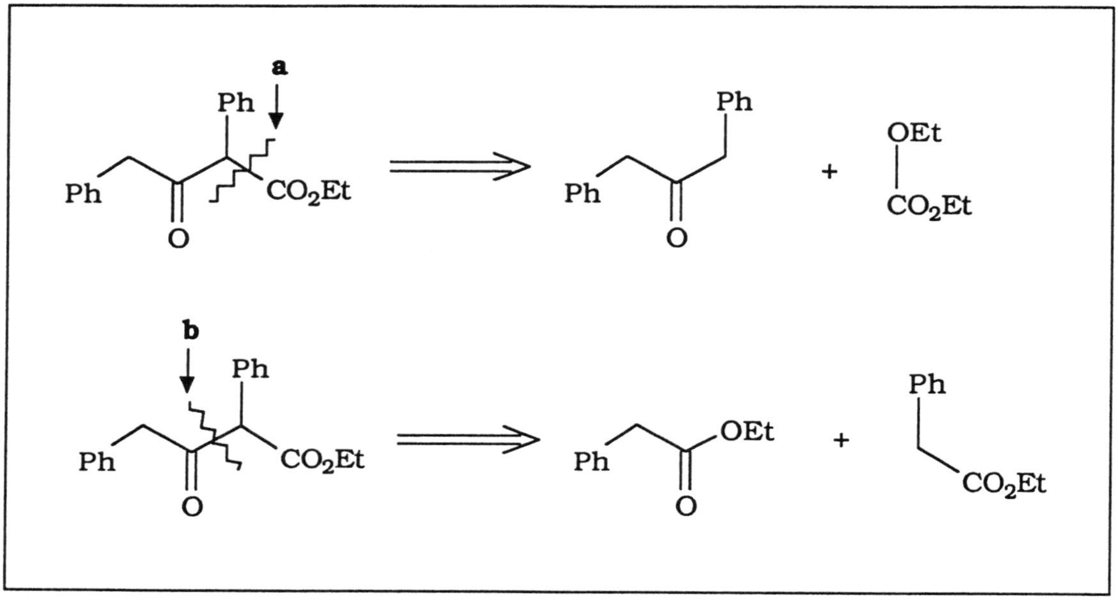
a
Ph
Ph
CO2Et
O
Ph
Ph
O
+
OEt
CO2Et
b
Ph
Ph
CO2Et
O
Ph
OEt
O
+
Ph
CO2Et

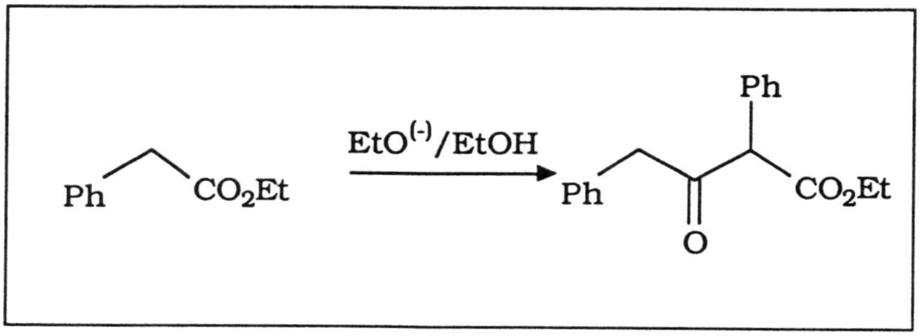
Ph
CO2Et
EtO(-)/EtOH
Ph
Ph
CO2Et
O

The Claisen condensation is the first 'great' reaction introduced here that allows for synthetic elaboration. Let us consider the target molecule indicated in the adjacent scheme.

If the disconnection—in a retrosynthetic sense—is carried out according to **a**, it produces the two indicated precursors which can undergo a Claisen condensation to form the target molecule. The advantage of the disconnection **b** is that the two precursors are in fact the same molecule. In practice, it will be sufficient to allow ethyl phenylacetate to react under basic conditions, providing the target molecule together with regenerated ethoxide anion, which had served originally as the base, and is thus a catalyst.

---

**Note**: The symbol used for retrosynthetic transformations is the double arrow, with the meaning to logicians of 'implication'. In this manner, a formal distinction is made between the real transformation and the formal inverse transformation. In general, synthesis elaborates more and more complex chemical structures with increasing numbers of carbons. Retrosynthesis, on the other hand, breaks down the molecule into modules or *synthons* with a smaller number of carbon atoms. A synthon, in the retrosynthetic sense, corresponds to a reaction intermediate in the synthetic sense.

**Some Important Carbonyl Derivatives**

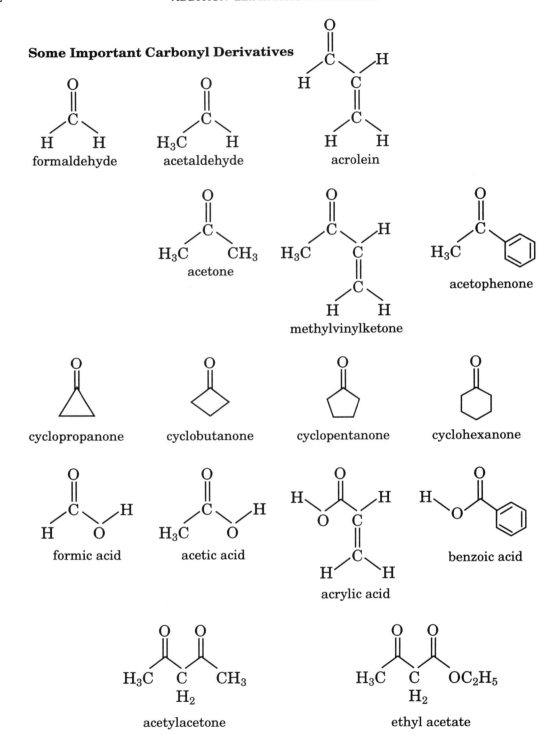

formaldehyde

acetaldehyde

acrolein

acetone

methylvinylketone

acetophenone

cyclopropanone

cyclobutanone

cyclopentanone

cyclohexanone

formic acid

acetic acid

acrylic acid

benzoic acid

acetylacetone

ethyl acetate

Since departure of the leaving group Y can be triggered by the presence of a nucleophile, in an addition–elimination reaction molecules of the type $R \cdot CO \cdot Y$ ($T = OR'$, halogen) are equivalent to an acyl (or acylium) synthon: $R \cdot CO^{(+)}$. Thus, formal analysis by retrosynthetic disconnections shows the importance of addition–elimination reactions, and particularly of the Claisen condensation: combination of an acyl cation synthon $R \cdot CO^{(+)}$ with a carbanion $R^{(-)}$. In the case of the Claisen condensation the carbanion is an activated methylene group.

So far, this chapter has presented several examples of carbon–carbon bond formation between an electron-donating and an electron-accepting center. The donor is a nucleophile, the acceptor the electrophilic carbon of the carbonyl group in additions which provide alcohols from aldehydes and ketones. The donor is a carbanion and the acceptor is (formally) an acylium cation in the Claisen condensation. We will end this chapter by considering reductions with metal hydrides where the acceptor is again an aldehyde or ketone and the donor is formally a hydride anion $H^{(-)}$.

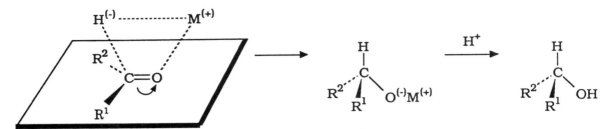

|  | $k_2 \times 10^4/\mathrm{M^{-1}\,s^{-1}}$* |
|---|---|
| $C_6H_5\ CHO$ | 12 400 |
| $(C_6H_5)_2\ CO$ | 1.9 |
| $C_6H_5 \cdot CO \cdot CH_3$ | 2.0 |
| $(H_3C)_2\ C{=}O$ | 15.1 |
| $(H_2C)_3\ C{=}O$ | 264 |
| $(H_2C)_4\ C{=}O$ | 7 |
| $(H_2C)_5\ C{=}O$ | 161 |

*$\mathrm{NaBH_4}/i-C_4H_9OH$; 0 °C.

# 10   Reduction with Metal Hydrides

The simplest nucleophile as already mentioned is the hydride anion $H^{(-)}$. It is provided by metal hydrides such as $NaBH_4$ or $LiAlH_4$. Its addition to the carbonyl group followed by exchange of the metal cation with a proton yields an alcohol (only one of the enantiomers is indicated): a primary alcohol from an aldehyde, a secondary alcohol from a ketone.

Examination of the kinetic data in the adjacent table provides a first example of a structure–activity relationship. The aldehyde (first line) has a reactivity by far superior to that of the ketones. This is universal for additions to carbonyl compounds. This higher reactivity of aldehydes compared to ketones goes parallel with the sequence of stability of the carbocations: tertiary carbocations are more stable than secondary carbocations, themselves more stable than primary carbocations. Since reactivity and stability vary in opposite directions, and based on the analogy between a carbocation and the positively charged carbon of a carbonyl group, it is logical to find higher electrophilicity for aldehydes than for ketones (the stability sequence of the carbocations results from hyperconjugation of the C—H bonds in the groups $R^1$, $R^2$, $R^3$ with the vacant $2p_z$ orbital at the positively charged carbon (cf. next page). Benzophenone and acetophenone, on the next two lines, show a relatively low reactivity relative to the examples below because their carbonyl groups are stabilized by conjugation. The opposite is true for cyclobutanone where the initial state is more destabilized than the tetrahedral intermediate: it is more costly, from an energy point of view to force a trigonal carbon into a four-membered ring—the C—C—C angle is constrained to a value of ~90° instead of ~120°—than to do the same with a tetrahedral carbon (~109.5° → ~90°). In the case of cyclobutanone, there is hence a reduction in the angular strain when passing from the initial to the transition state.

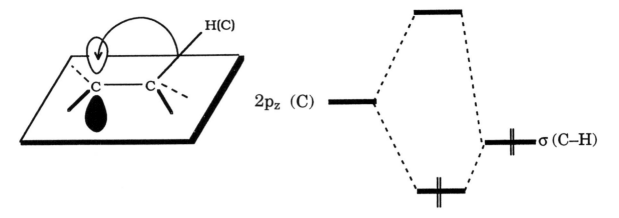

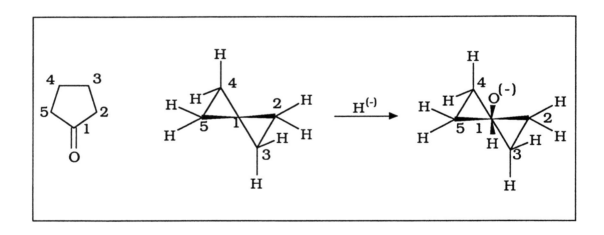

Cyclopentanone is less reactive than cyclohexanone: the trigonal arrangement at carbon 1 is accompanied by a partially eclipsed arrangement along the C(1)–C(2) and C(1)–C(5) bonds.

---

### Victor Grignard (1871–1935)

Victor Grignard, who had a modest family background, was born in Cherbourg. His excellent progress at the local lycée enabled him to attend the Ecole Normale Supérieure Spéciale in Cluny. This school was founded by Victor Duruy in 1866 in order to provide a professional education. The closing of Cluny resulted in the transfer of Grignard to the Faculté des Sciences in Lyons. He obtained there a B.S. in Mathematics and Chemistry. He then entered the laboratories of Ph. Barbier, a man of a noble and full character, who let him explore the reaction bearing his name, by replacing zinc with magnesium. In this way Grignard discovered (1899–1900) the organomagnesium compounds:

'I think that I can attribute the formula RMgI or RMgBr to the organometallic compounds which I have obtained, R being a fatty alcohol or an aromatic residue (. . .). I will continue finding new applications for these new organometallic compounds'.

Organomagnesium compounds, immediately adopted worldwide, brought Grignard the Nobel Prize in Chemistry (1912), which he shared with Sabatier, at the age of 41.

At the beginning of World War I, the French Army was a bit slow to take advantage of his competence: till July 1915, Grignard served as reserve corporal in the Cherbourg area and guarded railroad tracks. Finally a laboratory was given him at the Sorbonne. Grignard gathered there a small group for the analysis of combat gases: they found the structure of yperite only four days after its first use on a battlefield. After the war, Grignard resumed his professorship in Nancy. A steadfast provincial, he turned down twice a chair at the Collège de France but accepted to succeed Barbier in Lyon in 1919. His influence on French chemistry was deep-seated and durable, to such an extent as to become sterilizing at length, long after his passing: in the Fifties and in the Sixties, a number of French organic chemists still worked in the prestigious and overvalued area of organomagnesium compounds.

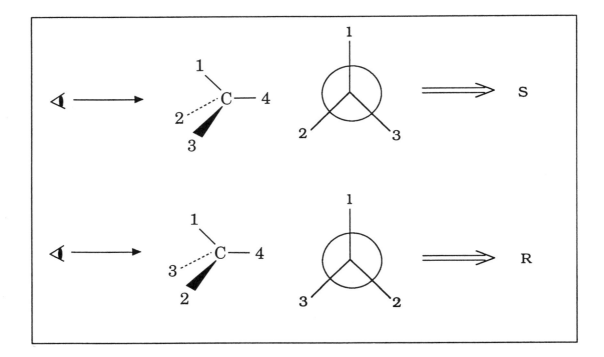

# 11  Notation of Configuration of an Asymmetric Carbon

Let us consider an asymmetric carbon C (1234), with $1 \neq 2 \neq 3 \neq 4$, by definition. It is always feasible to attribute a priority 1, 2, 3, 4. After establishing this priority, one examines the relationship of the groups 1, 2, 3 along the axis of the bond from carbon to group 4, the latter pointing backwards by convention: if the sequence 1, 2, 3 is clockwise, the carbon has a configuration R (Latin *rectus*); otherwise it has a configuration S (Latin *sinister*).

The rules that define the priorities are:

(1) One considers first only the atoms directly bonded to the carbon (or more generally to the chiral center). That with the highest atomic number has the highest priority and so on, following the atomic number.

(2) If two atoms directly bonded to the chiral center share the same priority, one examines the next atoms linked to them in turn, two bonds away from the chiral center; and so on.

(3) The heavier isotope has priority; e.g. tritium $^3H$ is followed by deuterium $^2H$, and then by $^1H$.

(4) Atoms with multiple bonds are formally considered as atoms with the corresponding number of single bonds. For example a substituent $-CHO$ is from this point of view equivalent to $-CH(O)_2$.

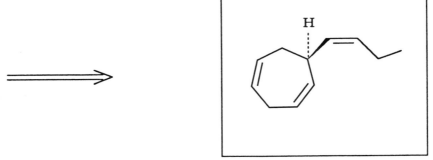

## Application of these Rules on an Example

The female gametes of the brown alga *Ectocarpus silicosus* emit a pheromone which attracts the male gametes at astonishingly low concentrations (20 picomolar solutions are active which corresponds to a response of the sperm to only twenty individual molecules!). The pheromone molecule ectocarpene with the formula $C_{11}H_{16}$ is hydrogenated with diimide in the side chain, exclusively. Scission of the two C=C bonds of the seven-membered ring by ozonolysis affords a known diacid the asymmetric carbon of which has an R configuration. Consequently, ectocarpene has the configuration indicated in the box.

Reference: L. Jaenicke and W. Boland, *Antigen Chem. Int. Ed. Engl.*, **21**, 643–710 (1982).

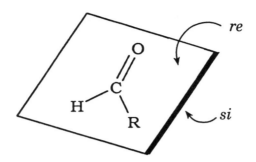

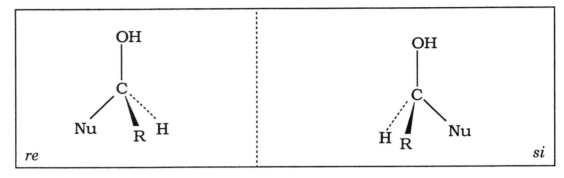

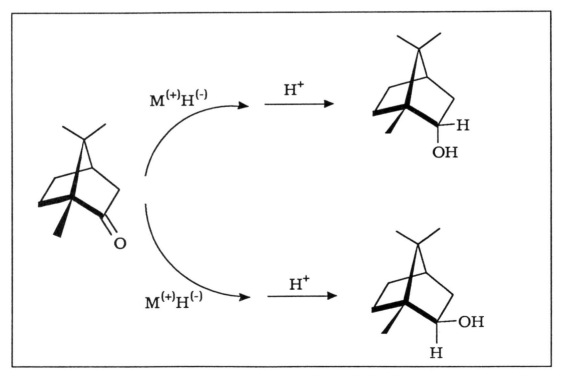

## 12  Enantiotopic and Diastereotopic Faces

Let us consider the example of an achiral aldehyde molecule RCHO. The plane of the carbonyl group C=O bisects two half spaces. These are called *si* or *re* according to the rotational sense of the three groups at the trigonal carbon (oxygen, carbon of R and hydrogen, ranked here in decreasing priority): counterclockwise (*si*) and clockwise (*re*).

Addition of any nucleophile: Nu to this molecule consequently forms one or the other enantiomer (i.e. optical antipodes) depending on whether the attack occurs from the *si* or the *re* face. Hence the two faces of the carbonyl group in this first example have a so-called *enantiotopic* relationship.

Consider now a chiral aldehyde molecule R*CHO. Addition of a nucleophile results in diastereoisomeric adducts, depending on the side of attack. The two distinct, non-equivalent faces are thus called *diastereotopic*. This terminology can be generalized to any double bond C=C, C=O, C=N, etc. For instance, addition of a hydride anion to the carbonyl group of camphor (=reduction) after acidic hydrolysis (exchange of $M^+$ by $H^+$) leads to diastereoisomeric alcohols: by definition, stereoisomers are either enantiomers or diastereoisomers.

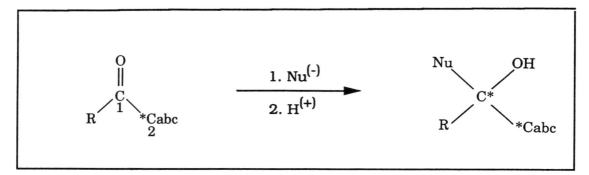

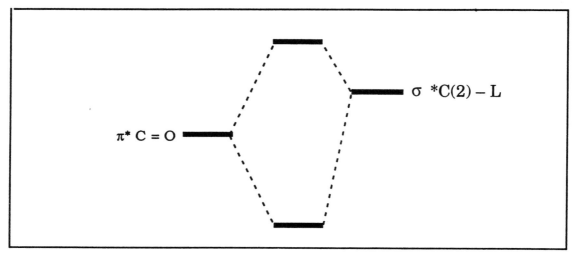

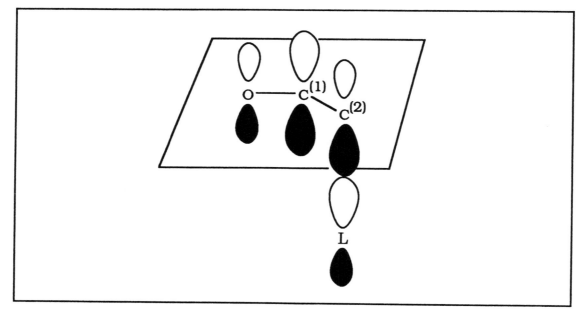

# 13   Asymmetric 1,2-Induction

An important practical consequence of the difference between the two faces of a carbonyl group bearing an asymmetric carbon —*C (abc) is the preference by a nucleophile for the attack on one of the two faces. Since the presence of an asymmetric carbon induces the chirality of the new asymmetric center formed by addition of the nucleophile, this preference is called asymmetric induction.

Let us consider a ketone $R \cdot CO \cdot C$ (abc). We will represent it in the Newman projection along the axis $C(1)-C(2)$: How will the groups a, b, c be arranged during the attack of a nucleophile? The predominant interaction (two electrons) is that between the HOMO of the nucleophile: $Nu^{(-)}$ and the LUMO of the ketone: the LUMO of the ketone is energetically higher. Consequently, any interaction that lowers the energy of the LUMO of the ketone will reduce the HOMO–LUMO gap and stabilize the transition state. Now the LUMO of the ketone can be constructed from the vacant $\pi^*$ orbital of the carbonyl group perturbed by the overlap with the $\sigma^*$ orbital of one of the three bonds $C(2)-a$, $C(2)-b$, $C(2)-c$. That one of these three bonds that has the $\sigma^*$ level closest in energy to the $\pi^*$ level of the carbonyl group will have the largest interaction: let us call it L (for large); the bond (substituent) with the weakest interaction will be called S (for small); the third substituent will be called M (for medium).

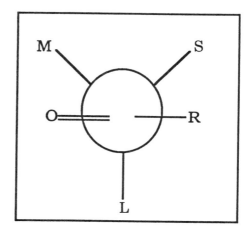

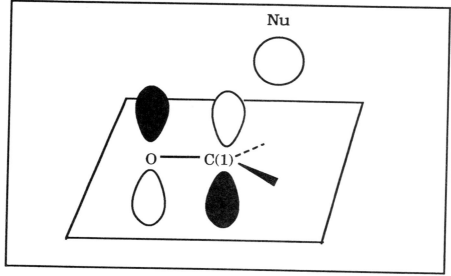

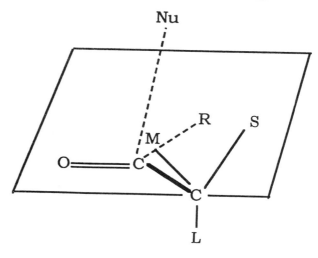

This reasoning dictates the orthogonality of the bond C(2)–L to the plane of the carbonyl group in the Newman projection during the approach of the nucleophile. (It is often the size of the groups a, b, c that determines their sequence large, medium, small.) The privileged geometry for the transition state is that having the group S *syn* to R, with a torsion angle S,C(2),C(1),R of about 30°.

It now remains to determine the position of the nucleophile to characterize completely the geometry of the transition state. The LUMO of the ketone resembles the $\pi^*$ orbital of the carbonyl group and the nucleophile attacks at the carbon with an obtuse angle O–C(1)–Nu in such a way as to maximize the bonding overlap Nu, C(1) and to minimize the antibonding overlap Nu,O. So the nucleophile Nu attacks the carbonyl group in the half-space occupied by the M and S groups, and opposite to the half-space with the L group. During the Nu–C(1) bond formation there is pyramidalization of the C(1) carbon, the bonds C(1)–O and C(1)–R move down to the L half-space so that the torsion angles O,C(1), C(2),L and R,C(1),C(2),L assume values of about 60°.

References: N. T. Anh and O. Eisenstein, *New J. Chem.*, **1**, 61–70 (1977).
E. P. Lodge and C. H. Heathcock, *J. Am. Chem. Soc.*, **109**, 3353–3361 (1987).

## Summary: Carbonyl Group—Addition and Addition–Elimination Reactions

Whereas the HOMO is a non-bonding lone pair at the oxygen, the LUMO is of $\pi^*$ type with a large coefficient at carbon. This enforces the angle of attack of the nucleophile at carbon (106°, Bürgi–Dunitz). The carbonyl group is strongly polarized ($C^+O^-$) with an electric dipole moment between 1.7 and 3.7 D.

The reversible addition of nucleophiles leads to a tetrahedral intermediate: example of cyanohydrins.

Reductions with metal hydrides provide alcohols. Aldehydes are more reactive. The reactivity of ketones is increased by ring strain, and decreased by conjugation.

The C—X bond in the tetrahedral intermediates is weakened (interaction $\sigma^*$ (C—X)/$n$(O)), hence eliminations.

The addition of the conjugated base of compounds with activated methylene groups $XCH_2Y$ (X,Y attracting groups) is called the Claisen condensation. It allows for the formation of new carbon–carbon bonds.

# 1.II
# Carbonyl Group: Additions and Addition–Elimination Reactions

## 1 Chemical Weapons of Insects

Ants are predators of the millipede *Apheloria corrugata*. The latter emit a defensive secretion rich in mandelonitrile, containing an enzyme that promotes hydrolysis of mandelonitrile to benzaldehyde and hydrogen cyanide:

Thus the millipede protects itself by spraying its attackers with hydrogen cyanide!

## 2 Biological Activity and Chirality

As a general rule, whenever a molecular system is chiral and there are thus two coexisting enantiomers, only one has biological activity, irrespective of the test used. In fact, most receptors are proteins (for example enzymes) and the chiral binding site fits one of the enantiomers but not the other: this is analogous to the difference between putting the left hand in the left glove (or pocket) or in the right glove (or pocket).

Even though olfaction is still not well-understood, the smell receptors are likely to be proteins. They are certainly chiral: an inference consistent with the difference in odor between two enantiomers. The example below is that of the enantiomers of carvone. Both have the same boiling point (231 °C). The (−) enantiomer, that causes the plane of polarized light to rotate to the

(−)                                                    (+)

left, has a mint odor. The (+) enantiomer, that rotates the plane of polarized light in the other direction, has the characteristic odor of caraway seeds, *Carum carvi* or of cumin.

## 3  Chemistry and Finance

John Emsley, lecturer in the Chemistry Department at King's College, London, was intrigued in the spring of 1990 by the large number of requests for recommendation letters from accounting firms. When he interviewed the 80 second-year students about their future plans, a majority indicated that they intended to follow a financial career such as accountancy, insurance or banking. A minority of them, about 20, expressed a wish to continue in chemistry.

When he asked the former the reason for their choice, the answer was 'MONEY' and Emsley commented: 'I could have guessed it!' A few days later, Emsley asked the same question from first-year university students: less than a quarter had the intention of continuing in chemistry. Pursuing his query, Emsley asked one of his former students, who had switched to a financial job, the reason for his initial choice of chemistry. He answered: 'Chemists are not afraid of data, numbers and their units, and . . . do not hesitate to work hard'. Emsley then reflected that indeed whoever could convert energies in $cm^{-1}$ to kilojoules per mole should have no problem converting pounds into pesetas.

## 4  Manufacture of Plexiglas

The Rohm and Haas process forms methyl methacrylate by:
  (i) obtaining the cyanohydrin through addition of HCN to acetone at 40 °C, in the presence of basic catalysts such as hydroxides or alkaline carbonates

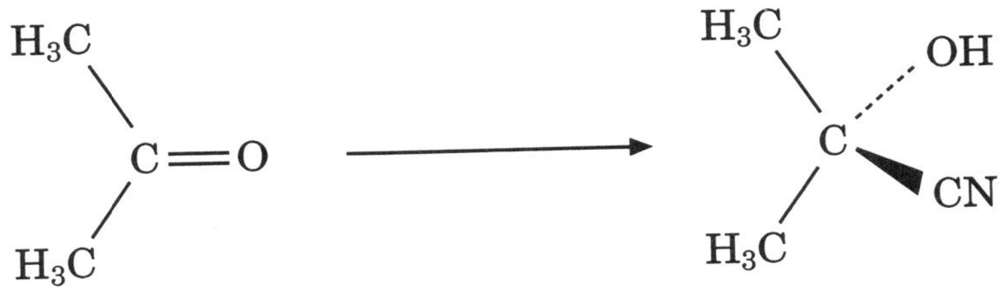

(ii) transformation of the cyanohydrin to the hydrogenosulfate of methacrylamide, between 80 and 140 °C, catalyzed by sulfuric acid, responsible both for dehydration of the tertiary alcohol and for hydrolysis of the nitrile function to an amide

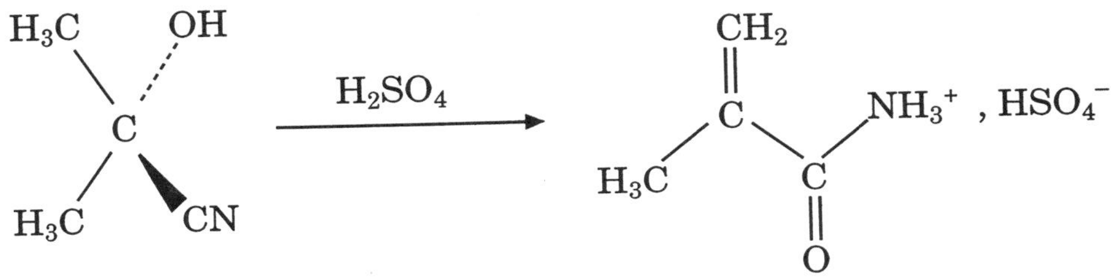

(iii) obtaining methyl methacrylate, by a reaction with methanol at 80 °C

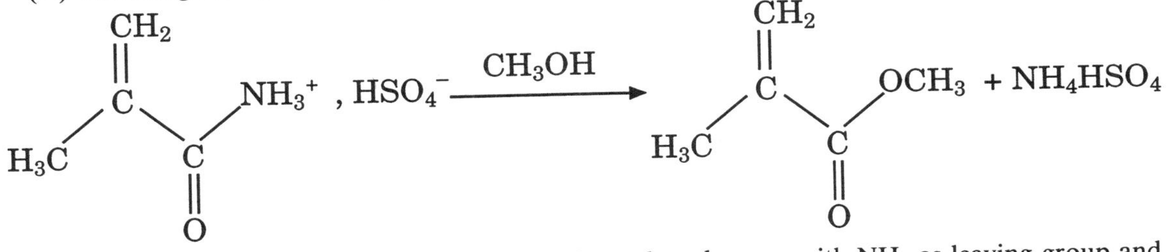

This step is an addition/elimination reaction at the carbonyl group with $NH_3$ as leaving group and $CH_3OH$ as nucleophile.

The overall selectivity for methyl methacrylate, based on acetone, is superior to 77%. Methyl methacrylate is the monomer for making Plexiglas, with the formula:

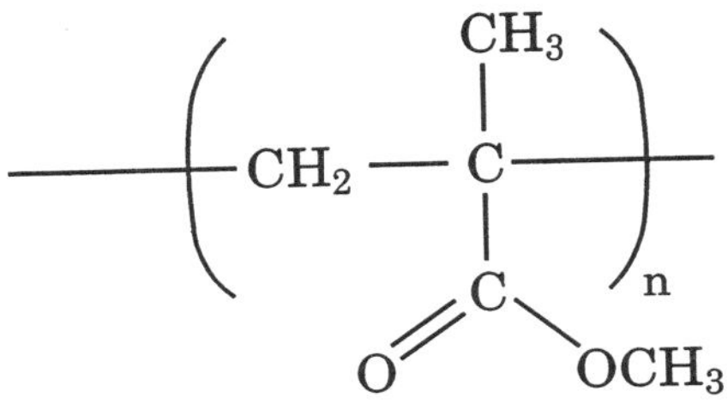

It is a plastic of high transparency with strong mechanical and chemical resistance. Methyl methacrylate is also used in a variety of copolymers. The world production exceeds a million T per year, with a substantial rate of increase. Thus the US production increased by 245 000 T per year (+6% from 1979 to 1989).

## 5   Accessibility and Selectivity

One of the cedrol syntheses—a tricyclic tertiary terpene alcohol, found in the essential oil of cedars and cypresses—illustrates a general principle: taking advantage of the masking of one of the two faces of a double bond, here a carbonyl group, by one of the two twin methyl groups at one of the quaternary carbon atoms. Accordingly, attack of the methyllithium nucleophile is feasible only from the opposite face left totally unhindered and thus accessible. This addition of $CH_3Li$, followed by acidic hydrolysis exchanging $Li^+$ with $H^+$, leads stereospecifically to cedrol.

This strategy is very general, using a hindering group (if no such group is present, an auxiliary group cleaved afterwards can be introduced), so that one of the faces of a C=C double bond, a carbonyl group C=O, and so on, is selectively blocked.

For this purpose, very often transition metal complexes are used. Such $\pi$ complexes, for instance complexes of butadiene or, as in the following Scheme, of the pentadienyl cation, mask the half-space occupied by the metal. When this auxiliary group has fulfilled its role, it is easily cleaved.

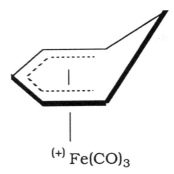

$^{(+)}$ $Fe(CO)_3$

References: (synthesis of cedrol): G. Stork and F. H. Clarke, *J. Am. Chem. Soc.*, **77**, 1072 (1955); **83**, 3114 (1961).
    ($\pi$ complexes): G. R. Stephenson, R. P. Alexander, C. Morley, and P. W. Howard, *Phil. Trans. R. Soc. (London)*, **A326**, 545–556 (1988).

## 6 Biochemical Use of Schiff Bases

Addition of an amine to a carbonyl group, followed by elimination of a molecule of water, produces imines, also referred to as Schiff bases (see the following Scheme).

Since formation of a Schiff base occurs through a sequence of consecutive equilibria, it is a reversible reaction. Biochemistry frequently makes use of this property: Schiff bases play a role in glycolysis (action of an aldolase), in biosynthesis of tryptophan from indole and serine, in binding of retinal to opsin (mechanism of vision), and in the action of transaminases.

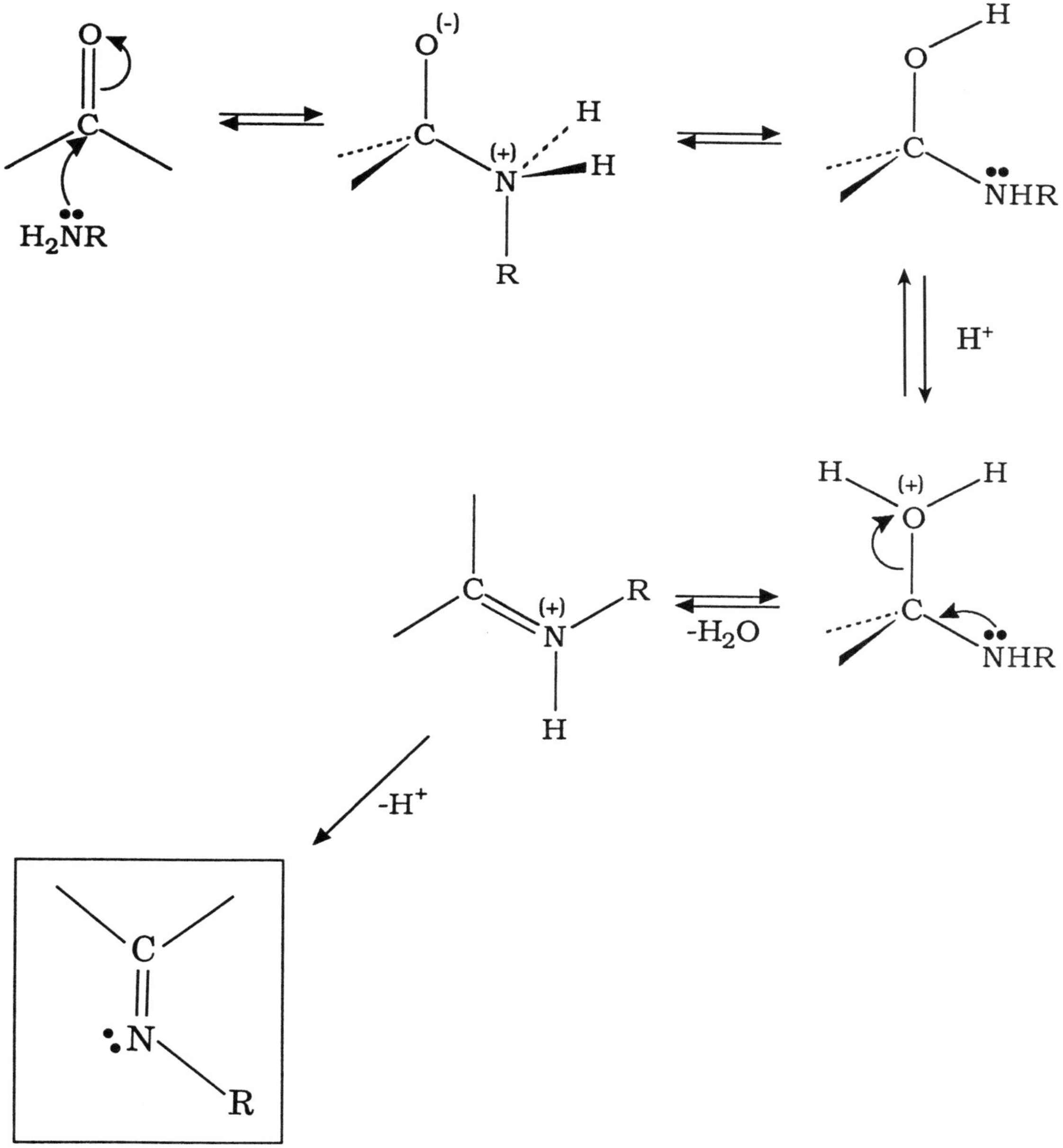

We shall focus on this last example, for its historical importance: it was the first co-enzyme whose mechanism was elucidated by the group of A. S. Braunstein in Moscow in 1945. A co-enzyme—also referred to as prosthetic group—is the non-polypeptidic part of an enzyme.

Vitamin $B_6$ or pyridoxin is the triol form (depicted in the adjacent Scheme) of a pyridinium cation (= conjugate acid of pyridine). The corresponding aldehyde or pyridoxal (see box below) has also $B_6$-type biological activity.

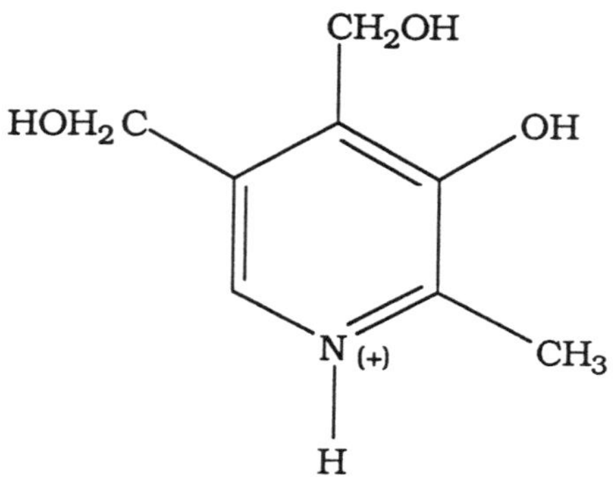

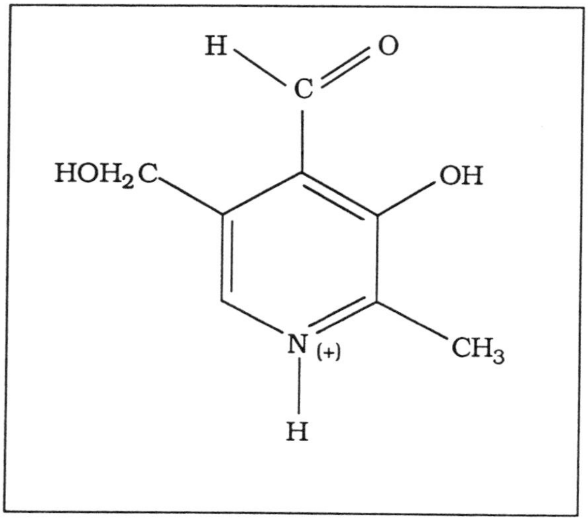

For a transaminase (= enzyme) in the resting state, the co-enzyme shown in the Scheme at the top of the facing page is pyridoxal phosphate. Union of the enzyme and of this co-enzyme is due to the $\epsilon$-amino group of a lysine residue forming a Schiff base with the aldehyde function of pyridoxal. In this way, the co-enzyme is attached to the active site of the transaminase (E denotes the enzyme, P stands for the pyridoxal phosphate group).

How does the transaminase function? An amino acid binds to the active site of the enzyme. The first Schiff base undoes itself to reform anew with the $\alpha$-amino group of the amino acid. This regenerates the side chain of lysine.

The reaction then proceeds by conversion of this aldimine into a ketimine. The driving force is electronic migration to the positively-charged nitrogen of the pyridinium cation. The co-enzyme has the function of an 'electron waste bin'.

The ketimine, still at the active site of the enzyme, is then hydrolyzed to the phosphate of pyridoxamine and to an $\alpha$-ketoacid $R^1 \cdot CO \cdot COOH$. The overall balance for the first half of the reaction can thus be written:

amino acid 1 + pyridoxal phosphate $\rightarrow \alpha$-keto acid 1 + pyridoxamine phosphate.

The second part of the reaction, conversely, sets the enzyme–pyridoxamine phosphate complex to react with *another* $\alpha$-ketoacid:

$\alpha$-ketoacid 2 + pyridoxamine phosphate $\rightarrow$ amino acid 2 + pyridoxal phosphate.

Hence the overall result shows up as:

amino acid 1 + $\alpha$-ketoacid 2 $\rightarrow$ amino acid 2 + $\alpha$-ketoacid 1.

This is what is meant by transamination.

## 7 Mechanism of Vision

The retina contains retinal, a substance related to vitamin A. It is attached to a protein called opsin (MW ~ 38 kD). Retinal has a conjugated chain allowing the molecule to absorb visible light. In the carbon side chain of retinal, all double bonds have *E* (or *trans*) configuration.

Neoretinal *β*, the photosensitive pigment in the rods, is the 11-*cis* isomer of retinal. It is bound by opsin as the Schiff base formed with the lysine 53 residue, a complex known as rhodopsin. This Schiff base upon protonation forms a hydrogen bond with an acceptor site A, another amino acid of the protein.

Rhodopsin is the main photosensitive pigment in the retina (the so-called retina purple). Absorption of a photon triggers a concerted migration of the double bonds, the π electrons being drawn to the positive charge; the latter thus goes to carbon-5 (a tertiary allylic carbocation, hence stabilized). In this way, the 11–12 bond is made into a single bond and internal rotation can occur. Simultaneously with the electronic migration, the proton H⁺ is transferred to the nitrogen: this proton transfer occurs within about 10 picoseconds.

As a consequence of rotation around the 11–12 double bond, the molecule isomerizes into the more stable *trans* form. This isomer, with a geometry different from that of the *cis* isomer, is unable to form a stable complex with the protein. Then follows a series of conformational changes leading to hydrolytic cleavage of the Schiff base and to bleaching of the protein. The *cis* → *trans* isomerization occurs in less than 200 femtoseconds!

References: Y. A. Ovchinnikov, *Pure Appl. Chem.*, **58**, 725–736 (1986),
          Bernstein, Law, and Rando, *Proc. Nat. Acad. Sci. USA*, **84**, 1849 (1987).
          Bridges and Alvarez, *Science*, **236**, 1678 (1987).
          J. Nathaus, D. Thomas, and D. S. Hogness, *Science*, **232**, 193–202 (1986).

$\lambda_{max}$ , nm

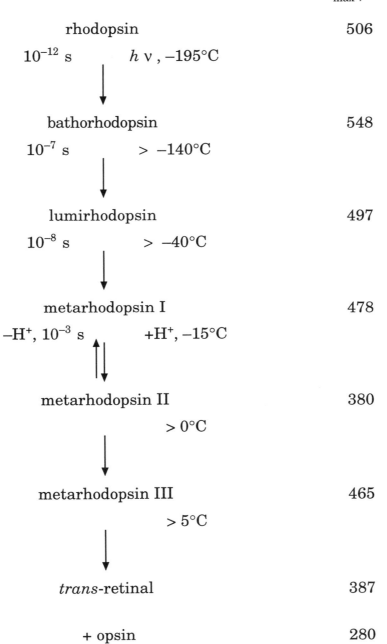

| | | $\lambda_{max}$, nm |
|---|---|---|
| rhodopsin | | 506 |
| $10^{-12}$ s    $h\nu$, −195°C | | |
| bathorhodopsin | | 548 |
| $10^{-7}$ s    > −140°C | | |
| lumirhodopsin | | 497 |
| $10^{-8}$ s    > −40°C | | |
| metarhodopsin I | | 478 |
| −H⁺, $10^{-3}$ s    +H⁺, −15°C | | |
| metarhodopsin II | | 380 |
| > 0°C | | |
| metarhodopsin III | | 465 |
| > 5°C | | |
| *trans*-retinal | | 387 |
| + opsin | | 280 |

Rhodopsin, with an absorption maximum at 506 nm and a large extinction coefficient of 40 000, is adapted to solar light. The polypeptide chain of rhodopsin crisscrosses seven times the membrane of the retina. The carbon and nitrogen ends of this chain are separated by the membrane. The lysine-53 residue, that binds retinal, is located near the C-terminal fragment, on the inside of the membrane.

After bleaching of the protein (= cleavage to *trans*-retinal and opsin), the aldehyde is reduced to the alcohol *trans*-retinol. It was only in 1987 that the enzyme membrane responsible for isomerization of this *trans*-retinol to *cis*-retinol in the dark was discovered: scientists had been searching for more than 30 years for this enzyme! Other enzymes convert *cis*-retinol to 11-*cis*-retinal.

## 8   Vision of Colors

Thomas Young (1802) postulated three types of rod cells involved in color vision. Each of them has a maximum absorbance for a specific wavelength:

|        | $\lambda_{max}$, nm |
|--------|---------------------|
| blue   | 455                 |
| green  | 530                 |
| red    | 625                 |

The opsins corresponding to these three pigments are related to rhodopsin. The DNA sequences coding for them have been established recently: 41% identical with rhodopsin. The green and red pigments are 96% identical with one another, but only 43% identical with the sequence for the blue pigment.

Each of these three types of rods (blue, green and red) contains 11-*cis*-retinal with an absorption maximum of 376 nm in the absence of the protein. The three pigment types arise from structural variations in the protein, genetically determined (see above) and identified by Nakanishi to a simple and convincing model of point charges: the absorption maximum of 11-*cis*-retinal shifts around depending on the precise location of a negative charge in the protein, at about 3 Å distance from carbons-12 and 14 of retinal.

Carotenoid pigments with structures related to those of retina pigments are responsible for the yellow color of rose petals.

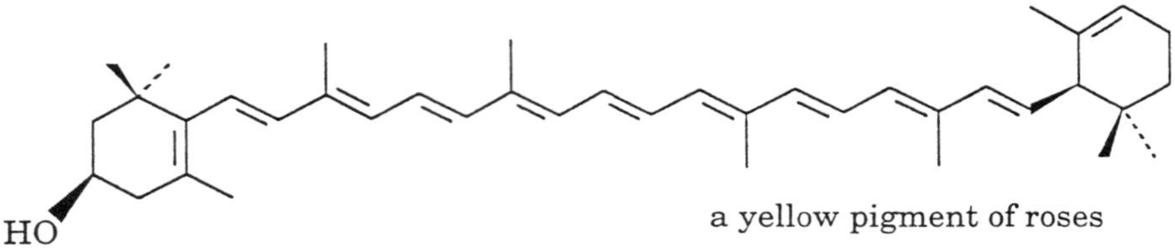

a yellow pigment of roses

References: K. Nakanishi, et al., *J. Am. Chem. Soc.*, **101**, 7082, 7084, 7086 (1979).
           K. Nakanishi, *Pure Appl. Chem.*, **57**, 769–776 (1985).
           M. Neitz, J. Neitz, and G. H. Jacobs, *Science* **252**, 971–974 (1991).
           R. W. Schoenlein, L. A. Peteanu, R. A. Mathies, and C. V. Shank, *Science*, **254**, 412–415 (1991).
           C. H. Eugster and E. Märki-Fischer, *Angew. Chem. Int. Ed. Engl.*, **30**, 654–672 (1991).

## 9   Stealth, in New Style

To escape radars, a military vehicle can be coated with a reflecting material (ferrites) stratified so that the reflections interfere destructively with one another.

Another possibility is conversion of the photons of the incoming beam to thermal energy. Retinal forms a Schiff base with the ε-amino group of the lysine moiety in opsin: as explained above, their assembly (rhodopsin) is responsible for vision. A molecular system related to retinal (see the following Scheme) shows much promise as a candidate for furtivity. There are two conformations of almost the same energy. In both the perchlorate anion is stabilized by electrostatic pairing to the organic cation, in which the underlined atoms all bear partial positive charges: either with an electrostatic bond between one oxygen and C-15 or with bonds between two oxygens and C-13 *and* C-15. Let us assume that the ground state consists of an ion pair of the latter type. Absorption of a photon in the microwave range will transform it within a short time to an ion pair of the first type. The perchlorate anion then reverts back to its former position with minimal emission of (thermal) energy, thus dissipating away the energy of the incoming, searching radar beam.

It remains only to synthesize a sufficient number of molecules of this type to absorb a large range of radio frequencies.

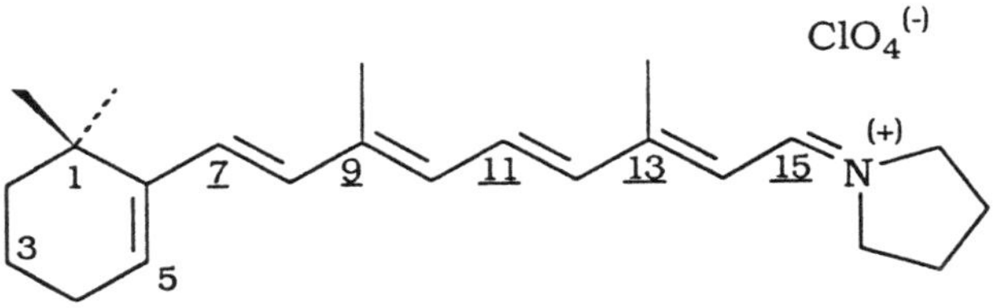

Reference: R. R. Birge, L. P. Murray, R. Zidovertzky, and H. M. Knapp, *J. Am. Chem. Soc.*, **109**, 2090–2101 (1987); see also *International Herald Tribune*, August 20, 1987.

## 10 Polyamides

Wallace H. Carothers (1896–1937), one of the best organic chemists in the United States, was hired away from Harvard to Wilmington by the research director of Dupont de Nemours with the promise that he would be able to continue doing fundamental research there. Some years later, in the course of an administrative reorganization, Carothers inherited a new boss. He was not as broad-minded as his predecessor. He insisted from the beginning that Carothers should confine himself to applied research. The first discovery to Carothers's credit were the polyamides (or nylons). A few months later, he committed suicide in a hotel room: he had been very much affected by the death of a sister to whom he felt closely attached; his depression was likely to have been worsened by his extreme dissatisfaction with the change of direction in his scientific career.

Let us consider an example of a polyamide: nylon 6,6 results from condensation of adipic acid (a $C_6$ compound) with hexamethylene diamine (another $C_6$ compound) at 280 °C. It is used in synthetic fibres for clothing, in coatings for tires, and in bearings.

In recent years, polyamides formed from aromatic residues or *aramides* have become very popular. Thus condensation of terephthalic acid dichloride with *para*-phenylene diamine affords *Kevlar*. Such polymers have excellent thermal stability due to the presence of aromatic rings. Their mechanical properties are excellent: Kevlar is used for bulletproof vests, hulls of competition sailing boats, ultralight ice skates, tire treads, radial tires, construction of ultra-light aircrafts such as the 'Gossamer-Albatros', etc. Kevlar has a mechanical resistance five times that of steel, and 10 times that of aluminum, at equal weight. Its resistance to stretching goes with smaller heat evolution, hence the widespread application to radial tires. The Formula 1 driver Nelson Piquet had his life saved by his Kevlar helmet.

This fiber has been produced in the United States since 1972, its development having started already in the sixties. A spinning factory for Kevlar opened at Maydown, Ulster in mid-1988, with a capacity of $7\,kT\,year^{-1}$ to serve Europe.

Reference: *Chem. in Britain*, **24**, 18 (1988).

# Carbonyl Group: Enolization—
# Michael Addition

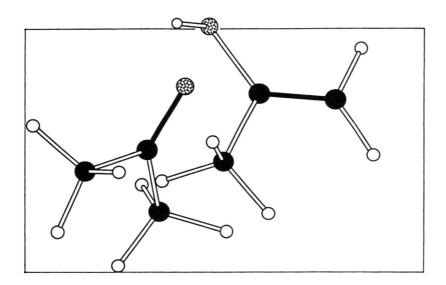

A remarkable feature of the carbonyl group is that it makes more acidic the adjacent carbon—hydrogen bonds. Such *enolization* is catalyzed by acids and bases. The easy formation of carbanion enolates is one of the major assets of the carbonyl group. Enols and their fixed form, enol ethers, are often-used synthetic intermediates. They owe their importance both to the high reactivity of their C=C double bonds, and to the high nucleophilicity of the terminal carbon.

Conjugation of a carbonyl group with a C=C double bond forms a so-called Michael acceptor. The Michael addition is the second carbon–carbon bond formation mentioned in this book. It consists simply in the union of a nucleophile (an electron pair donor) with a Michael acceptor. When the Michael addition is followed by a nucleophilic addition to a carbonyl group, it constitutes the Robinson annulation, an elegant method for constructing medium rings, especially six-membered ones.

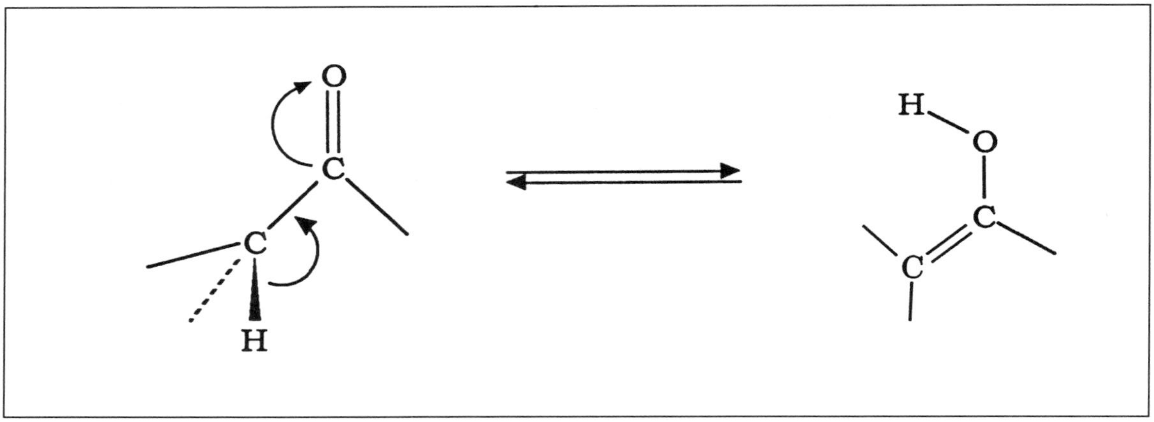
O
C
C
H
H
O
C
C

# 1 Keto–Enol Equilibrium

Such tautomerism, in which a ketone is transformed into an enol, occurs by migration of a proton, from one of the carbons $\alpha$ to the carbonyl group, to oxygen: it is accompanied by the concomitant electron transfer.

The equilibrium favors usually the keto form. The enthalpy difference, of the order of 16 kcal mol$^{-1}$, favors the ketone. The predominant factor is the greater bond energy of the C=O bond (179 kcal mol$^{-1}$) as compared to a C=C bond (142 kcal mol$^{-1}$). For simple ketones, see Table 12, the equilibrium constant is in a range between $10^{-2}$ and $10^{-6}$.

**Table 12**  Equilibrium constants of simple ketones

| Ketone | $K = [\text{enol}]/[\text{ketone}]$ |
|---|---|
| acetone | $2.5 \times 10^{-6}$ |
| methylethylketone | $1.2 \times 10^{-3}$ |
| diisopropylketone | $3.7 \times 10^{-5}$ |
| cyclobutanone | $0.55 \times 10^{-2}$ |
| cyclopentanone | $4.8 \times 10^{-5}$ |
| cyclohexanone | between $1.2 \times 10^{-2}$ and $4.1 \times 10^{-6}$ depending on authors |

'My respect for the miniscule takes on gigantic proportions' (Karl Kraus).

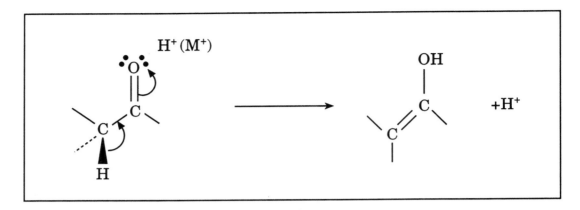

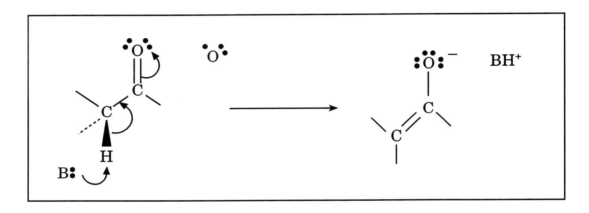

## 2 Catalysis of the Keto–Enol Equilibrium

Acids catalyze the interconversion of a ketone and the corresponding enol. The attachment of a proton or a Brønsted acid or, more generally, of an electrophile (Lewis acid) to the carbonyl oxygen enhances the polarization of the C=O bond. An electron migration $\sigma$ (C—H) $\rightarrow \pi$ (C=C) affords the conjugated acid of the enol.

In a similar way, bases catalyze the transformation of ketone into the corresponding enolate anion. The base removes a proton from the $\alpha$ carbon. The resulting enolate carbanion is stabilized by migration of an electron lone pair from carbon to oxygen, where the negative charge gains additional stabilization with formation of an ion pair.

'It is often the abstract that is bad and stupid'. Alain, *Propos* (October 21, 1923).

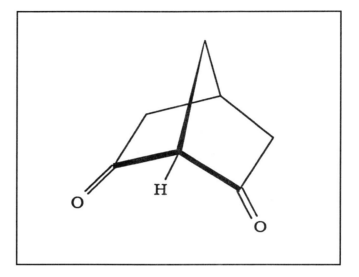

# 3 Geometry Required for Enolization

The C—H bond broken in the ketone has to be orthogonal to the C—C=O plane, for maximum overlap ($\sigma$ C—H/$\pi$* C=O) as will be explained below. The enol has a C=C bond. Its formation however can be prohibited (e.g. by Bredt's rule which forbids double bonds at bridgehead positions in strained bicycles). Thus the diketone in the adjacent Scheme is not enolizable at the bridgehead.

Enolization is accompanied by a lengthening of the C—O bond and a shortening of a C—C bond.

There is a great deal of experimental evidence in support of these statements.

---

'A physical theory is above all a hypothetico-deductive system built a priori according to artistic preferences for the algebraic formulation to be tried on the phenomenon. In order for the physical data and the algebraic conclusions to coincide, it is by no means necessary that the principles of the theory be true. As shown by formal logic, although true premises never lead to wrong conclusions, the wrong can however generate the true'. J. Maritain, *Réflexions sur l'intelligence*, Desclée de Brouwer, Paris, 1930, p. 242.

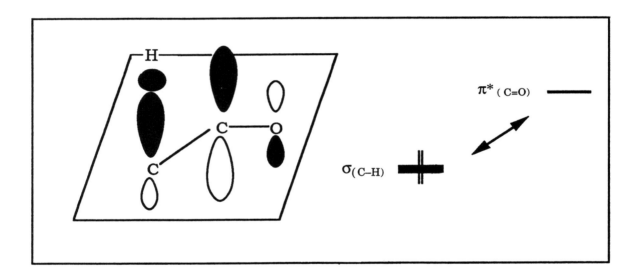

# 4  Weakening of the H—Cα Bond in a Ketone

The increase of acidity is considerable in going from a hydrocarbon to the corresponding ketone. Propane has a $pK_a$ of the order of 50; that of acetone is about 20.

In the resonance formalism, this acidity gain is pictured as a hybrid. There are three main contributing forms: one with a bond between carbon and hydrogen and two lacking that bond. They are the uncharged keto form with a C—H bond; and the two forms, without a carbon—hydrogen bond, with a positive charge at hydrogen and a negative charge either at carbon or oxygen.

In equivalent manner, the MO method describes the weakening of the hydrogen α carbon bond as a two-electron interaction between the $\pi^*$ level of the carbonyl as acceptor and the $\sigma$ level of the *anti* C—H bond as donor. This explains the required orthogonality of the C—H bond and the carbonyl plane: this $\sigma$–$\pi^*$ two-electron stabilizing interaction is at its maximum with the C—H bond parallel to the axes of the $2p_z$ atomic orbitals at carbon and oxygen.

'I leaped from peak to peak: my discourse should not follow a single path', Empedokles, fragm. 24.

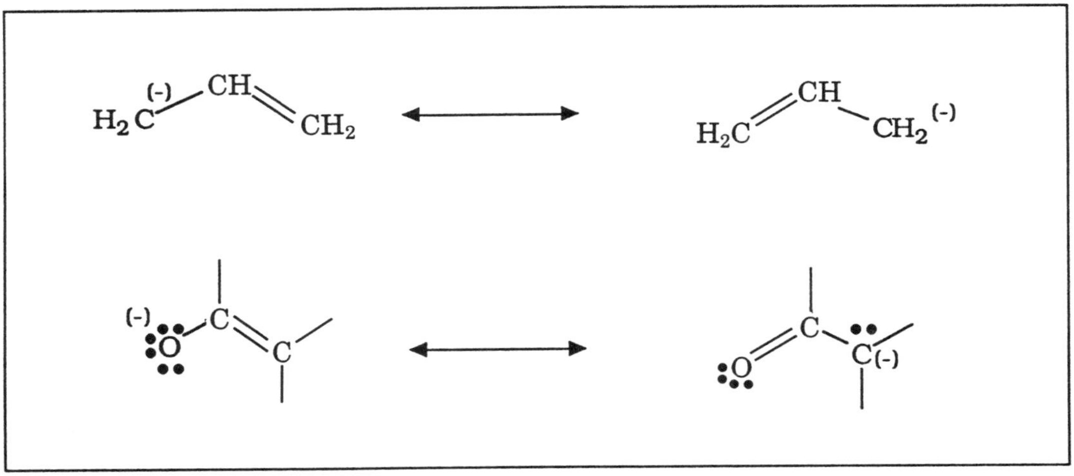

(-)
H2C
CH
CH2
H2C
CH
CH2 (-)
(-)
O
C
C
O
C
C(-)


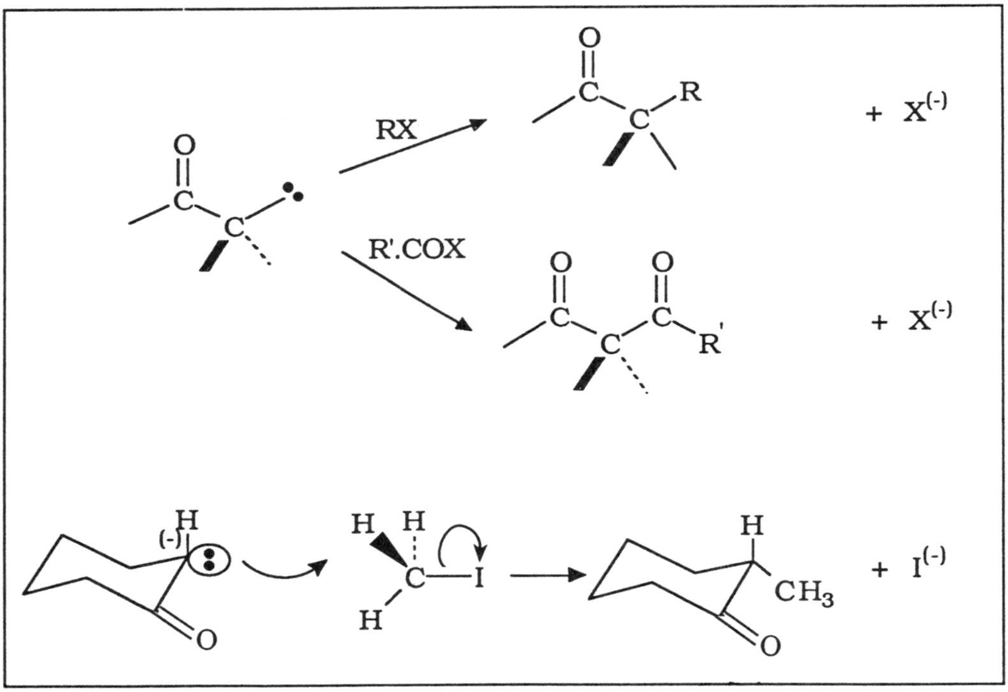

O
C
C
RX
R'.COX
R
+ X(-)
R'
+ X(-)
H
(-)
H
H
H
C
I
H
CH3
+ I(-)
O


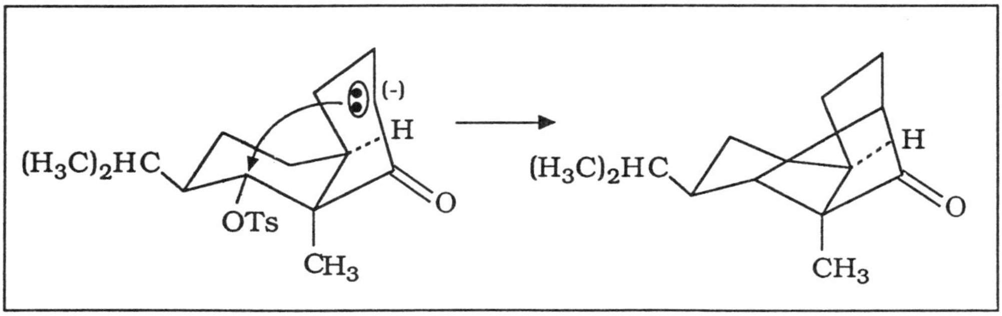

(H3C)2HC
(-)
H
OTs
O
CH3
(H3C)2HC
H
O
CH3

# 5  Enolates and their Formation

The conjugate base of a ketone, resulting from ionization of the $H-C_\alpha$ bond, is an enolate anion. This anion is stabilized by resonance, the negative charge residing either on carbon or on oxygen: the enolate anion is isoelectronic to the allyl anion. In obvious contrast to the allyl anion, the two limiting forms of the enolate anion are not equivalent. That with the negative charge on the more electronegative oxygen has greater weight. Conversely, the more reactive form is that with negative charge on carbon. In other words, in an enolate, carbon is a better nucleophile than oxygen.

Since enolates possess this carbanion reactivity, they lend themselves to easy formation of new carbon—carbon bonds at the $\alpha$-carbon, either by alkylation or acylation. It is sufficient to let the enolate react with an entity R.X or R'.CO.X, in which X is a good leaving group.

A recent synthesis of sativene, a hydrocarbon produced by the lichen *Helminthosporum sativum* provides an intramolecular equivalent of such a substitution reaction at the $\alpha$-position of a carbonyl group, through the intermediate formation of an enolate. Only the key-step is represented. The tricyclic product is eventually transformed to sativene by a Wittig reaction ($C=O \rightarrow C=C$).

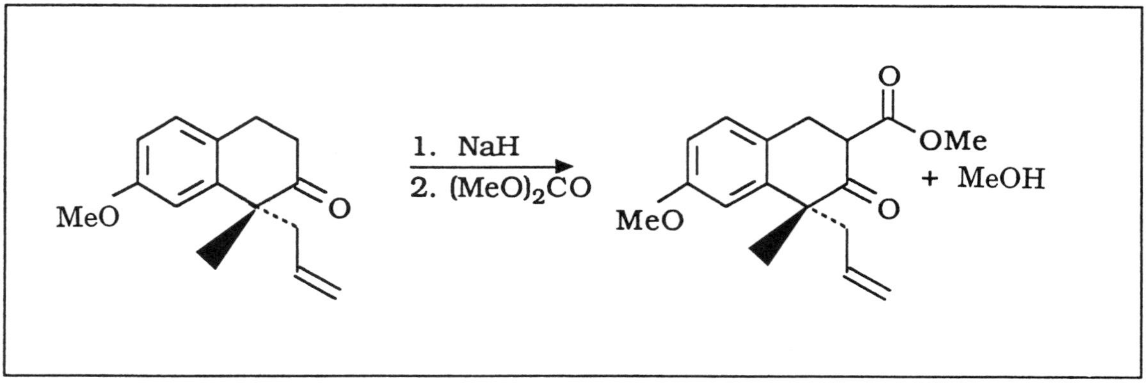
MeO
O
1. NaH
2. (MeO)2CO
O
OMe
+ MeOH
MeO
O

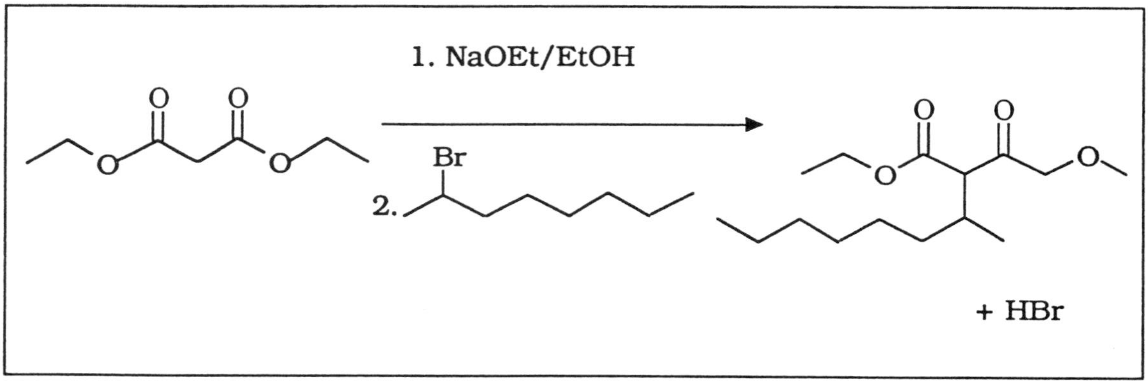
1. NaOEt/EtOH
O
O
O
O
Br
2.
O
O
O
O
O
+ HBr

As indicated above, enolate anions tend to react as nucleophiles from the carbon end rather than from the oxygen end. Let us be more precise. The general rule is that the reactivity at carbon prevails for those enolates, devoid of any special stabilization, when the $pK_a$ of their conjugate acid is at least 24. In the adjacent example, sodium hydride forms the enolate anion in the $\alpha$-position to the carbonyl group and its reaction at carbon rather than at oxygen leads to the indicated ketoester (D. C. Palmer and M. J. Strauss, *Chem. Rev.*, **77**, 1 (1977)).

Use of protic solvents SH (e.g. ethanol in the example indicated) is another means to induce reactivity of an enolate at carbon rather than at oxygen. The reason is stabilization of the resonance form of the enolate bearing a negative charge on oxygen by a hydrogen bond $RO^{(-)} \dots HS$; which reduces the reactivity of this form (E. Le Goff, S. E. Ulrich and D. B. Denney, *J. Am. Chem. Soc.*, **80**, 662 (1958)).

Formation of a $\beta$-ketoester (in the upper box) is an example of an acylation via an enolate. In the same way alkylation of a 1,3-diester with *i*-octyl bromide occurs via the enolate generated with sodium ethoxide.

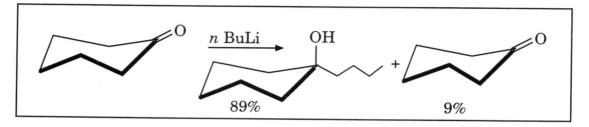

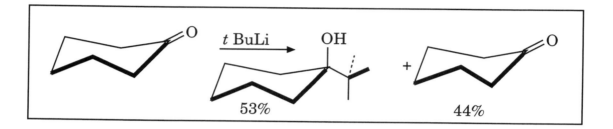

If a reagent is both a strong base and a good nucleophile, one can expect two types of reactivity. Addition of the nucleophile to the carbonyl group competes with formation of the enolate anion, due to attack by the same reagent as a strong base at one of the $\alpha$-carbons. Thus reaction of $n$-butyl lithium—a strong base and a very reactive nucleophile—with cyclohexanone affords a mixture of products: the tertiary alcohol in 89% yield and cyclohexanone in 9% yield (reprotonation by exchange of $Li^+$ by $H^+$ during isolation). This experiment is due to J. D. Buhler, *J. Org. Chem.*, **38**, 904 (1973).

If instead of $n$-butyl lithium, $t$-butyl lithium is the reagent, the part accounted for by enolization increases to 44% (with 53% addition). This shows that $t$-butyl lithium is a stronger base than $n$-butyl lithium. This observation can be generalized thus: organolithium compounds RLi have increased basicity with increasing substitution of the R substituent. $t$-BuLi is more basic than $s$-BuLi, itself a stronger base than $n$-BuLi. Thus $t$-BuLi is capable of deprotonating acids with a $pK_a$ up to 54 (saturated hydrocarbons), $s$-BuLi goes up to $pK_a \sim 50$, whereas $n$-BuLi is limited to $pK_a < 46$. The other organometallic reagents, organomagnesium compounds and organocuprates have weaker basicities.

$$HN\,(CH(CH_3)_2)_2 + Li\,CH_2CH_2CH_2CH_3 \xrightarrow{\text{THF}} Li\,N\,(CH(CH_3)_2)_2 + H_3CCH_2CH_2CH_3$$

Let us come back to the acidity of a carbon–hydrogen bond $\alpha$ to a carbonyl group with a p$K_a$ of the order of 20. This acidity is rather weak compared to that of an alcohol (p$K_a$ approx. 16) or by far weaker than a carboxylic acid (p$K_a$ approx. 5). Therefore strong bases are required to transform a carbonyl compound such as an aldehydes or a ketone into the corresponding enolate.

Strong bases commonly used for this purpose are sodium hydride NaH, sodium amide NaNH$_2$ and lithium diisopropylamide LiN($i$-C$_3$H$_7$)$_2$ (abbreviated as LDA). The latter base has the advantages of high basicity (since isopropylamine has a p$K_a$ of 40), of good solubility in organic solvents, and particularly of a weak nucleophilicity due to the steric hindrance of nitrogen by the isopropyl groups. Thus, there is no longer risk of competition between addition to the carbonyl group and deprotonation at the $\alpha$-carbon. In practice, LDA is prepared by reaction of the amine with an organolithium compound (such as butyl lithium) in an ether solvent, e.g. tetrahydrofuran (THF).

‘There is chemistry from red to blue, effervescence and violet profusion in the test tubes, in mauve colored paper filters’, F. Ponge, *Le Grand Recueil (Pièces)*, Gallimard, Paris, 1961, p. 135.

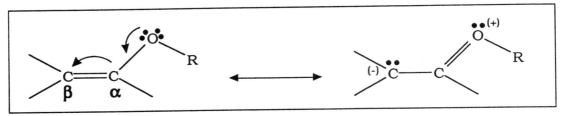

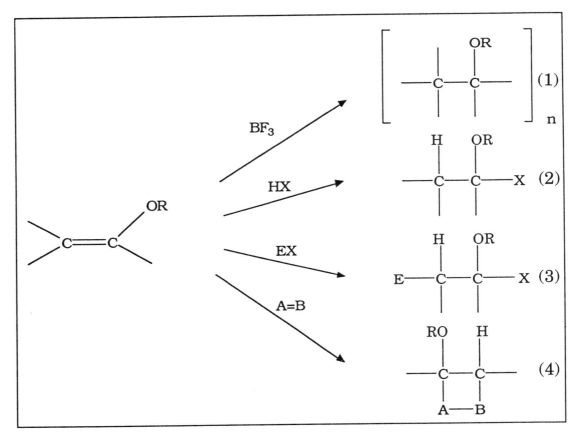

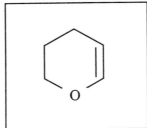

# 6   Enol Ethers

Like the isoelectronic enamines, enol ethers have an 'electron rich' C=C double bond. In the formalism of resonance, this is expressed by the delocalization of a lone pair at oxygen towards the $\beta$-carbon.

Enol ethers can undergo four different types of reaction:

    (1) polymerization with Lewis acids;
    (2) reaction with protic reagents, with regeneration of the parent carbonyl compound or formation of acetals by addition of a molecule of alcohol to the $\alpha$-carbon;
    (3) attack of electrophiles at the $\beta$-carbon; this leads to additions (which can be followed by eliminations) and/or substitutions;
    (4) cycloadditions.

An enol ether often used due to its low cost is dihydropyran (DHP). An acid catalyzed reaction of type (2) with alcohols provides the corresponding acetals; this is an easy way of protecting an alcohol function:

$$ROH + DHP \xrightarrow{\;H^{(+)}\;} RO\,(THP) + H^{(+)}$$

The alcohol is regenerated by an acidic hydrolysis

$$RO(THP) + H_3O^{(+)} \longrightarrow ROH + DHP + H_2O$$

(THP = tetrahydropyran)

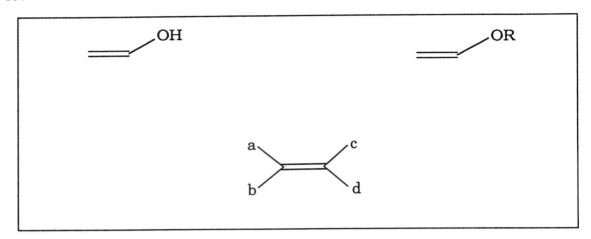

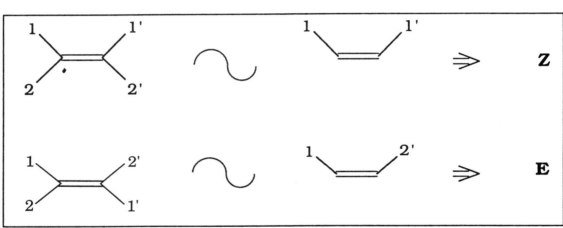

# 7 Notation of the Configuration: Enols and Enol Ethers

Enols and enol ethers are to be considered in like manner as alkenes.

If one examines the priorities of the substituents at the two ends of the C=C double bond, the two possibilities are for 1 and 1′ to be *cis* to one another (i.e. on the same side of the C—C axis) or *trans* to one another (i.e. on different sides of the C—C axis). In the first case the configuration is named Z (German *zusammen*), in the second case the configuration is named E (German *entgegen*). The priorities of the substituents follow the same rules already explained for the R and S configuration of an asymmetric carbon.

This is applied to the Z enol of butanone. One of the carbons bears a hydrogen and a methyl group: the latter has priority. The other carbon bears a hydroxy group OH and a methyl group: the oxygen function has priority, oxygen having an atomic number greater than that of carbon.

---

'There are figures that are clear and demonstrative; but others do not appear natural and convince only those who are already convinced', Pascal, *Pensées*.

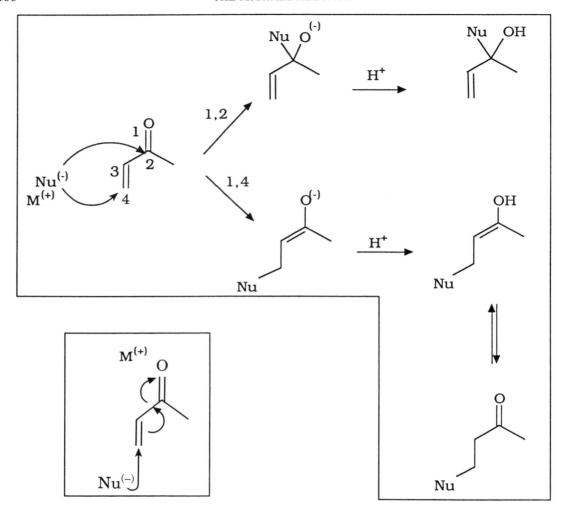

# 8 The Michael Addition

The substrate in this reaction is a conjugated aldehyde or ketone, the Michael acceptor. The Michael donor can be any nucleophile—most commonly an enolate anion, or a carbanion, e.g. derived from an activated methylene compound. A new carbon–carbon bond is formed between the donor and either carbon 2 or carbon 4 of the acceptor (we shall return to this regioselectivity problem). A first possible outcome is simple addition to the carbonyl group. Michael addition, also known as conjugate addition, is the name of the process when carbon 4 is the acceptor site.

The great utility of this reaction lies in the formation of a new C—C bond. Thus Michael additions allow extensions of a carbon chain length.

During the reaction at carbon 4, an electron lone pair migrates from the donor carbon to the more electronegative oxygen of the acceptor. The driving force is the formation of a new carbon–carbon bond (ca. $80 \, kcal \, mol^{-1}$ bond energy). This enthalpy factor is more than sufficient to compensate for the loss of entropy attendant upon joining two particles. This addition, invented by Michael in 1887, illustrates well his theoretical concepts.

Reference: A. Michael, *J. Prakt Chem.*, **35**, 349 (1887).

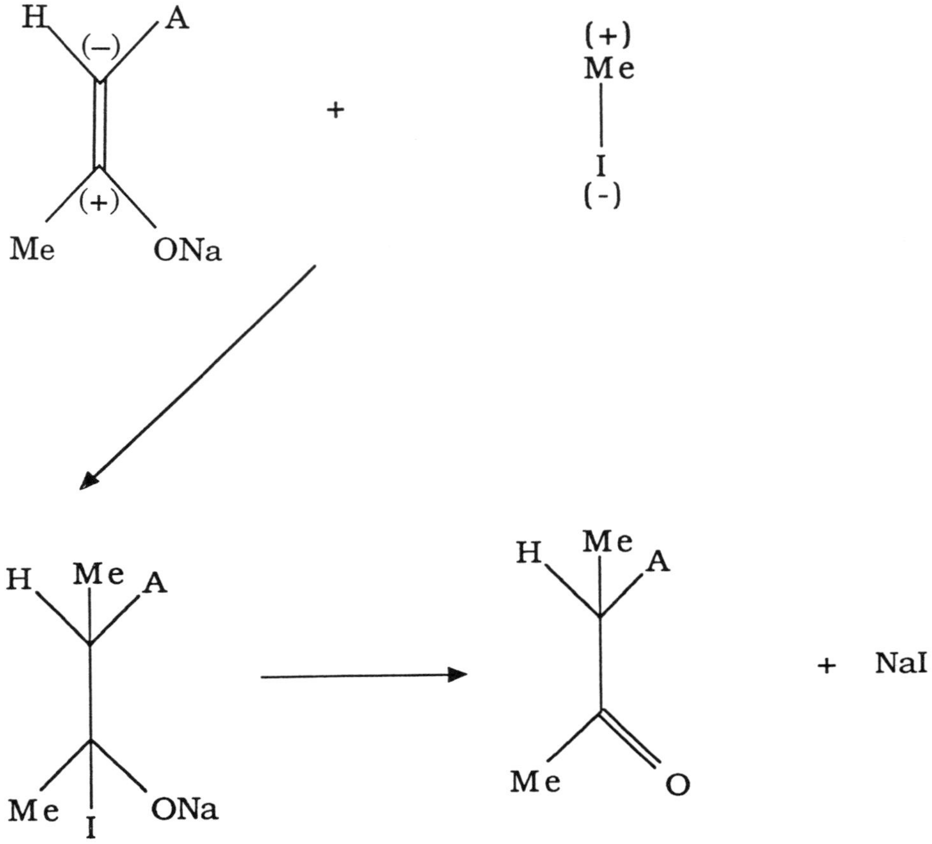
H
A
(–)
(+)
Me
ONa
+
(+)
Me
I
(-)
H
Me
A
Me
I
ONa
H
Me
A
Me
O
+ NaI

# 9 Who was Michael?

Arthur Michael (1853–1942) was a great American chemist. Born in Buffalo on August 7, 1853, and having completed his secondary school education in this city, he went to study with Hofmann at the University of Berlin in 1871. A year later he spent two years with Bunsen in Heidelberg. From 1875 to 1878 he was again with Hofmann in Berlin; then after 1879 he spent some time in the laboratories of Wurtz in Paris and of Mendeleyev in Russia. After these years of study, concentrated mainly in Germany, he became professor at Tufts College from 1880 until 1891 and from 1894 until 1907. From 1912 until his retirement he was a professor at Harvard. Michael died in 1942. Michael considered himself basically a theoretical organic chemist. He is to be credited with the introduction of physico-chemical concepts, especially thermodynamic, in organic chemistry: in his view, it was where German chemistry made a large contribution. Physical chemistry was much favored by the German universities in the last quarter of the 19th century. Michael was one of the propagators of this new chemistry in the United States.

For Michael, influenced by chemical evolutionism, everything followed from the second law of thermodynamics. He interpreted entropy as 'chemical neutralization', i.e. as mutual neutralization of free energies associated with reactive atoms during a chemical transformation. Michael took up Kekulé's idea: the first step in the reaction $A + B$ is the formation of a super-molecule A,B. He interpreted the second law as follows:

'Every chemical system adjusts itself to the configuration leading to a maximum of chemical neutralization' (*J. Prakt. Chem.*, **60**, 286 (1899)).

And Michael went on to deduce the 'positive–negative' rule: the maximum neutralization obtains if the most electronegative atom (or group) from one of the partners combines with the most electropositive atom (or group) of the other partner, or vice versa (*J. Prakt. Chem.*, **40**, 171 (1899); **46**, 189 (1892)).

For Michael, the reaction that bears his name, which he invented in 1887, was such an example of the 'positive–negative' rule. He formulated it as the union of a positive methyl, methyl iodide, with the most negative unsaturated carbon whereas the attraction of sodium for iodine was stronger than that for the oxygen ($A = COOC_2H_5$).

---

Reference: A. B. Costa, *J. Chem. Ed.*, **48**, 243–246 (1971).

'In case of doubt, choose what is correct', Karl Kraus.

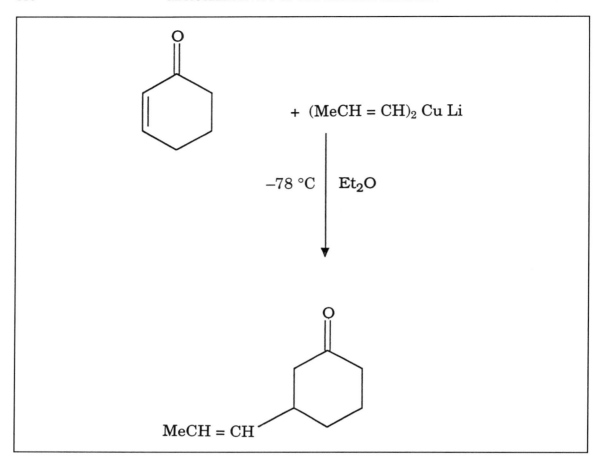

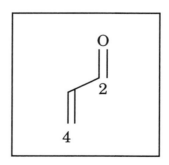

# 10   Regioselectivity of the Michael Addition

The nature of the metal cation determines which pathway predominates, between 1,2-addition or 1,4-addition. Organoalkali reagents ($M^+ = Li^+$, $Na^+$, $K^+$) give preferentially 1,2-addition. Organomagnesium and organoaluminum compounds lead to mixtures of 1,2-addition or 1,4-addition products. A regioselectivity favorable to 1,4-addition—i.e. what is usually meant by 'Michael addition'—is obtained by means of organocadmium compounds and, especially, organocuprates. Why?

The formalism of hard and soft acids and bases provides an explanation. It would appear that orbital control is in charge whether 1,2-addition (naively described by charge control) or 1,4-addition occurs. In the latter case, the soft metal cation ($Cd^+$, $Cu^+$) induces softness in the anionic nucleophile. Under orbital control, the new C—C bond is formed at the carbon having the larger coefficient of the atomic orbital: for the free ketone, uncomplexed by the metal, at C(4) (oxygen=hard base; metal=soft acid, hence a weak interaction). Let us consider the opposite case of a hard metal ($Li^+$) strongly coordinating oxygen: now the largest coefficient is at C(2). Thus the maximum overlap also occurs at C(2) and therefore the new C—C bond is formed with this carbon.

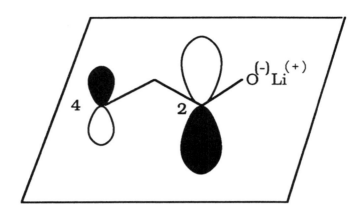

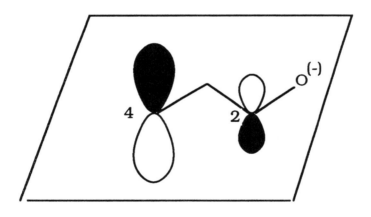

Indeed the predominant interaction involves the frontier orbitals: a doubly occupied level of the nucleophile combines with the LUMO (lowest vacant orbital) of the Michael acceptor.

This has been proved experimentally. If cyclohexenone is reduced with lithium aluminum hydride, 98% of 1,2-addition is accompanied by 2% 1,4-addition: lithium coordinates with oxygen with the largest coefficient being at C(2); thus the hydride anion adds to this center. If the lithium cation is removed from coordination with oxygen, by encapsulation into one of Jean-Marie Lehn's cryptands (Nobel Prize for Chemistry, 1987), that with a cavity of appropriate size (the '2.1.1'), the LUMO becomes that of the free ketone. One can predict then a predominant addition of the $H^{(-)}$ to C(4). This is indeed observed: 77% of 1,4-addition *vs.* 23% of 1,2-addition.

Let us repeat the rule: a soft (hard) reagent adds preferentially at 4 (2).

The Michael acceptor has carbocationic character due to the $C^+O^-$ polarization of the carbonyl group. Thus the C(2), C(3), C(4) segment resembles an allyl cation. Therefore the LUMO of a Michael acceptor (adjacent scheme) is isomorphous to the non-bonding level of an allyl cation.

References: A. Loupy and J. Seyden-Penne, *Tetrahedron Lett.*, **1978**, 2571.
N. T. Anh, *Topics in Current Chemistry*, **88**, 146–162 (1980).

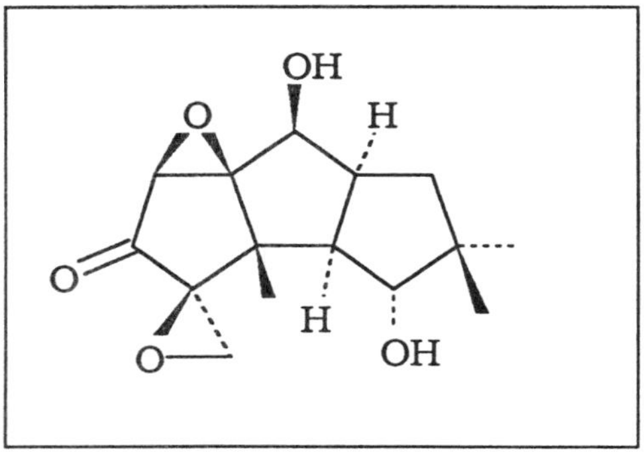
OH
O
H
O
H
OH
O

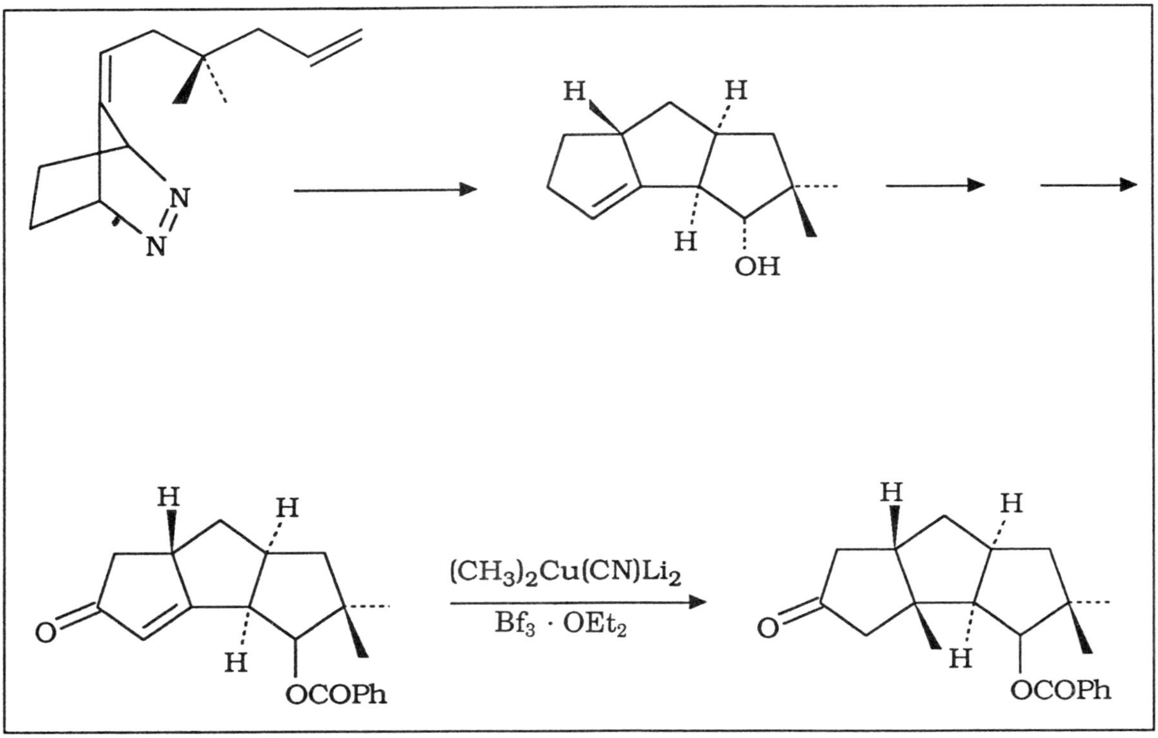
N
N
H
H
H
OH
H
H
H
OCOPh
$(CH_3)_2Cu(CN)Li_2$
$Bf_3 \cdot OEt_2$
H
H
H
OCOPh
O
O

# 11  Use of Organocuprates

Gilman reagents $R_2CuLi$ are obtained by reaction of an organolithium compound RLi with a cuprous halide according to the equation $2RLi + CuX = R_2CuLi + LiX$, $X = Br$, I. They allow for a regioselective addition, cf. regiospecificity in chapter 4.

Another type of mixed organocuprate of the general formula $R_2Cu(CN)Li_2$ obtained according to the reaction $2\ RLi + CuCN = R_2Cu(CN)Li_2$ is also used.

The example is a step in the synthesis of coriolin, an antitumor agent (see upper box). It was the intention of Van Hijfte and Little to introduce a methyl group (by way of a Michael addition) at the strongly hindered 4-position. For this purpose, they used $BF_3$ as an activator (coordination of the carbonyl oxygen) and an organocuprate: the Michael addition takes place very smoothly at $-50°C$ in tetrahydrofuran with an isolated yield of 80% after three hours.

References: C. Fréjaville and R. Jullien, *Tetrahedron Lett.*, 2039 (1971).
C. P. Casey and R. A. Boggs, *Tetrahedron Lett*, 2455 (1971).
L. van Hijfte and R. D. Little, *J. Org. Chem.*, **50**, 3940 (1985).
B. H. Lifshutz, *Synthesis*, 325–341 (1987).

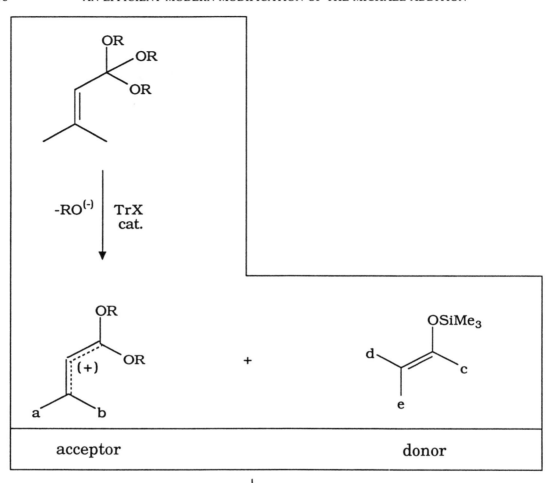

# 12 An Efficient Modern Modification of the Michael Addition

In this modification, the Michael donor is neutral: instead of an enolate, a silylated enol ether is used. The Michael acceptor is charged, an allyl cation, stabilized by two terminal ether functions; it is formed from an $\alpha$, $\beta$-unsaturated orthoester, with a trityl salt TrX as catalyst (Tr=CPh$_3$; X=ClO$_4$). Typically, the reaction is carried out in methylene chloride overnight followed by hydrolysis with NaHCO$_3$ in aqueous solution. The yields are very good (58–90%).

One can similarly catalyze the reaction between an $\alpha$, $\beta$-unsaturated ketone (Michael acceptor) and a silylated enol ether.

References: T. Mukaiyama, M. Tamura and S. Kobayashi, *Chem. Lett.*, 1817 (1986).
S. Kobayashi and T. Mukaiyama, *Chem. Lett.*, 1183–1186 (1987).

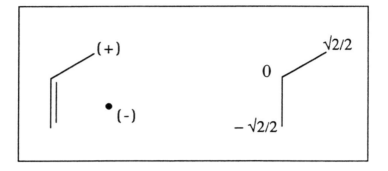

# 13  Conjunction or Disjunction

Attack by a nucleophile of an $\alpha$, $\beta$-unsaturated carbonyl compound leads to a 1,2- or 1,4-addition and thus to different reaction products. One could ask the title question: why do not 1,2- and 1,4-addition occur at the same time?

Reduced to its simplest expression, the Michael addition combines a carbanion $^{(-)}CH_3$ with an allyl cation $H_2C{=}CH{-}^{(+)}$. In the perturbational approach of Dewar the variation in the energy is

$$\Delta E = \sqrt{2}.\beta < 0$$

if the carbanion is attached to one or the other terminal carbons of the allylic system. Thus, on this simple approximation level, the addition is predicted to occur indifferently: 1,2 or 1,4 due to the same coefficients at the two terminal carbons in the non-bonding molecular orbital of the allyl cation.

If, however, two bonds between the carbanion and the allyl cation are formed, the gain of perturbational energy is zero: in the non-bonding MO of the allyl cation, the signs of the coefficients of the terminal carbons are opposite. This is the major reason for the non-existence of a double addition $(1,2) + (1,4)$.

To put it in other terms, the transition state is isoelectronic with 1,3-butadiene in the first case (stabilization by extension of conjugation) and with cyclobutadiene (antiaromatic destabilization) in the second case.

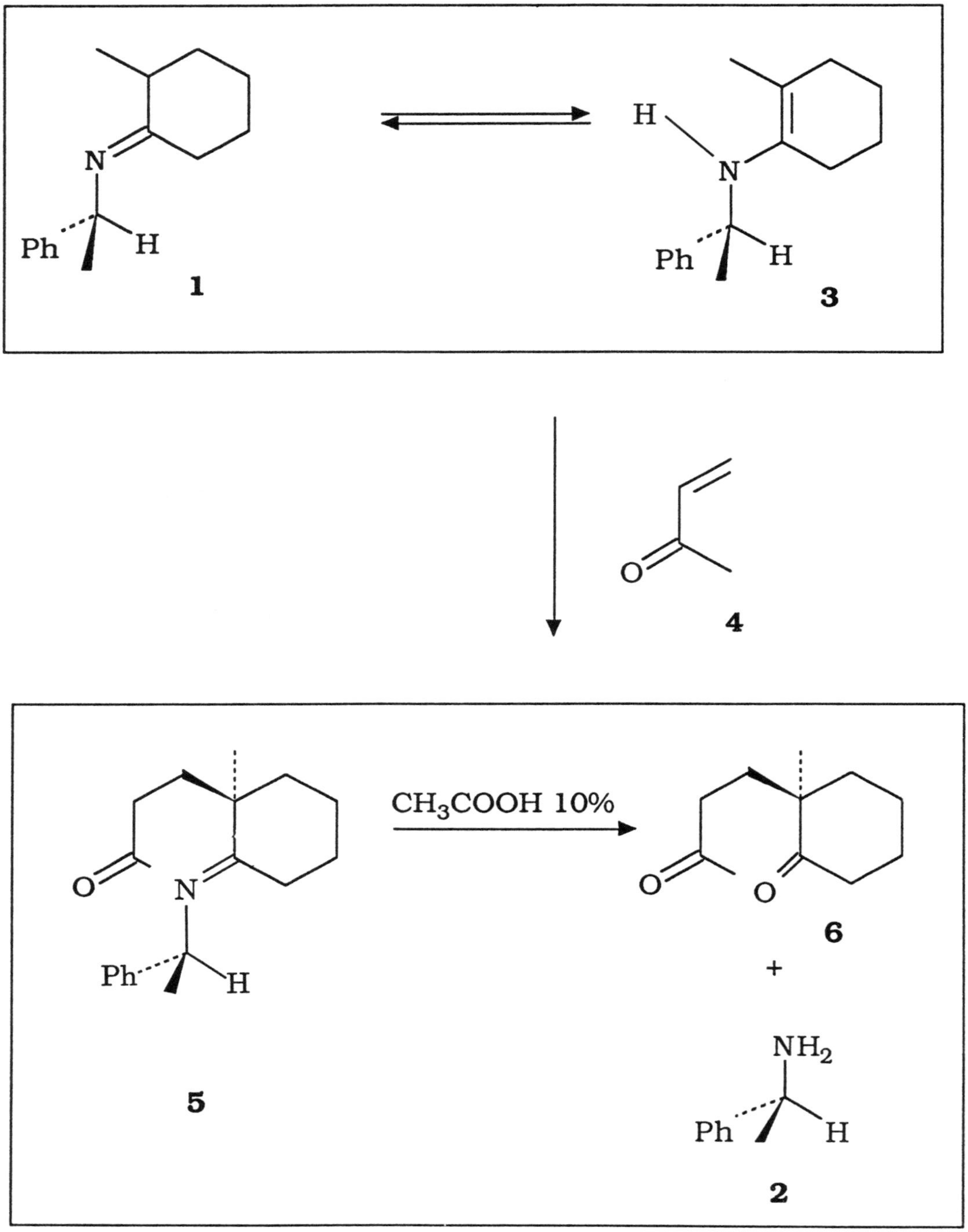
Ph
H
N
1
H
N
Ph
H
3
O
4
CH3COOH 10%
O
N
Ph
H
5
O
O
6
+
NH2
Ph
H
2

# 14  Michael Addition with Asymmetric Induction

Imine 1 results from the reaction between racemic 2-methylcyclohexanone and (S)-(−)-1-phenylethylamine 2. It undergoes a Michael addition (1 and 3 are in a tautomeric equilibrium) with methylvinylketone 4 to afford the adduct 5. Hydrolysis of 5 with acetic acid quantitatively regenerates the chiral auxiliary 2 and provides the alkylated cyclohexane 6 in good yield (88%) and with an enantiomeric excess of 91%.

Reference: M. Pfau, G. Revial, A. Guingant, J. J. d'Angelo, *J. Am. Chem. Soc.*, **107**, 273–274 (1985).

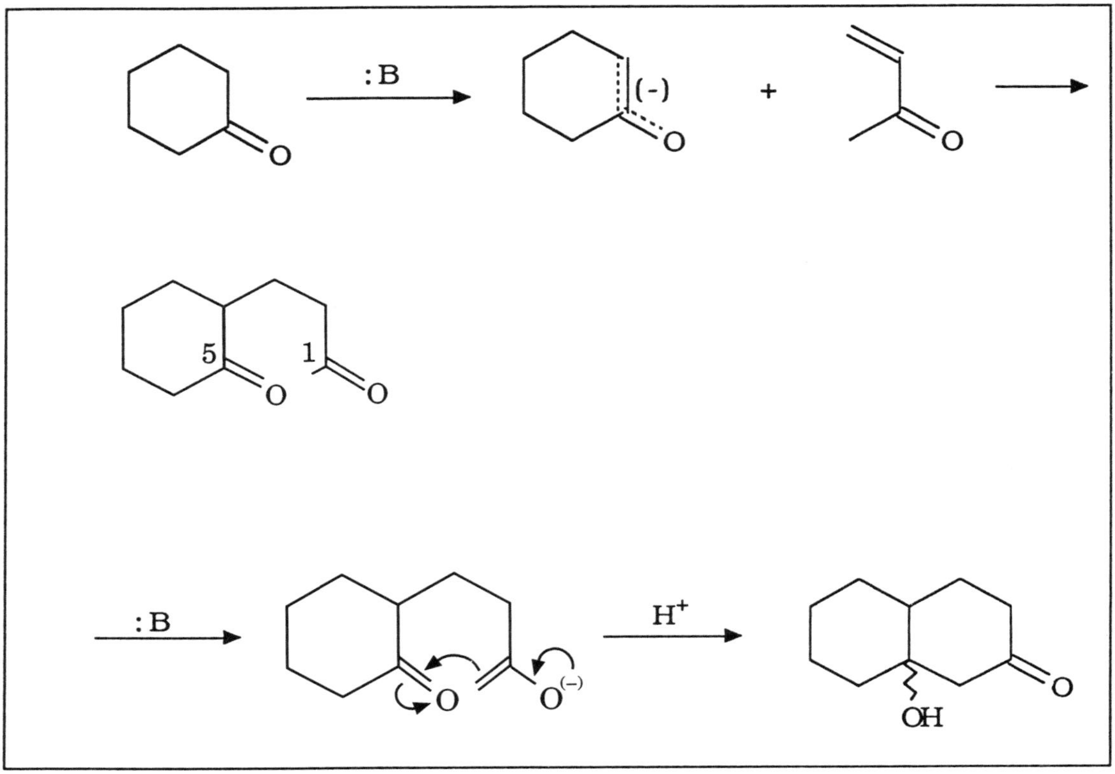

:B
(-)
O
+
5
1
H+
OH


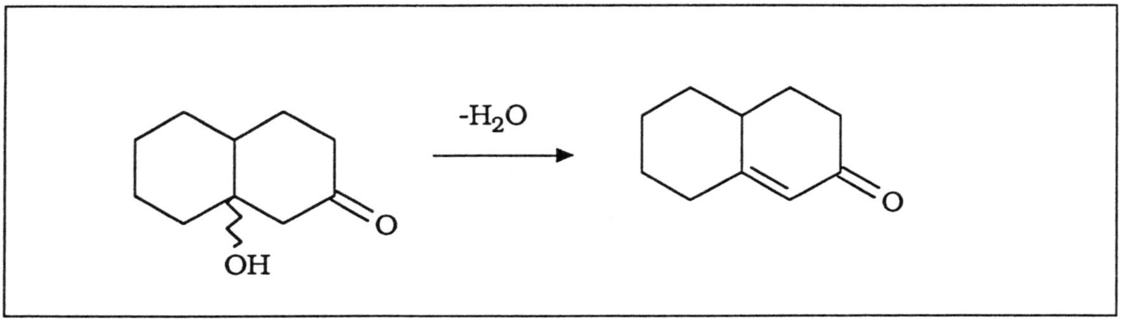

-H2O
OH
O

# 15   Robinson Annulation

The reaction sequence known under this name is a useful application of the Michael addition. Let us use an enolate anion as a Michael donor. The carbon end—a better nucleophile than the oxygen end, as one will recall—adds to position 4 of the Michael acceptor. The result is a 1,5-dicarbonyl compound.

It is again possible to enolize one of the two carbonyl groups. Under kinetic control, a base removes a proton from the methyl group of the methylketone. The enolate thus formed is trapped by the other carbonyl group in position 5. After this aldol condensation provides a six-membered ring, dehydration of the tertiary alcohol—an extremely smooth and usually spontaneous reaction—leads to the conjugated ketone, this conjugation being the driving force for the overall reaction.

The global process is called the Robinson annulation (Latin *annulus* = ring). This reaction sequence was discovered by Robert Robinson in 1937.

'What combinations of ideas, nonsense alone can wisely form!', James Clark Maxwell.

# 16 Sir Robert Robinson (1886–1975)

He was born on September 13, 1886, in Rufford Farm near Chesterfield. His father, William Bradbury Robinson, an inventor of textile machines, proved to be as productive as he was prolific: eight children from his first marriage, five from the second, among which Robert. He became educated in a religious school (Moravian sect): the good fathers gave him an unquenchable desire for the chess game and convinced his father to send him to university (Manchester, Chemistry Faculty).

The decisive moment was the second year of his studies when he discovered Perkin's lectures; he wrote later that the clarity of the exposition touched upon the miraculous. After his BSc (1905) he entered Perkin's private(!) laboratory. In 1909 he defended his thesis, working on the structure of brasiline, a natural pigment. During his years in Manchester, Robert Robinson was hyperactive: in the laboratory until three o'clock in the morning, playing the piano; and above all mountain climbing where he met his first wife Gertrude, also an ardent alpinist.

The university career of Robinson can be summed up as follows: professor in Liverpool, in Sydney (Australia) (1912–1914), at St. Andrews and at Manchester (1916–1928), then at University College London (1928–1930), and eventually in Oxford where he succeeded Perkin (1930–1955). His work dealt with structure and synthesis of natural products: dyes, steroids and mainly alkaloids. Robinson, who became Sir Robert in 1939, received the Nobel Prize in chemistry in 1947 for his work on alkaloids.

He also made another career as chess champion (twice winner of the Oxfordshire championship) and was president of the English Federation from 1950 to 1953. He wrote a chess manual.

An outstanding alpinist, he started his climbing career in New Zealand (Mount Sealy, Coronet Peak, Mount Meeson) and added some hundred ascents in the Swiss Alps (Eiger, Jungfrau, Mönch) to his record. At the age of 70, he was still able to climb Piz Julier (3 385 m) near St. Moritz. His last ascent (Table Mountain, South Africa) dates from 1966, when he was 80.

Robinson was an impulsive and emotional man with an enormous power of concentration, an amazing memory and an impressive analytical capacity.

Reference: Lord Todd, *Natural Products Reports*, 4, 3–11 (1987).

A=COOC$_2$H$_5$

# 17   Exercise: Obtaining the Target Molecule Starting from Simple Precursors

The combination of the three structural elements (six-membered ring, carbonyl group, and conjugated C=C bond) suggests a Robinson annulation. Let us break down mentally (reasoning in retrosynthetic sense) the C=C bond, in order to show the elimination of $H_2O$ that can give rise to it; thus one gets the idea of an intramolecular condensation of a 1,5-diketone (of the aldol condensation type). It is itself a Michael from the addition of methylvinylketone (Michael acceptor) with ethyl acetoacetate (Michael donor, via the enolate).

The importance of this sequence lies also in its selectivity. The presence of the ester group A results in the preferential enolization of the activated methylene group. This thermodynamic enolate serves as a Michael donor. If necessary the auxiliary group A can be removed at the end of the synthesis by hydrolysis to the corresponding carboxylic acid which eliminates $CO_2$ by simple heating.

A = COOR

The reason for the ease with which this decarboxylation can take place is that the target molecule is *vinylogous* (by interposition of the C=C double bond between the keto carbonyl group and the ester carbonyl group of A) to a $\beta$-keto ester. A general property of the latter is that the corresponding acid formed by acidic hydrolysis readily undergoes thermal decarboxylation, according to a concerted six center–six electron mechanism.

---

# Summary: Michael Addition—Enolization

The *Michael reaction* is a conjugated 1,4-addition to an $\alpha,\beta$-unsaturated carbonyl compound O=C−C=C, a so-called 'Michael acceptor'. The electrophile (the metal counter ion) is attached to the oxygen (atom labeled 1); the nucleophile forms a new bond with the terminal carbon (labeled 4). The LUMO of the Michael acceptor which resembles an allyl cation is characterized by a large coefficient at carbon 4, the reaction site of the acceptor. The preference of 1,4- over 1,2-addition can be enhanced by use of a mixed lithium cuprate. In difficult cases a previous activation of the Michael acceptor is required which occurs by means of a strong Lewis acid, e.g. $BF_3$, complexing the oxygen atom (example: **coriolin** synthesis). The Michael addition can be made selective: *regioselective* by using an auxiliary group if the nucleophile can remove a proton from the $\alpha$- or from the $\alpha^1$-position of the ketone (example: **helminthosporal** synthesis); stereoselective, especially by way of the dielectric constant of the solvent controlling the electrostatic interaction in the transition state.

The use of the Michael addition in a multi-step synthesis is illustrated by the key step in the great classic of organic syntheses, the synthesis of **longifolene** by Corey (Nobel Prize in Chemistry 1990).

*Enolization* of a ketone or aldehyde is acid ($\rightarrow$ enol) or base ($\rightarrow$ enolate) catalyzed. A dual description of the enolate anion is given, in terms of molecular orbitals (MO) and as a resonance hybrid. The enolate and allyl anions are isoelectronic. The orthogonality of the C($\alpha$, $\alpha$*)–H bond to the carbonyl plane determines the feasibility of enolization. Involvement of a Michael addition, enolate formation from the adduct and intramolecular trapping by the other carbonyl group constitutes the *Robinson annulation*, a useful method of ring formation.

# 2.II
# Carbonyl Group:
# Enolization—Michael Addition

## 1  Smelling an Enol

β-Ionone, with a carbonyl group conjugated with a butadiene, smells like violets.

In contrast, β-damascone, an isomer of β-ionone, displays a much different fragrance, comprising components of fruit, exotic spices and chrysanthemum.

The β-diketone shown is subject to keto–enol equilibrium. The two enolic tautomers resemble β-ionone and β-damascone, respectively. In fact this particular diketone combines the two fragrances.

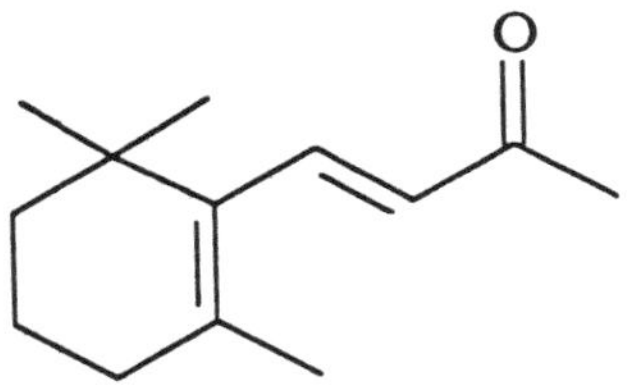

β-ionone

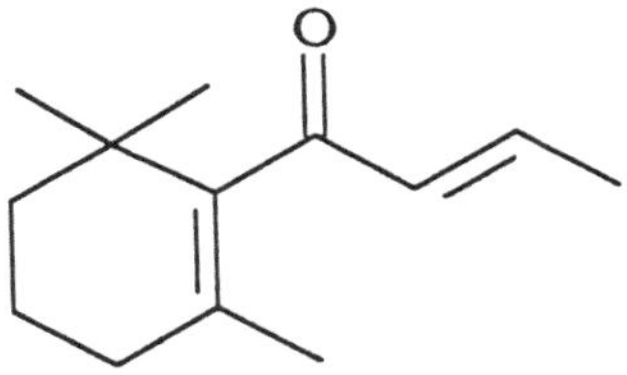

β-damascone

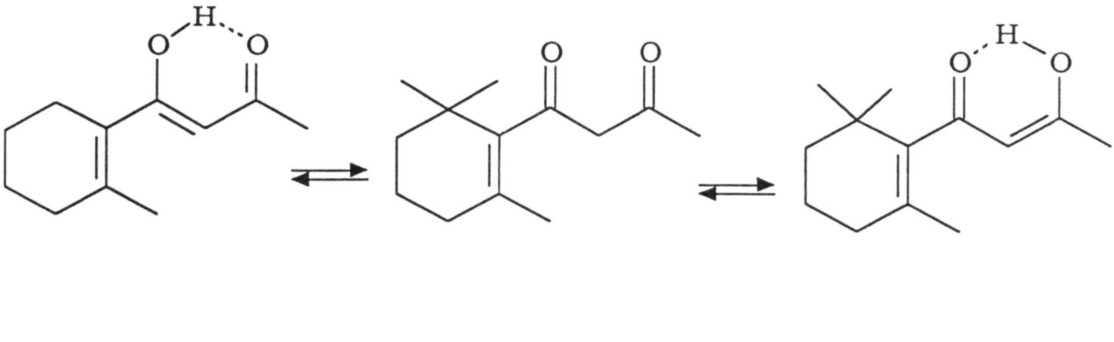

Reference: G. Ohloff, *Experientia*, **42**, 271–279 (1986).

## 2   In the Field of Enolates

Squaric $C_4O_4H_2$, croconic $C_5O_5H_2$ and rhodizonic acids $C_6O_6H_2$ belong to a family of strong organic diacids, studied by R. West, of the University of Wisconsin. Their acidity constants $pK_1$ and $pK_2$ are 1.7 and 3.2 (squaric acid), 0.5 and 1.5–2.0 (croconic acid), and 3.1 and 4.9 (rhodizonic acid). Their high acidities stem from the high stability of their conjugate bases, depicted in the adjoining Scheme.

These conjugate bases are aromatic; they fulfil the Hückel rule with two electrons in a planar monocycle for the squarate, croconate and rhodizonate dianions. It is possible to write a large number of resonance forms for each of these dianions. A further indication in favor of their aromaticity is provided by the bond lengths: 1.469 Å ($C_4$); 1.457 Å ($C_5$) and 1.488 Å($C_6$).

'The real truths are those that can be invented', Karl Kraus.

## 3    Chemicals of Major Importance for Industry: Formaldehyde

Formaldehyde—besides being an interstellar molecule with a major role in the origin of life—is manufactured mainly by catalytic oxidation of methanol. Its annual production in the United States is in the order of 3 MT (1991).

Its major uses include urea–formaldehyde resins (30%), and phenol–formaldehyde (25%) resins. They are used as adhesives for wood and plywood, and they serve as plastics.

## 4    The Miracle of San Gennaro

In Naples since 1389, a phial containing the precious relic of the blood of this Saint (Gennaro = January) regularly shows a surprising phenomenon: whereas the blood would appear to be coagulated, it liquefies in about half-an-hour without any need for shaking or overturning the container; then it appears to boil and foam. During liquefaction, the faithful pray fervently and when the prayers fail to induce the physical change, the people gathered there break out in lamentations and start cursing. Whenever the expected miracle of Saint January failed to occur, Naples was struck by disasters: in 1527 by a plague epidemic (40,000 deaths); in 1836 by the cholera (24,000 deaths) and more recently in 1980 by an earthquake that claimed 3,000 victims.

The Church has so far objected to analyzing the sample. Amongst the theories about the physico-chemical origin of the phenomenon, let us mention thyxotropy of gels arising from aqueous solutions of chalk and ferric chloride, mixtures of blood and honey, or mixtures of other natural products.

References: L. Garlaschelli, F. Ramaccini, and S. Dalla Sala, *Nature*, **353**, 507 (1991).
G. A. F. Hendry and A. J. E. Lyon, *Nature*, **354**, 114 (1991).
J. M. Dunlop, *Nature*, **354**, 114 (1991).

## 5    Bakelite

One of the major uses of formaldehyde is to manufacture polymers, the ancestor of which is bakelite dating back to the beginning of this century. At that time, only celluloid was known; it is a thermoplastic derived from nitrocellulose and invented by John Wesley Hyatt. The drawback of this plastic was that, for instance, billiard balls made of this material would sometimes explode . . .

Leo Hendrik Baekeland (1863–1944), a chemist of Belgian origin, settled in the United States. In 1907, he discovered bakelite by chance, while looking for an elastic lacquer to replace natural lacquers. Bakelite is a resin prepared from phenols and formaldehyde. Baekeland observed that, upon heating these under pressure, they formed a soft solid which could be molded and which hardened upon cooling. Bakelite, first manufactured by the company founded by Baekeland (Bakelite Corporation, 1910–1939), continues to be used in a variety of applications: adhesives, molded plastics, e.g. thermoabrasion cones for missile heads.

'One shouldn't always name names. What should be said is not that someone has done it, but that it was possible to do it'. Karl Kraus

# 6  A Natural α,β-Unsaturated Carbonyl Compound

In the species *Utetheisa ornatrix*, a moth, larvae of both sexes live on plants known as crotolaria. Alkaloids produced by these plants are accumulated by the insects, which protects the adults against predators such as spiders. Furthermore, the males metabolize these alkaloids to the bicyclic products shown. They use them as sexual pheromones.

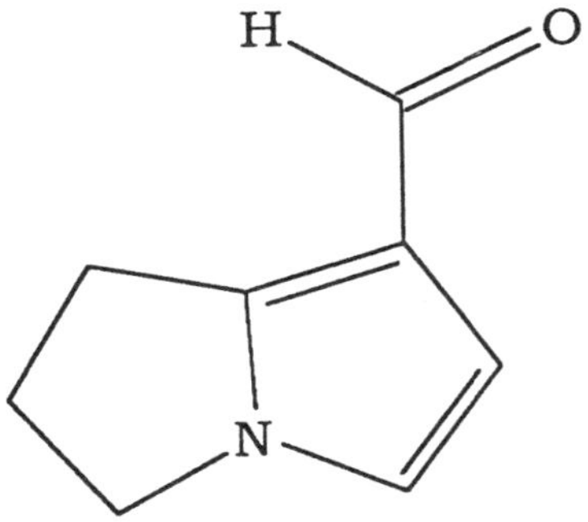

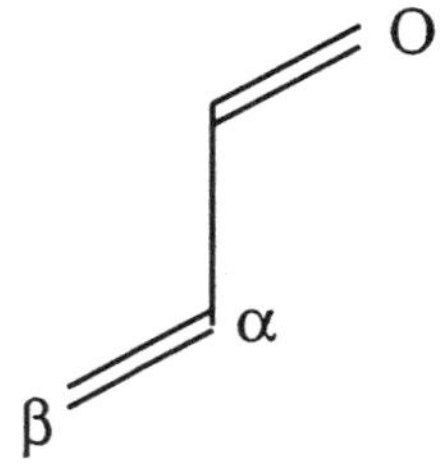

Why such dual use of the same alkaloids? During copulation, the males give the alkaloid to the females who accumulate it in their eggs, which protects them against predators (all the control experiments were done). Thus the bicycle shown in the adjacent Scheme serves as a chemical message: it tells the females that the male will be a good father, since he provides them with the alkaloid needed for protection of their offspring. The double use makes sense, in the context of evolutionary pressure.

Reference: J. Meinwald. *Ann. NY. Acad. Sci.* **471**, 197–207 (1986).

# 7  Study in Red

Conjugation of a quinone—formed by oxidation of a benzene ring bearing two OH functions in positions 1 and 4—with other unsaturated groups often forms strongly colored molecules.

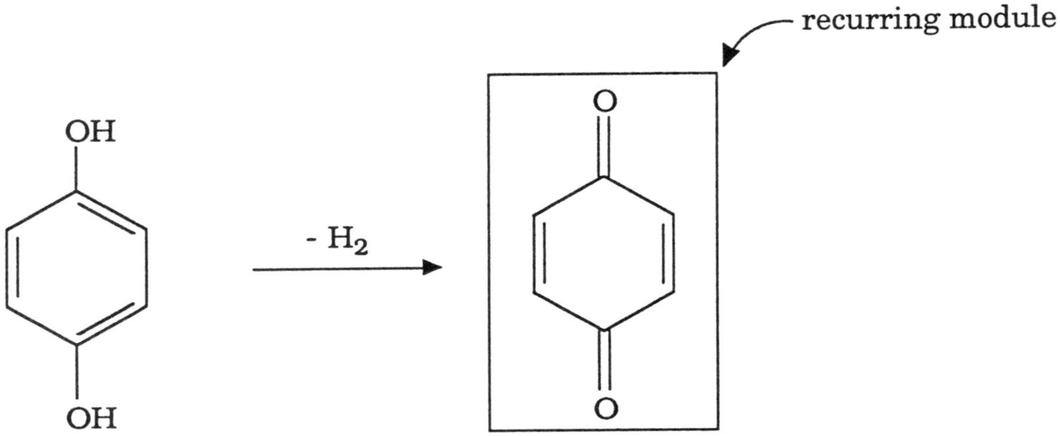

— recurring module

$- H_2$

A first example is the orange red alizarin. This pigment is traditionally extracted from madder, *Rubia tinctorum*. It was used as a dye for cotton ('Turkish red'), with an aluminium mordant. The conspicuous red trousers of the French soldiers dyed with it made superb targets! The decision of the French Army to abandon this dye, hence reducing its demand, plus its first industrial synthesis led to the ruin of the cultivators of madder in Southern France.

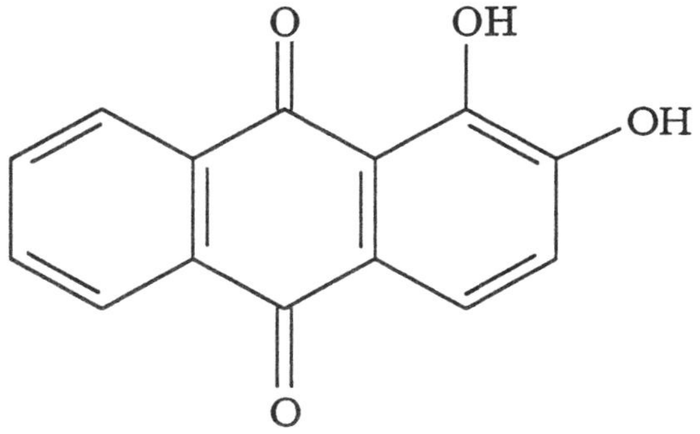

alizarin

The brown red color of henna, *Lawsonia inermis*, comes from lawsone. The nut shell dyes skin brown.

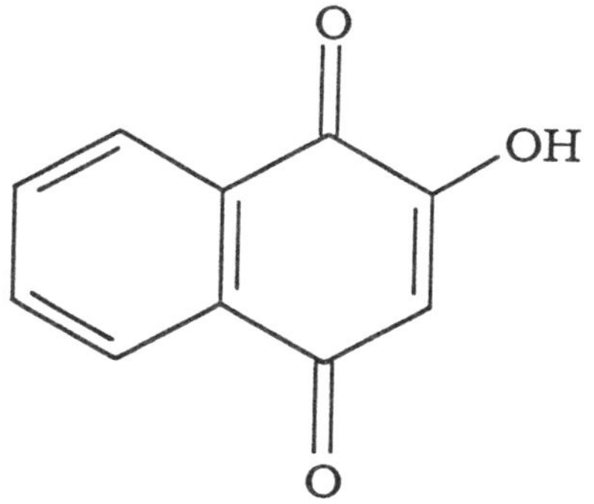

lawsone

The substance that causes this is juglone (nut tree = *Juglans regia*), resulting from air oxidation of a hydroquinone.

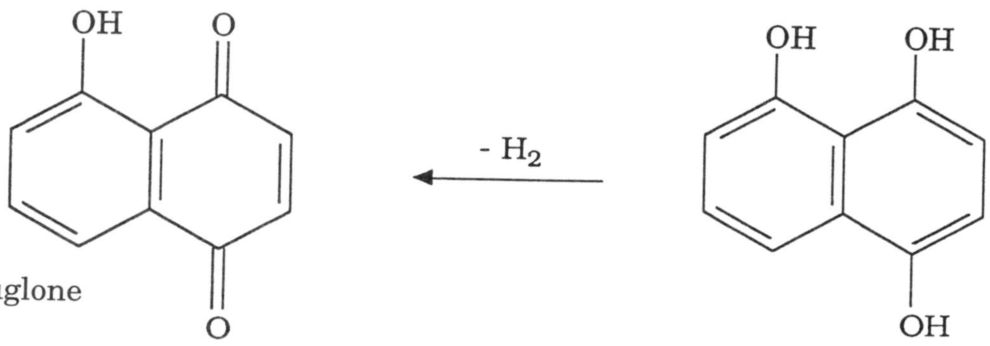

juglone

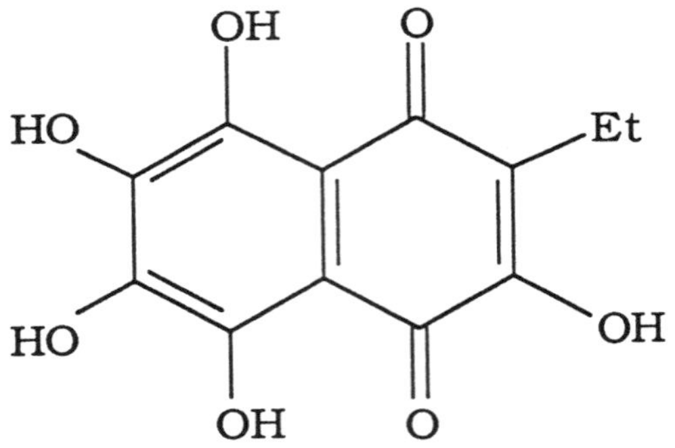

echinochrome A

Echinochrome A is the red pigment of the urchin eggs, *Arabacia pustosa*.

The vitamins $K_2$ ($n = 1$–$13$) or menoquinones with antihemorrhagic properties are produced by several symbiotic intestinal bacteria.

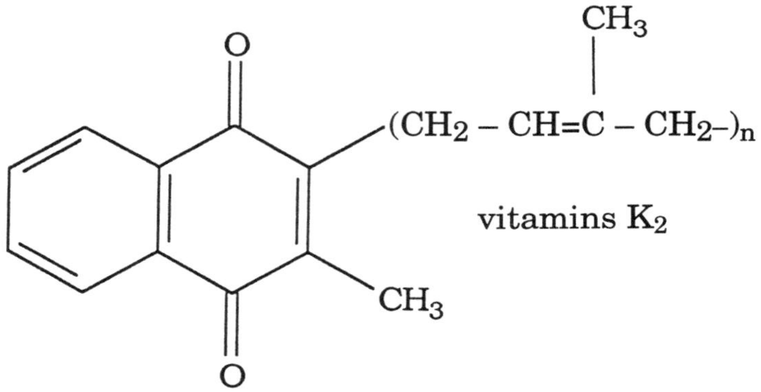

vitamins $K_2$

Kermes is one of the oldest dyes known (for instance 'Venetian pink') and is obtained from the dried insect *Coccus ilici*. It basically consists of kermesic acid ($R = H$). Carminic acid ($R = C_6H_{11}O_5$, glucose) is the main constituent of carmine, the extract of cochineal—a preparation similar to that of kermes but from the Mexican homopter *Coccus cacti* found on certain cacti (*Coccus ilici* haunts the islands in the whole Mediterranean).

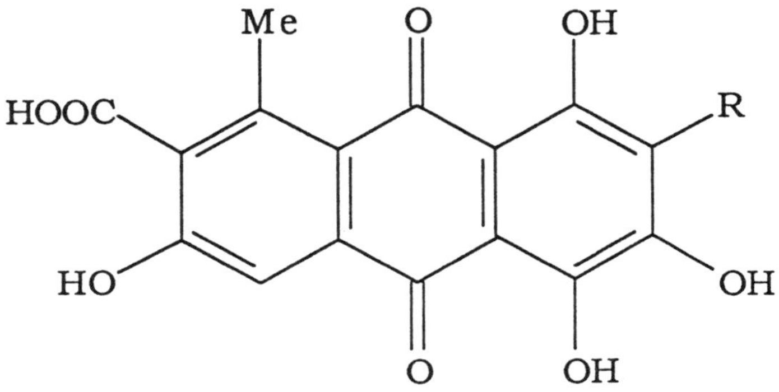

kermes

Isolation of alizarin from madder: Colin and Robiquet, *Ann. Chim. (Phys)*, **34**, 225 (1927).
First synthesis of alizarin: Graebe and Liebermann, *Liebigs Ann. Supp.*, **7**, 257 (1870).

# 8 How Nature Improves Basicity

Vitamin K is important to blood-clotting. It is involved in the biosynthesis of prothrombin, of factors VII, IX and X as well as of the C and S proteins which inhibit coagulation: in each of these proteins, vitamin K transforms 10–12 glutamate residues into $\gamma$-carboxyglutamate residues. The latter chelate calcium ions, binding them to the protein, and this plays an important role in the cascade of reactions giving rise to a blood clot.

Carboxylation of glutamate requires a base with $pK_a$ of 26–28 for enolization $\alpha$ to the carboxyl group. However ionization of a hydronaphthoquinone—the reduced form of vitamin K—leads to a base with a $pK_a$ of 9, insufficient. This is where an oxidation takes place: the intermediates shown—a dioxetane followed by an epoxide—provide the required increase in basicity, an alkoxide being a stronger base than a phenoxide.

NB: Anticoagulants such as dicumarol or warfarin (a rat poison) function by interfering with the mechanistic pathway indicated; they inhibit the reduction step to the hydroquinone.

References: E. M. Arnett, P. Dowd, R. A. Flowers II, S. W. Ham., and S. Naganathan, *J. Am. Chem. Soc.*, **114**, 9209–9210 (1993).
S. Stinson, *Chem. Eng. News*, November 16, 18–20 (1992).

# 9   Use of Michael Addition in Steroid Syntheses

Michael addition of the enolate of 2-ethylcyclopentane-1,3-dione to the conjugated ketone indicated affords the 1,4-adduct in 97% yield. The enolate is formed with potassium methoxide.

This Michael adduct is a precursor of steroid hormones, in the biomimetic syntheses devised by Professor W. S. Johnson of Stanford: a double aldol type condensation allows formation of two new carbon—carbon bonds, leading to the characteristic tetracyclic systems of steroids, more precisely of the estrone series.

Another access to steroids, in the same estrone series (see Box) is provided by the Torgov modification of the Michael reaction: the Michael acceptor is an allylic alcohol instead of a conjugated ketone. Departure of $OH^{(-)}$ as the leaving group forms an allyl cation. Indeed, a Michael acceptor is formally an allyl cation. This reaction, where the Michael donor is the same as in the previous reaction and where the enolate is also formed with potassium methoxide, leads to the adduct in 70% yield.

References: H. Smith, D. H. P. Watson, et al. J. Chem. Soc., 4472 (1964).
S. N. Anachenko and I. V. Torgov, Tetrahedron Lett., 1553 (1963).

## 10   An Elegant Solvent Effect in the Robinson Annulation

The Michael donor here is the C=C double bond of a 'push–pull' olefin, strongly polarized by a donor nitrogen and by an attracting carbonyl substituent. Robinson annulation occurs in the presence of sodium acetate and aqueous acetic acid (as reagents), hydrolyzing the amine at the end of the reaction.

When benzene with its low dielectric constant is the solvent, the stereoselectivity is high: 10:1 favoring the bicycle with the methyl groups in a *trans* arrangement. In contrast when dimethyl formamide, with a much larger dielectric constant, is used as solvent, the stereoselectivity vanishes: equimolar amounts of the *cis* and *trans* isomer are formed (only one of the enantiomers is represented here).

|        |       |        |
|--------|-------|--------|
| C₆H₆   | 6%    | 59%    |
| DMF    | 25%   | 25%    |

The *cis* or *trans* stereochemistry is determined during formation of the first carbon—carbon bond in the Michael addition. The two partners lie in parallel planes, to maximize overlap between the donor (d) and acceptor (a) centers. Their mutual orientation also tends to increase to a maximum Coulomb attraction of the two electric dipoles. In the conjugated ketone, the dipole moment points from the positive carbon 4 to the negative oxygen 1. In the enaminoketone, the dipole moment points from the positive nitrogen to the negative oxygen.

This Coulomb attraction of the two dipoles organizes the transition state for the Michael addition in such a way as to place them in a tight antiparallel arrangement for both formation of the *cis* (see adjacent Scheme, top) as well as the *trans* products (Scheme, bottom). As compared to the *trans* transition state, the *cis* transition state is destabilized by steric interaction of the two methyl groups.

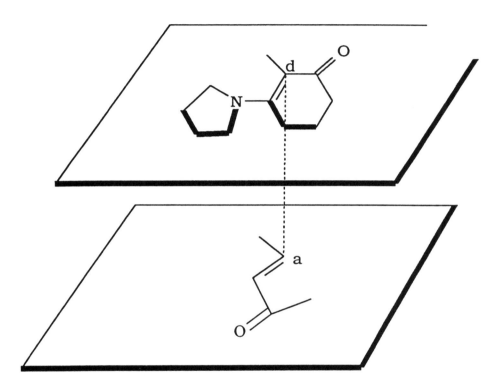

Therefore, in a solvent with small dielectric constant (benzene), where the Coulomb attraction is strong and the transition state is held tightly, the isomer with *trans* methyl groups forms preferentially. If the dielectric constant D is increased by an order of magnitude, in going from benzene to dimethyl formamide, the Coulomb attraction (proportional to $D^{-1}$) is substantially reduced: the partners have a looser mutual orientation and therefore the *trans* product preference goes away.

Reference: R. M. Coates and J. E. Shaw. *Chem. Comm.*, 47 (1968).

## 11　Stereoselectivity in the Robinson Annulation

The cyclohexanone indicated reacts with methylvinylketone to form the decalone with a *cis* junction. Enolization of this cyclohexanone (with sodium ethoxide in ethanol at $-10°C$) precedes the Michael addition. The thermodynamically controlled more stable enolate (at the methyl-substituted position) is the Michael donor, which accounts for the regioselectivity observed.

As to stereoselectivity, the cyclohexanone (also under thermodynamic control) has the methyl and isopropylidene groups both equatorial (e); this directs the methylvinylketone to the axial position (a).

In the second step of the Robinson annulation, the reagent EtONa-EtOH enolizes the methyl ketone and the carbanion adds to the carbonyl group of the cyclohexanone, i.e. from the most accessible face, in the same half-space. Hence the formation of a single *cis*-decalone.

Reference: J. A. Marshall, G. L. Bundy, and W. I. Fanta, *J. Org. Chem.*, **33**, 3913 (1968).

## 12   Application of the Robinson Annulation to the Industrial Synthesis of Estrone (Roussel–Uclaf)

The enolate from 2-methyl-1,3-cyclopentadienone is formed regioselectively in the position 2. Addition to the Michael acceptor occurs as indicated. The resulting triketone is not isolated but is immediately cyclized to afford the Robinson annulation product.

'You'd be surprised how hard it can often be to translate an action into an idea', Karl Kraus.

After several steps, the female sex hormone estrone is formed, which controls estrus and nesting of the ovum; furthermore childbirth and milk rise is controlled by its metabolites, such as estradiol and estriol.

## 13   The Sacrilege of the Bloody Host

Holy wafers had traces of blood. This is now known to be due to a strongly red colored pigment, a metabolite of the very common non-pathogenous bacterium *Serratia marcesens*, also known as *Chromobacterium prodigiosum*. This 'miracle' excited the imagination of our ancestors. In times of religious intolerance, to quote from the book by F. Mayer and A. H. Cook, *Chemistry of Natural Coloring Matters*, Reinhold, New York, 1943, p. 269, it was believed that:

'sacramental wafers had been defiled by the Jews and were bleeding. The offenders were quickly found and usually burned or simply put to death. Hecatombs of men have met their end through *Prodigiousus*'.

The pigment responsible has been isolated. It is called prodigiosin. Its chemical structure is that of a tripyrrole derivative, strongly conjugated, hence its color. I choose a reaction from its multistep synthesis which illustrates particularly well some of the concepts presented in the first two chapters of this book.

The initial reaction in the synthesis involves a Michael addition: the donor is a negatively charged nitrogen atom (conjugate base of an $\alpha$-amino ester); the acceptor is a strongly polarized olefin with an EtO-donor substituent at the $\alpha$ carbon and two electron-attracting ester functions (A=—COOEt) at the $\beta$ carbon. This Michael adduct **M** is deactivated by re-formation of the C=C double bond and cleavage of an ethoxide fragment EtO$^{(-)}$. (Such a fragment is a good leaving group since the negative charge is located on the electronegative oxygen.) The enamine **N** is in equilibrium with its conjugate base **O**, since the methylene group in **N** between the nitrogen and the ester function is activated and readily loses a proton, i.e. under the action of a base such as EtO$^{(-)}$. The carbanion **O** thus formed is trapped by the nearby carbonyl group and the addition–elimination reaction thus triggered leads to the cyclic molecule **P**, again with loss of the fragment EtO$^{(-)}$. The transformation **O**→**P** is a Claisen condensation. The molecule **P** bears an enolizable C—H bond: the acidity of this bond is due to the two adjacent C=O groups of the cyclic ketone and the ester. Thus, **P** is in equilibrium with the enol form **Q**, stabilized furthermore by its aromaticity (six electrons in a monocyclic plane). Due to this stabilization, the reactive site of **Q** is no longer the ring but the ester group on nitrogen, the carbonyl group of which can react with the previously expelled fragment EtO$^{(-)}$, affording **R**. The relatively unstable intermediate **R** then immediately cleaves off ethyl carbonate giving the conjugate base of the final product, the pyrrole **S**.

Reference: H. Rappoport and K. G. Holden, *J. Am. Chem. Soc.*, **84**, 635–642 (1962).

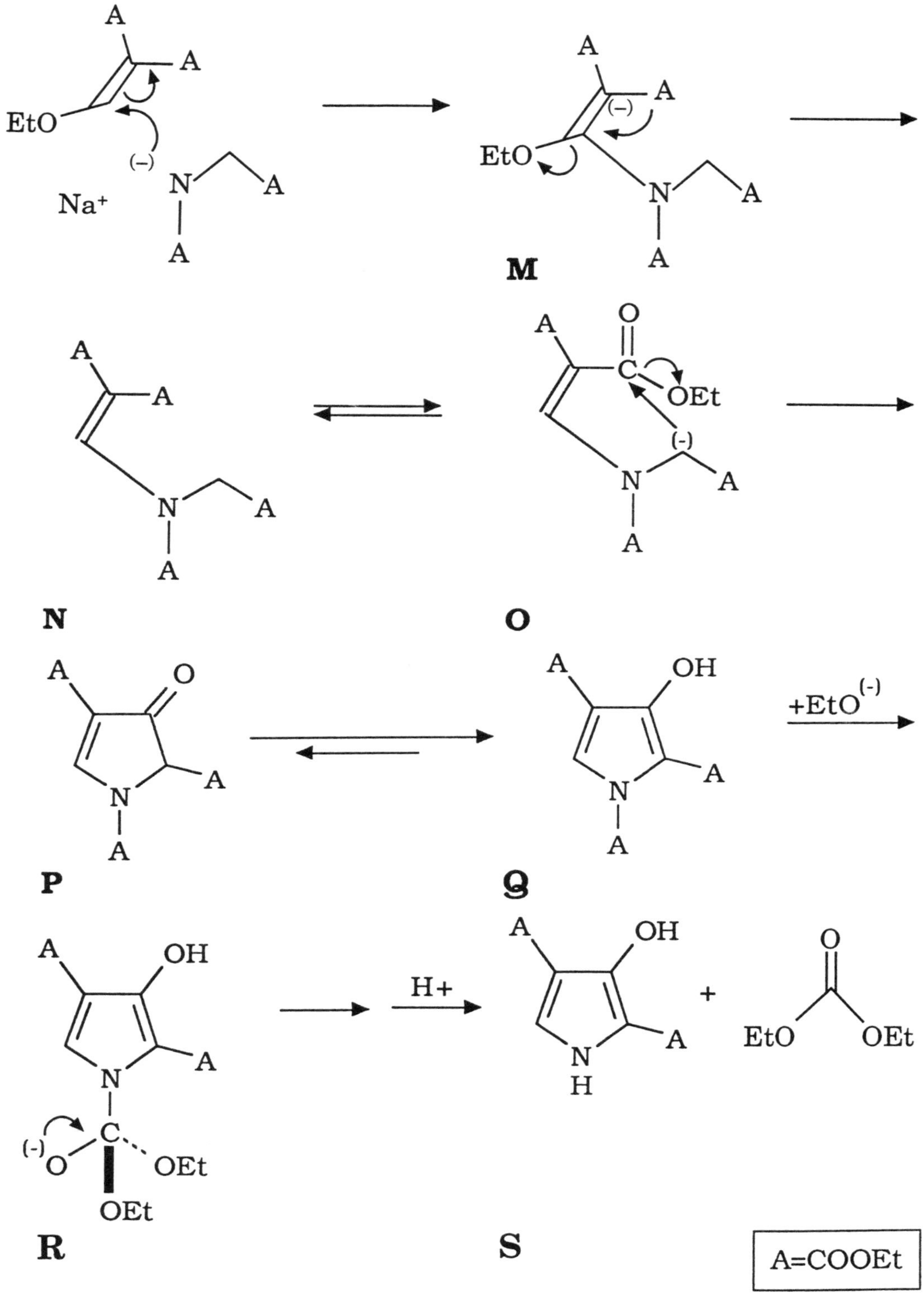
A
A
EtO
(–)
Na+
N
A
A
A
A
(–)
EtO
N
A
A
M
A
A
N
A
A
N
A
O
C
OEt
(-)
N
A
A
O
A
O
N
A
A
P
A
OH
N
A
A
Q
+EtO(-)
A
OH
N
A
(-)
O
C
OEt
OEt
R
H+
A
OH
N
H
A
S
+
O
EtO
OEt
A=COOEt

# 3.I
# Carbonyl Group:
# Aldol Condensation

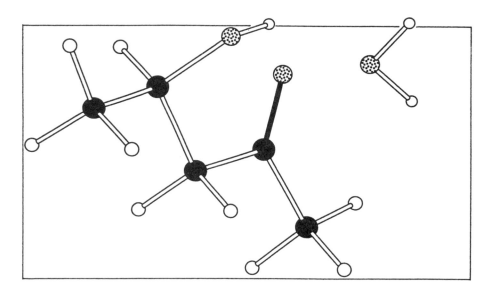

Addition of an enolate carbanion as the donor to the acceptor electrophilic carbon of a carbonyl group defines the aldol condensation: it is the third great reaction for the formation of a new carbon–carbon bond that we come to. Its major asset is its simplicity; treatment with a base suffices to transform a carbonyl compound into the corresponding enolate. The major disadvantage of the aldol condensation is its lack of selectivity. For instance, the enolate can add to the carbonyl group of its precursor.

Thus this chapter focuses on control of the aldol condensation to make it more selective.

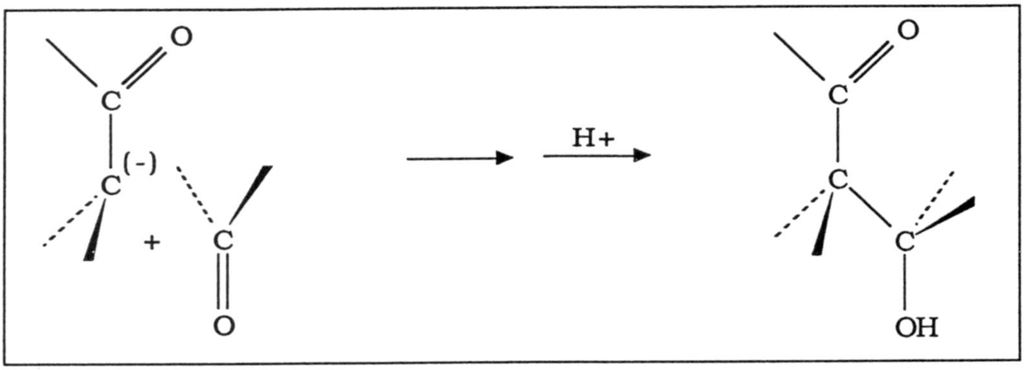
O
C
C(-)
+
C
O
H+
O
C
C
C
OH

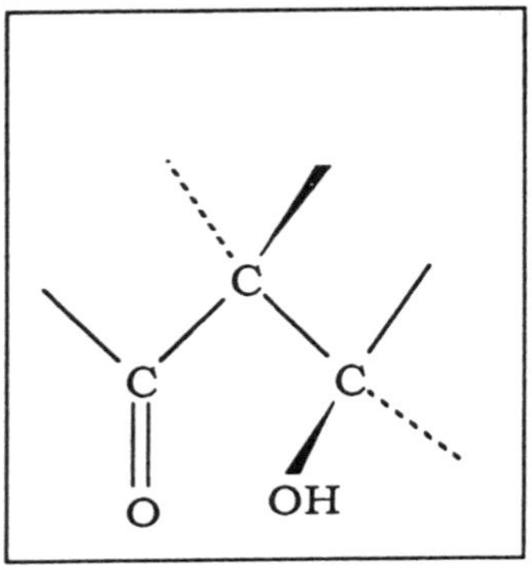
C
C
C
O
OH

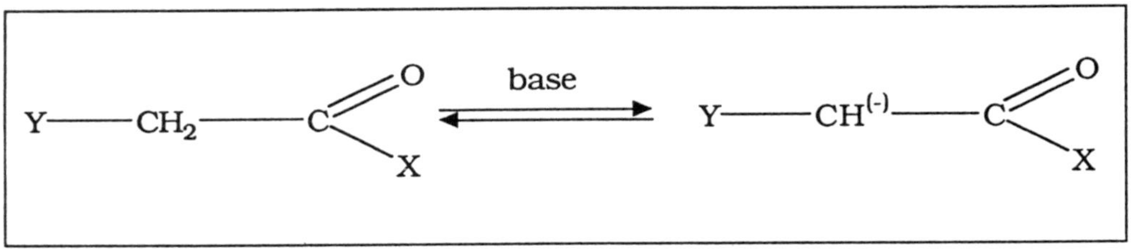
Y—CH2—C
O
X
base
Y—CH(-)—C
O
X

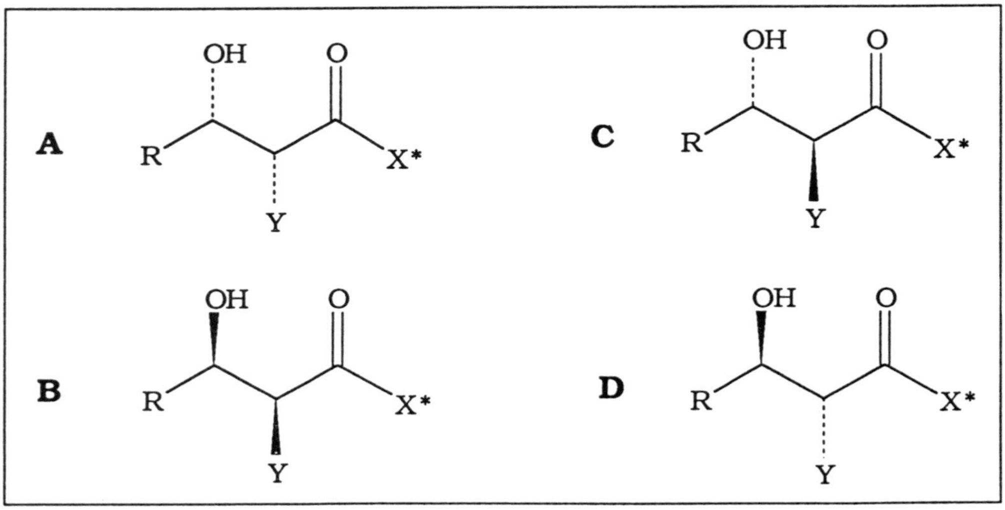
A
OH
O
R
X*
Y
C
OH
O
R
X*
Y
B
OH
O
R
X*
Y
D
OH
O
R
X*
Y

# 1  Aldol Condensation

As in numerous other reactions, a new carbon–carbon bond results from the union of an electron donor and an electron acceptor; the donor is the negatively charged carbon of an enolate; the acceptor is the carbonyl carbon, with its positive partial charge.

The adduct (box) has two oxygen-containing groups, a ketone and an alcohol, in a 1,3 mutual relationship (a $\beta$-ketol). A chain of the type $-O-C-C-C-O-$ is the typical result of an aldol condensation.

This reaction is catalyzed by bases: a base removes a proton from a ketone and thus forms the enolate anion, which is the conjugate base of this ketone.

In order to avoid an exchange of the roles (donor and acceptor) of the two reaction partners, an aldehyde is frequently used as acceptor. Aldehydes are better electrophiles than ketones. Thus the risk of trapping the enolate by the corresponding ketone is reduced.

Addition of the enolate of a ketone $YCH_2 \cdot CO \cdot X$ to an aldehyde $RCHO$ forms four stereoisomers because two asymmetric carbons are present at the ends of the new carbon–carbon single bond. Two of these products are *syn* stereoisomers, the other two *anti*. Use of several procedures, especially that of chiral auxiliary agents X* allows for selective access to either of these diastereoisomers.

---

The terms *syn* and *anti* refer to the relative position of the OH and Y group in the four adducts: *syn* on the same side, *anti* on different sides of the carbonyl plane. **A** and **B** are *syn* stereoisomers, **C** and **D** are *anti* stereoisomers.

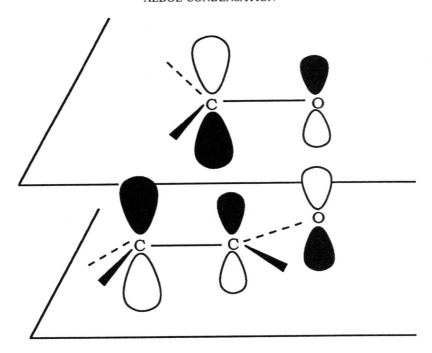

The aldol addition can be represented by the predominant interaction of the frontier molecular orbitals, the highest occupied orbital (HOMO) of the enolate and the lowest unoccupied orbital (LUMO) of the aldehyde. You will note the bonding overlap of the carbonyl carbon with the two carbons of the enolate as well as the bonding overlap of the $2p_z$ orbitals at the two oxygens, somewhat unexpected because of their Coulomb repulsion (but the metal cation present, i.e. the counterion of the base used, also contributes to their linkage by forming a chelate). The relevant frontier orbitals are the $\pi^*$ (C=O) with a large coefficient at carbon and a small coefficient at oxygen and the non-bonding n orbital of the enolate with a large coefficient, but with opposite phases at the terminal atoms and a small coefficient at the central carbon.

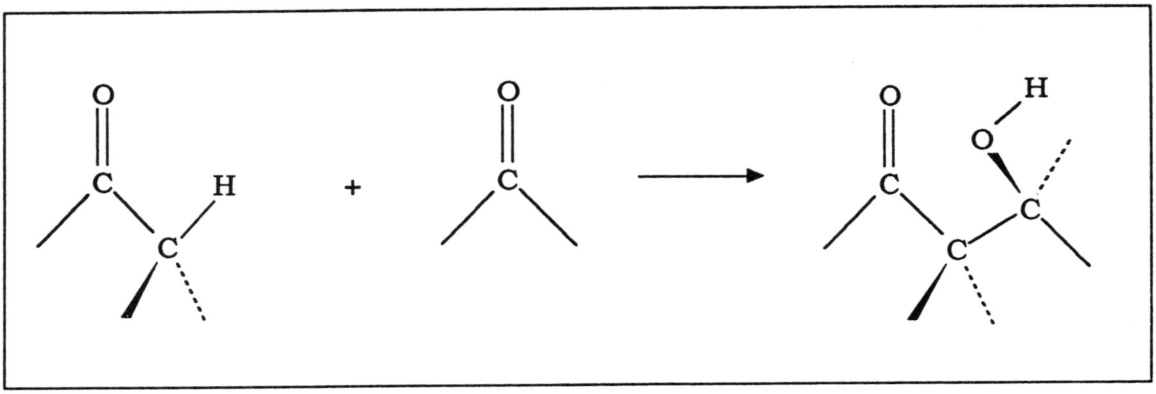
O
C
H
+
O
C
O
C
H
O
C
C

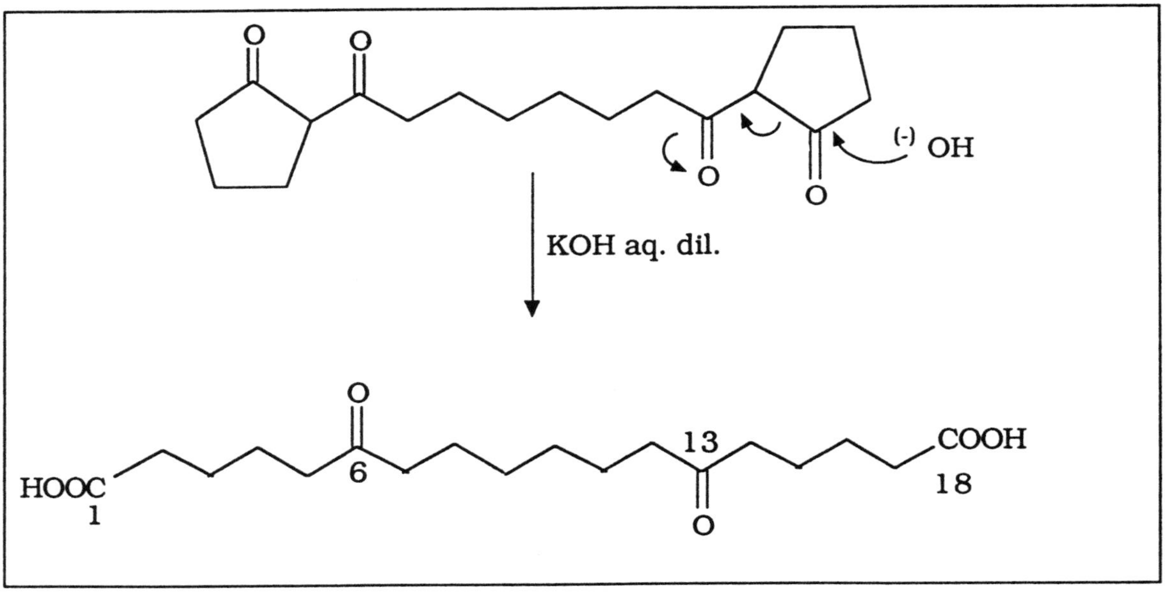
O
O
(-) OH
O
O
KOH aq. dil.
O
HOOC
1
6
13
COOH
18
O

## 2   Product Stability in the Aldol Condensation

The purely thermochemical balance is loss of a $\pi$ C=O bond (about 93 kcal mol$^{-1}$) and a $\sigma$ C—H bond (98 kcal mol$^{-1}$), almost compensated for by formation of a new $\sigma$ C—C bond (83 kcal mol$^{-1}$) and a $\sigma$ O—H bond (103 kcal mol$^{-1}$). Very often, the equilibrium is shifted to the side of the adduct, the negative charge migrating from the carbon to the more electronegative oxygen atom.

Nevertheless when the conditions are favorable, the equilibrium can on the contrary favor the inverse reaction; such a fragmentation is known as the retroaldol reaction. In the reaction indicated, the two cyclopentanones thus open under base catalysis (S. Hünig, 1969).

The product of an aldol condensation, typically a $\beta$-ketol, is labile and loses readily a molecule of water by elimination. This can be observed in the last step of the cantharidin synthesis by Stork, one of the examples for this chapter (page 177).

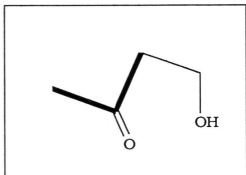

# 3   Steering of Aldol Reactions: Donor and Acceptor

The aldol addition is valuable not only for its aptitude at creating a new carbon–carbon bond, but also by providing two functions (an alcohol and a ketone), readily susceptible to further elaboration on the way to a target molecule in a multistage synthesis. These assets go together with the disadvantage of leading a priori to several products. Accordingly, it is important to steer aldol reactions so as to render them univocal.

Some are selective from the outset; take for example the target molecule in the box, where the characteristic β-ketol pattern suggests an aldol addition. In fact, the formation of secondary products is not so critical here since only cyclohexanone can enolize and the 1,2-diketone is more electrophilic (the mutual repulsion of the C=O dipoles enhances their reactivity).

A second, more general case is those aldol reactions involving an aldehyde and a ketone. Wittig's rule is applicable to these: aldehydes have more reactive carbonyl groups (better electrophiles) than ketones, whereas enolates derived from ketones have more nucleophilic carbons than those derived from aldehydes.

Reference: G. Wittig and H. Reiff, *Angew. Chem. Int. Ed. Engl.*, **7**, 7 (1968).

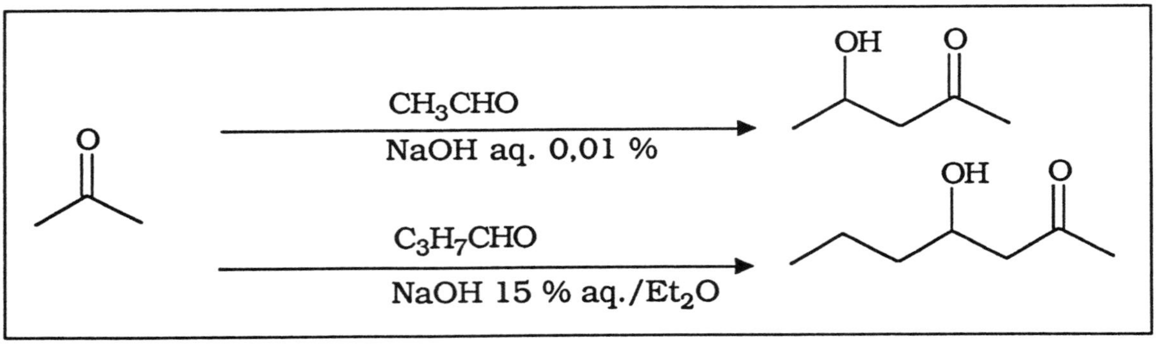
O
$CH_3CHO$
NaOH aq. 0,01 %
OH
O
$C_3H_7CHO$
NaOH 15 % aq./$Et_2O$
OH
O

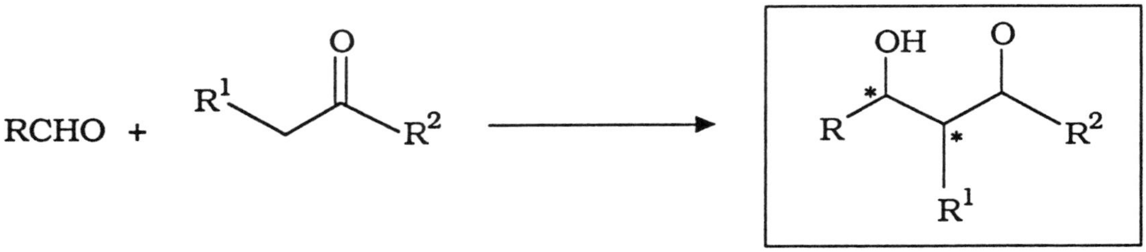
RCHO +
O
R1
R2
OH
O
R
R2
R1

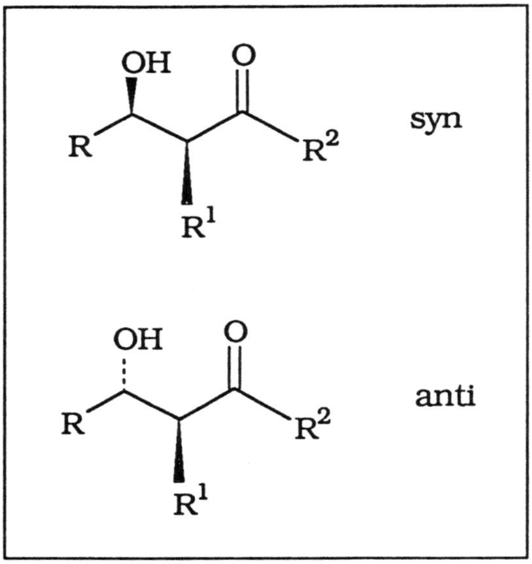
OH
O
R
R2
R1
syn
OH
O
R
R2
R1
anti

Therefore condensation of acetone with an aldehyde, acetaldehyde or butyraldehyde in the example indicated, leads exclusively to $\beta$-hydroxyketones with yields of 80 and 84%, respectively; self-condensation of the aldehydes does not occur. This chemoselectivity is obtained, provided that the aldehyde is slowly added, drop-by-drop to the solution of the ketone.

# 4   Steering of Aldol Reactions: *Syn* and *Anti*

Let us consider again an aldol condensation between an enolate from a ketone and an aldehyde; the products can belong to the *syn* series if the groups $R^1$ and OH are in the same half-space or to the *anti* series ($R^1$ and OH being in different half-spaces); the half-spaces are defined by the plane of the carbonyl group.

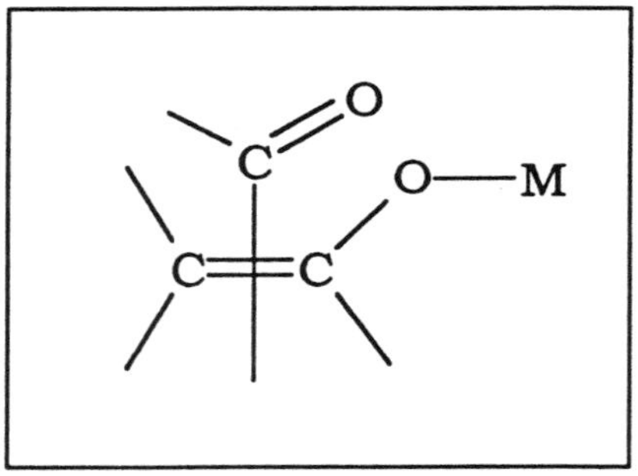
C
O
O—M
C═C

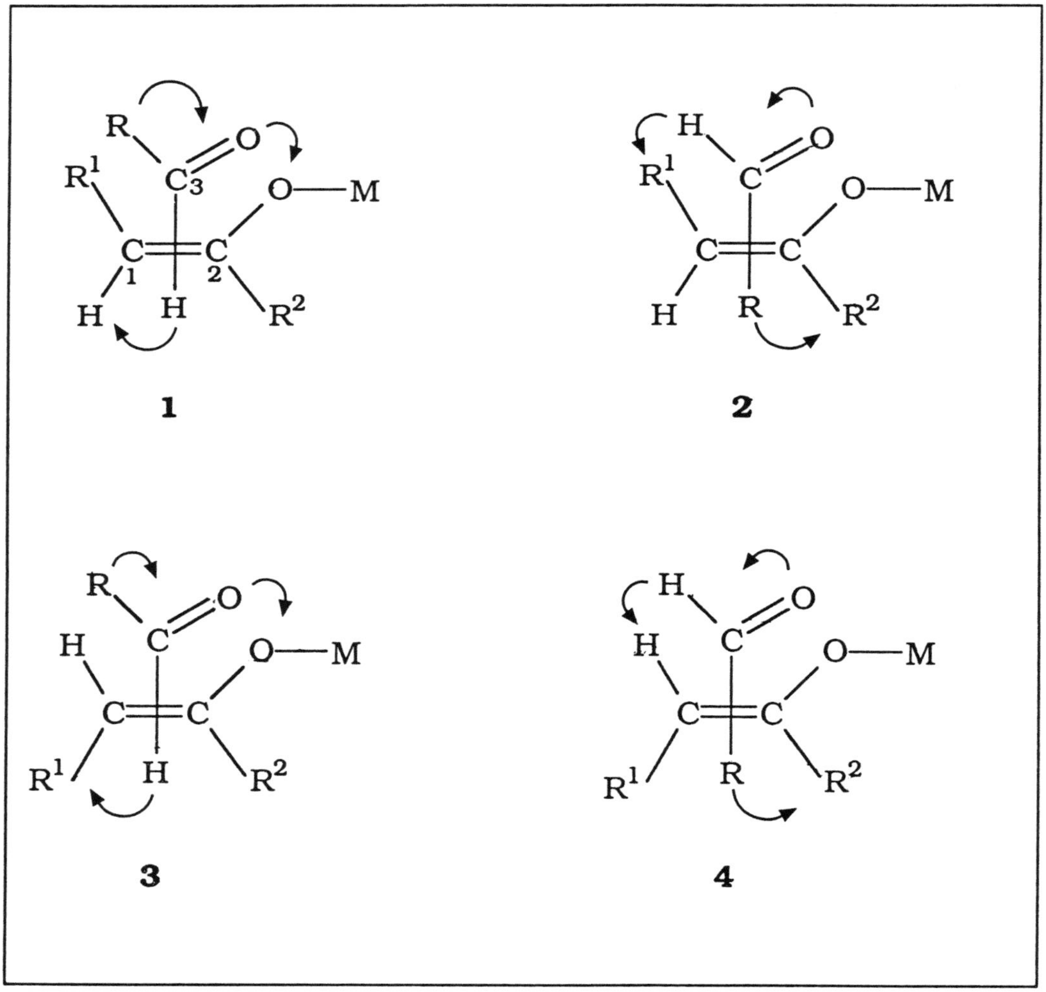
1
2
3
4

It is possible to predict the geometry of the product from the geometry of the enolate. For this purpose, let us consider once more the predominant orbital interactions between the LUMO ($\pi^*$) of the aldehyde and the HOMO (n) of the enolate. We have seen that these interactions tend to place the aldehyde and the enolate planes parallel to each other, with the oxygens face-to-face. This model requires still further refinement since attack of any nucleophile at a carbonyl group occurs in an obtuse Bürgi–Dunitz angle and closer to the hydrogen atom than to the R group of the aldehyde. The new C—C bond is formed closer to the C—H bond than to the C—R bond of the aldehyde: for this to take place a rotation of the aldehyde in its plane by about 30° is sufficient. Let us examine the consequences of this rotation for each of the four transition states of Nguyen Trong Anh and Bui Tho Thanh: four, because the aldol addition creates two asymmetric centers, at both ends of the new carbon–carbon bond.

Transition state 1, for one of the enolates, requires such a clockwise rotation of 30°. This rotation decreases the unfavorable eclipsing of the R and $R^1$ groups attendant upon formation of the $C_1$—$C_3$ bond.

Transition state 2 also gives rise to an enolate of $Z$ configuration. In this case, rotation by 30°, now required counterclockwise, increases the already unfavorable steric interaction of R and $R^2$.

Transition state 1 is preferred; it leads to a *syn* ketol. In fact, predominant formation of *syn* products from $Z$-enolates is observed. An analogous argument makes transition state 3 preferable to 4: formation of the anti product from $E$-enolates. To summarize the rule: the $Z(E)$ configuration of the enolates determines the *syn* (*anti*) stereochemistry of the aldols.

References: C. H. Heathcock and L. A. Flippin, *J. Am. Chem. Soc.*, **105**, 1667 (1983).
T. A. Nguyen and B. T. Thanh, *Nouv. J. Chim.*, **10**, 681–683 (1988).
E. P. Lodge and C. H. Heathcock, *J. Am. Chem. Soc.*, **109**, 3353 (1987).

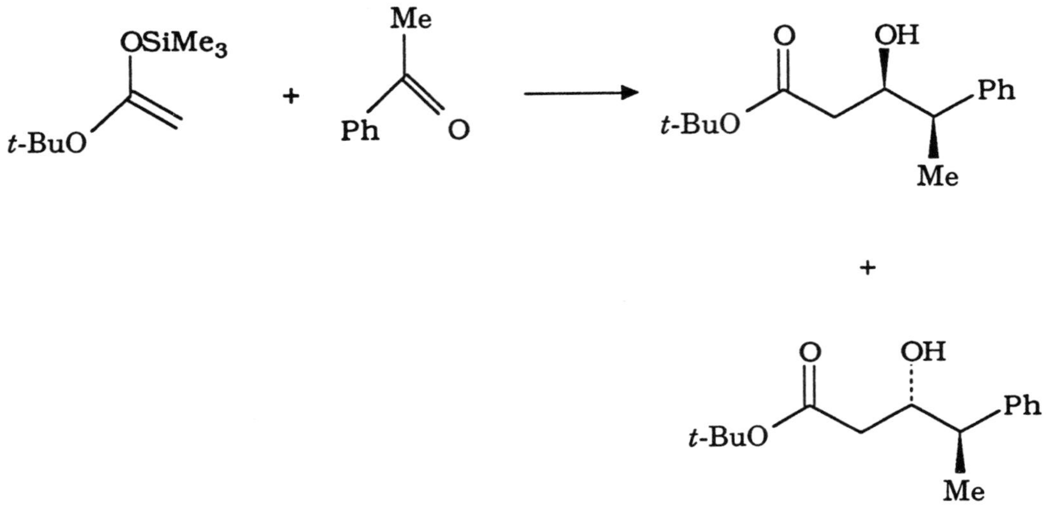


97

3

# 5  Stereoselective Aldol Reactions

If one of the reaction partners in the aldol reaction is chiral, e.g. if the acceptor is a chiral aldehyde, and if a silylated enol ether is used as donor, catalysis with a Lewis acid such as $BF_3$ enhances diastereoselectivity. $BF_3$ coordinates with the aldehyde oxygen *anti* to the chiral group R* ($= CHMePh$), which is more bulky than the hydrogen; this favors a transition state of type 4 in the model of Nguyen and Thanh. One takes advantage thus of the *repulsive* interaction between $BF_3$ and the nucleophile ($=$ silylated enol ether).

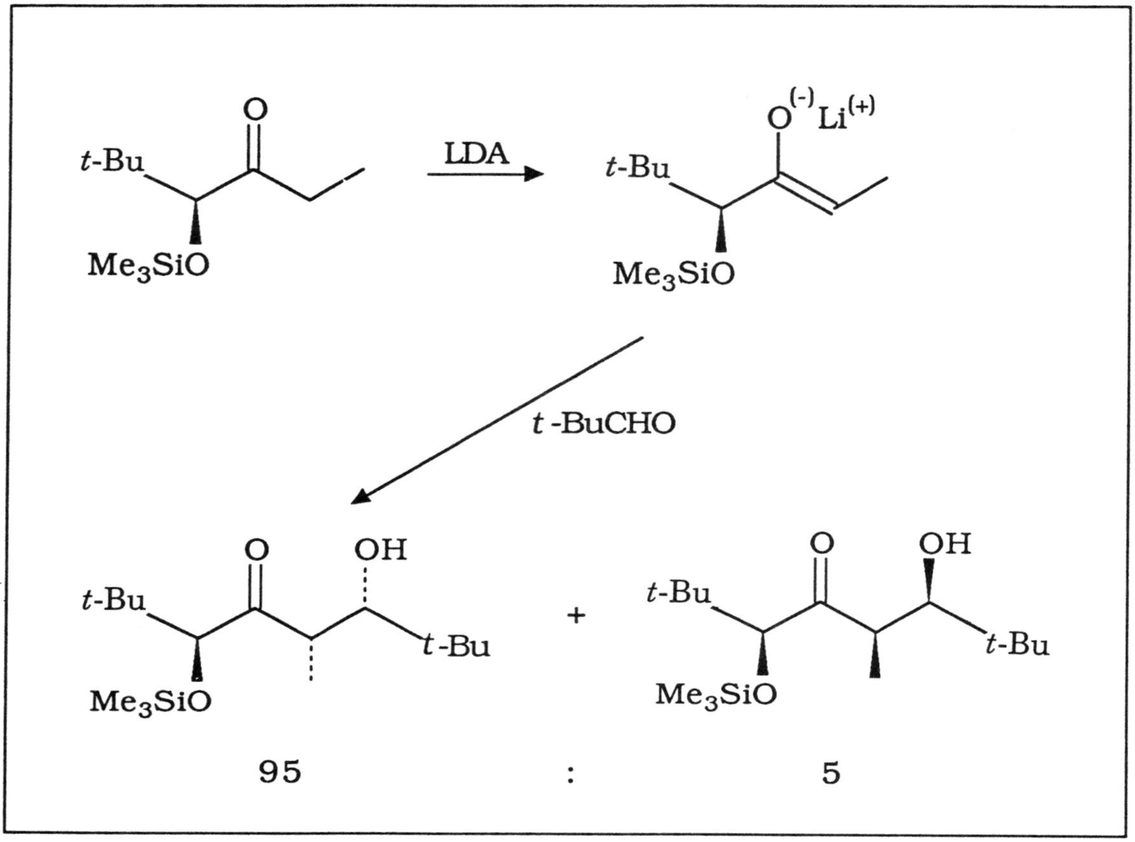
t-Bu
O
Me3SiO
LDA
O(-) Li(+)
t-Bu
Me3SiO
t-BuCHO
O
OH
t-Bu
t-Bu
Me3SiO
+
O
OH
t-Bu
t-Bu
Me3SiO
95
:
5

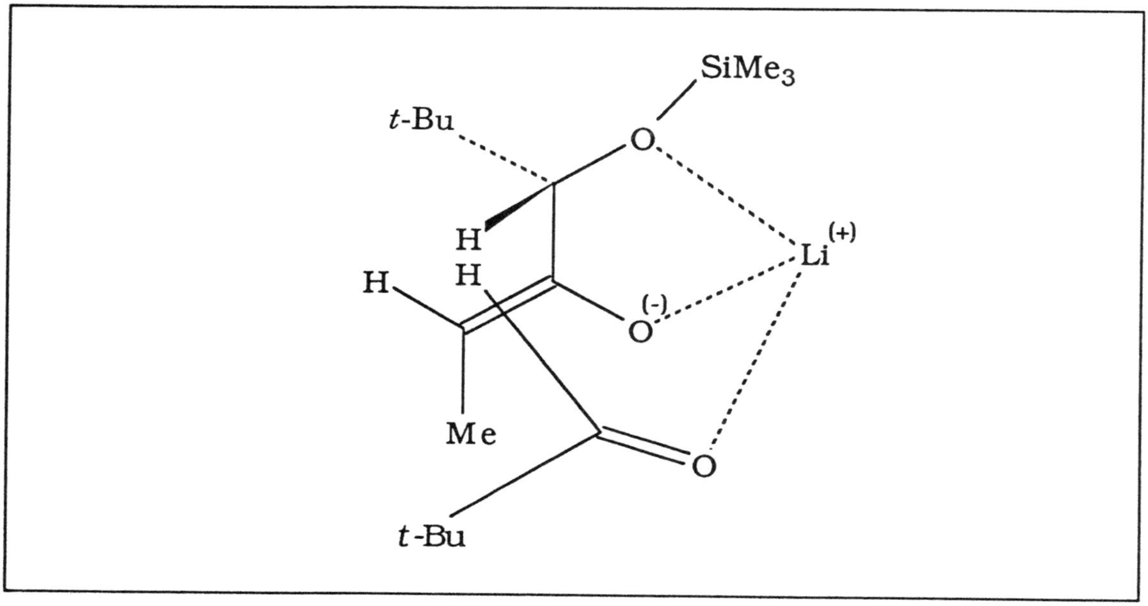
SiMe3
t-Bu
O
H
Li(+)
H
H
O(-)
Me
O
t-Bu

It is also possible to organize the transition state of an aldol addition by an *attractive* interaction and thus control the diastereoselectivity. Thus chelation of donor and acceptor by a lithium cation directs the attack of the donor to the more accessible side of the acceptor RCHO. Diastereoselectivity is excellent ($>95:5$ for $R = i\text{-Pr}$, $t\text{-Bu}$, Ph).

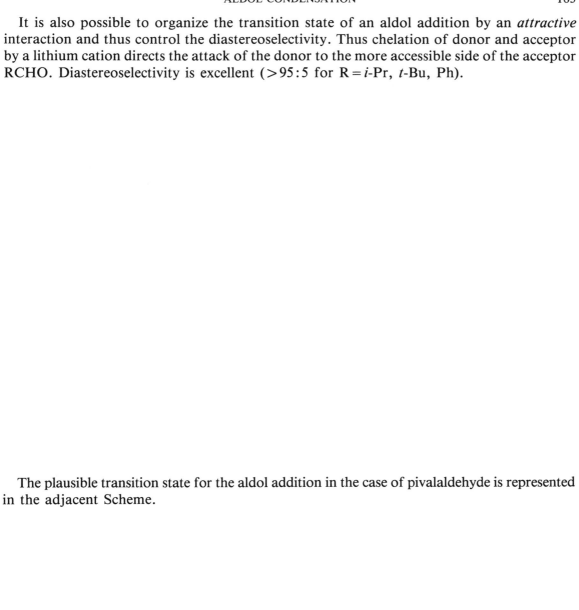

The plausible transition state for the aldol addition in the case of pivalaldehyde is represented in the adjacent Scheme.

Reference: C. H. Heathcock, *Aldrichimica Acta*, **23**, 99–111 (1990).

Ar = $p$-CH$_3$-C$_6$H$_4$-

Let us consider an aldol addition between an aldehyde RCHO acceptor and an enolate derived from a ketone $*Y \cdot CH_2CO \cdot X$ as donor. The idea consists in using a substituent Y in $\alpha$ position as a chiral auxiliary group. It has to fulfil two conditions: to induce a high stereoselectivity in the aldol addition step; and to be cleaved readily after the addition has proceeded to completion.

One of the solutions to this problem has been provided by Solladié and Mioskowski, in Strasbourg. They formed the conjugate base of an $\alpha$-sulfinylester using a Grignard reagent. This carbanion, stabilized by the sulfur group in position $\alpha$, adds diastereoselectively to aldehydes. The $\alpha$-sulfinyl function after serving as chiral auxiliary group is cleaved by reduction with aluminum amalgam. The yields are good (50–85%) and the enantiomeric excesses are high (80–91%).

The explanation for the observed diastereoselectivity again involves chelation with the metal, probably in a transition state with magnesium coordinated with the three oxygen atoms of the enolate, of the sulfoxide, and of the aldehyde. The enolate attacks the aldehyde from the *si* face. The enolate presents preferably its face with the lone pair of the sulfur rather than the hindered bulky Ar group; the substituent R of the aldehyde is placed *anti* to the sulfur substituent.

Reference: C. Mioskowski and G. Solladié, *J. Chem. Soc. Chem. Comm.*, 162 (1967).

*Reminder*:

1. If a trigonal atom, e.g. carbon, bears three different groups, its two faces have an enantiotopic relationship. The three substituents (a, b, c) are ranked according to their relative priorities: large > medium > small. If the substituents (a, b, c) in this order are arranged anticlockwise, the face is named *si* (from the Latin *sinister*); if (a, b, c) are arranged clockwise, one speaks of a *re* face (from the Latin *rectus*).

The particular case of aldehydes RCHO:

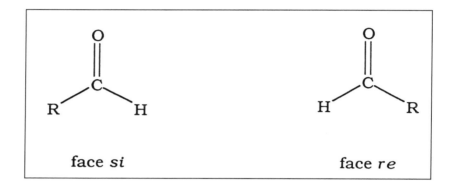

2. The enantiomeric excess, indicated ee, is the difference between the percentages of the two enantiomers formed in a reaction, the sum of which is arbitrarily set to 100%, irrespective of whether the reaction is quantitative or not.

| Percentage difference | ee |
|---|---|
| 99:1 | 98 |
| 98:2 | 96 |
| 97:3 | 94 |
| 95:5 | 90 |
| 90:10 | 80 |
| 80:20 | 60 |

# 6  Musical Interlude

Chapter 3.II presents a minibiography of Borodin to whom we owe the aldol reaction. His opera *Prince Igor* illustrates the inadequate place that music occupied in the life of the eminent chemist. He started to write *Prince Igor* in 1869. The opera was still uncompleted 18 years later when Alexander Borodin died! It consisted of a string of unrelated fragments and of the memories of the friends to whom the composer had played informally on the piano, the overture for instance. After Borodin's death, Rimsky-Korsakov and Glazounov established the version of *Prince Igor* known to us, from the manuscripts which Borodin had left and from their own memories. Borodin had been engaged too much in his research work in chemistry to have had enough time to set the scene, the orchestration and the numerous episodes of his opera!

References: D. Brown, *Times Literary Supplement*, February 16–22 1990, p. 173.
         G. B. Kauffman and K. Bumpass, *Leonardo*, **21**, 429–436 (1988).

## Summary 'Aldol Addition'

*The aldol addition* is the condensation of a carbanion enolate to a carbonyl group. It was discovered by Borodin, whose contributions to chemistry compare favorably with those of his monumental musical achievements.

A biochemical application is the 'shikimate pathway'. By this pathway microorganisms and plants synthesize natural products such as ubiquinone and morphine as well as the three aromatic amino acids phenylalanine, tyrosine, and tryptophan: a key step is an *intramolecular aldol addition*.

*Selective* aldol addition is possible: using an enolate as donor derived from a ketone, thus having a more nucleophilic carbon; and by resorting to an aldehyde as acceptor, as a better electrophile (Wittig's rule).

The principal problem presented by the aldol addition is stereoselectivity. In fact, a priori, a mixture of two $\beta$-ketol diastereoisomers is formed. The frontier orbital description guides the reasoning. We deal with the HOMO (analogous to the non-bonding level of the allyl anion) of the enolate; and the $\pi^*(C=O)$ of the carbonyl. For both partners the angle of attack is close to the tetrahedral angle. The model by Nguyen Trong Anh accounts for predominant formation of a *syn* (*anti*) ketol from a Z(E)-enolate.

Use of *auxiliary groups* allows for improvement of stereoselectivity; e.g. resorting to a silylated enol ether allows for an impressive *anti* selectivity. A sulfoxide as chiral auxiliary group similarly allows for equally good stereoselectivities.

The aldol addition is reversible: its attractiveness in a multi-step synthesis (example: the longifolene synthesis by Oppolzer) is thus yet more enhanced.

## 1 Recognition of Molecular Fragments Synthesizable by an Aldol Reaction

Rifamycine S, an antibiotic, was the target for several great recent syntheses. The handle in its structure can be considered as equivalent to the thioether shown. The latter has eight asymmetric carbon atoms for a total of no fewer than 256 stereoisomers!

Let us focus on the near-symmetry of the C(18)—C(28) chain in the thioether (or in the corresponding ether): to make it appear, oxidation of the alcohol function at C(23) to the ketone, and hydration of the C(18)=C(19) double bond are sufficient.

In retrosynthetic mode, one can now dissect this chain with bilateral symmetry into the enantiomeric moieties A and A' together with the B and C fragments. This crucial observation leads to the notion of accessing both the A and A' enantiomers pure by way of an aldol reaction: they consist of the

Reference: S. Masamune, B. Imperiali, and D. S. Garvey, *J. Am. Chem. Soc.*, **104**, 5528–5531 (1982).

**characteristic moiety** $-O-C-C-C=O$. Furthermore, the $A' \supset B \supset A$ entity is also available, in a stereoselective manner, by an aldol reaction. Masamune and his group (MIT) have thus synthesized the thioether in 18 steps with an overall yield of 30% and an overall stereoselectivity of 80% (as compared with the earlier, more laborious syntheses of Kishi at Harvard: 51 steps, 0.7% yield, 50% stereoselectivity; 48 steps, 3.1% yield, 74% stereoselectivity; 45 steps, 4.3% yield, 75% stereoselectivity). The take-home lesson is the importance of the initial, analytical and imaginative hard look at the structure of the target molecule.

## 2 Musical Paternity of the Aldol Reaction

One of the inventors (another is the Frenchman Wurtz) of the aldol reaction is the Russian composer Alexander Borodin (1838–1887). It is not well known that he was also an excellent chemist. For his PhD thesis (1858) under the direction of Nicolai Zinin—who had synthesized aniline in 1842 by reduction of nitrobenzene—Borodin had studied the biological chemistry of arsenic and phosphoric acids, set in parallel.

Zinin had warned Borodin: You have to make a choice between chemistry and music, the latter would interfere with your scientific activity. In parallel manner, musical historians lament the low productivity of Borodin and blame it on his professional activity as a chemist!

Borodin succeeded Zinin in 1864 to the Chair in organic chemistry at Saint-Petersburg. Besides his work on the aldol reaction, he studied aromatic amines, organic fluorine compounds (he was the first to prepare benzoyl fluoride $C_6H_5 \cdot CO \cdot F$) and elucidated the bromodecarboxylation of silver salts:

$$RCOOAg + Br_2 \rightarrow RBr + CO_2 + AgBr$$

A. Borodin, *Ann.*, **119**, 121 (1861), a reaction now referred to as the Hunsdiecker reaction.

Besides Rimsky-Korsakov, another great musician friend of Borodin was Franz Liszt, whose acquaintance he made in Jena in 1877; on this occasion he was accompanied by two of his students about to defend their dissertations in this university.

References: A. Borodin, *Bull. Acad. Sci. St. Petersburg*, **7**, 463 (1864); *J. Pract. Chem.*, **93**, 413 (1864); *Zhur. Russ. Khim. Obsch.*, **1**, 214 (1869); **2**, 90 (1870); **4**, 207–209 (1872); *Zhur. Khim. Fiz. Obshch.*, **5**, 386–410 (1873); *Ber.*, **3**, 423–552 (1870); **5**, 480–481 (1872); **6**, 982 (1873); *J. Chem. Soc.*, **27**, 145 (1874).
I. D. Rae, *Chem. in Australia*, 144 (1987).
G. B. Kauffman, I. D. Rae, Y. I. Solov'ev, and C. Steinberg, *Chem. Eng. News*, 28–35 (1987).
G. B. Kauffman, Y. I. Solov'ev, and C. Steinberg, *Ed. in Chem.*, **24**, 138–140 (1987).
I. D. Rae, *Leonardo*, **23**, 156 (1990).
C. A. Wurtz, *Compt. Rend.*, **74**, 1361 (1872); **76**, 1165 (1873).

## 3   Industrial Uses of Benzene

(a) Transformation to ethylbenzene, followed by dehydrogenation to styrene which then serves as the monomer leads to polystyrene. This is its major use (55%) with an annual production in the order of 4.5 MT.

(b) Transformation to cumene (= *i*-propylbenzene): alkylation with propylene catalyzed by $AlCl_3$ or $H_3PO_4$.

The Hock process first oxidizes cumene to a hydroperoxide, which is then cleaved to a molecule of phenol and one of acetone. This second use accounts for 24% of the production of benzene: about 2.3 MT of cumene are manufactured annually.

(c) Hydrogenation to cyclohexane, subsequently transformed into nylon, according to: cyclohexane → cyclohexanol → cyclohexanone → cyclohexanone oxime → ε-caprolactam, by way of a Beckmann rearrangement of this oxime. This pathway represents 13% of the annual production of benzene.

Reference: B. F. Greek, *Chem. Eng. News*, December 10 1990, pp. 10–11.

## 4   An Industrial Application of the Aldol Reaction

Aldol condensation of acetone is catalyzed by bases, and the product upon acid-catalyzed dehydration affords mesityl oxide.

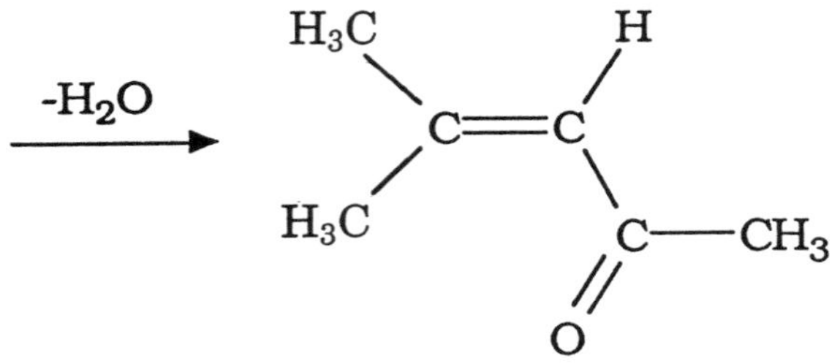

$$\xrightarrow{-H_2O}\quad \begin{array}{c} H_3C \\ \phantom{H_3C} \\ H_3C \end{array} C{=}C \begin{array}{c} H \\ \phantom{} \\ C{-}CH_3 \\ \parallel \\ O \end{array}$$

Hydrogenation of mesityl oxide, catalyzed by several metals (Cu,Ni) affords isobutylmethylketone. The various processes used allow for conversion of acetone to reach 30–50% with a 83–95% selectivity in formation of isobutylmethylketone.

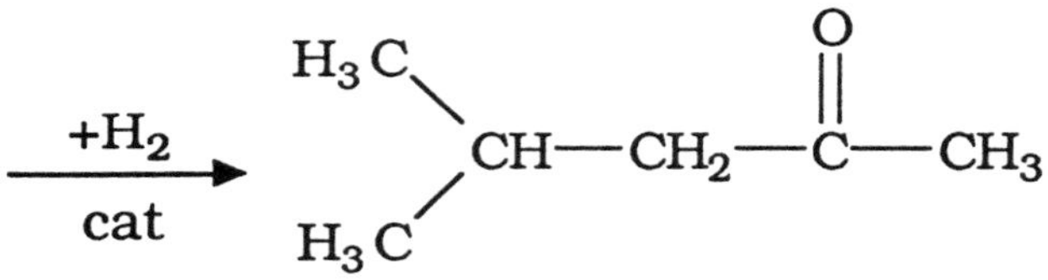

$$\xrightarrow[\text{cat}]{+H_2}\quad \begin{array}{c} H_3C \\ \phantom{} \\ H_3C \end{array} CH{-}CH_2{-}\overset{\displaystyle O}{\overset{\parallel}{C}}{-}CH_3$$

Isobutylmethylketone, like the other aldolization products of acetone, is used as solvent in the following applications:

—cellulose acetate
—cellulose acetobutyrate
—acrylic resins
—purification of organic products by liquid–liquid extraction
—removal of paraffins from mineral oil
—paint industry
—extraction from niobium and tantalum minerals.

The German Texaco (Deutsche Texaco) has a production unit in Germany with a capacity of about 25 kT year$^{-1}$.

# 5  Another Great Industrial Application of the Aldol Reaction

Condensation of formaldehyde and isobutyraldehyde under basic conditions affords a $\beta$-aldol intermediate the hydrogenation of which leads to neopentyl glycol.

The latter is produced yearly in an amount of about 20 kT. It is used for lubricating oils, in paints, in some polyesters and in plasticizers. The first step of aldol addition uses formaldehyde as acceptor, with the enolizable isobutyraldehyde as donor. The reason for such selectivity in a reaction between two aldehydes is twofold: formaldehyde can't enolize (absence of $\alpha$-carbons); its carbonyl group is more accessible (steric hindrance from the isopropyl groups).

'Socrates: this wonderful definition, in which the beautiful consists in what is advantageous and in what is useful and capable of producing some good has nothing wonderful about it ( . . . ) it is even, if it were possible, more ridiculous than the previous one', Plato, *Hippias Major*.

# 6 Aldol Reaction in Major Organic Syntheses

This condensation is one of the major processes used by the wizards of organic synthesis: it allows for formation of a new carbon–carbon bond; and it presents the challenge of four possible stereoisomeric products where only one is required.

The example shown is from Oppolzer's longifolene synthesis. Longifolene is a sesquiterpene. The two key steps are:

(i) a photochemical $2\pi + 2\pi$ cycloaddition between two cyclopentenes—one of which results, in a pre-equilibrium, from enolization of a cyclopentanone. This cycloaddition affords a tetracyclic system, comprising the bicyclo[2.2.1]heptane part of longifolene.

(ii) a ring opening: the strained cyclobutane formed in the preceding step opens; thus a seven-membered ring, also present in longifolene, obtains.

The pathway followed in this opening is a spontaneous retro-aldol reaction. In fact, the photoadduct possesses the **fragment** −O−C−C−C=O **characteristic** of an aldol. Since the aldol addition is an equilibrium, the inverse reaction is also possible. In this specific case, the driving force for the retro-aldol reaction is the opening of a small strained ring. This opening is likely to be triggered off by the hydrogenolysis of the ester function.

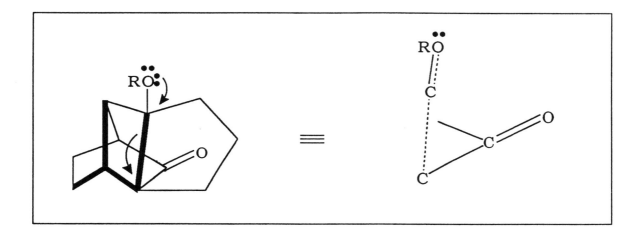

From a retrosynthetic viewpoint, the tricyclic diketone is equivalent (from the point of view of the aldol reaction) to the tetracyclic β-ketol (see box on the right hand side, R=H).

The latter can again be disconnected, in the manner indicated, to a cyclopentene and a cyclopentenol (R=H). A photochemical $2\pi + 2\pi$ addition will join them all the more easily because they are already linked together by a carbonyl bridging group.

This longifolene synthesis is extremely elegant because of its simplicity (see the note below).

Corey had given a splendid longifolene synthesis, 14 years earlier, which remains one of the epochal achievements of organic synthesis. Oppolzer, having opened a new route, was inviting comparison with his great predecessor. Esthetic considerations are not absent from this field. In like manner as Brahms and his 'Variations on a Theme of Haydn' or as Ravel with his 'Tombeau de Couperin', Oppolzer has chosen to present posterity with his own masterpiece, in implicit honor of Corey, but at the same time to display his own prowess.

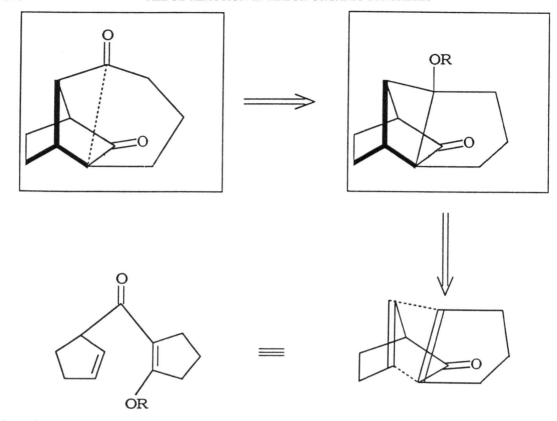

Complementary note:

The initial bicycle, bridged by a carbonyl group, is obtained by addition of an enamine derived from cyclopentanone to the acid chloride of cyclopentene-2-carboxylic acid. The final steps of this synthesis are:

   (iii) on the more accessible carbonyl group of the cycloheptanone ring: use of the Wittig reagent $Ph_3P{=}CH_2$, followed by methylene iodide $CH_2I_2$ in the presence of the copper–zinc couple (Simmons–Smith conditions); finally hydrogenolysis of the cyclopropane with molecular hydrogen, catalyzed by platinum under acidic conditions;

$$\diagup\hspace{-6pt}\diagdown C{=}O \rightarrow \diagup\hspace{-6pt}\diagdown C{=}CH_2 \rightarrow \diagup\hspace{-6pt}\diagdown C\overset{CH_2}{\underset{CH_2}{\diagup\hspace{-4pt}\big|\hspace{-4pt}\diagdown}} \rightarrow \diagup\hspace{-6pt}\diagdown C(CH_3)_2$$

   (iv) introduction of an *endo* methyl group at the carbon $\alpha$ to the cyclopentanone carbonyl group: enolization with $LiNR_2$ followed by alkylation with methyl iodide $CH_3I$;

   (v) final transformation to longifolene: the Wittig reagent $Ph_3P{=}CH_2$ again transforms the carbonyl group to a vinyl group

$$\diagup\hspace{-6pt}\diagdown C{=}CH_2$$

Reference: W. Oppolzer and T. Godel, *J. Am. Chem. Soc.*, **100**, 2583 (1978).

# 7   Synthesis of an Aphrodisiac

Cantharidin, the active principle (vesicant) of the Spanish fly *Cantharis vesicatoria* was isolated by the pharmacist Robiquet in 1810. Its synthesis was achieved by Woodward in 1941. We present here the Stork synthesis in the Büchi modification: (i) Diels–Alder addition in ethanol at 100 °C followed by (ii) a reduction with lithium aluminum hydride: —COOMe → —CH$_2$OH and (iii) esterification with methanesulfonyl chloride MsCl: —CH$_2$OH → —CH$_2$OMs; then (iv) a substitution with EtS$^{(-)}$K$^{(+)}$ is carried out: $-CH_2OMs \rightarrow -CH_2SEt$. The tricyclic compound, bearing the two sulfur-containing side-chains, is subjected to reaction with osmium tetroxide in the dark, and then hydrolyzed with $Na_2SO_3$ in water/ethanol (v and vi), which transforms the cyclohexene into a 1,2-cyclohexadiol. Desulfuration with Raney nickel affords the tricyclic glycol in which the two fragments $-CH_2SEt$ have been transformed into methyls. Cleavage of the glycol by periodic acid (viii) affords the intermediate dialdehyde. The latter is submitted, without isolation, to the action of piperidinium acetate, a relatively weak base (ix) which promotes an aldol addition followed by dehydration of the primary aldol product to cantharidin.

Note that the intermediates of the synthesis are symmetric (the symmetry plane passing through the bridging oxygen). Only in the last step (ix) dissymmetry is introduced by the aldol condensation, which necessarily provides a racemic product.

References: Robiquet, *Ann. Chim.* [1], **76**, 307 (1810).
            G. Stork, E. E. Van Tamelen, L. J. Friedman and A. W. Burgstahler, *J. Am. Chem. Soc.*, **75**, 384 (1953).
            G. Büchi and G. A. Carlson, *J. Am. Chem. Soc.*, **90**, 5336 (1968).

# 8   Aldol Condensation as Source of Mechanical Resistance of the Ligaments

Collagen, the most common protein in mammals (to the extent of about 25%), is an assembly of fibrous proteins. They can be found in bones, cartilages, teeth, tendons, ligaments, skin, blood vessels, etc.

The mechanical resistance of collagen fibers is increased by their reticulation, i.e. the bridges formed between polypeptide chains. Thus, in rats, the resistant collagen in the Achilles' tendon is much more reticulated than that in the more flexible tail.

In a first step, $\epsilon$-amino groups of the lysine residues are oxidized into the corresponding aldehydes by means of lysyl oxidase (1):

1.

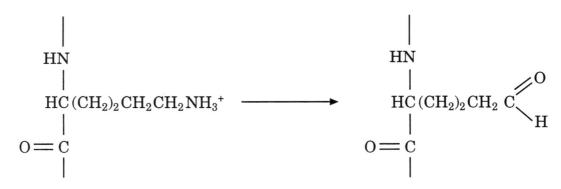

Then a reticulation by way of an aldol condensation can occur (2):

2.

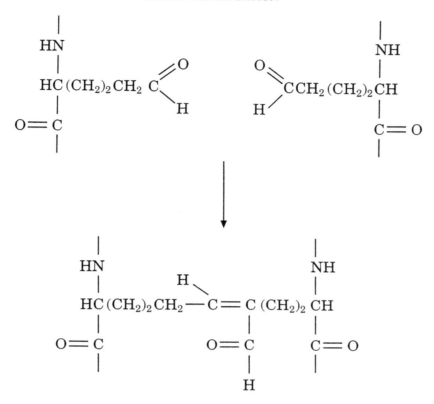

Reticulation can then continue in two ways:

(i) addition of an imidazole of a histidine residue to the C=C double bond of the aldol bridge (3):

3.

(ii) the aldehyde group still present after this double aldol–histidine connection, can react with another side chain, e.g. hydroxylysine to afford a Schiff base.

In the same way, up to four side chains can attach to one another, which increases the rigidity and the solidity of the collagen fibers.

The corresponding pathology *lathyrism* is caused by ingestion of some foodstuffs, especially beans of various sorts. The toxic agent present in these seeds is β-aminopropionitrile which inhibits oxidation of lysine ε-amino groups into aldehydes.

---

'Wretched people, wretched people, take your hands away from the bean', Empedocles, fragment 141.

## 9  Chemical Products of Major Importance to Industry: Vinyl Acetate

Vinyl acetate $H_3C \cdot CO \cdot O \cdot CH = CH_2$ is produced mostly by a gas-phase reaction between acetic acid $H_3C-COOH$, ethylene $H_2C=CH_2$ and dioxygen $O_2$. Its production in the United States is in the order of a million tons per year.

Its principal derivatives are polymers (55%), polyvinylalcohol (20%) and several copolymers with butyraldehyde, ethylene or vinyl chloride (20%).

The three major uses are:

1. Adhesives (35%)
2. Paints: interior paints based on latex (30%)
3. Paper and textile coatings

## 10  Thiamine Deficiency and the Korsakov Syndrome

The first description of this partial amnesia in 1887 is due to the Russian Korsakov:

'Only the memory of recent events is perturbed. The most recent impressions seem to be erased first; the patient correctly remembers more ancient memories. Accordingly, his intelligence, his spiritual freedom and his ability to adapt himself to a situation are practically not impaired'.

This neurological disorder found amongst a minority of alcoholics and of people having a genetic predisposition or suffering from chronic malnutrition, originates from the destruction of mamillary bodies in the brain, caused by a thiamine deficiency in food.

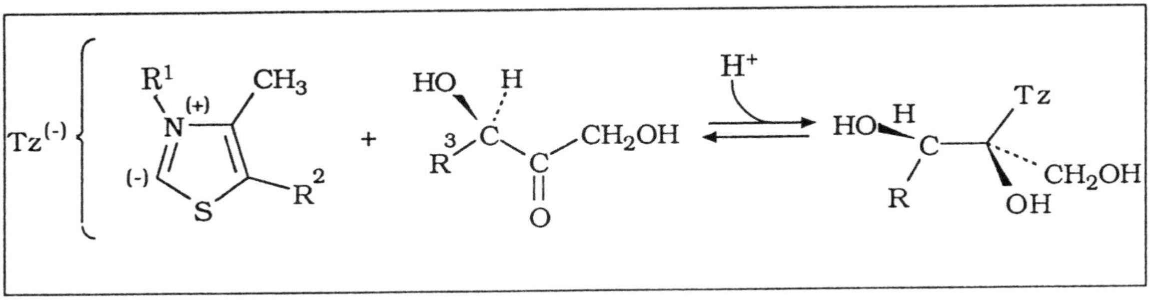

Thiamine, in the form of its pyrophosphate and as its conjugate base symbolized here as $Tz^{(-)}$, serves as the prosthetic group (non-proteic part) of an enzyme, transketolase. It promotes condensation of a ketose-type sugar with an aldehyde. In a first step, nucleophilic addition of $Tz^{(-)}$ to the carbonyl group of the ketose takes place.

The addition compound thus formed is cleaved (in a second step) to an aldehyde molecule $R^3CHO$ and to an activated glycoaldehyde module. The activation is due to the entity $C=C=N$, which would be more stable with a linear geometry. Thiamine thus acts as a carrier for the two-carbon fragment $-CO-CH_2OH$. The second cleavage step *resembles* a retroaldol reaction.

This activated glycoaldehyde entity can then condense with another aldehyde RCHO, which forms a new ketose, set free by the enzyme in a reaction inverse to the first addition step, with regeneration of $Tz^{(-)}$.

Transketolase in patients with the Korsakov syndrome combines with thiamine pyrophosphate $Tz^{(-)}$ ten times less efficiently than the enzyme in a normal person.

Reference: L. Stryer, *Biochemistry*, 2nd ed., W. H. Freeman, San Francisco, 1981, pp. 340–343.

'—This is, said Frère Jean, an incongruous and non pertinent comparison. Because I heard it told a long time ago that the snake that had got inside the stomach does not cause any trouble and readily goes out if by the feet the patient is suspended, and via the mouth when a dish filled of warm milk is presented to it.
—You have heard it told, said Pantagruel: likewise for the people who told it to you. But such a remedy was neither seen nor read. Hippocrates (*Epidemics*, Book V) writes that such a case happened in his time, and the patient died by spasm and convulsion.' Rabelais, *Quart-Livre*, XLIV

## 11   From Underwater Exploration to Biotechnologies

Microorganisms known as hyperthermophiles are found near hydrothermal springs, on ocean floors, where the local temperatures can reach 150 C or higher. The conjecture (authors: Eric J. Toone and Michael Shelton, Duke University, in Durham, North Carolina; Michael W. W. Adams, University of Athens, Georgia) is that robust enzymes, more versatile than those functioning at 37 C in standard organisms could survive under such conditions. In fact, it has been possible to isolate aldolases 100 times more active and more substrate-selective than normal aldolases.

These enzymes* catalyze the addition of pyruvic acid to aldehydes such as glyceraldehyde-3-phosphate, affording 2-keto-4-hydroxybutyrate derivatives.

The latter are the precursors to numerous natural products: 2-deoxyaldoses, after decarboxylation; β-hydroxycarboxylic acids, after oxidative decarboxylation; α-amino-γ-hydroxycarboxylic acids, by stereospecific reductive amination; and α,γ-dihydroxycarboxylic acids, by reduction.

$$RCHOHCH_2CHO$$

$$RCHOHCH_2COO^{(-)} \longleftarrow RCHOHCH_2\overset{\overset{O}{\|}}{C}COO^{(-)} \longrightarrow RCHOHCH_2CHOHCOO^{(-)}$$

$$RCHOHCH_2CH(NH_3{}^+)COO^{(-)}$$

Reference: S. Borman, *Chem. Eng. News.*, January 11 1993, pp. 26–36.

# 4.I
# Some Solutions to the Problem of Selectivity
# (Chemo-, Regio-, Stereo-)

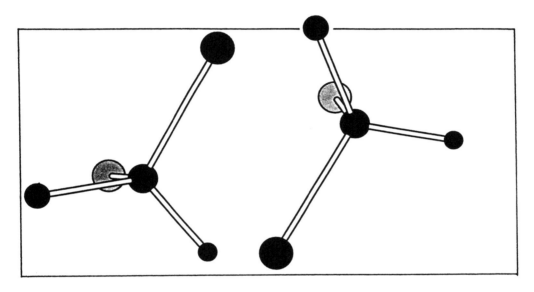

To synthesize unambiguously the target molecule, without being sidetracked due to the multiplicity of reaction sites: this is what is at stake. This is the great difference between present and past syntheses.

In this chapter, we shall refrain from dealing with localized solutions or with specific tricks. Our focus will remain on general principles. Taking the most desirable goal, that of enantioselectivity, how do we discriminate between the two faces *si* and *re* of a double bond? We can do it by inducing the reagent to attack exclusively from one side or by preventing it from attacking from the other.

The examples provided again describe formation of carbon–carbon double bonds, especially by way of Michael addition. This chapter also presents two major selective alkylation methods, the elegant route through enamines (Stork), a classic, and another avenue through R-AMP or S-AMP (Enders) which even though much more recent is no less superb in its conception and in its effectiveness.

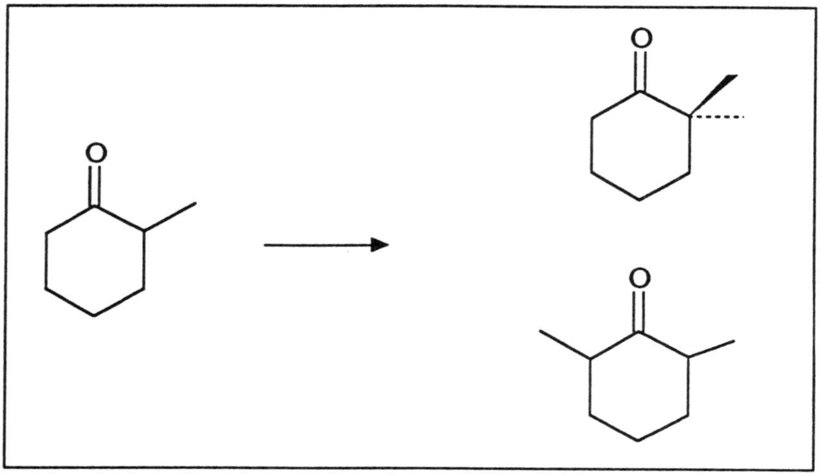
O
O
O

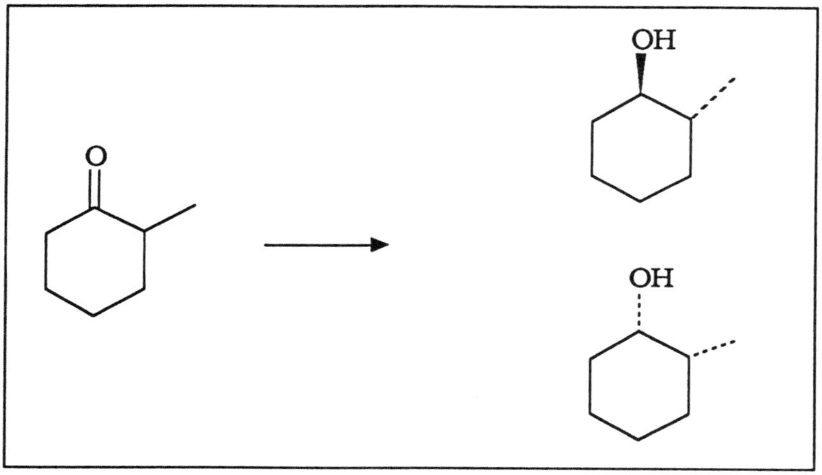
O
OH
OH

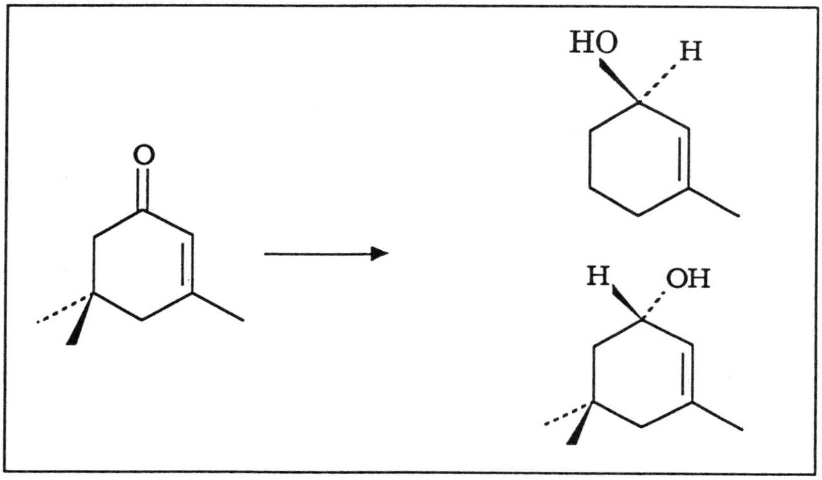
O
HO
H
H
OH

# 1 The Vocabulary of Selectivity

A reaction is selective if it affords predominantly a single product. It is specific if this is the only product.

Chemoselectivity is associated with difference in reactivity between different groups, aldehydes and ketones for instance.

Regioselectivity involves preferential reaction at one site rather than at another. An example is the alkylation of 2-methylcyclohexanone at C(2) or at C(6).

Stereoselectivity is the preferential formation of one among the various stereoisomers (example: reduction of 2-methylcyclohexanone to cis- or trans-2-methylcyclohexanol).

Enantioselectivity is the preferred formation of one enantiomer rather than its mirror image (example: reduction of isophorone to the R or to the S alcohol). Use of enzymes (if feasible) is one means for achieving enantioselectivity.

$(H_3C)_3 Si \text{———} H \qquad 90$

$(H_3C)_3 Si \text{———} CH_3 \qquad 89$

$(H_3C)_3 Si \text{▬▬} OCH_3 \qquad 127 \ (\text{kcal mol}^{-1})$

## 2 Regioselectivity in the Formation of Enolates

Use of silyl enol ethers solves this problem, because an enolate can be formed from the $\alpha$- or the $\alpha'$-carbon, adjoining a carbonyl group.

The concept is attack at the most accessible site by a hindered base such as lithium diisopropylamide (LDA). This affords the kinetically controlled enolate if the proton is removed at low temperature ($-78\,°C$) so fast that the enolate does not have the time to equilibrate.

It only remains to trap this kinetic enolate through formation of a silyl enol ether. The rationale is the exceptional strength of the silicon–oxygen bond: this high bond energy is the driving force for the formation of the silylated enol ether (Gilbert Stork, 1968 and Herbert O. House, 1969).

The explanation for the high Si—O bond energy is twofold: the difference in electronegativity between silicon (electropositive) and oxygen (electronegative) displaces the electron density to the internuclear bonding region; the polarizability of silicon helps to stabilize the non-bonding lone pairs on oxygen.

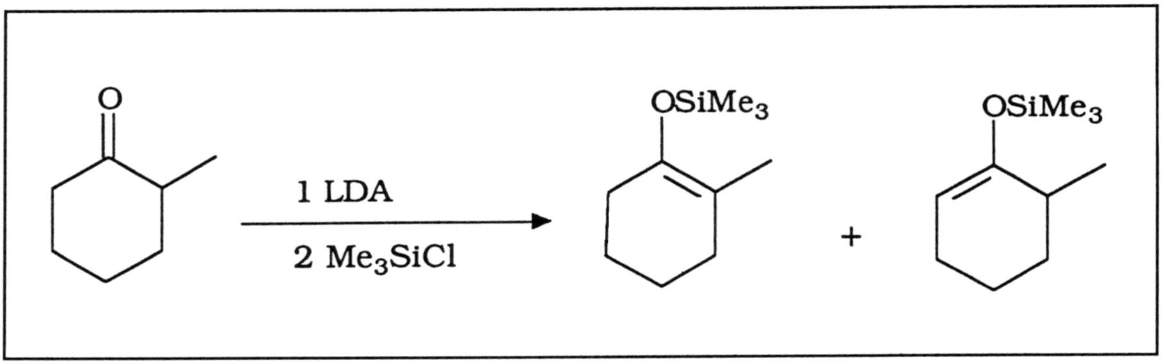
O
OSiMe3
OSiMe3
1 LDA
2 Me3SiCl
+

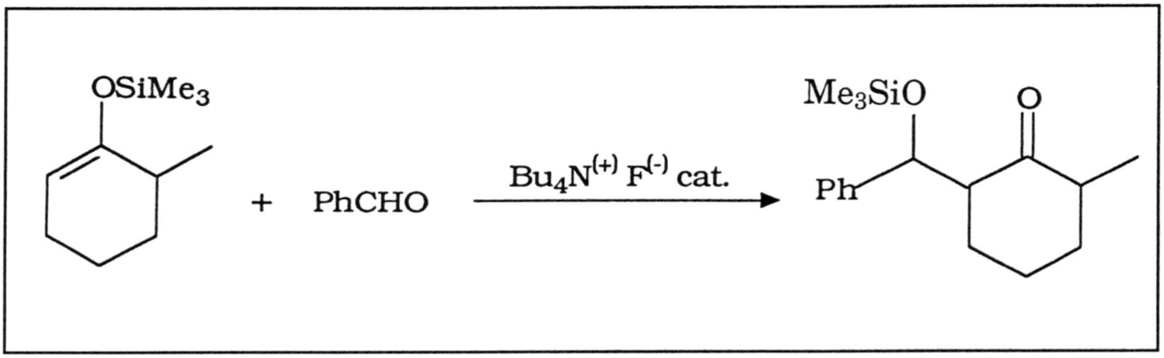
OSiMe3
+
PhCHO
Bu4N(+) F(-) cat.
Me3SiO
O
Ph

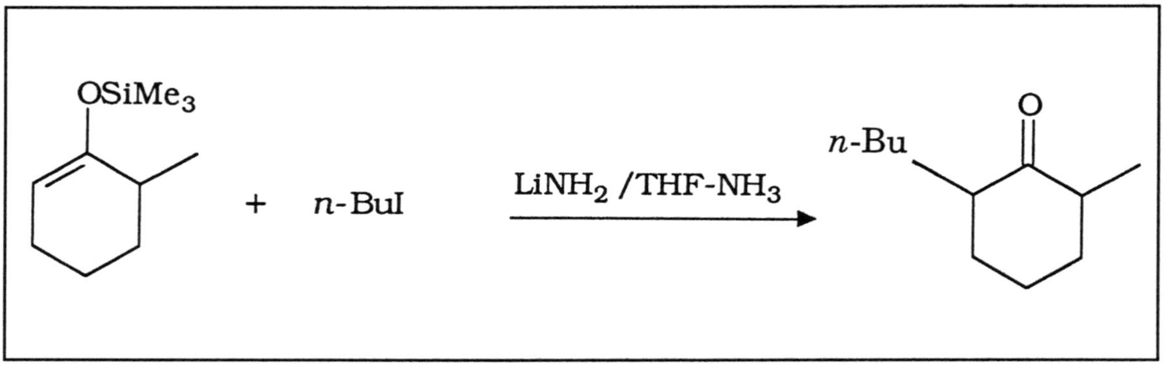
OSiMe3
+
n-BuI
LiNH2 /THF-NH3
n-Bu
O

Formation of the kinetic enolate and its trapping as a silyl enol ether have just been described as specific reactions (selectivity 100%). Actually, with 2-methylcyclohexanone as substrate, a small amount (1%) of the silyl enol ether resulting from the thermodynamically-controlled enolate, is obtained from treatment with LDA.

These silyl enol ethers react just like enolates in aldol additions. The trimethylsilyl group is cleaved at the end of the reaction, usually by means of fluoride anions, taking advantage of the high bond energy of the Si—F bond.

These silyl enol ethers can also be alkylated: this is the equivalent of a regioselective alkylation of a ketone.

References: G. Stork and P. F. Hudrlik, *J. Am. Chem. Soc.*, **90**, 4462–4464 (1968).
H. O. House, L. J. Czuba, M. Gall, and H. O. Olmstead, *J. Org. Chem.*, **34**, 2324–2336 (1977).
R. Noyori, K. Yokoyama, J. Sakata, I. Kuwajima, E. Nakamura, and M. Shimizu, *J. Am. Chem. Soc.*, **99**, 1265–1267 (1977).

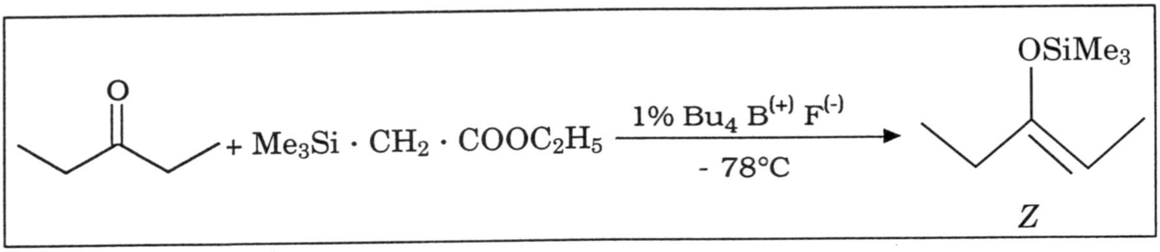
O
+ Me3Si · CH2 · COOC2H5
1% Bu4 B(+) F(-)
- 78°C
OSiMe3
Z

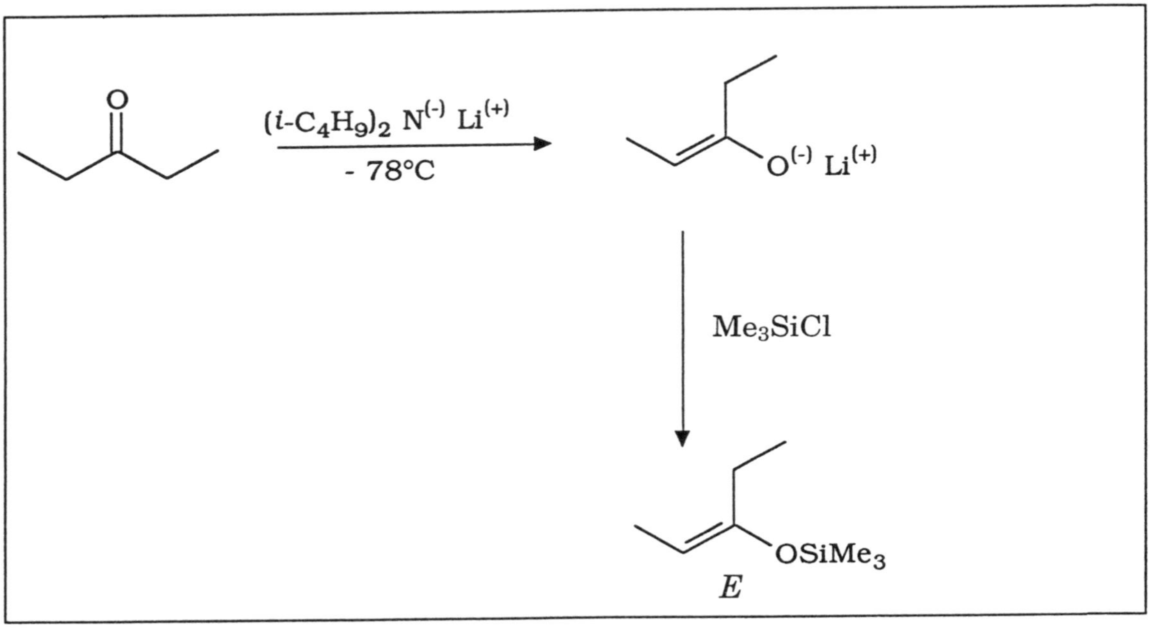
O
(i-C4H9)2 N(-) Li(+)
- 78°C
O(-) Li(+)
Me3SiCl
OSiMe3
E

# 3 Selective Access to *E*- and *Z*-Enolates

One tactic is use of silyl enol ethers. Ethyl trimethylsilyl acetate forms stereoselectively enol ethers of *Z*-configuration in quantitative yield (99.5%).

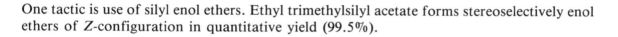

Silyl enol ethers with *E*-configuration are preferentially formed with trimethylsilyl chloride with LDA (lithium diisopropylamide) as the base.

References: E. Nakamura, T. Murofushi, M. Shimizu, and I. Kuwajima, *J. Am. Chem. Soc.*, **98**, 2345–2348 (1976).
E. Nakamura, K. Hashimoto, and I. Kuwajima, *Tetrahedron Lett.*, 2079–2082 (1978).
R. E. Ireland, R. H. Mueller, and A. K. Willard, *J. Am. Chem. Soc.*, **98**, 2868–2877 (1976).

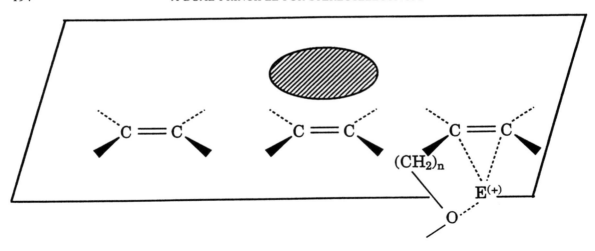

# 4    A Dual Principle for Stereoselectivity

Attack of a double bond from one side only is the aim. There are two solutions to this problem. The first is negative, by previous occupation of the site: if one of the two faces is blocked by a bulky group, then the reagent will attack preferably from the other face. The second way is positive, an invitation to stay: by means of a temporary bond with an auxiliary group, which binds the reagent preferentially to one of the faces. This is frequently an oxygen atom, coordinating an electrophile $E(+)$.

Let us examine each of these two tactics:

(A)  Steric hindrance:  (A.1) enantioselective Michael reaction
                        (A.2) monoalkylation of ketones
(B)  Attraction:        (B.1) enantioselective alkylation of ketones via the enamine
                        (B.2) enantioselective alkylation of ketones with the SAMP method

---

'The proposition is an image of reality. The proposition is a transposition of reality and thought', L. Wittgenstein, *Tractatus Logico-philosophicus*, 4.01.

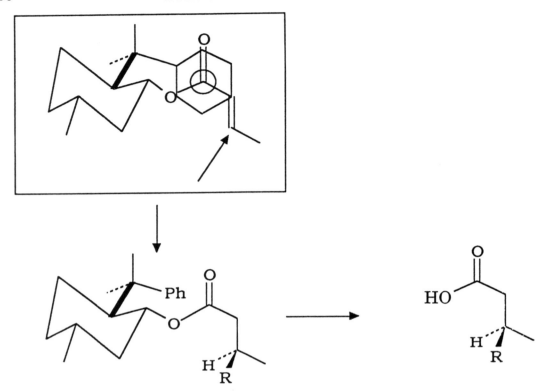

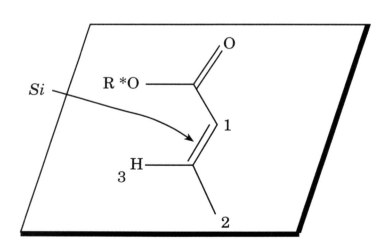

# 5   Enantioselective Michael Addition

Use of organocuprates complexed with $BF_3$ can lead to some elegant asymmetric inductions. In the case at hand, the $\alpha,\beta$-unsaturated ester shown, having one of the two faces of its double bond masked, reacts with several organocuprates with a large enantiomeric excess:

| Organocuprate | ee (%) |
|---|---|
| $BuCu \cdot BF_3$ | 99.5 |
| $MeCu \cdot BF_3$ | 78 |
| $MeCu \cdot BF_3 \cdot PBu_3$ | 87 |

Hydrolysis of the Michael adduct provides a chiral acid.

Reference: W. Oppolzer, R. Moretti, T. Godel, A. Meunier, and H. Lother, *Tetrahedron Lett.*, **24**, 4971–4974 (1983).

'In the state of things the objects are connected to each other like the links of a chain', L. Wittgenstein, *Tractatus logico-philosophicus*, 2.03.

$I^{(-)}$

# 6  Regioselectivity in Alkylation of Ketones: The Stork Enamine Method

The goal is monoalkylation of a ketone at the carbon adjacent to the carbonyl group. A direct reaction is not the answer. Reaction of a ketone, e.g. cyclohexanone, with an alkylating agent, e.g. methyl iodide, leads to a mixture of mono-, di-, tri- and tetramethylation products.

A solution is the enamine method of Stork. Addition of pyrrolidine to cyclohexanone forms the enamine (together with a molecule of water). The enamine can be described as a resonance hybrid with a dipolar limit form that manifests the donor character of the olefinic carbon (2).

Accordingly, electrophiles attack this donor center in regioselective manner. They also attack in a stereoselective manner, hence avoiding polyalkylation.

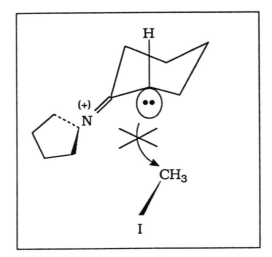

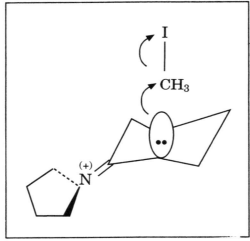

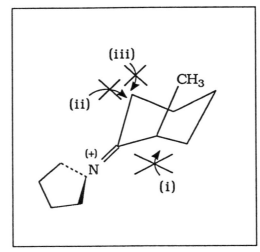

Let us consider the first alkylation. The $S_N2$ transition state aligns carbon (2) of the donor with the carbon of the methyl group and the iodide leaving group. In the equatorial position of the cyclohexanone chair conformation, such an arrangement is disfavored due to the strong steric repulsion from the *syn* methylene group of the pyrrolidine; furthermore, this form of the enamine is deconjugated.

This alkylation can only take place in the axial position. The intermediate resulting from the alkylation thus possesses an *axial* alkyl group before its hydrolysis to the corresponding cyclohexanone.

This alkylated iminium ion intermediate behaves from now on as a carbonyl compound and no longer as an enolate; thus the problem of alkylation in the $\alpha'$-position vanishes.

Reference: G. Stork, R. Terrell, and J. Szmuszkovisz, *J. Am. Chem. Soc.*, **76**, 2029–2030 (1954).

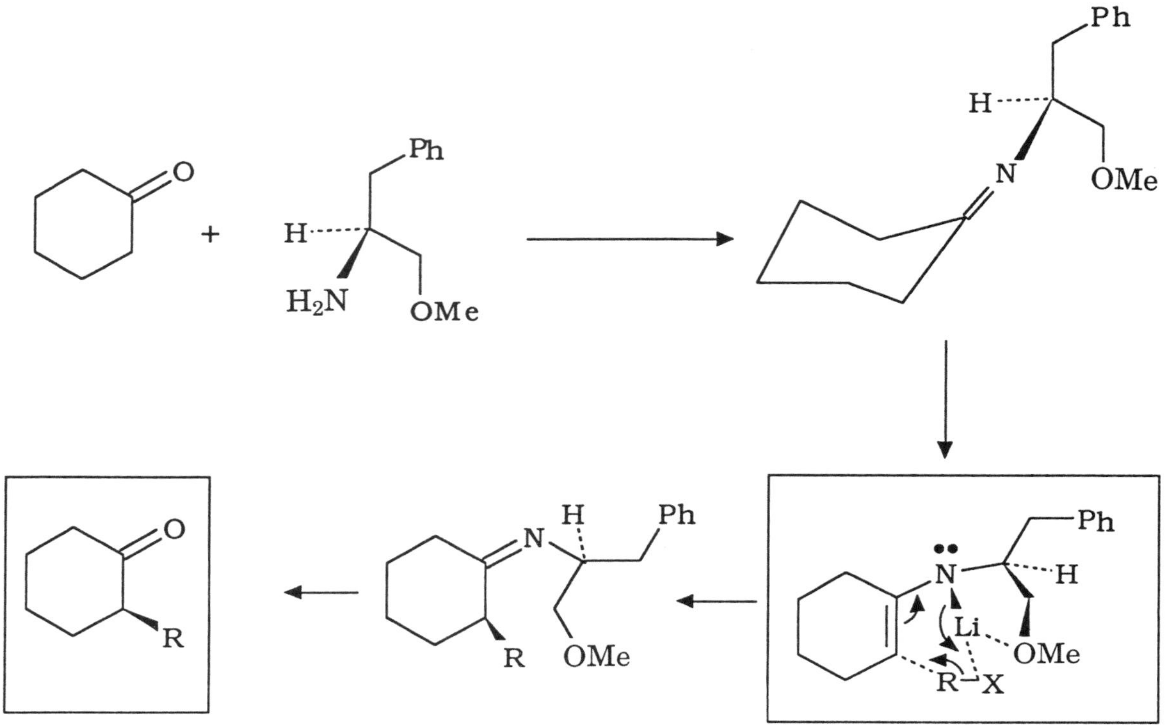
O
+
Ph
H
$H_2N$
OMe
Ph
H
N
OMe
Ph
H
N
R
OMe
Ph
N
H
Li
OMe
R
X
O
R

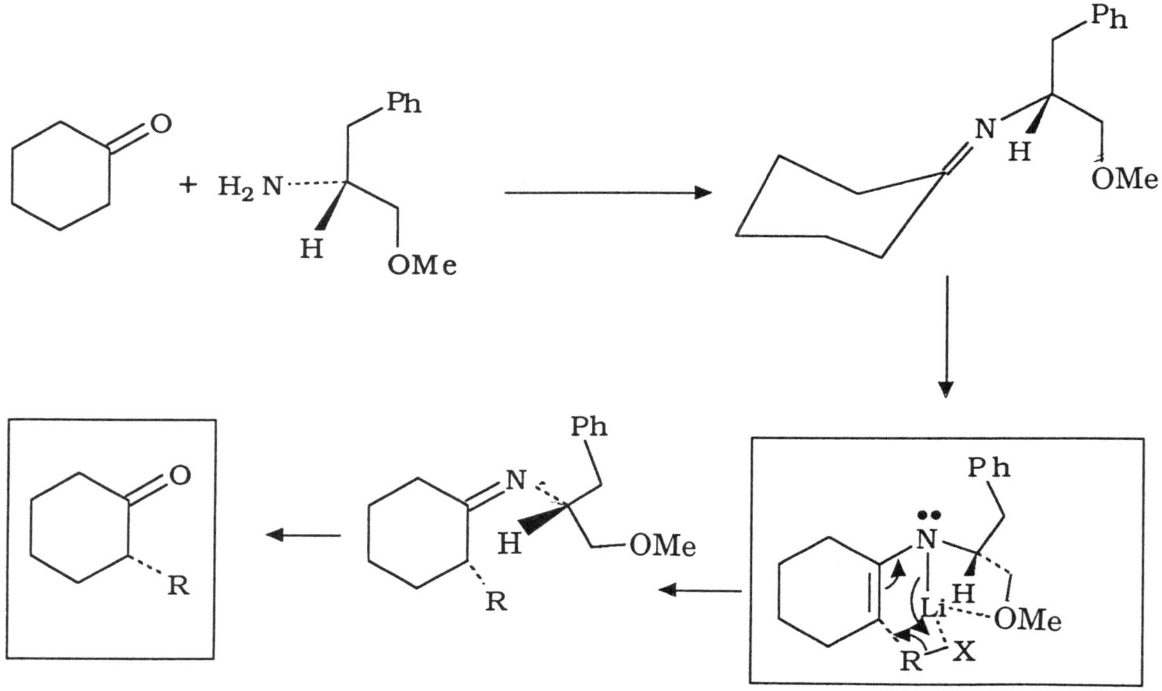
O
+
Ph
$H_2N$
H
OMe
Ph
N
H
OMe
Ph
N
H
OMe
R
Ph
N
H
Li
OMe
R
X
O
R

# 7   Formation of a Chiral Carbon Center by Alkylation of Enamines

A method for enantioselective alkylation of a ketone involves an enamine obtained from a chiral amine. Metallation of the enamine by lithium is achieved by reaction with lithium diisopropylamide (LDA) in THF at $-20\,°C$. Alkylation with a compound RX is then carried out at $-78\,°C$. In the transition state (cf. box), the lithium cation is chelated by the oxygen substituent present in the starting chiral amine. If the oxygen atom belongs to the half-space above (below) the C=C bond of the enamine, the lithium ion is confined also into the same above (below) half-space. The X leaving group also coordinates to the lithium ion: thus alkylation by the R group occurs from the same above (below) side of the C=C bond of the enamine. Thus, 2-alkylcyclohexanones of the S-(R) configuration are obtained with enantiomeric excesses ranging between 87 and 100%. A further advantage of this method is that the starting chiral amine can be regenerated.

Reference: A. I. Meyers, D. R. Williams, and M. Druelinger, *J. Am. Chem. Soc.*, **98**, 3072–3033 (1976).

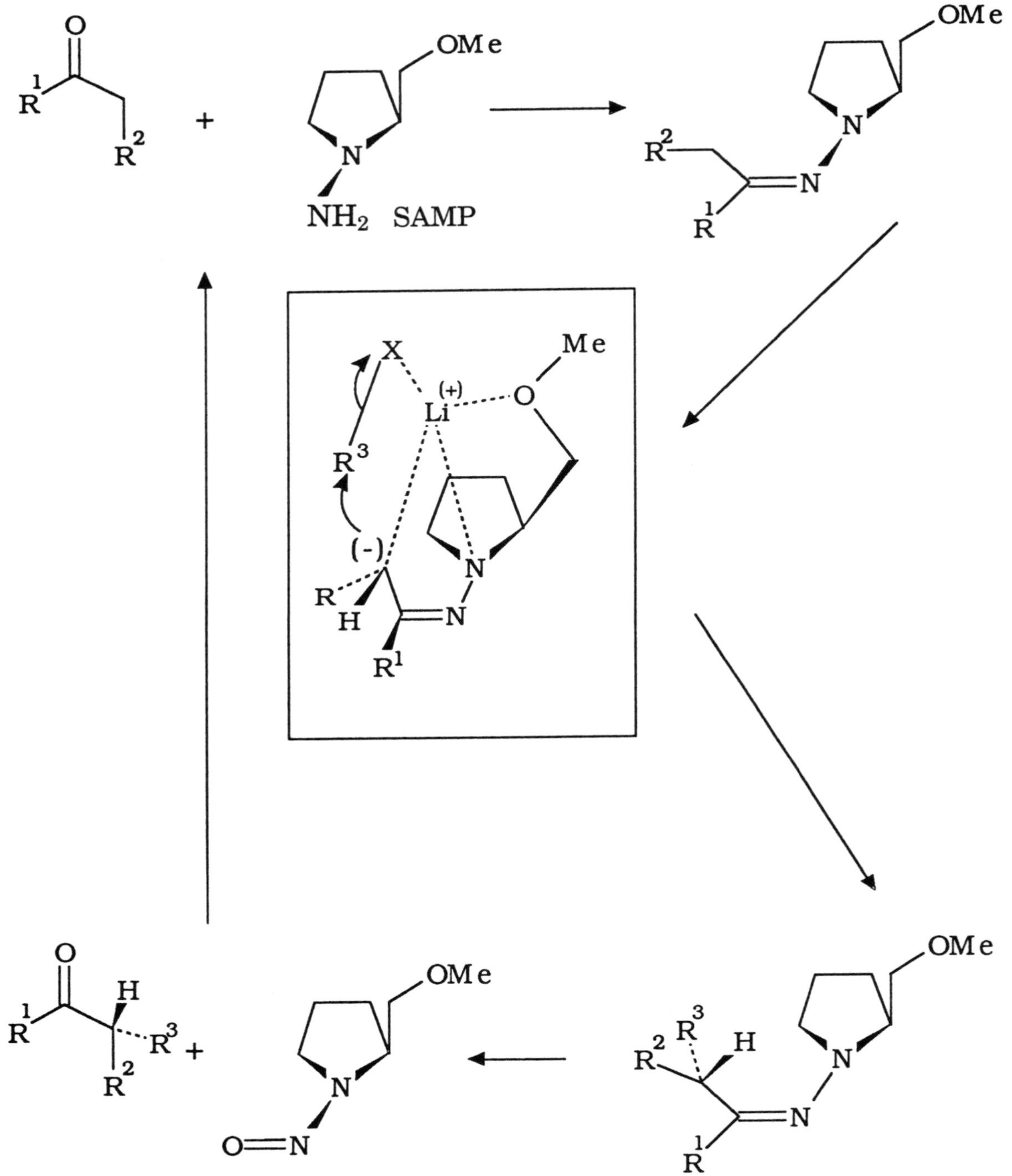
OMe
NH2 SAMP
Me
X
Li (+)
O
R3
(-)
R
H
N
R1
R2
H
O=N

# 8 Enantioselective Alkylation of Enamines

There is a tendency now to replace the enamines by imines and by hydrazones, metallation of which leads to anions: yields and selectivities are improved. The principle is the same, viz. to prepare the organolithium compound and to coordinate the metal by an oxygenated group. In this way the lithium cation is fixed to one side of the C=N double bond: this in turn orientates the attack of the alkylating reagent RX to the same side (by Coulomb attractions $R^{(+)}X^{(-)}$ and $X^{(-)}Li^{(+)}$).

One starts from the ketone and (S)-amino-1-methoxymethyl-2-pyrrolidine (or SAMP) which is enantiomerically pure. The hydrazone of SAMP is obtained in a yield of 80–100%. Treatment with LiN $(i\text{-Pr})_2$ in THF (yield 95–100%) followed by addition of the alkylating agent $R^3X$ (X=I, $OSO_3R$) affords the alkylation product (yield 75–90%, enantiomeric excess 80–100%). The transition state for the alkylation is depicted in the box. Cleavage of the C=N bond is effected by ozone $O_3$ in methylene chloride at $-78\,^\circ$C. The SAMP can be regenerated and recycled by reduction (recovery about 80%).

If the other enantiomer is required, (R)-amino-1-methoxymethyl-2-pyrrolidine (RAMP is the acronym) is used instead of SAMP as the chiral inductor.

References: D. Enders and H. Eichenauer, *Chem. Ber.*, **112**, 2933–2960 (1979).
D. Enders, H. Kipphardt, and P. Fey, *Org. Synth.*, **65**, 183–202 (1987).
D. Enders, H. Kipphardt, and P. Fey, *Org. Synth. Coll.*, **8**, 403–414 (1993).

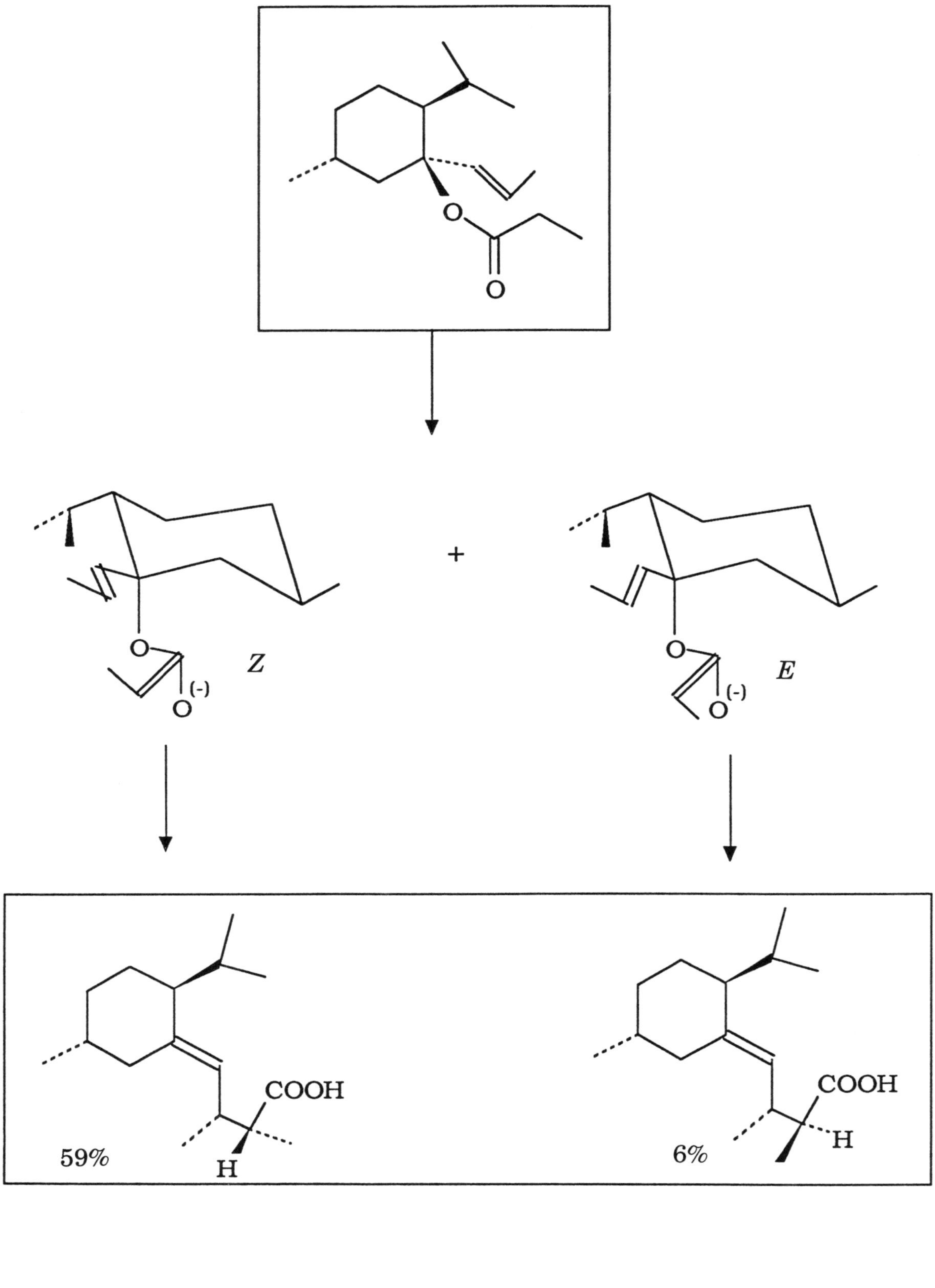
O
O
O
(-)
O
Z
+
O
(-)
O
E
COOH
59%
H
COOH
H
6%

# 9  Transfer of Chirality

Numerous reactions take advantage of the pre-existence of a chiral center for the selective formation of another chiral center. The example given is that of a sigmatropic [3,3] shift, a Claisen rearrangement. The starting ester is enolized by a strong base such as lithium cyclohexylisopropylamide. The strong steric hindrance makes the nucleophilicity negligible.

The two enolates thus formed can arrange the two unsaturated chains originating at the same carbon in a chair-like transition state. The Claisen rearrangement results in a mixture of two diastereoisomeric acids in yields of 59% (left-hand side in the adjacent scheme) and 6% (right-hand side), respectively. The difference in yields stems from preferred formation of the Z-enolate. The starting ester, the propionate, is formed from menthone; addition of several organometallic reagents R-M is diastereoselective and the axial alcohol is formed preferentially.

This alcohol is then esterified with propionic acid or a derivative.

Reference: S. E. Chillous, D. J. Hart, and D. K. Hutchinson, *J. Org. Chem.*, **47**, 5418–5420 (1982).

'Basic science
has to do with isotopes and ions
sols and gels
inorganic and organic smells
and variously differentiated cells.
In this scientific mélange
Plus c'est la même chose, plus ça change.
What people write
Was out of date on the previous night.
No sooner do you see data neatly analysed
Than BOOM comes another research
and the facts are changed.' *Journal of Irreproducible Results*, **13**, 5 (1964).

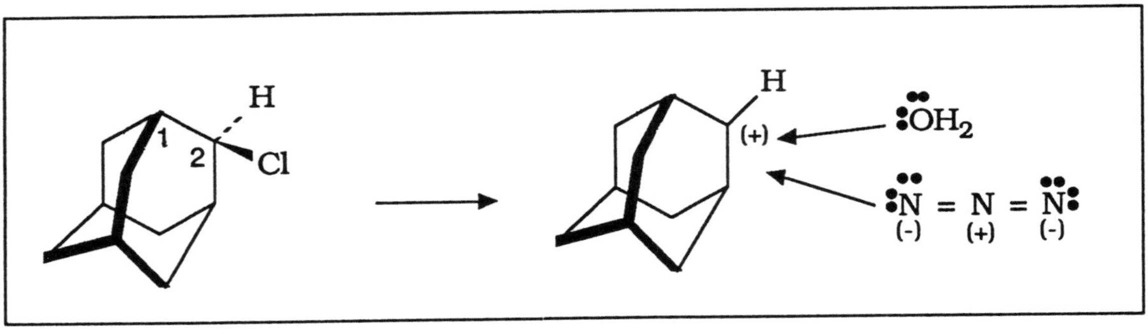
H
1
2
Cl
H
(+)
OH2
N = N = N
(-) (+) (-)

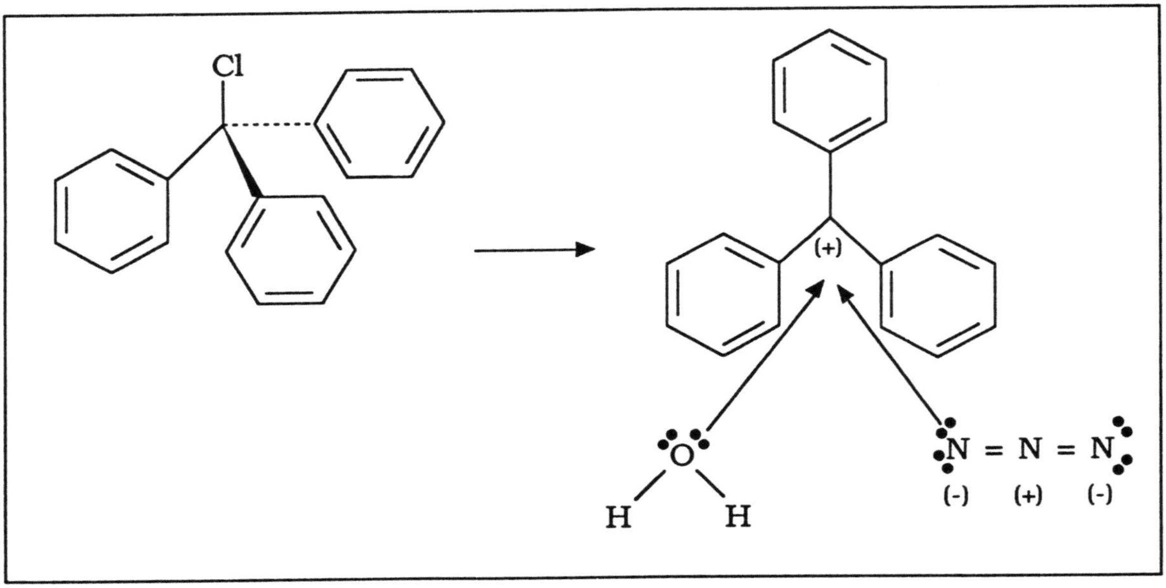
Cl
(+)
O
H H
N = N = N
(-) (+) (-)

# 10  The Reactivity–Selectivity Principle

This is an intuitive, relatively useful postulate according to which the more reactive a chemical species is, the less selective it will be. Let us compare for example, the selectivity of two carbocations with a large difference in stability towards two nucleophiles, azide anion and water: the 2-adamantyl and the trityl (triphenylmethyl) cations. If the solvolysis of the corresponding chlorides is carried out in 80% aqueous acetone, $Ph_3CCl$ has a greater reactivity by a factor $10^{15}$ than the secondary chloride. In fact, the carbocation $Ph_3C^{(+)}$— very stable and hence less reactive—is much more selective than its congener $Ad^{(+)}$: it traps azide anions much faster than water; the ratio ($k_{N_3}/k_{H_2O}$) is almost $10^3$ larger.

The reactivity–selectivity principle, and this is important, is, however, often invalid. In particular, one should note that the selectivity is not necessarily mediocre if the reactivity is high. The converse correlation of a high selectivity with a low reactivity seems to be better supported by experiment as well as by theory.

$S_N2$ reactions in particular provide numerous counter-examples to the principle of reactivity–selectivity. When the reactions of the RBr and $PhCOCH_2Br$ bromides with nucleophiles are compared, the more reactive reagent $PhCOCH_2Br$ is also the more selective toward reaction with variously substituted pyridines.

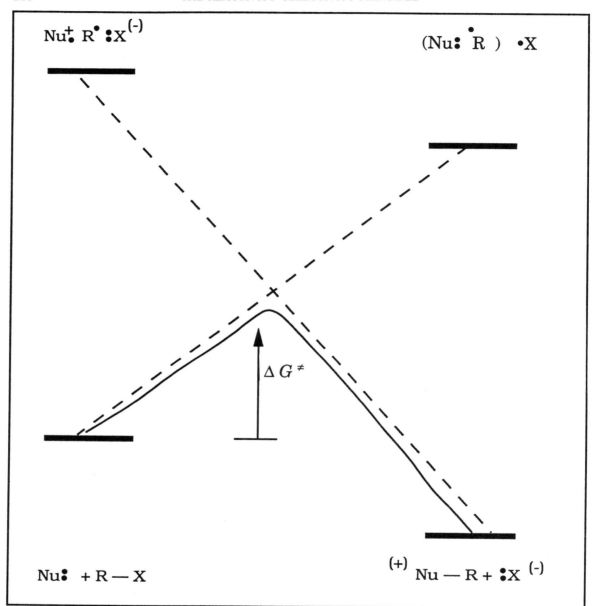

An explanation can be given, in the language of resonance. Its merit is in showing that—far from being limited to the transfer of electron pairs as the curved arrows formalism would suggest—chemistry also proceeds by transfer of single electrons. Let us provide here an introduction to Pross diagrams.

To construct the diagram on the adjacent scheme, we start with the resonance forms Nu: and R—X for the reactants and $^{(+)}$Nu—R and :X$^{(-)}$ for the products of the $S_N2$ substitution. It is more of a problem to write the rather heterodox form for the transition state of this reaction. It is described as the hybrid of the two resonance forms:

$$\text{Nu:R} \cdot \ \cdot \text{X and Nu} \cdot ^{(+)} \ \cdot \text{R:X}^{(-)}$$

The first is an excitation of the R—X bond by homolysis. The second form transfers an electron of the nucleophile to the radical R of the substrate R—X. To avoid the forbidden cross-over between the ground and the excited states (dashed lines), the reaction pathway bypasses it below.

If the substrate RX is PhCOCH$_2$Br, a third resonance form should be added to the resonance hybrid representing the transition state: Nu$^+$, PhCOCH$_2$$^{(-)} \cdot$Br, therefore, stabilized. This stabilization gives an explanation for the increased reactivity. Two of the three resonance forms accommodate a positive charge on the nucleophile (the nitrogen atom of the pyridines) which is thus more selective in donating its electrons.

References: R. D. Levine, *Israel J. Chem.*, **26**, 320–324 (1985).
  E. Buncel and H. Wilson, *J. Chem. Ed.*, **64**, 475–480 (1987).

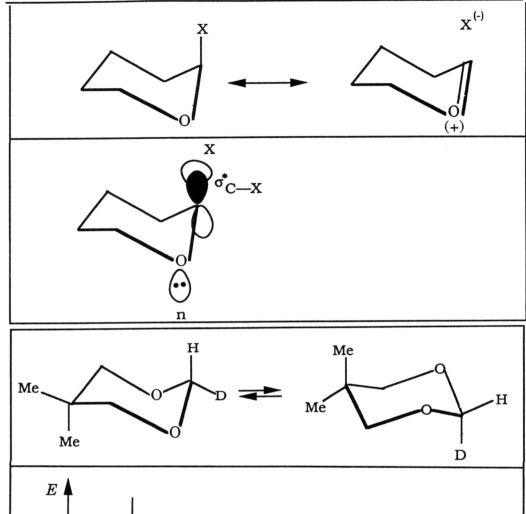

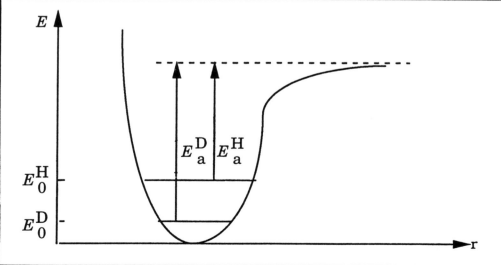

# 11 Stereoelectronic Effects

## Anomeric Effect

Preference of the conformation with an axial position of the C—X bond (X = attracting atom, halogen, oxygen, etc.) in a six-membered ring if the C—X bond is adjacent to an oxygen atom. Moreover, compared with the conformation with an equatorial C—X bond, a shortening of the C—O bond and a lengthening of the C—X bond is observed. This is explained by resonance theory in terms of the indicated hybrid. The MO explanation implies a two-electron interaction between the vacant antibonding $\sigma^*$ (C—X) level and the 'axial' non-bonding $n$ electron pair on oxygen with parallel axes so that the lateral $\pi$ type overlap is significant.

Let us now give a spectacular example of an anomeric effect: the deuterium of $d_1$-5,5-dimethyl-1,3-dioxane in position 2 prefers the equatorial position ($\Delta G = 52 \pm 10$ cal.mol$^{-1}$ at 203 K). The explanation is:

(i) in the language of resonance, the no-bond resonance form has greater weight with hydrogen (protium) than with deuterium since, due to the primary isotope effect, a C—H bond is easier to break than a C—D bond ($E_a^H < E_a^D$); this favors the conformation with an axial position of the hydrogen;

(ii) in the language of MO theory, the C—H bond is weaker than the C—D bond, the $\sigma^*$ (C—H) level is therefore energetically lower than the $\sigma^*$ (C—D) level, consequently an n-$\sigma^*$ (C—H) interaction is stronger than an n-$\sigma^*$ (C—D) interaction.

Reference: F. A. L. Anet and M. Kopelevich, *J. Am. Chem. Soc.*, **108**, 2109–2110 (1986).

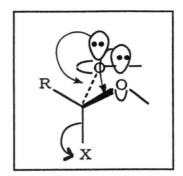

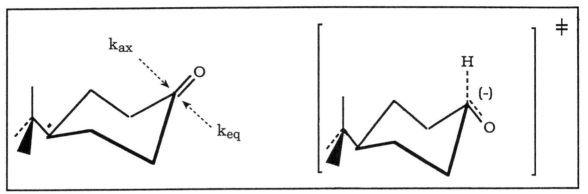

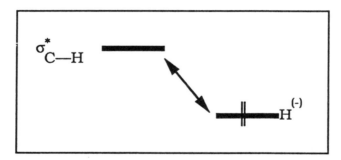

## Reactivity of Tetrahedral Intermediates

As we have already seen with addition–elimination reactions, the anomeric effect facilitates breaking of a C—X bond, provided that there is an antiperiplanar disposition of the lone pairs of the oxygen atoms bonded to the same carbon. The criterion given by Deslongchamps is the presence of at least two such lone pairs antiperiplanar to the leaving group.

## Stereoselective Reduction of Cyclic Ketones

The reduction of 4-$t$-butyl-4-cyclohexanone with lithium aluminum hydride in ether at 0 °C affords the product of axial attack, i.e. from the more hindered face, in 91% yield. The ratio of the rate constants $k_{ax}/k_{eq}$ is of the order of 10! Likewise, reduction of the tetracyclic ketone, shown on the adjacent scheme, occurs from the more hindered side, with $k_R/k_H$ of the order of 4 (R = $i$-Pr). The transition state is characterized in all these cases by an antiperiplanar arrangement of the C—H bond during the formation of one or several C—H bond(s) at the adjacent carbon. The explanation for this stereoselectivity remains the same: a stabilizing two-electron interaction H$^{(-)}$, $\sigma^*$ (C—H) (one should note that an author— invoking the electrophilic character of the carbon–hydrogen bond in the course of its formation—prefers to speak of an interaction between a vacant orbital of this C—H bond and a doubly occupied $\sigma$(C—H) orbital).

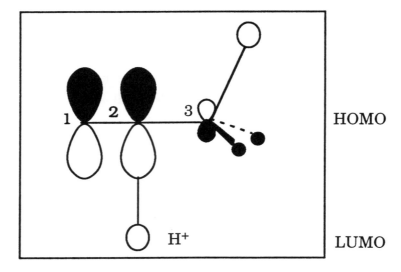

HOMO

LUMO

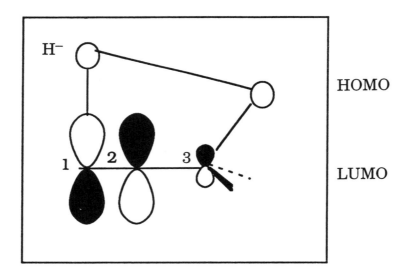

HOMO

LUMO

## Additions to C=C Double Bonds

Let us consider the frontier orbitals of propene (conformation with a dihedral angle (C−1, C−2, C−3/C−2, C−3, H) of 90°) interacting with either a proton H$^+$ (model for an electrophile) or a hydride anion H$^-$ (model for a nucleophile). It is a question of either the HOMO of propene combining with the LUMO of the electrophile or the LUMO of propene combining with the HOMO of the nucleophile. The secondary orbital interactions disfavor the *syn*-periplanar approach of an electrophile at C−1 or C−2. Antiperiplanar attack is thus preferred. *Syn*-periplanar attack of a nucleophile is disfavored at C−2 but favored at C−1.

References: A. J. Kirby, *The Anomeric Effect and Related Stereoelectronic Effects at Oxygen*, Springer Verlag, Berlin, 1983.

F. A. L. Anet and M. Kopelevich, *J. Am. Chem. Soc.*, **108**, 2109–2110 (1986).

D. G. Gorenstein, *Chem. Rev.*, **87**, 1047–1077 (1989).

N. T. Anh and O. Eisenstein, *Nouv. J. Chim.*, **1**, 61 (1977).

A. S. Cieplak, *J. Am. Chem. Soc.*, **103**, 4540–4552 (1981).

P. Caramella, N. G. Rondan, M. N. Paddon-Row, and K. N. Houk, *J. Am. Chem. Soc.*, **103**, 2438–2440 (1981).

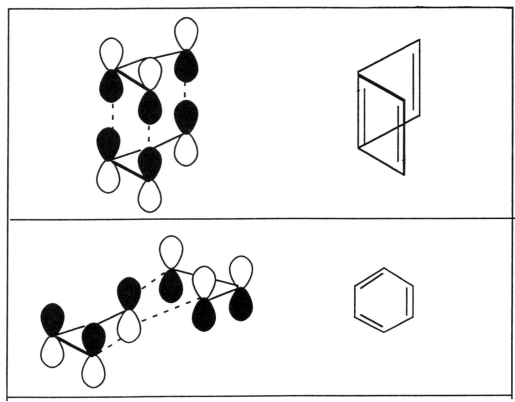

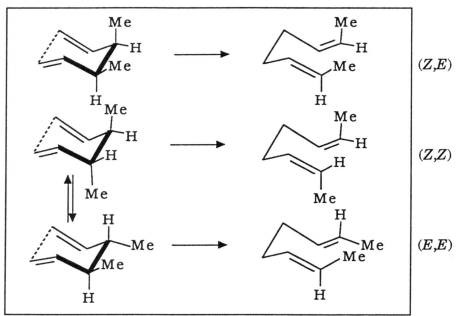

# 12   Chair and Boat Transition States

We will treat this subject only very briefly. If we consider a sigmatropic [3,3] shift such as the Cope rearrangement, we can depict chair and boat transition states. The former is, according to the Dewar–Zimmerman rule, analogous to benzene, i.e. aromatic. The latter, with three instead of two interactions of the *2p* orbitals of two allyl fragments is isoelectronic to butalene, i.e. doubly antiaromatic.

As a general rule, chair-like transition states prevail in concerted six-center six-electron reactions. The classic example is that of the Cope rearrangement of 3,4-dimethyl-1,5-hexadienes. Heating of *meso*-3,4-dimethyl-1,5-hexadiene at 225 °C gives exclusively 2,4-octadiene with the *E,Z*-configuration. The rearrangement products of (±) 3,4-dimethyl-1,5-hexadiene at 180 °C are 2,4-octadienes of *E,E*- and *Z,Z*-configuration (ratio 9:1).

References: W. von E. Doering and W. R. Roth, *Angew. Chem. Int. Ed. Engl.*, **2**, 115 (1963).
            M. J. S. Dewar and C. Jie, *J. Am. Chem. Soc.*, **109**, 5893–5900 (1987).

## Summary 'Selectivity'

**Regioselectivity** denotes control of which position to be attacked: substitution of a benzene ring at the ortho, meta or para position; or to give another example, attack by peracids of one rather than another ethylenic C=C bond in a polyene. In the same way as regioselectivity favors *one* regioisomer, **stereoselectivity** denotes preferential formation of *one* stereoisomer. For instance, the introduction of an alkyl group $\alpha$ to a carbonyl group can be directed to one or the other face of the C=O double bond.

**Enantioselectivity** is a special case of stereoselectivity: preferential formation of *one* of the two enantiomers. Enzymes are capable of enantioselectivity; organic synthesis uses them routinely for this purpose. More generally, enantioselectivity requires a source of chirality, frequently a chiral auxiliary group that can be cleaved and regenerated.

The enamine method devised by G. Stork allows for a high degree of selectivity of alkylation $\alpha$ to a carbonyl group. The latter is transformed in the first step to an enamine. The reagent ensures monoalkylation by axial attack (six-membered rings). It is sufficient to consider the transition state in the realistic chair conformation for the explanation for this stereoselectivity.

Enantioselective alkylation can be achieved by the enamine method (A. I. Meyers) or by the hydrazone method (D. Enders). It makes use of the preferential occupation of a half-space due to attractive interactions in the half-space: metallation by lithium occurs with coordination to another 'hard' atom, namely oxygen. The enamine method (Meyers) places the alkylating agent in the same half-space as the lithium previously attached by the oxygen. The hydrazone method (Enders) uses the same effect and affords one or the other enantiomer, depending on the (regenerable) chiral auxiliary agent, *R*- or *S*-AMP.

# 4.II
# Some Solutions to the Problem of Selectivity (Chemo-, Regio-, Stereo-)

## 1 Regarding the Active Site of Fumarase

A step in the Krebs citric acid cycle is transformation of fumarate to malate. When carried out in heavy water, the only reaction product is ($-$)-malate of *erythro* configuration.

This experiment demonstrates that addition of water occurs in *anti* fashion ($H^+$ and $OH^{(-)}$) and is also enantiospecific: the enzyme distinguishes between the two faces of the C=C double bond; attack of the electrophile ($H^+$) occurs exclusively from the *re* face whereas the nucleophile ($OH^{(-)}$) approaches only the *si* face.

This stereochemical information translates into these characteristics of the active site of the enzyme: positioning of positive charges stabilizing the two carboxylate groups; presence of an acidic A—H site to effect the *re* face protonation; probable existence of a basic site :B to deprotonate a molecule of water attacking the substrate from the *si* face.

References: J. C. Chottard, J. C. Dépezay, and J. P. Leroux, *Chimie Fondamentale*, Hermann, Paris, 1982, III–137.
L. Stryer, *Biochemistry*, W. H. Freeman, San Francisco, 2nd ed., 1981, p. 288.

*Note*: Since the whole book is replete with examples of selectivity, only a small number will be presented here.

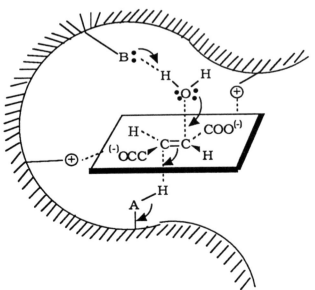

## 2 How to Manufacture a Zombie

During his second circumnavigation, on the *Resolution*, Captain Cook discovered New Caledonia. An unknown fish, caught by the Kanakas, was brought aboard on September 8, 1774. Before it was handed over to the cook, the biologist of the expedition, Forster, made a description and drawing of it. The rest of the story is known from Cook's diary:

'luckily for us the operation of describing and drawing took up so much time till it was too late so that only the Liver and Roe was dressed of which the two Mr Forsters and my self did but just taste. About 3 or 4 o'Clock in the Morning we were siezed with an extraordinary weakness in all our limbs attendant with a numness or Sensation like to that caused by exposeing ones hands or feet to a fire after having been pinched much by frost, I had almost lost the sence of feeling nor could I distinguish between light and heavy bodies, a quart pot full of Water and a feather was the same in my hand. In [the morning] one of the Pigs which had eat the entrails was found dead ( . . . ) In the Morning when the Natives came on board and saw the fish hanging up, the immidiately gave us to understand it was by no means to be eat, expressing the utmost abhorance of it . . .'

*The Journals of Captain James Cook on his Voyages of Discovery. The Voyage of the* Resolution *and* Adventure *1772–1775.* J. C. Beaglehole, Ed., Cambridge University Press, London, 1961.

In his report, Forster states that the fish belonged to a sub-class, referred to by Linnaeus as *Tetraodon*, of which several species are known to be poisonous.

The Japanese relish this fish which they call *fugu*. In spite of the care which they take to clean it and because the guts have the reputation of being a delicacy, poisoning by *fugu* is not uncommon: 715 cases during the period 1956–1958, 420 of which ended in death. The toxin responsible, *tetrodotoxin*, is also found in certain salamanders (in the eggs, in embryos and adults), particularly in those of the genus *Taricha*.

Let us quote this testimony:

'The Japanese gourmets who savour the fugu pufferfish, as delicious as it is potentially fatal, nowadays have to spend more in order to risk their lives for their tastebuds. The price of this fish has doubled, due to less abundant catches and because of increased demand. It can be found now at prices up to ¥ 15 000 the kilogram (which translates to US$57/lb). One restaurant in Tokyo specialized in fine fugu dishes charging ¥ 25 000 a helping.

Arthur Higbee, *Int. Herald Trib.*, December 1 1988.

Table 13 compares the toxicities of some poisons as measured on mice.

This data was so impressive as to make Ian Fleming use it in his James Bond series: in *From Russia with Love*, the secret agent becomes paralyzed and loses consciousness after a minor injury with a concealed knife. The writer revealed later, in *Doctor No*, that the blade of the knife had been poisoned with tetrodotoxin.

The action potential of nerves is connected, as the reader probably knows, to the selective permeability of axons to sodium and potassium ions. Tetrodotoxin is a powerful poison because it obstructs sodium channels in the membranes. This blocks transmission of the nervous impulse along the neuron. This also happens in the excitable membranes of the muscle fibres, hence leading to paralysis of breathing. As indeed described by Captain Cook, anesthesia adds itself to the neuromotor block. Like curare, tetrodotoxin has become a tool for physiologists and neurochemists. For instance, by this means it could be established that a membrane of an axon contains on average about twenty sodium channels per square micron.

An interesting question is that of the biological function of this poison for the fish that secretes it. These fish belong to the subclass *Tetraodon* and to the genus *Sphaeroides*: the latter name comes

**Table 13**   Toxicities of some poisons as measured on mice

| Substance | Minimum lethal dose ($\mu g\ kg^{-1}$) |
|---|---|
| sodium cyanide | 10 000 |
| muscarine | 1 100 |
| (*Amanita muscaria*) | |
| curare | 500 |
| strychnine | 500 |
| tetrodotoxin | 8 |

from their characteristic behavior. When attacked, they swallow large quantities of water, blowing themselves up and becoming spherical, which intimidates or renders ingestion by their predators more difficult. In addition, recourse to the powerful toxin would seem to ensure better survival for the species. The novelist John Steinbeck wrote of the *Sphaeroides lobatus* of low California, which he referred to as the *botete*:

> '*Botete* is sluggish, fairly slow, unarmored, and not very clever at either concealment, escape, or attack. (. . .) Did he develop poison in his flesh as a protection in lieu of speed and cleverness, or being poisonous and quite unattractive, was he able "to let himself go", to abandon speed and cleverness? The protected human soon loses his power of defense and attack. Perhaps *botete*, needing neither brains nor tricks nor techniques to protect himself except from a man who wants to poison a cat, has become a frump.' John Steinbeck and Edward F. Ricketts, *Sea of Cortez. A leisurely journal of travel and research*. The Viking Press, New York, 1941, p. 125.

R.B. Woodward elucidated the structure of tetrodotoxin at the beginning of the Sixties; chemists had worked on it since 1910! It is a zwitterion, i.e. a molecule in which anionic ($O^-$) and cationic ($H_2N^+$) poles coexist. This tetracyclic molecule is an *ortho*-ester (a species of the type $RC(OR')_3$ with an oxygen function in the form of an alkoxide ($-O^-$). The tetracyclic skeleton comprises five alcohol functions besides this alkoxide.

tetrodotoxin

To conclude this presentation of one of the most toxic natural products, we shall describe its use as a traditional drug by the *bokors*, the sorcerers in voodoo cults of Haïti.

The sorcerer starts to prepare his poison. The ingredients are numerous: human bones, fragments of lizards, spiders, toads, etc. Invariably, these preparations also contain the poison of the *Sphaeroides* puffer fish. They are reminiscent of a witches'-brew:

'Round about the cauldron go;
In the poison'd entrails throw.
Toad, that under cold stone
Days and night hast thirty-one
Swelter'd venom sleeping got,
Boil thou first i' the charmed pot.
Double, double toil and trouble;
Fire burn and cauldron bubble.
(. . .)
Eye of newt, and toe of frog,
Wool of bat, and tongue of dog,
Adder's fork, and blind-worm's sting,
Lizard's leg, and howlet's wing,
For a charm of powerful trouble,
Like a hell-broth boil and bubble.''
    Shakespeare, *Macbeth*, IV.i.4

To become a zombie in Haïti punishes those who have infringed the law set by a secret society. These secret societies, similar to the Mafia in Sicily, hold great power on the rural areas. They go back to the clandestine organization of colored slaves at the time of Toussaint-Louverture and his uprising against the French.

The poison is applied directly to the skin after a cut or abrasion. Then the *'apparent'* death of the poor fellow occurs. His burial is the most spectacular part of zombification. In the evening following the burial, the *bokor* digs him out. He is then beaten-up and administered another mixture to drink, extracted from the 'zombie cucumber', *Datura stramonium*, rich in alkaloids of the tropane family. This provokes a psychotic delirium.

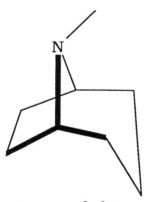

tropane skeleton

Afterwards, the zombie is enslaved. He has lost his soul, he believes that he is dead: in so doing, he forsakes his personal identity and autonomy. The power of the *bokors* lies in the terror that they inspire: most Haïtians are terrified of having to undergo such suffering and of becoming a zombie. The case of Clairvius Narcisse, who 'died' like this in 1962 in the Albert Schweitzer hospital in

Deschapelles, has been documented. His death and inhumation certificate exist, yet he remained a zombie, working as a slave on a sugar plantation for the following 18 years. When his master died, Narcisse and the other slaves were able to escape.

References: J. C. Beaglehole, Ed. *The Journals of Captain Cook*. Vol. 2 *Voyages of the Resolution and Adventure 1772–1775*. Cambridge University Press, London, 1961, pp. 534–535.
J. R. Forster and G. Robinson, *Observations Made During a Voyage Around the World*. London, 1778, pp. 643–649.
H. S. Mosher, F. A. Fuhrman, H. D. Buchwald, and H. G. Fischer, *Science*, **144**, 1100–1110 (1964).
L. Stryer, *Biochemistry*, W. H. Freeman, San Francisco, 2nd ed., 1981.
J. Steinbeck and E. F. Ricketts, *Sea of Cortez*, Viking, New York, 1941.
R. B. Woodward, *Pure Appl. Chem.*, **9**, 49–79 (1964).
W. Davis, *Passage of Darkness: The Ethnobiology of the Haitian Zombie*, University of North Carolina Press, 1988; *Amer. Scientist*, **75**, 412–417 (1987).

# 3   Natural/Unnatural

### Just a Little Bit Unnatural (*by Roald Hoffmann*)

Human beings have been put on this earth to create. Some write poems. Others build additions to houses, draft new civil rights legislation, dig ditches. Some make molecules—these are the chemists. All—poets, builders, lawmakers, ditchdiggers, chemists—either create something new (call it man or woman-made, synthetic, artificial, or unnatural) or they modify a product of nature.

Is the natural different from the unnatural? Yes, on the spiritual level, as the designers of food labels know too well. No, on a material level. All stuff—natural or unnatural—is at the microscopic level molecular. And the observable macroscopic properties—color, toxicity, strength, conductivity—follow from that micro-structure. Synthetic molecules, carefully made, can replace natural ones. Your MSG headache is equally well induced by synthetic or natural MSG; your pneumonia cured by an antibiotic made by a mold or in the laboratory.

Chemists do have a special way of playing with the natural. First they see it as a challenge to make any molecule nature can. And they can do it, those master builders of tiny structures. They may manage their synthesis less efficiently than nature, but then nature has had a few more million years to optimize most any process.

Second, chemists want to make molecules that aren't there in nature. Why not a molecule that looks like an icosahedron ($B_{12}H_{12}^{2-}$)? Or one like a soccer ball ($C_{60}$)? This is real fun.

Third, they want to make molecules resembling natural ones, but better in this or that specific respect. There are polymers stronger than steel, or fats in which you can fry your onion rings but which are calorie-free, because they are not digested.

Fourth, chemists want to make molecules that are sort of like natural ones, but a little different. Why? To fool bacteria and viruses, of course. There is profit in this.

Fifth, chemists make synthetic molecules to understand nature—its highways and byways, how it got to be the beautiful way it is.

Men and women seem to find ever more ingenious ways to confound the natural/unnatural dichotomy. Here is a recent sample of minor and not-so-minor changes made to molecules critical to life.

(1) Nucleic acids, including DNA and RNA, are the information carriers of life. A typical DNA double helix is shown below. Note the constituent components of this biopolymer: a backbone of phosphate groups and five-membered carbohydrate (sugar) rings, called riboses, and the 'bases'— guanine, cytosine, adenine, uracil (in RNA) or thymine (in DNA), which engage in pairing.

There was no greater chemical achievement (even if it was accomplished by two non-chemists!) in midcentury than the recognition of this structure. And it has taken much ingenious work to explore fully the consequences and workings of what Watson and Crick had divined.

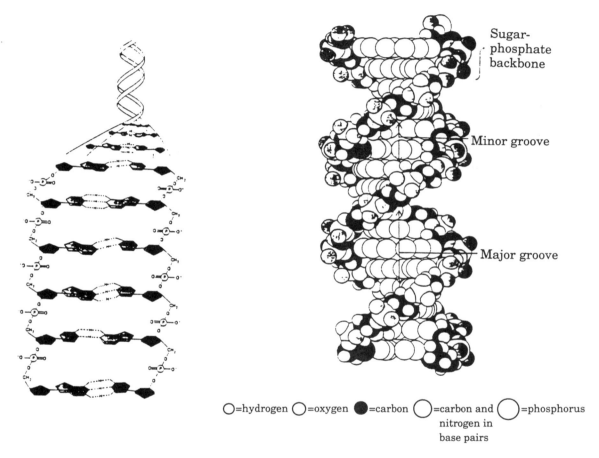

Sugar-
phosphate
backbone

Minor groove

Major groove

◯=hydrogen ◯=oxygen ●=carbon ◯=carbon and ◯=phosphorus
nitrogen in
base pairs

But now chemists are curious. Why this structure and not all others? Albert Eschenmoser, of the Eidgenössiche Technische Hochschule in Zürich, one of the deepest thinking chemists of our time, has focused in on the sugars. He argues quite convincingly that hexoses (another kind of sugar, actually a more common one, with a six-membered ring, see below) were just as likely to form under prebiological conditions. And then Eschenmoser asks: 'Why, then, did Nature choose pentoses (ribose is an example of these) and not hexoses as the sugar building blocks of nucleic acids?'

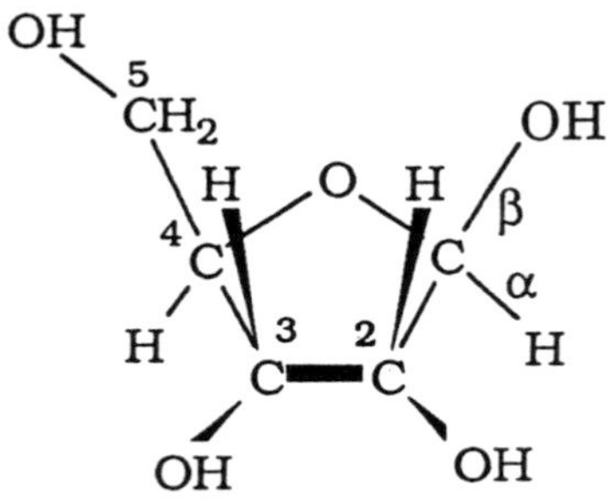

β-*D*-ribose: a pentose

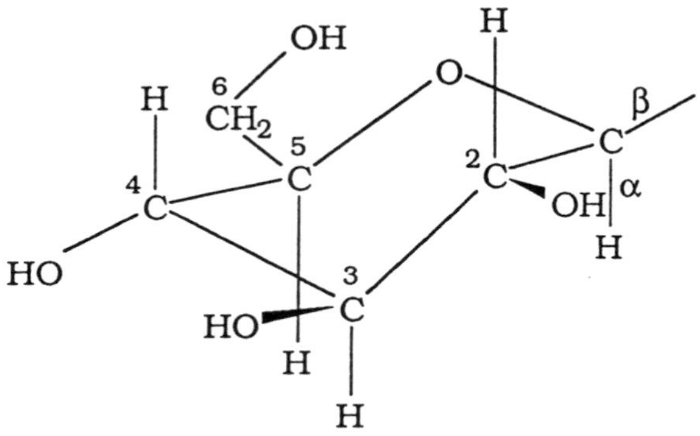

β-glucose: a hexose

Eschenmoser is not a philosopher but a natural philosopher. So he follows up the question by . . . synthesis. He and his able co-workers build up an entire alternative universe—not just the sugars, but single-stranded phosphate—containing chains built from them; they add the bases. They do what nature chose not to do, build a 'hexose-NA'. And then differences emerge. Those hexose-NA strands pair differently, pair more strongly. They are not helical. Their stability is fine, but the hexose-NA is not capable of the ready pairing–unpairing that is characteristic of normal pentose-DNA. It does not form helices. The alternative universe is not good enough, so it seems, to do what has been done.

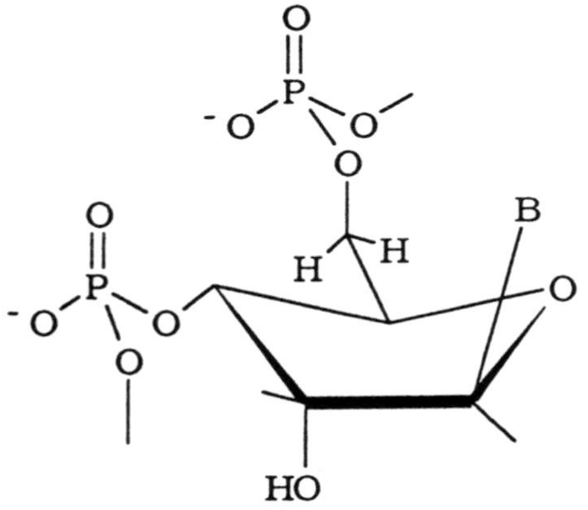

RNA

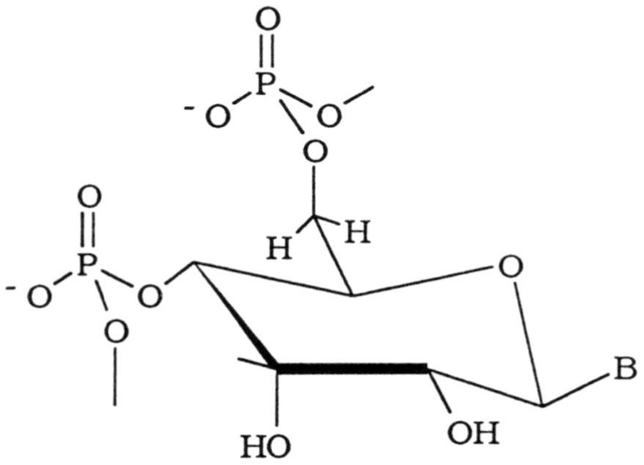

hexose – NA

(2) Not too far from Eschenmoser, in the same laboratories in Zürich, Steven Benner is playing with nature in another way. Recall the four bases, G, C, A, T(U), C. These couple, through so-called hydrogen bonds, in specific A–T(U) and G–C pairs shown below.

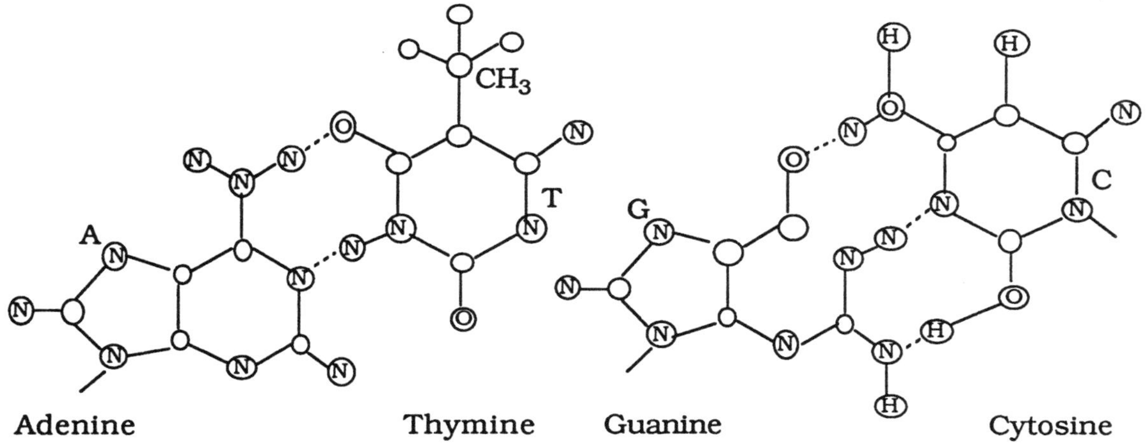

Adenine                    Thymine        Guanine                        Cytosine

Benner's group thought that Nature's alphabet might be extended. So they designed another base pair, κ–χ (structure shown below)

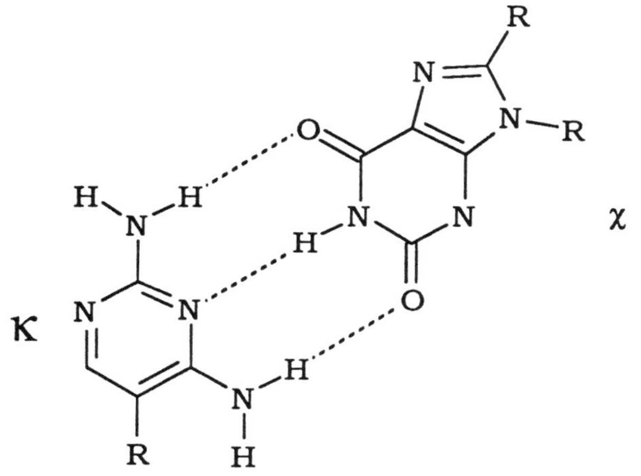

which fits into the double helix. Neither κ nor χ pairs with known bases, so the new 'letters' do not sow confusion, they just extend the alphabet from 4 to 6 letters. Benner and his co-workers say, 'This expanded genetic alphabet should allow more diversity in the functional groups available to RNA . . . [It] should provide RNA molecules with the potential for greatly increased catalytic power, including perhaps RNA molecules that catalyze their own replication.'

There is a kind of *hubris* in this game, but it sure is pretty.

(3) Still more molecular games, now with proteins. Proteins are biopolymers built up from amino acids. The amino acid building block and the piece of the protein (called a peptide) assembled from several such blocks is shown below.

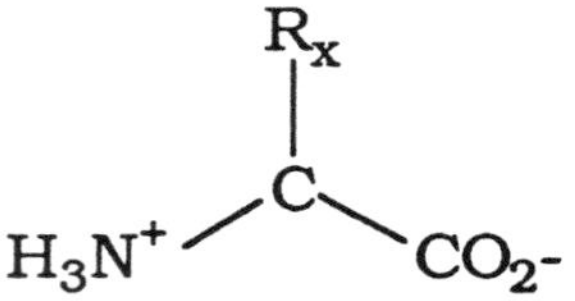

normal amino acid

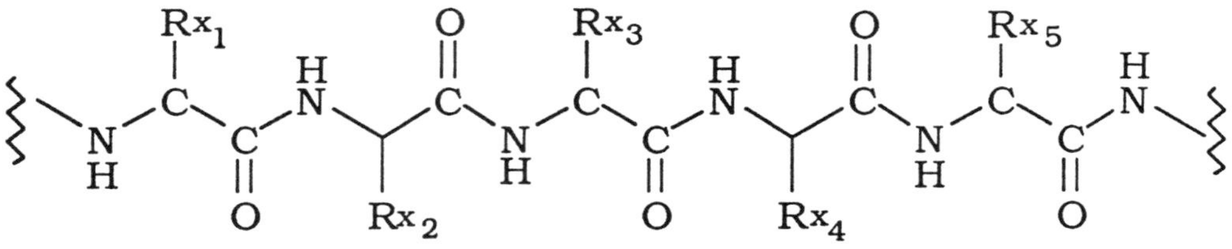

normal peptide

Stuart Schreiber at Harvard University and John Clardy at Cornell University have designed a set of stretched or 'vinylogous' peptides. You see the same amino ($NH_2$) and acid (COOH) handles that confer function (and the ability to form a chain) at the ends. But in the middle there is something different, two Rs, perhaps providing more variety than the single R group of a normal amino acid.

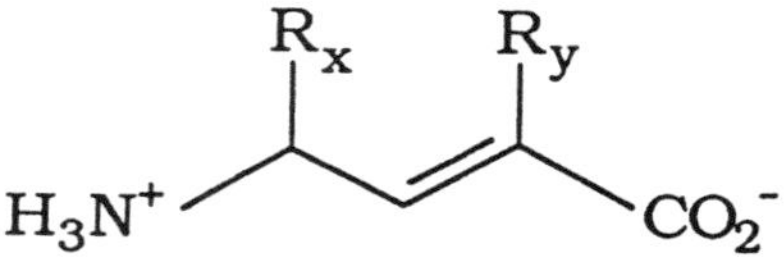

vinylogous amino acid

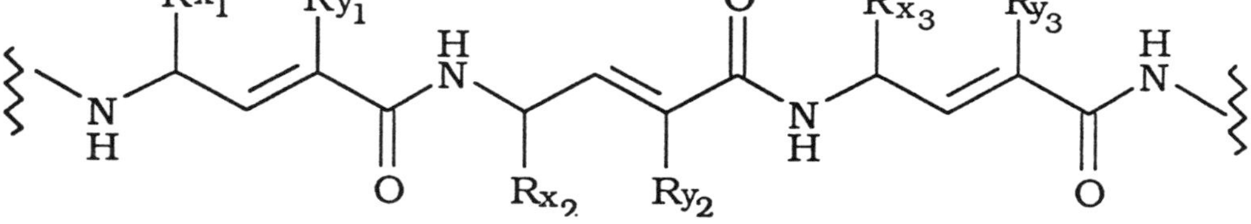

vinylogous peptide

Incredibly one of these vinylogous peptides has turned up in a marine sponge and acts as a thrombin inhibitor. What will other modified peptides do?

(4) Meanwhile a cohort of 16 chemists at Chiron Corporation, Emeryville, Calif., the Whitehead Institute in Cambridge, Mass., and two branches of the University of California, at San Francisco and Berkeley, has made an entirely new family of molecules based on the amino acid building block. They rang their changes just by substituting one of the hydrogens on a nitrogen by a range of chemical groups. That way the variation inherent in amino acids, a variability responsible for their biological versatility, has been shifted from the carbon of the amino acid to its nitrogen.

normal amino acid                    modified amino acid

It's easy to build polymers from these new building blocks. These the cohort calls 'peptoids'. Their structure is contrasted with a normal peptide below.

Peptides:

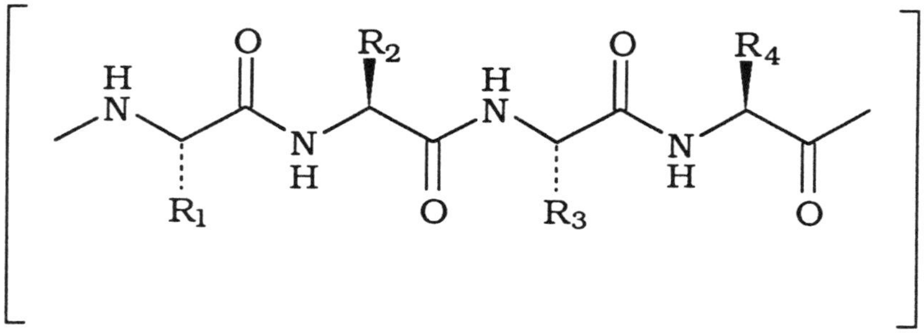

The idea is that a library of such peptoids might be of value in drug design. They would be the same and not the same, and among them some would fool and subdue the pathogens that threaten our well-being.

The motivations of the chemists engaged in this work are clear. First, there is the sense of wonder. Why did nature do it this way and not another? What if I changed things just a little? The power to effect such change is in our hands. Then there is benefit and its sidekick, profit. Some of these molecules may be useful drugs, just because they are small variations on a natural theme. When curiosity and benefit find themselves on the same trail, the hands and minds of human beings seem to quicken. You can see this in the current flurry of activity in the making of slightly unnatural molecules.

Yet some people are scared by these molecular games, as they are by genetic engineering. The range of concern is wide: even if there be safeguards against letting loose possible new pathogens, even if these molecules save lives or improve our standard of living, what right do we have to tamper with a God-given universe? I do not have a full answer to these concerns, but I would begin a dialogue as follows: We *have* tampered irreversibly with nature, from the time we became a species, from the moment prehistoric man used a copper axe to modern times, to the 10 billion chickens that now share the world with us. Before us, cataclysmic natural events, as destructive of other species as we are, shaped this world. We did not invent species extinctions. The difference about our transformation of nature is its scale and pace. But there is another difference, the potential of repair. There is a human creation (fully as unnatural as all those new molecules)—ethics—which makes every chemist think about the consequences, the possible harm, the potential ill use, of the new molecules he or she makes. We have no choice but to make these molecules, for curiosity drives us relentlessly on. But we also have no choice but to worry about what we do. And act upon it.

References: R. Hoffman, 'How should chemists think', *Sci. Amer.*, **268**, 66 (1993).
R. Hoffmann, 'Natural/Unnatural', *New England Rev.*, **XII**, 323 (1990).
A. Eschenmoser and E. Loewenthal, *Chem. Soc. Revs*, 1 (1992).
A. Eschenmoser and M. Dobler, *Helv. Chim. Acta*, **75**, 218 (1992).
J. A. Piccirilli, T. Krauch, S. E. Moroney and S. A. Benner, *Nature*, **343**, 33 (1990).
L. E. Orgel, *Nature*, **343**, 19 (1990).
M. Hagihara, N. J. Anthony, T.-J. Stout, J. Clardy and S. L. Schreiber, *J. Am. Chem. Soc.*, **114**, 6568 (1992).
R. J. Simon *et al.*, *Proc. Natl Acad. Sci. USA*, **89**, 9367 (1992).
S. Borman, *Chem. Eng. News*, 31 August, 27 (1992).

# 5.1
# Protection and Regeneration. Cases of the Alcohol and Carbonyl Group. Inversion of Polarity

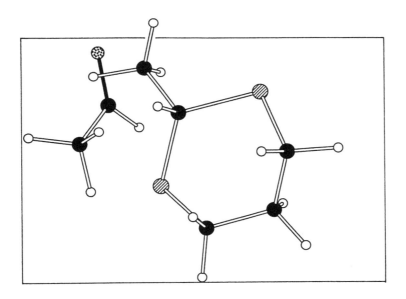

A major objective of organic synthesis remains to synthesize the large polyfunctional molecules of nature in order to best take advantage of their biological activity. The problem is to modify one part of the molecule without affecting the remainder: if a group is likely to be affected by a step that transforms another group elsewhere in the molecule, then the former has to be protected. Obviously, protection and regeneration have to be carried out by simple means and quantitatively. This chapter presents one or two procedures for each of the key groups: alcohols and carbonyl compounds. A related synthetic operation is inversion of polarity: how can a given dipole A—B, polarized as A$^+$ and B$^-$, be modified temporarily in such a way that A becomes negative? The dithiane method of Corey and Seebach illustrates conversion of a normally electrophilic carbon, that of an aldehyde carbonyl group into a nucleophilic center.

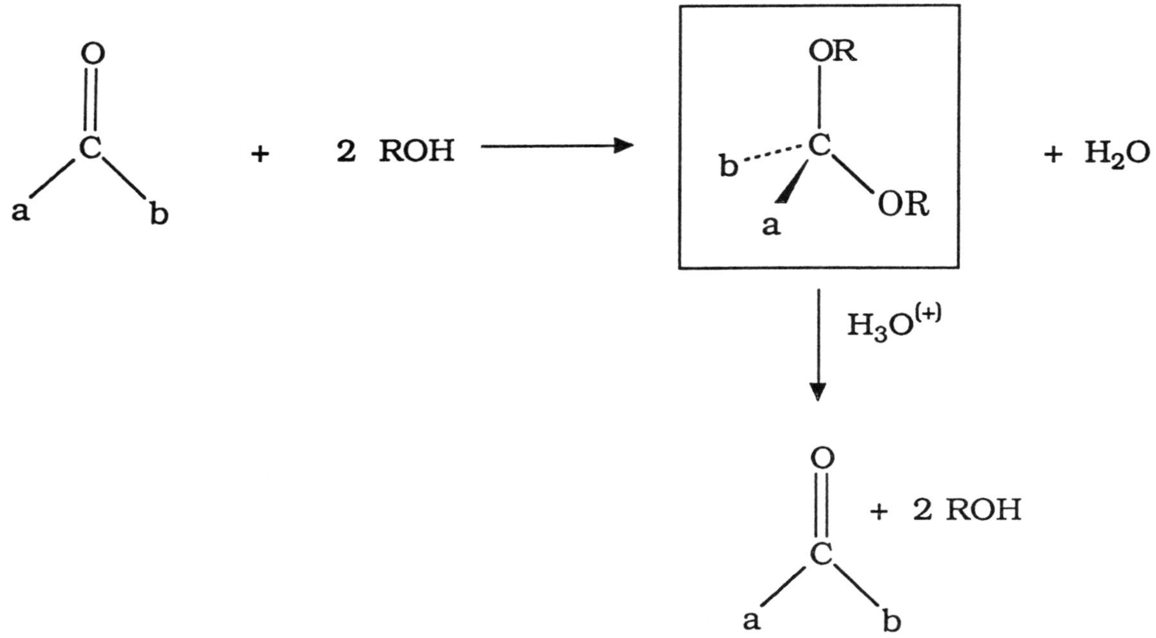
+ 2 ROH
OR
OR
+ H₂O
H₃O(+)
+ 2 ROH

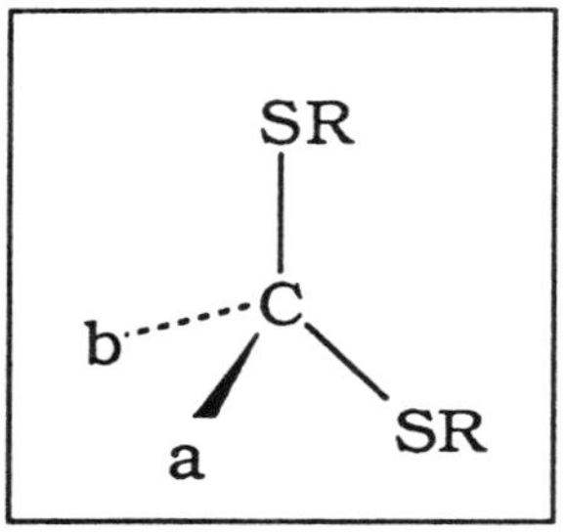
SR
SR

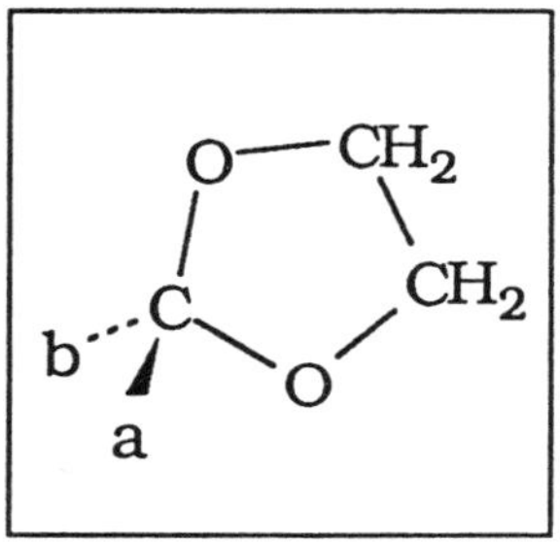
CH₂
CH₂

# 1  The Ketal Protecting Group for Carbonyl Compounds

Advantage is taken of the reversible formation of a ketal from a carbonyl compound and two molecules of an alcohol R—OH (Section 1.5). Thus, a carbonyl group is protected by formation of the corresponding acetal or bisthioketal etc.; if protection is no longer necessary, the carbonyl group can be regenerated by acidic hydrolysis or by another method.

Classically, this is carried out using ethylene glycol, catalyzed by traces of p-toluenesulfonic acid, in a solvent such as benzene. Thus an ethyleneketal is obtained.

Best Witchcraft is Geometry
To the magician's mind—
His ordinary acts are feats
To thinking of mankind.

Emily Dickinson, *The Complete Poems*, T. H. Johnson, ed., Little, Brown and Co., Boston, 1955, No. 1158.

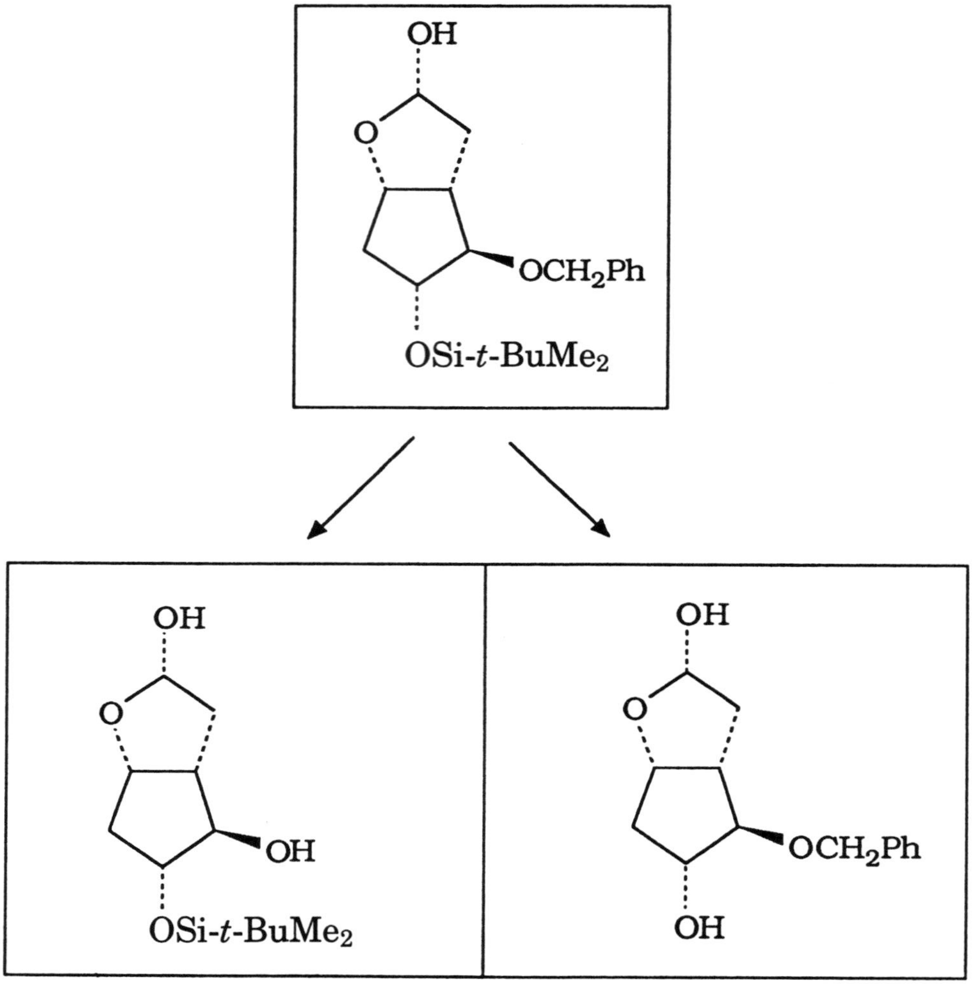
OH
O
OCH2Ph
OSi-t-BuMe2
OH
O
OH
OSi-t-BuMe2
OH
O
OCH2Ph
OH

# 2   Protection of Alcohols: Dimethyl-*t*-butylsilyl Ethers

Protection of alcohols R—OH can be effected by way of trialkylsilyl ethers R—OSiR$_3'$. The simplest way is to use the trimethylsilyl ether (R=CH$_3$). These derivatives are very useful and because of their high volatility are readily separable by gas chromatography. They are, however, relatively labile and undergo solvolysis in protic solvents.

*t*-Butyl-dimethylsilyl ethers are on the other hand about 10$^4$ times more stable, being easily formed from the corresponding chlorides.

The regeneration of the alcohol can be carried out with tetra-*n*-butylammonium fluoride. The example given here is that of the selective regeneration of an alcohol function blocked either by a benzyl or a *t*-butyl-dimethylsilyl group. The former can be cleaved selectively by catalytic hydrogenation (Pd/C) at room temperature in a quantitative way (96%). The latter is removed as follows: treatment with tetra-*n*-butylammonium fluoride for 5 min at 0 °C, then for 1 h at room temperature; yield 92%.

Reference: E. J. Corey and A. Venkateswarlu, *J. Am. Chem. Soc.*, **94**, 6190–6191 (1972).

= RO(THP)

# 3  Protection of Alcohols: Enol Ethers

Enol ethers (examples: dihydropyran, methylvinyl ether) have electron rich double bonds due to the interaction of the $\pi$ electrons of the C=C double bond and the n electrons at the oxygen. The carbon $\beta$ to the oxygen has an increased electron density. In the presence of catalytic amounts of a protic acid, which forms the conjugate acid by protonation of the $\beta$-carbon, a molecule of the alcohol ROH is attached to dihydropyran. Thus the tetrahydropyranyl ether of this alcohol is obtained, often written in the abbreviated form RO(THP).

The alcohol protected in this way can be readily regenerated by treatment with dilute acid, when the protection is no longer required.

The acetal RO(THP) is stable under basic conditions. It remains inert to the action of alkylating agents, organometallic reagents or LiAlH$_4$. Nevertheless, it can be hydrolyzed under mild conditions using aqueous acetic acid or 0.1 M hydrochloric acid.

---

'No, no! The adventures first', said the Gryphon in an impatient tone: 'explanations take such a dreadful time'.
Lewis Carroll, *Alice in Wonderland*, X.

# 4   Inversion of Polarity: Definition

*Inversion of polarity* consists in the inversion of the sign of the electric charge (partial or full) that an atom bears.

*Notes*

(i) We will confine ourselves to the case of carbon.

(ii) The transformation of a nucleophile into an electrophile (or vice versa), that of a donor synthon into an acceptor synthon (or vice versa) are inversions of polarity.

(iii) The limiting case would be: carbanion $\leftrightarrow$ carbocation;

(iv) In practice, we speak of inversion of polarity only if we deal with a simple and generally applicable procedure.

(v) One finds frequently the German term 'Umpolung' to name this operation.

*Relevance*:

The asset of a polarity reversal is that it bypasses the 'normal' reactivity expected for an atom, thus tuning into another register of reactivity normally inaccessible to that atom; for instance the normally electrophilic carbon of a carbonyl is made into a nucleophile.

---

'Methinks the Chymists, in their searches after the truth, are not unlike the navigators of Solomon's Tarshish fleet, who brought home from their tedious voyages, not only gold, and silver and ivory, but apes and peacocks too: for so the writings of several of your hermetick philosophers present us, together with diverse substantial and noble experiments, theories, which either like peacock feathers make a great show, but are neither solid nor useful; or else like apes; if they have some appearance of being rational, are blemished with some absurdity or other, that, when they are attentively considered, make them appear ridiculous'. Robert Boyle, *The Skeptical Chymist*.

# 5  E. J. Corey

The beginning of the scientific and academic careers of Elias J. Corey, born in 1928 in Methuen, Massachusetts, is that of a highly gifted student. Fond of mathematics in secondary school, he prepared himself to enter electrical engineering at MIT. In 1944 it seemed unrealistic to him to find a job as mathematician. At MIT, after the wonders of chemistry had been revealed to him, he decided to pursue a chemical engineering curriculum. He had entered MIT in 1946, he obtained the B.S. in 1948 and completed his PhD thesis in 1951—after only five years of study, in total! He was then appointed as an instructor in chemistry at the University of Illinois, in Urbana. He continued to climb rapidly the academic ladder, being offered a professorship at the very young age of 28. From 1959 on, he has held a chair at Harvard University.

Corey's vast scientific output (more than 600 publications) is focused on organic synthesis. Besides the syntheses of some hundred target molecules—some of which (e.g. longifolene) have become classics—Corey contributed considerably to the development of methodology in organic synthesis. The retrosynthetic view, the synthon approach as well as innumerable new reagents are his inventions. One owes to him for example the inversion of polarity of the aldehyde using the dithiane method.

A number of Corey's works on leukotrienes and prostaglandins in the past two decades have revolutionized our knowledge of the molecular origins of inflammation, immunology and other biomedical sectors. E. J. Corey has received numerous scientific awards: the Wolf prize for chemistry in 1986 which he shared with Albert Eschenmoser of the ETH in Zurich; and above all the Nobel prize in 1990.

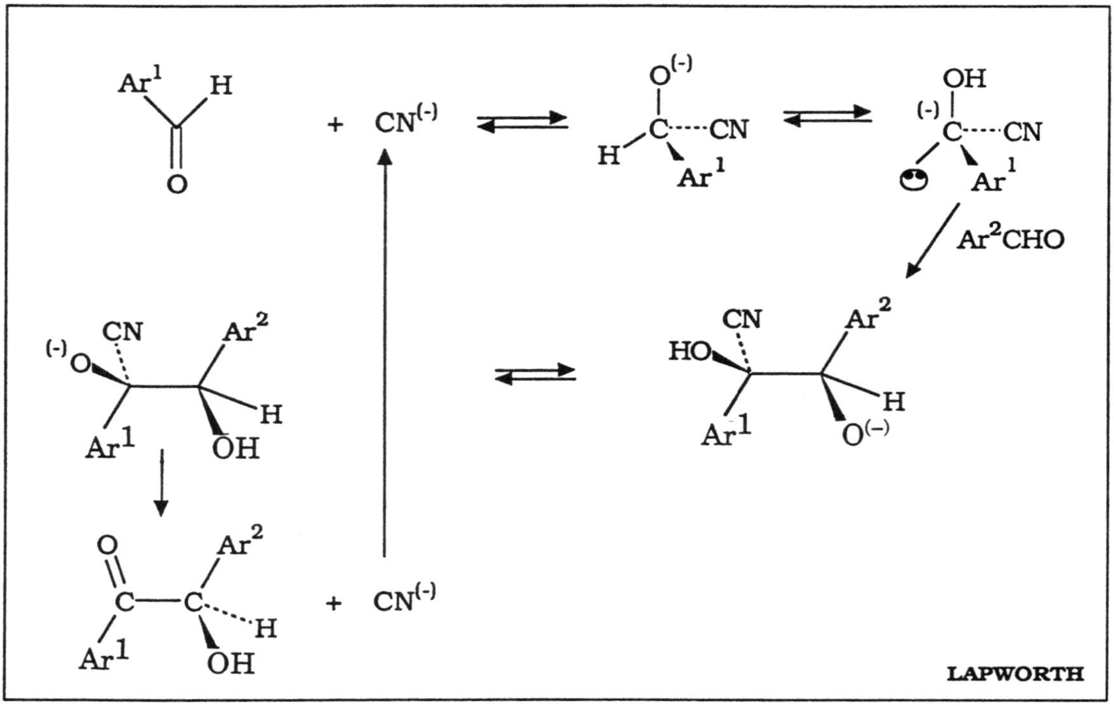
Ar^1CHO + CN(-)
Ar2CHO
LAPWORTH

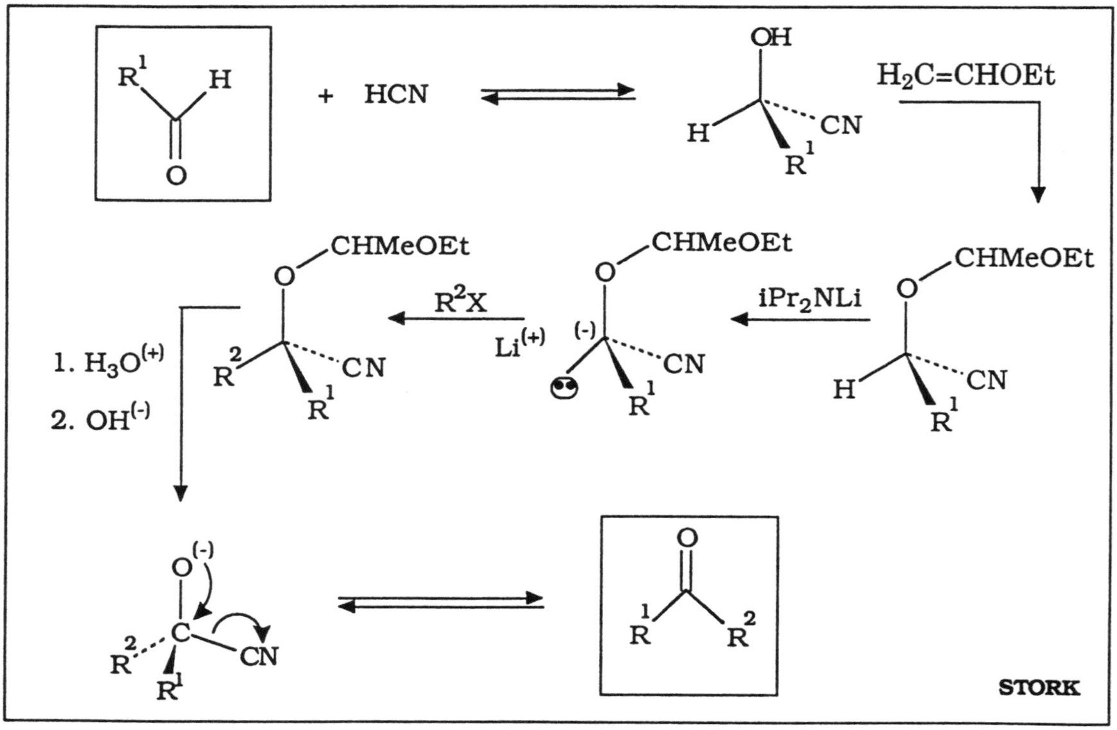
+ HCN
H2C=CHOEt
CHMeOEt
iPr2NLi
R2X
Li(+)
1. H3O(+)
2. OH(-)
STORK

# 6 Inversion of Polarity According to Lapworth–Stork

Arthur Lapworth (1872–1941) in 1903 proposed a mechanism for the benzoin condensation catalyzed by cyanide anions.

From the viewpoint of polarity inversion, the key step is migration of the aldehydic proton from carbon to oxygen. This prototropic change leads to an inversion of polarity since the carbon, initially electrophilic in the starting aldehyde, has been transformed to a carbanion, i.e. to a nucleophilic center (or donor). Migration of the negative charge from oxygen to the carbon is made easier by the presence of the strongly attractive cyano group on the latter.

In the Seventies, Gilbert Stork (1921– ) of Columbia University refined this inversion of polarity postulated by Lapworth and created a general alkylation method that allows conversion of an aldehyde $R^1CHO$ to a ketone $R^1R^2CO$. The OH group of the cyanohydrin produced in the first step is *protected* by reaction with an enol ether such as ethylvinyl ether. Thus only the acidic proton at carbon is removed by a strong base. The resulting carbanion, stabilized both by the cyano substituent and by the lithium counterion, reacts with an alkylating agent $R^2X$: a new bond between $R^2$ and the carbon atom is formed with the $X^{(-)}$ group leaving as the lithium salt. It remains only to cleave the protecting group of the OH function under acidic conditions, to displace the cyano group with a base and to provide the required ketone $R^1R^2C{=}O$.

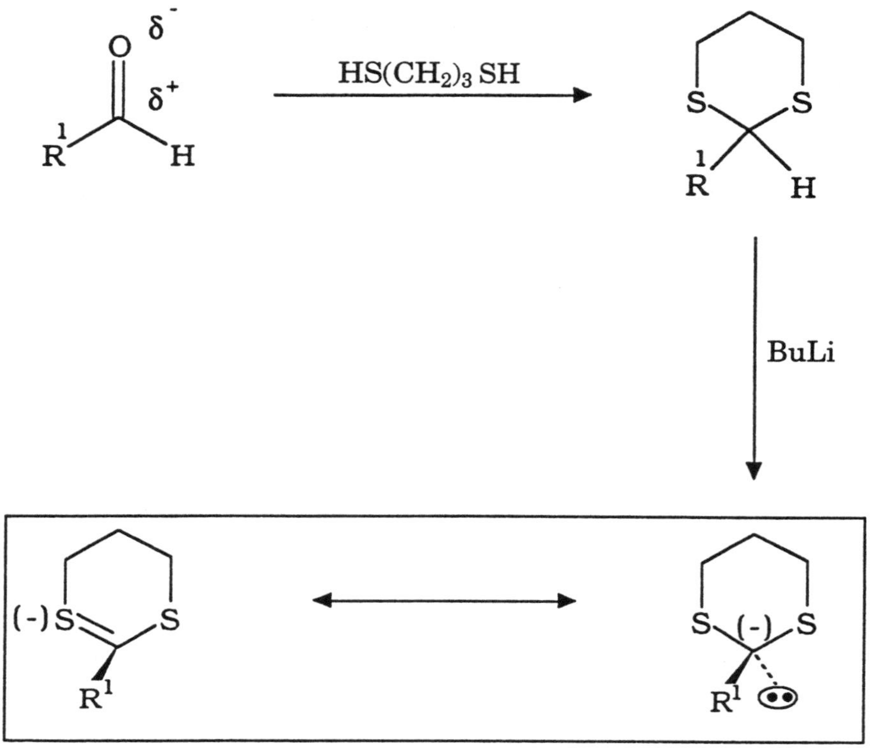
δ⁻
δ⁺
O
R1
H
HS(CH2)3 SH
S
S
R1
H
BuLi
(-) S
S
R1
S (-) S
R1

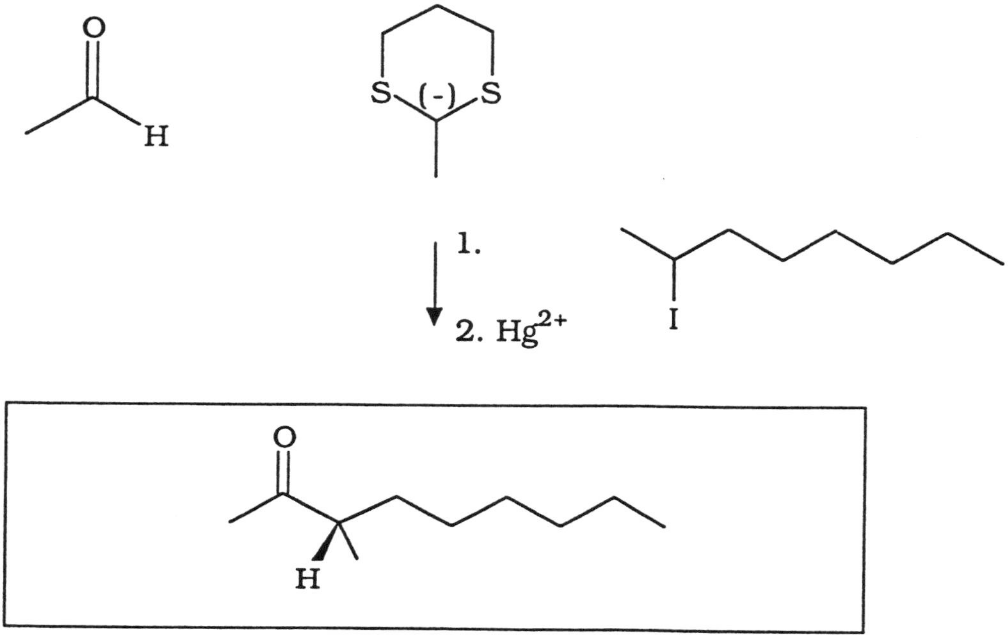
O
H
S (-) S
1.
I
2. Hg2+
O
H

# 7   Dithiane Method (Corey and Seebach)

This method is applicable to aldehydes R'CHO. The first reaction with $HS(CH_2)_3SH$ transforms them to dithianes: this is the cyclic equivalent of a bisthioacetal. The second step is metallation of the 1,3-dithiane: it is carried out by means of a strong base such as $n$-butyllithium which removes the acidic proton. The conjugate base is stabilized in two ways: by the inductive effect of the two attracting sulfur atoms, but especially by their mesomeric effect.

The result of this operation is the transformation of a carbon atom, electrophilic in the original aldehyde, into a nucleophilic center: this newly acquired donor character of the carbon can then be turned to advantage for alkylation or acylation at this carbon center. After it has served its use, the dithiane group can be discarded; the carbonyl group is regenerated by hydrolysis catalyzed, for instance, by a mercuric salt $Hg^{2+}$.

References: D. Seebach, *Synthesis*, 1969, 17.
            D. Seebach, M. Kolb, and B. T. Gröbel, *Chem. Ber.*, **106**, 2277 (1973).
            D. Seebach and E. J. Corey, *J. Org. Chem.*, **40**, 231–237 (1975).

## Summary:
## 'Protection, Regeneration,
## Inversion of Polarity'

The **protection** and **regeneration** of a sensitive functional group during a multistep synthesis has some analogy with some of the procedures followed by a dentist or a surgeon. We examine them as applied to the carbonyl group, as the most important example.

Let us assume that an $S_N2$ reaction has to be carried out with an aldehyde bearing a leaving group; this requires a protection of the very reactive carbonyl group. Therefore the latter is transformed to an ethylene acetal and is regenerated as soon as the substitution is accomplished. This prevents secondary reactions such as nucleophilic attack on the carbonyl group, aldol coupling, an $S_N2$ reaction with the enolate as nucleophile, and so on, from taking place. It is easy to transform the carbonyl compound to an ethylene acetal by an acid catalyzed reaction with ethylene glycol. Its regeneration is carried out, for instance, by acidic hydrolysis.

The **inversion of polarity** was first illustrated in 1903 by the cyanide-catalyzed benzoin condensation (Lapworth); addition to the carbonyl group affords the corresponding cyanohydrin. With an aromatic aldehyde as substrate, the cyanohydrin is sufficiently acidic at carbon that a proton can shift from the carbon to oxygen atom; this produces a carbanion which can be trapped by a second molecule of the aromatic aldehyde. Expulsion of the cyanide anion from the adduct affords benzoin. There has been thus a polarity inversion: the electrophilic carbon of the carbonyl group has been transformed into the nucleophilic carbon of the carbanion. In the seventies, Gilbert Stork extended the principle of the benzoin condensation to a similar reaction that converts aldehydes to ketones.

The enzymatic formation of acyloins is catalyzed by thiazolium salts as demonstrated by Ronald Breslow in 1957. Their use is illustrated by one of the syntheses of dihydrojasmone, used in perfumes because of its jasmine odor.

The link between protection and inversion of polarity is manifested by the dithiane method (E. J. Corey and D. Seebach, 1965); an aldehyde adds ethanedithiol to afford a 1,3-dithiane (six-membered ring with two sulfur atoms in position 1 and 3). The carbon bearing the two sulfur atoms can be readily deprotonated into the nucleophilic carbanion. In contrast, the starting aldehyde has an electrophilic center. The method is very general; mercury (II) salts cleave dithianes and regenerate the aldehyde function.

# 5.II
# Protection and Regeneration. Examples of the Alcohol and Carbonyl Group. Inversion of Polarity

## 1 Nucleophilic Displacement in a Molecule with a Very Reactive Carbonyl Group

The trivial solution, $S_N2$ reaction with methoxide $CH_3O^{(-)}$ as the nucleophile, and with bromide $Br^{(-)}$ as the leaving group, is excluded: the reagent would add to the carbonyl group, and it could also provoke aldol coupling between $Br(CH_2)_3CHO$ as donor and another molecule of aldehyde, with the $-CHO$ group as the acceptor.

A protection of the carbonyl group is required, for instance as shown in the Scheme.

---

'I subscribe totally to the judgement that Your Excellency holds of chemists, and I am convinced that they keep saying words outside of the common use, to make themselves appear as knowledgeable on topics they know nothing about.' Descartes, letter to the Marquess of Newcastle, November 23 1646.
'The trouble with chemists is that chemistry is too hard for them.' Albert Einstein, quoted by J. J. Zuckerman, *J. Chem. Ed.*, **63**, 829–833 (1986).

## 2   The Beginning of Corey's Longifolene Synthesis

The starting point of the synthesis is the bicyclic diketone, usually referred to as the Wieland–Miescher ketone. We will see later on (Chapter 16) how readily it is made. The first step requires the selective transformation of the carbonyl group conjugated to an ethylenic system.

A protecting function solves both problems, of reactivity and of selectivity. The more reactive cyclohexanone is converted by reaction with one equivalent of ethylene glycol, with *para*-toluenesulfonic acid as catalyst, in benzene as solvent, to the corresponding ethylene ketal. Then the functional group interchange can be carried out at the conjugated carbonyl group (C=O → C=CH·CH₃) by a Wittig reaction (see Chapter 12).

Finally, the carbonyl group is regenerated from its protected form (2N HCl in aqueous ethanol).

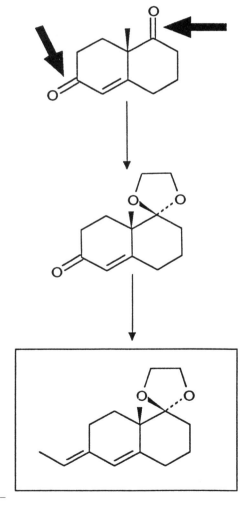

Reference: E. J. Corey, M. Ohno, P. A. Vatakencherry, and R. B. Mitra, *J. Am. Chem. Soc.*, **83**, 1251 (1961); **86**, 478 (1964).

# 3 Protection of a Carbonyl Group in an Industrial Synthesis

The methylketone, more reactive than the conjugated ketone, can thus be protected selectively: the ethylene ketal is formed in a reaction with ethylene glycol, catalyzed by *para*-toluenesulfonic acid. It is therefore possible to alkylate selectively at the free $\alpha$-position of the carbonyl group: a base (NaOCH$_3$) removes a proton from the $\alpha$-face at C-11, forming a vinylogous enolate, so as to set-up an antiperiplanar arrangement of the broken bond (C—H) and the newly-formed C—C bond (with the CH$_3$I reagent). In this way, the angular methyl group C-19 is introduced at carbon C-10 (to use steroidal numbering). It only remains to regenerate the second carbonyl group by an acid catalyzed hydrolysis reaction.

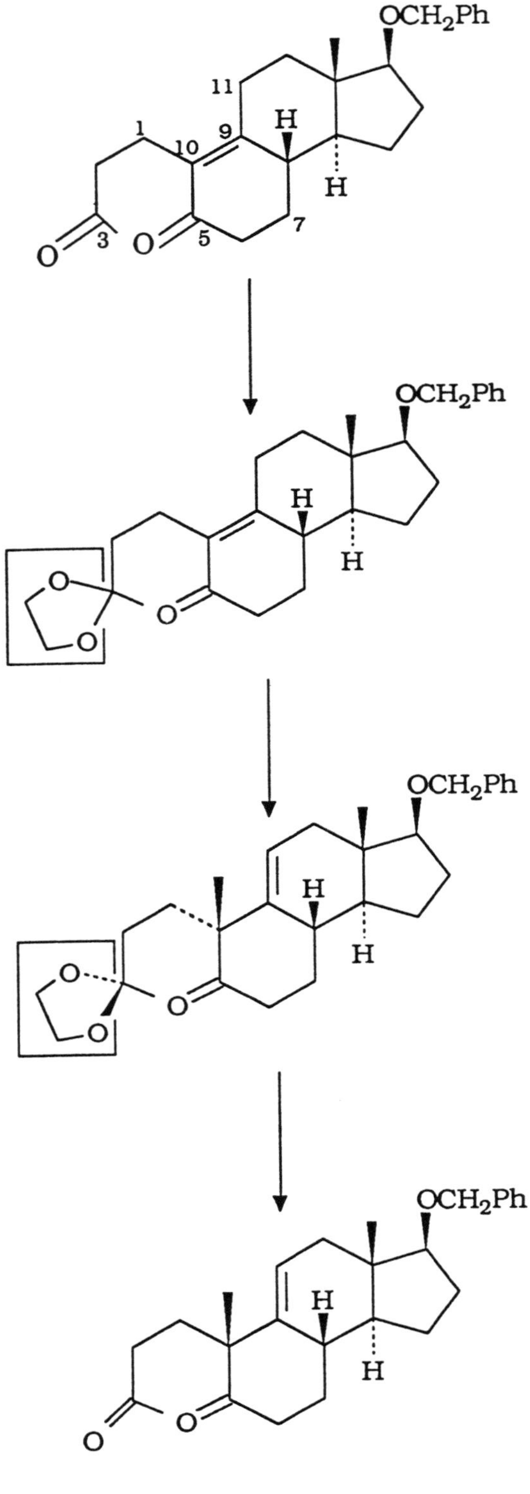
OCH2Ph
11
H
1
9
10
H
3
5
7
O
O
OCH2Ph
H
H
O
O
O
OCH2Ph
H
H
O
O
O
OCH2Ph
H
H
O
O

This process is interesting because it involves alkylation of a vinylogous enolate with isomerization of the $C=C$ double bond. It is part of the Roussel–Uclaf steroid synthesis devised in the Sixties.

## 4   Mechanism of the Enzymatic Formation
## of Acetyloin

The cyano group is not the only means for polarity inversion of an aldehyde, preferably catalytic, as in the Lapworth–Stork style. Another is recourse to thiazolium salts, whose position 2—between sulfur and positively-charged nitrogen—which makes it even more electron-attracting, is acidic. In 1957 Ronald Breslow (Columbia) deduced the mechanism of action of the co-enzyme thiamin pyrophosphate from this observation. We show here the simple case of self-condensation between two molecules of benzaldehyde (see adjacent Scheme).

---

'Science always comes from life, and returns to it by a by-road'. J. W. Goethe.

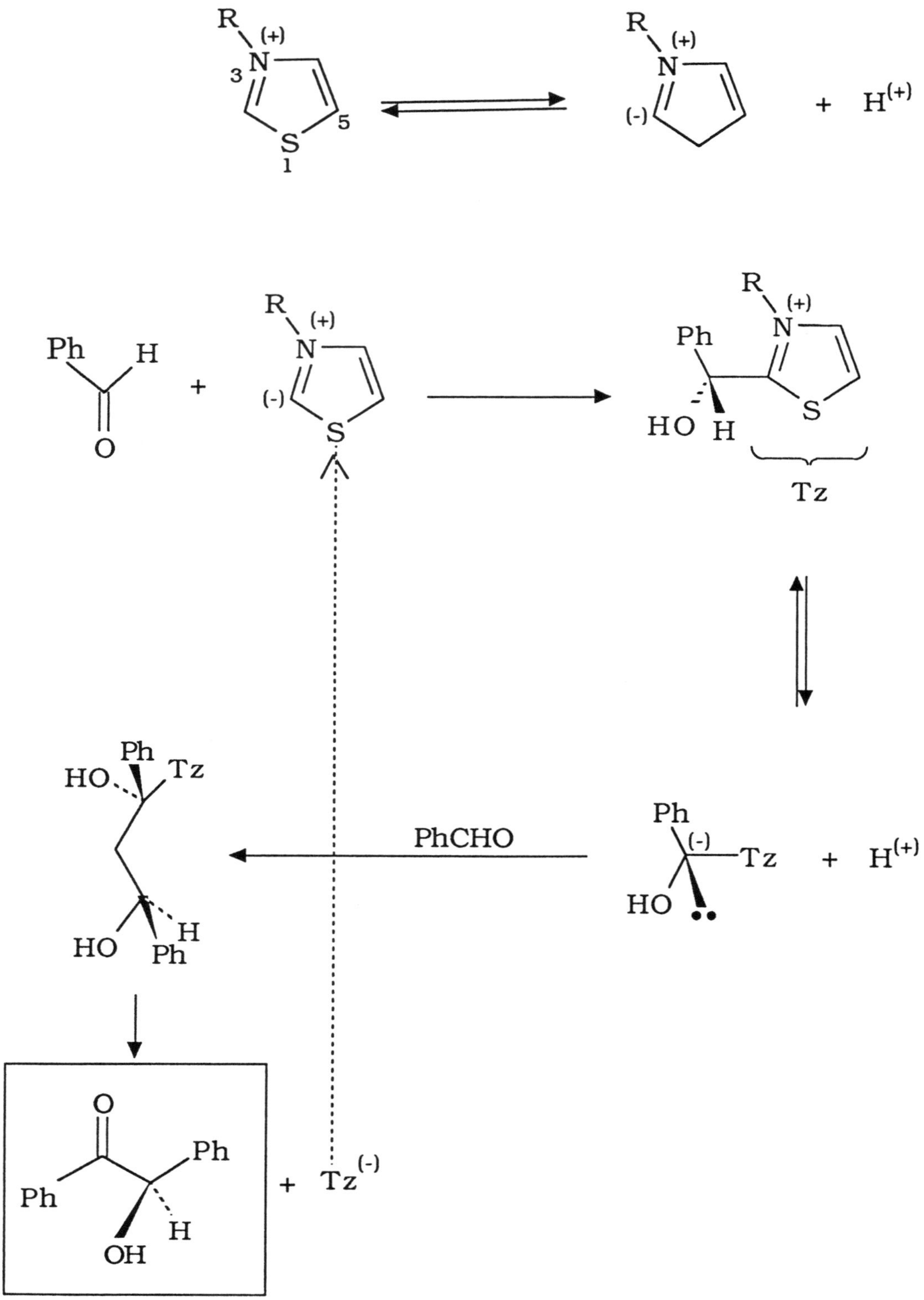
R
(+)
N
3
5
S
1
+ H(+)
(-)
Ph
H
O
PhCHO
HO
Tz
Tz(-)
OH

# 5 The Fragrance of Jasmine

*cis*-Jasmone is the active constituent of the jasmine essential oil used in perfumes; it has an exotic quality (Ohloff) that *trans*-jasmone lacks. Synthetic dihydrojasmone is used in perfumery as a replacement for the natural *cis*-jasmone even though it is inferior in olfactive quality. Dihydrojasmone is obtained by base-catalyzed cyclization of 2,5-undecadione.

*cis*                    *trans*

This dione is synthetized elegantly by condensation of heptanal with methylvinylketone, acting as Michael acceptor. Triethylamine is used to deprotonate the thiazolium salt at C-2. The conjugated base $Tz^{(-)}$ of the latter serves as the catalyst for the inversion of polarity of the aldehyde: *n*-heptanal is replaced by the carbanionic synthon $H_3C(CH_2)_5\text{-}C(Tz)(OH)^{(-)}$. The carbanion adds to the carbon end of methylvinylketone in a Michael reaction and affords the required dione.

References: G. Ohloff, *Experientia*, **42**, 271–279 (1986).
        H. Stetter, H. Kuhlmann, and W. Haese, *Org. Synth.*, **65**, 26–31 (1987).

## 6    Natural/Artificial

'Organic foods, untouched by chemical fertilizer', 'a natural product, not a chemical product': such assertions, with a strong whiff of vitalism, bring a smile. In the same way as for Gertrude Stein 'a rose is a rose is a rose', a chemical substance is the same whatever its origin. Acetone resulting from the Hock oxidation of cumene, or from a microbiological process established by Chaim Weizmann for the British Government during World War I (which rewarded him with the Balfour Declaration) is the same substance. Is this really true? A counter-example would appear to be that of vanillin, a natural product of nevertheless simple structure. The natural substance, originating from *Vanilla planifolia* pods costs kFF 25 kg$^{-1}$, whereas synthetic vanillin (from guaiacol (Rhône-Poulenc) or from lignin (Monsanto) is sold for 100 FF kg$^{-1}$. Same stuff, 250-fold difference in price! The amounts sold are enormous because it is the most common flavor, used mainly in vanilla ice cream. Certain legislations (USA) explicitly distinguish between natural and synthetic products and inform the consumer by way of labeling. In the former case, the selling price is allowed to reflect the difference in manufacturing costs. Such huge cost differences are an invitation to fraud. But it ain't easy, to sell synthetic vanillin as natural vanillin. In the Eighties, governmental agencies specializing in fraud detection made use of mass spectrometry to reveal the minute differences in mass distribution between the two vanillins, stemming from non-identical isotopic distributions. Interpol was intrigued by massive deliveries of $^{13}$C-labeled methyl iodide ($I^{13}CH_3$) to a small chemical concern in Surrey: these people methylated the OH of the catechol precursor, in order to adjust the mass distribution of synthetic vanillin to that of the natural product. Since then, a better spectroscopic method has ruled out this particular fraud: the two vanillins differ in distribution of the H/D hydrogen isotopes at six different positions and it is easy to determine which is which by NMR spectroscopy ($^2$H, $^{13}$C, $^{17}$O). The natural and synthetic substances are thus told apart by their isotopic signatures.

vanillin

References: Maubert et al., *Analusis*, **16**, 434 (1988).
　　　　　 Caer et al., *Anal. Chem.*, **63**, 2306 (1991).
　　　　　 G. J. Martin, *Biofutur*, March, 33–40 (1992).
　　　　　 B. Lacaze, *Technologies*, 62–68 (1992).

## 7    Synthesis of Cyclic Ketones by the Dithiane Method

This dithiane is prepared from formaldehyde. Its reaction with 1-bromo-3-chloropropane, after lithiation with *n*-butyllithium, allows introduction of a three-carbon chain. After lithiation of the intermediate,

---

'Philosophical analysis indeed starts from symbolism considered as a fact. But neither does it take from the manner of the philologist, who describes this factual state as such, nor of course from the manner of a scientist, who occasionally destroys and reforms natural symbolisms in order to set-up an abstract model for our world of objects, which are then associated in as exact a manner as possible to well-formed assertions in his new language. Philosophy sees it as its task to be explicit in which way and how symbolisms, including those that science deliberately builds, both *are* and *represent* an experience that they organize, whether in latent or in manifest manner.'
　　G. G. Granger, *Pour la connaisance philosophique*, O. Jacob, Paris, 1988, p. 195.

a second alkylation occurs, this time with chloride as leaving group. This affords a bisthioketal of cyclobutanone.

Hydrolysis of this bisthioketal leads to the 'free' ketone. This is a general method for synthesizing cycloalkanones.

# 6.I
# Activation

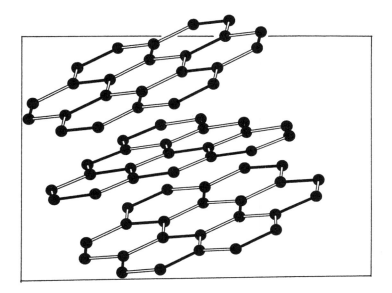

Activation of molecules is a very general and rather difficult problem. In actual industrial practice it includes activation of atmospheric nitrogen, and the challenge of direct functionalization of saturated hydrocarbons, to mention only two of the most prominent examples. There are three types of solutions: destabilization of the initial state, stabilization of the final state and stabilization of the transition state. We will examine each of these in turn.

'The rules of inference are completed by the *rules of defining*, which is indispensable if a given theory provides for definitions in an object language formulation'. K. Ajdukiewicz, *Pragmatic Logic*, PWN-Polish Scientific Publishers, Warsaw, and D. Reidel, Dordrecht, 1974, p. 220.

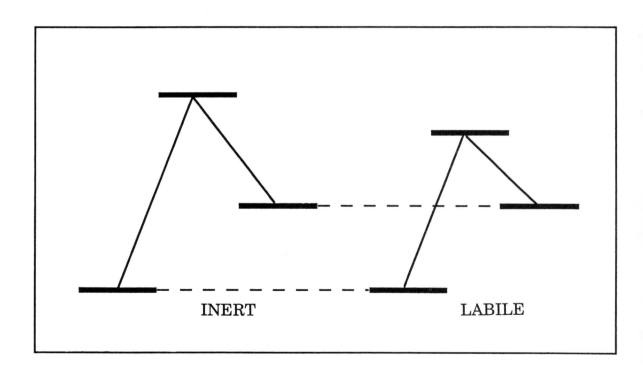

# 1 The Vocabulary

It is necessary to distinguish between the realms of thermodynamics and of kinetics; the term stability applies within the former. Only by extension—unfortunately rather common—can it be used for the latter. Yet, kinetics has its own terminology which we shall use.

Of the two energy states available to a system the lowest is the most stable; the higher state is the least stable, relative to the other, of course.

However, the operationally significant and thus determining factor in most cases is the rate of the transformation, stable ⇄ unstable, in either direction. If the energy barrier against the transformation is high (still in relative terms), the system is *inert*. If the energy barrier is low, the system is *labile*. We prefer these terms to those of 'kinetically stable' and 'kinetically unstable'. 'Stable' and 'unstable' thus belong to thermodynamics; while 'inert' and 'labile' refer to kinetics.

Therefore there are four logical combinations: stable and inert, stable and labile, unstable and inert, unstable and labile.

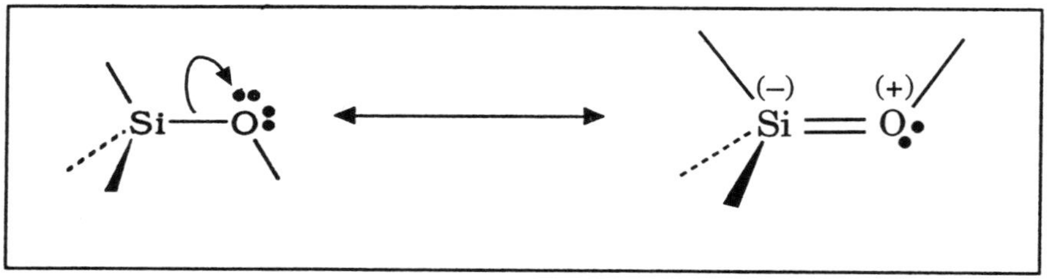
Si
O
(−)
(+)
Si
O

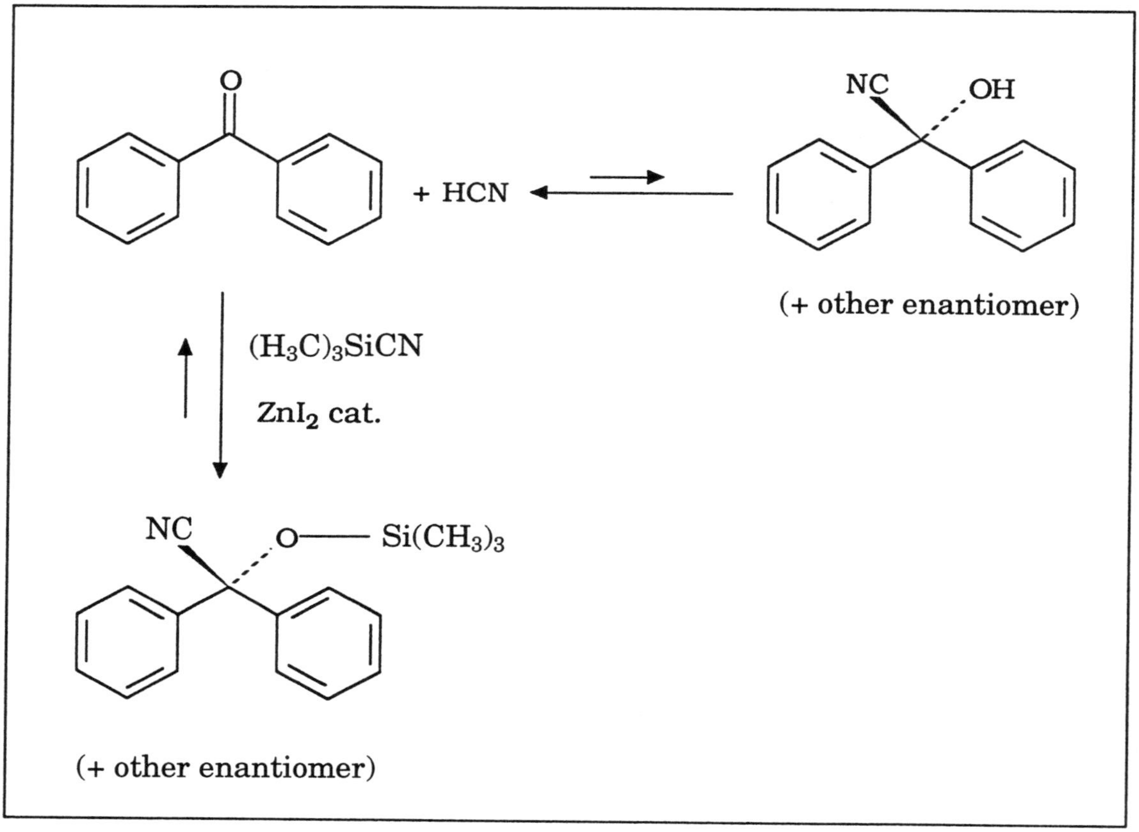
O
+ HCN
NC
OH
(+ other enantiomer)
(H3C)3SiCN
ZnI2 cat.
NC
O
Si(CH3)3
(+ other enantiomer)

# 2  Stabilization of the Final State

We will describe here the example of the silylated enol ethers discussed in the previous chapter. Their formation makes use of the very high Si–O bond energy:

|                    | $\Delta H^\circ$ (kcal mol$^{-1}$) |
|--------------------|-----------------------------------|
| $(H_3C)_3\,Si{-}H$    | 90                                |
| $(H_3C)_3\,Si{-}CH_3$ | 89                                |
| $(H_3C)_3\,Si{-}OCH_3$ | 127                              |

It is a very strong bond, for two main reasons: silicon is electropositive, oxygen electronegative, hence the strengthening of the bond from the difference in the electronegativities of the two bonded atoms; silicon is polarizable and can stabilize the non-bonding electron pairs on oxygen; this can be expressed in the language of resonance, as indicated in the adjacent scheme.

Let us consider another example (due to Professor David A. Evans, of Harvard University) that makes use of this high bond energy as a driving force. In this manner, one lowers the energy of the product which makes the reaction more exothermic. The addition of the nucleophile $CN^-$ to benzophenone as the starting compound does not occur readily. This substrate is too well stabilized by conjugation. However if the addition is coupled with formation of a silicon–oxygen bond, then the reaction can take place.

References: R. Walsh, *Accounts Chem. Res.*, **14**, 246–252 (1981).
            D. A. Evans, L. K. Truesdale, and G. L. Carroll, *J. Chem. Soc. Comm.*, **55–56**, (1973).

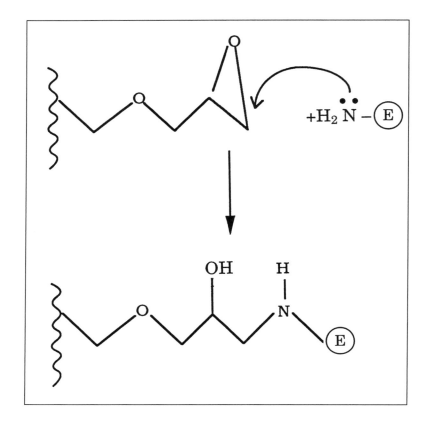

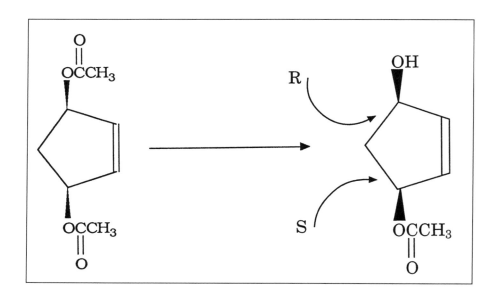

# 3   Destabilization of the Initial State

One way of reducing the activation energy is to raise the initial state: if, to first approximation, the energy of the transition state remains unaffected, then the reaction is accelerated.

(1)  This concept is very often used in the opening of strained small ring systems. The driving force for such reactions is the release of Baeyer ring strain. An example is the immobilization of an enzyme E on beads of a polyacrylic resin. To do so, one grafts onto the polymer chains with epoxy end groups. Their opening is the trigger for attachment of the resulting OH groups to carboxylic and other polar groups on the enzyme surface. The enzyme thus modified retains most of its activity. For instance, pork liver esterase retains 68% of the activity of the soluble form. In this form, the enzyme can be easily stored for several months in the refrigerator at 7 °C.

A practical use of this immobilized enzyme is to the specific hydrolysis of only one of the two acetate functions: aqueous solution, phosphate buffer 0.1 M, pH 7, 14 h at 32 °C; yield 68%, enantiomeric excess of R alcohol >98%!

Reference: K. Laumen, E. H. Reimerdes, and M. Schneider, *Tetrahedron Lett.*, **26**, 407–410 (1985).

## Jean-Marie Lehn (1939–   )

He originates in the small Alsatian village of Rosheim, where his father was both baker and church organist. He has retained much more than a mere flavor, or even a strong appeal for music: he is himself an excellent pianist, as evidenced by a grand piano in one of his offices; his favorite composers are Bach, Beethoven and Bartok. The influence of music on his scientific work, all in themes and variations, is obvious.

Besides giving him a problem of structural elucidation in the triterpenes series, his PhD supervisor, Guy Ourisson, also entrusted him with the responsibility for the nuclear magnetic resonance (NMR) spectrometer in his Strasbourg laboratory. This gave Lehn a strong background in NMR spectroscopy. He then continued as a postdoctoral fellow (1963–64) at Harvard with Robert B. Woodward, where he was involved in the synthesis of vitamin $B_{12}$. On returning to Alsace, he became a professor at Strasbourg in 1970. He was elected to the Collège de France in 1979 and to the Academy of Sciences in 1985. The Nobel Prize in Chemistry in 1987 crowned his work on supramolecular chemistry.

The period 1964–1968 can be referred to as the germinative period for his work: dynamic NMR and quantum chemical calculations applied to phenomena such as inversion of the nitrogen atom in organic molecules encouraged him to synthesize novel molecular objects interesting for the interactions amongst their components, in particular for their microdynamics. At that time (1967) Lehn borrowed his dominant research theme from the great German chemist Emil Fischer who had also worked for a relatively short period in Strasbourg. To describe the mutual fit of an enzyme and of the substrate embedded into it, Fischer had proposed the image of a key fitting into a lock. For Lehn, this theme has been truly seminal. He could find it embodied in ionophore antibiotics, with cyclic structures arranging themselves around a positive metal ion thus transportable across cellular membranes. Lehn embarked on the synthesis of similar artificial structures, which he called cryptands of cations. This was the founding act of the new supramolecular chemistry. It also formed a bridge towards biology, which had always held a special fascination for him.

(2) Let us consider an ion pair $A^-$, $C^+$. If the cation is stabilized by encapsulation in the cavity of a crown ether (Pedersen) or a cryptand (Lehn), the reduction of the Coulomb attraction $A^-$ $C^+$ activates the anion $A^-$.

*Notation* for crown ethers: $mCn$, with $n$ oxygen atoms and $(m - n)$ carbon atoms.

For instance, the 12C4 crown ether selectively encapsulates $Li^+$ ions whereas the 18C6 homolog selectively traps $K^+$ cations: the size of the cavity has to be matched to that of the encapsulated ion.

Take the example of the benzoin condensation, catalyzed by cyanide anions, as Lapworth had established in 1903. Addition of 7% of 18C6, which complexes the potassium ion of the KCN catalyst, increases the yield from 6 to 99%.

Reference: S. Akabori, M. Ohtomi, and K. Arai, *Bull. Chem. Soc. Japan*, **49**, 746–747 (1976).

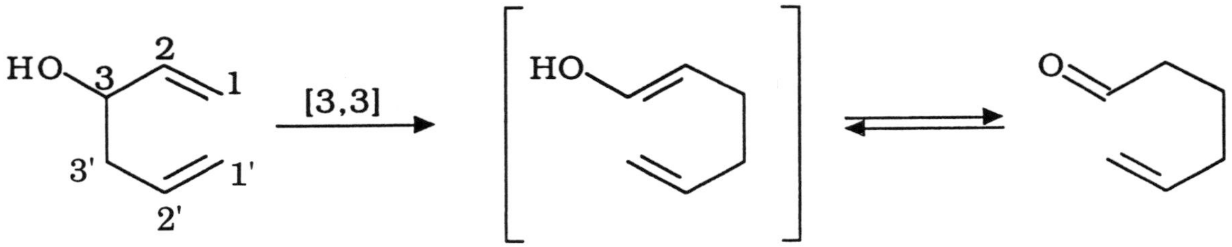
HO
3
2
1
3'
2'
1'
[3,3]
HO
O

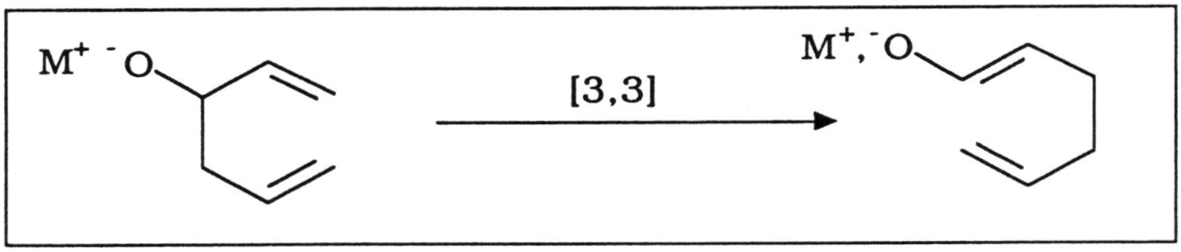
M⁺ ⁻O
[3,3]
M⁺,⁻O

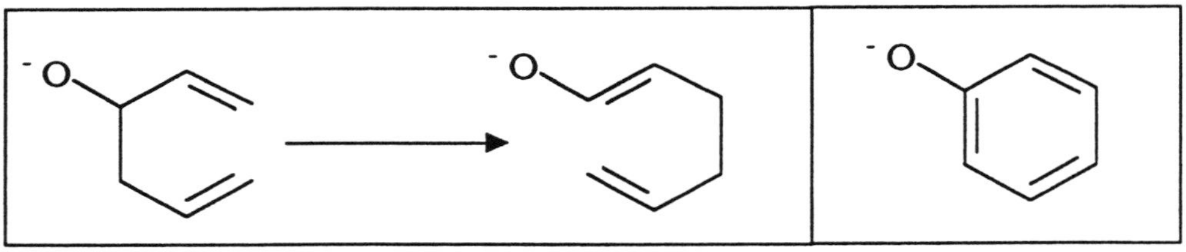
⁻O
⁻O
⁻O

# 4 Transition State Stabilization

Another way of lowering the energy barrier is to bring down the energy level of the transition state. The example given is that of a rearrangement termed (by the explicit but rather barbaric name), the 'anionic oxy-Cope rearrangement'. The oxy-Cope rearrangement is a sigmatropic [3,3] rearrangement of 1,5-cyclohexadienes bearing an alcohol group in position 3. The anionic oxy-Cope is the same rearrangement, thermally allowed (according to the Woodward–Hoffmann rules), the only difference being that it is applied to the alkoxide, rather than to the alcohol. The acceleration observed in this case is enormous: $10^{10}$–$10^{17}$, the highest factors being found in the presence of complexing agents of the $M^+$ cation such as crown ethers, which further increase the anionic character of the rearrangement.

Let us give an explanation for this spectacular activation in very simple terms. The transition state of the sigmatropic [3,3] rearrangement obeys the Dewar–Zimmermann rule, in that it is pseudoaromatic. In the same manner as the negatively charged oxygen atom of a phenoxide (the conjugate base of a phenol) is stabilized by conjugation with the benzene ring, that of an anionic oxy-Cope is stabilized by conjugation with the pseudo-ring present in the transition state, the energy of which is lowered. Another, more rigorous interpretation is weakening of the C—C bond and subsequent cleavage due to the $O^- - \sigma^*_{(C-C)}$ interaction.

Reference: D. A. Evans and A. M. Golob, *J. Am. Chem. Soc.*, **97**, 4765 (1975).

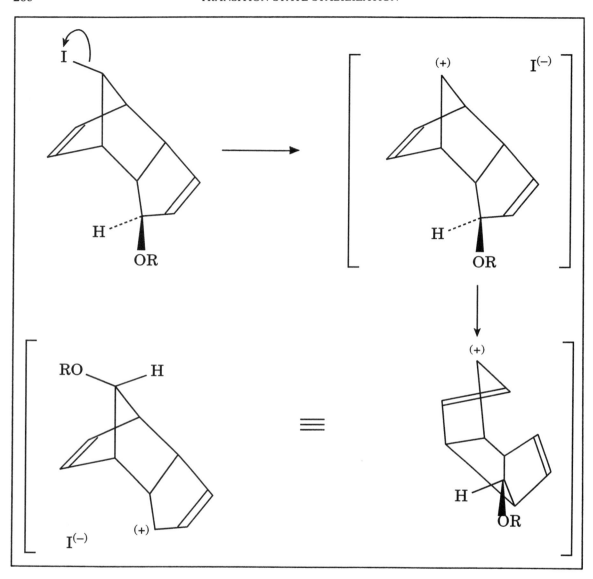

Since in the benzylic systems Ph-CH$_2^*$ (* = +, −, ·) the anion, the cation, and the radical are all stabilized, the generalization is obvious: a carbocationic Cope rearrangement should also be accelerated, by analogy with the stabilization of the benzyl cation.

This is indeed observed. The acceleration is considerable for the mechanism shown in the adjacent scheme.

We will not develop further this topic in this chapter: the conversion from inert to labile—by lowering the transition state with an auxiliary molecule, a mediator (just like the character of a 'Mittler' in Goethe's *Elective Affinities*)—is by definition *catalysis*. Catalysis of organic reactions is a field in full expansion.

---

Reference: R. Breslow and J. M. H. Hoffman, Jr., *J. Am. Chem. Soc.*, **94**, 2110–2112 (1972).

## Summary: 'Activation'

The language of activation distinguishes between the **thermodynamic** terms **stability/instability** and the *kinetic* terms *inertia/lability*.

The inclusion of the cation in a stabilizing cavity activates the anion and loosens the ion pair. The crown ethers, the cryptands of Jean-Marie Lehn, act as such selective complexing agents of cations via electrostatic bonding with the oxygen atoms of the cavity. The mutual adaptation of the sizes of the encapsulated ion and of the cavity are responsible for the selectivity. In this way, the 12C4 crown is selective for lithium, while the higher homolog 18C6 is selective for potassium. An application is to the benzoin condensation, catalyzed by cyanide anions as shown earlier in Chapter 5; with potassium cyanide as the catalyst, addition of 18C6 increases the yield from 6 to 99%! Even better the anionic oxy Cope rearrangement on a potassium alkoxide is accelerated by a factor $10^{17}$.

A possibility for activation is destabilization of the initial state. This is the reason for the frequent use of strained small rings (cyclopropane, epoxide) which can trigger off a sequence of reactions. Immobilization of an enzyme on a solid support, which allows for improved recovery of the enzyme, very often makes use of the ring opening reaction of an epoxide group, attached to the support, e.g. a polymer by a tether.

Stabilization of the transition state can be due to conjugation: an example is the anionic oxy-Cope (alkoxide) as compared to the oxy-Cope (alcohol); in the same way the cationic oxy-Cope rearrangement also displays spectacular accelerations. The intermolecular modification with regeneration of the auxiliary agent at the end of the reaction is **catalysis**. We give as an example the **abzymes** which are antibodies produced for catalytic use, capable of catalytic accelerations up to $10^6$.

In certain cases (luminol) the chemical activation of a system can be dissipated by the emission of photons; such photochemical deactivations are used for lights (rescue rafts), and attraction of fish where man is inspired by the bioluminescence frequently observed with marine organisms.

# 6.II
# Activation

## 1  Stable (by Roald Hoffmann)

Words are our enemies, words are our friends. We think in science that words are just an expedient for describing some inner truth, one perhaps ideally represented by a mathematical equation. Oh, the words matter, but they are not essential for science. We might admit there is a real question as to whether a poem is translatable, but we argue that it is irrelevant whether the directions for a synthesis of a molecule are in Japanese or Arabic or English—if the synthesis is described in sufficient detail, the same molecule will come out of the pot in any laboratory in the world.

Yet words are all we have, and all our precious ideas must be described in these history- and value-laden signifiers. Furthermore, most productive discussion in science takes place on the colloquial level, in simple conversation. Even if we know that a concept signalled by a word has a carefully defined and circumscribed meaning, we may still use that word colloquially. In fact, the more important the argument to us, the more we want to be convincing, the more likely we are to use simple words. Those words, even more than technical terms, are unconsciously shaped by our experience. Which may not be the experience of others.

I was led to reflect on this by the reaction a friend, a physicist, had to my use of the word 'stable'. I had said that an as yet unmade form of carbon was unstable with respect to diamond or graphite by some large amount of energy. Still I thought it could be made. My friend said 'Why bother thinking about it at all, if it's unstable?' And I said 'Why not?', and there we were off arguing. Actually, it was enjoyable, because we were friends. But perhaps we should have pondered why the simple English word 'stable' has different meanings for a physicist and a chemist.

First, a little background. Diamond (drawing **1**) and graphite (**2**) are the two well-known modifications (allotropes) of carbon. The carbon atoms are linked up in very different ways in

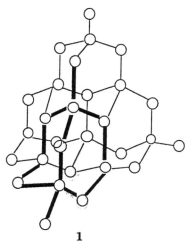

1                                                          2

the two: in diamond each carbon atom is tetrahedrally surrounded by four neighbors; in graphite a layer structure is apparent. Each layer is composed of 'trigonal' carbons, three bonds going off each carbon at a 120° angle from each other. The graphite layers are weakly held, not by real chemical bonds. They slip easily by each other, which is why graphite serves as a lubricant. Isn't it nice that black graphite is more stable (thermodynamically, more on this in a moment) than pellucid, hard diamond? Not by much, but so it is for carbon on the surface of the earth. Under high pressure the stability sequence, which is determined by a combination of energy and entropy, changes; the denser diamond becomes more stable, which is the basis of a commercial process for making industrial grade diamonds.

There are other forms of carbon.* The random and systematic incendiary activities of men and women have led to a multitude of pyrogenic materials. Most seem to be some form of graphite. There are a couple of rare, but well-established, allotropes of carbon, which are related to the diamond and graphite structures, and some others whose existence is disputed.* Carbon also turns up in short chains in the tails of comets and in flames, and recently hefty carbon clusters, $C_n$, $2 \leqslant n \leqslant 100$ have been produced in the gas phase. A most abundant cluster is one with 60 atoms, first detected by R. Smalley and co-workers at Rice. They suggested a soccer ball structure, 3, for this remarkable molecule. And a name, buckminsterfullerene.[†]

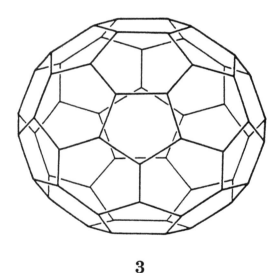

**3**

I digress, but that's just testimony to the fascination of allotropes. One day, I was trying to think up some alternatives to diamond and graphite. Why? Well, for one it was fun; for another, people have been squeezing elements which are not metals, trying to make them metallic. When you apply a megabar or so (some of my friends at Cornell do this routinely, using diamond anvils!) to most

References: *J. Donohue, *The Structure of Elements*, Wiley, New York, (1974).
      [†]H. W. Kroto, J. R. Heath, S. C. O'Brien, R. F. Curl, and R. E. Smalley, *Nature*, **318**, 162 (1985) and subsequent papers.
      [‡]Xe: D. A. Nelson and A. L. Ruoff, *Phys. Rev. Lett.*, **42**, 383 (1979).
      I$_2$: A. S. Balchan and H. G. Drickamer, *J. Chem. Phys.*, **34**, 1948 (1961); K. Syassen, K. Takemura, H. Tups and A. Otto, in *Physics of Solids under High Pressure*, J. S. Schilling and R. N. Shelton, Eds., North-Holland, Amsterdam, 1981, p. 125.
      O$_2$: S. Desgreniers, Y. K. Vohra and A. L. Ruoff, private communication.

anything, the atoms are forced so close together that their electron clouds overlap, the material becomes a metal. Xenon, iodine and oxygen have been made metallic;[‡] there's an argument whether hydrogen has been so transformed.

The interesting thing about both diamond and graphite is that they are, so to speak, full of nothing. They're not dense at all; a close-packed structure such as that of a typical metal would be much denser. Of course, there is a good reason why the density of the known C allotropes is so low—carbon atoms form bonds, and there is a lot of energy to be gained by forming those bonds directionally, trigonally or tetrahedrally. Carbon, with its four valence electrons, has better things to do than to try to shuffle its bonding among its 12 or 14 different nearest neighbors, as it might be forced to do in a close-packed structure.

Could there be, nevertheless, carbon networks filling space more densely than diamond or graphite, yet forming bonds along tetrahedral or trigonal directions? If there were, applying pressure to one of the known allotropes might be a way to make a hypothetical new form. To sum up a long story, there are *many* alternative space-filling structures.* But we haven't yet found one denser than diamond.

Peter Bird and I thought up one that is intermediate in density between diamond and graphite, and is quite special. This is illustrated in 4. It fills space with perfect trigonal carbon atoms. In the jargon of our trade, there are polyacetylene chains, needles of conjugation, running in two dimensions.

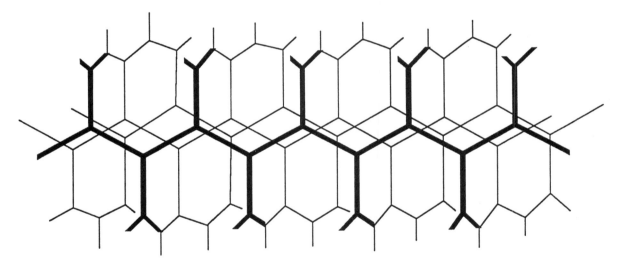

4

And no conjugation in the third. The most remarkable thing about this structure, something which emerged from calculations Tim Hughbanks did, is that it should be metallic. As it is, with no pressure applied to it.[†]

So there is a prediction of a metallic allotrope of carbon. If only one knew how to make it!

From structural reveries to matters of stability: the relatively unreliable calculations at our disposal indicated 4 to be a whopping 0.7 electron volt/carbon atom = 17 kilocalories/mole of carbons, *unstable* relative to graphite. This is what made my physicist friend react to my suggestion of the putative existence of this allotrope. But it didn't bother me at all.

References: *I. V. Stankevich, M. V. Nikerov, and D. A. Bochvar, *Russ. Chem. Rev.*, **53**, 640 (1984).
         [†]R. Hoffmann, T. Hughbanks, M. Kertesz, and P. H. Bird, *J. Amer. Chem. Soc.*, **105**, 4831 (1983).

Why the different reactions? Because the common English words 'stable' or 'unstable' had different meanings for the two of us!

To get at the source of our misunderstanding, let me go back to a scientific definition on which both my friend and I could agree. Stability, real stability, has to do with thermodynamics, the science of energy and entropy relationships, and with kinetics, the rates or speeds of imagined processes by which a system might be stabilized or destabilized. In chemistry we speak, in fact, of thermodynamic and kinetic stability. Suppose we have two molecules A and B which have the potential of changing one into the other:

$$A \rightleftharpoons B$$

Their relative thermodynamic stability is gauged by a marvelous function called the Gibbs free energy, which contains in it enthalpy (that's something very much like an energy, but with some specific conditions placed on it) and entropy terms. The natural, spontaneous way in which matter moves is to lower enthalpy and higher entropy, which means, in turn, greater disorder. The molecule which has the lower free energy is more stable, and the higher free energy molecule will transform spontaneously into the lower free energy one. To be specific: should it be B that has the lower free energy, then the spontaneous reaction will be A → B. One might represent this in a graph, **5**, in which the vertical axis is the free energy.

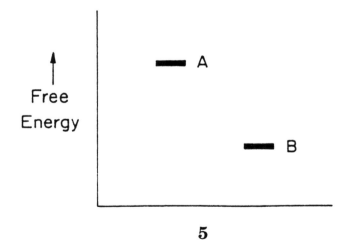

**5**

But now life is not so simple. Thermodynamics say what *must* happen, but not how fast it *will* happen. To go from molecule A (say my hypothetical metallic carbon, **4**) to molecule B (graphite) is no idle molecular promenade. Bonds have to break, many of them, and then to reform. Before poor A knows all the happiness that's waiting for it in those lovely graphite rings of B, it's got to suffer a lot of electronic indignity in the form of broken bonds. It resists. In general, molecules have barriers to their transformation. The situation is typically not that at left in **6**, but as at right.

There's a hill in the way. It's like a book that wants to fall under the force of gravity, but that has got a shelf under it. One might say that A is metastable, or, that A is a local minimum on some energy surface. And all of a sudden, it is no longer a question of falling, but of climbing hills!

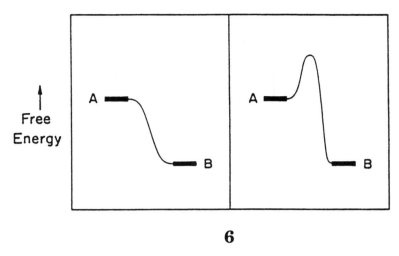

**6**

Will the reaction proceed after all? Well, it depends on the size of the hill, and on the temperature. Molecules don't sit still. In a gas, or in solution they are bouncing around at great speed, buffeted randomly by collisions with the $10^{20}$ other molecules in a typical flask. It's a crowded dance floor there. Some of the molecules acquire enough energy through collisions (here is where the temperature comes in, for the higher the temperature, the faster the molecules move) to pass over the hill. Others don't. If the hill be greater than about 30 kilocalories/mole high, then at room temperature A will remain A. Unless you wait a thousand years, for it is only then you might begin to see a little B.

A chemist would say A is thermodynamically unstable, kinetically stable, a physicist might call A metastable. These concepts are quite familiar to chemists and to physicists. So where is the problem? The difficulty is that our everyday discourse is perforce colloquial. We say 'stable' and not thermodynamically and/or kinetically stable. Some may label the colloquial characterization sloppy, and say it should be more precise. I would say that we wouldn't be human (and therefore have the potential of doing great science) unless we were often imprecise in just this way.

But now comes the crux of the matter. Into that word 'stable' goes the history of what we are or have done. When a chemist says 'stable' I think he or she means 90% kinetic and 10% thermodynamic. But a physicist, I would hazard a guess, means (he doesn't 'mean' in the sense of making a rational choice; I imagine he does this unconsciously) just about the converse extreme—90% thermodynamic, 10% kinetic.*

From the beginning of one's life in chemistry the importance of kinetic stability and relative unimportance of thermodynamic stability is highlighted. Every organic molecule in the presence of air (a typical laboratory and real life situation) is thermodynamically unstable with respect to $CO_2$ and $H_2O$. Think of methane ($CH_4$, natural gas), the essence of stability, having survived unchanged under the earth for thousands of years. Every time you light a gas stove, you demonstrate methane's thermodynamic instability. But it takes the complicated autocatalytic reaction set off by a match to take those $CH_4$ and $O_2$ molecules and get them over the hill, giving off light and heat along the way. Otherwise, methane is stable as a rock. Speaking of rocks, modern air pollution problems show that they are not that stable when strong acids come around.

---

Reference: *For some fascinating observations on the language of physics see C. F. Weizsäcker, *Die Einheit der Natur*, München, 1974, p. 61.

One amusing way to define synthetic chemistry, the making of molecules that is at the intellectual and economic center of chemistry, is that it is the local defeat of entropy, the construction of complex thermodynamically unstable molecules. In chemistry, a molecule that is strained, or otherwise thermodynamically disfavored by 1 electron volt/molecule relative to another molecule is not thought of as an occasion to throw up one's hands. It's a challenge, to be made, ingeniously . . .

I suspect there are a number of reasons why thermodynamic stability is set more firmly in the physicist's mind. First, a typical elementary physics course concentrates on mechanics, dynamics, electromagnetism in the absence of barriers or obstacles. Motion in the presence of barriers is too difficult to solve explicitly, so such problems are not mentioned. No one ever puts a shelf of variable permeability under that falling weight in Physics 100. Barrier penetration problems are probably first encountered in quantum mechanics courses.

Second, in thinking about the transformation of matter, physicists most of the time begin with motions governed by central forces, masses or charges moving around without hooks or directional valences. Entering the study of matter from the starting point of gases or close-packed metals one has few activated processes, only collisions, or balls sliding frictionlessly past balls, to reach the thermodynamically most stable point. Friction, barriers, the evolution in time of real systems in the end are just as important for physicists as they are for chemists. But the subtle weighting of concepts which shapes the colloquial language of science is fixed in scientific infancy. The early experiences matter; this is why I think the words stable and unstable mean different things to chemists and physicists.

Our metallic carbon allotrope is still waiting to be synthesized. I think it will be pretty stable, sorry, enduring, when it is made. If it is made.

## 2   Nitro Explosives

The explosion threshold is defined as the minimum pressure required to trigger off an explosion in 50% of the cases. This threshold varies with structure (indications in kbar):

| | |
|---|---|
| nitrobenzene | 17 |
| trinitrobenzene (TNT) | 21 |
| monoaminotrinitrobenzene | 30 |
| diaminotrinitrobenzene | 46 |
| triaminotrinitrobenzene | 75 |

Interpretation of these data provides an example of qualitative reasoning on substituent effects, as follows:

1. Nitro groups are strongly attractive and decrease the electron density of the benzene ring, and thus the stability of the molecule.

2. Amino groups are conversely electron donors: their presence to a certain extent compensates that of nitro groups.

3. Simultaneous presence of amino and nitro groups produces a polarization with partial positive charges on amino groups and partial negative charges on nitro groups; a network of intermolecular hydrogen bonds $-NH_2 \ldots O_2N^-$ is set-up, bringing additional stability to the system.

These intuitive considerations are confirmed by examination of the energy levels associated with the different molecules. The technique used is removal of electrons from the atomic inner shells of atoms by X-rays. The atom thus excited by a photon undergoes an Auger transition, i.e. an electron from an outer shell replaces the inner shell electron removed. In this way an Auger spectrum is obtained whose transitions indicate the molecular energy levels.

Reference: W. Worthy, *Chem. Eng. News*, August 10 1987, p. 25.

## 3  Destabilization of the Initial State: Phase Transfer Catalysis

Substitution reactions $A^{(-)} + BX \rightarrow AB + X^{(-)}$ do not proceed well with water as solvent. This is due to the anion $A^{(-)}$ being stabilized by coulombic dipolar interactions with solvent molecules. Solvation makes $A^{(-)}$ less reactive.

The remedy is to drag the nucleophile $A^{(-)}$ out of the aqueous phase into the organic phase, where the neutral substrate molecule BX prefers to stay. This can be done with an auxiliary agent: most frequently a quaternary ammonium cation $Q^{(+)}$ with one or several long hydrocarbon chains serving as bait. The hydrocarbon chain in the ion pair $A^{(-)}Q^{(+)}$ coils around the organic anion $A^{(-)}$, hydrophobic interactions mask the anion from solvent molecules. The ion pair $A^{(-)}Q^{(+)}$ can thus migrate continually into the organic phase. At the interface, where activation of the desolvated $A^{(-)}$ proceeds, substitution can take place, and the catalytic cycle closes up as shown.

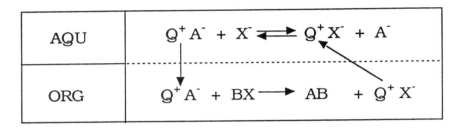

Let us give an application. Substitution of bromide with the rhodanide group occurs in 100% yield.

## 4  Bioluminescence or Communication in the World of Silence

Light intensity in the ocean has a maximum at 475 nm. Two-thirds of the marine organisms living at depths greater than 2000 m are bioluminescent and, coincidentally, emit also blue-green light. The biological role of bioluminescence is manyfold: communication; regrouping; anti-predator strategy; camouflage; differentiation of sexes; search and attraction of prey; love parade; . . .

Let us analyze this reaction in detail:

(1) luciferin (substrate) + luciferase (enzyme) → product*
(2) product * → product + $h\nu$

It boasts spectacular high efficiency: for fireflies, the overall yield is 80%! Now the overall quantum yield is the product of three terms: chemical yield of the first step, quantum yield for fluorescence of the electronically excited product (*), and proportion of the molecules promoted to the excited state.

In the first step, luciferin fixes a molecule of dioxygen. The negatively charged terminal oxygen undergoes a nucleophilic attack at the neighboring –COOR group (R=AMP): thus an addition–fragmentation reaction with $RO^{(-)}$ as leaving group, affording an intermediate with a strained four-membered ring. Strained enough to break open, the O–O bond is a weak bond: 34 kcal mol$^{-1}$ of bond energy. The four-membered ring has considerable angular strain, of the order of 25 kcal mol$^{-1}$ for cyclobutane. There is additional strain due to the carbonyl group: instead of the normal 120°, the angle at trigonal carbon is forced down close to 90°.

References: N. J. Turro, P. Lechtken, N. E. Schore, G. Schuster, H. C. Steinmetze, and A. Yekta, *Accounts Chem. Res.*, **7**, 97–105 (1974).
F. McCapra, *Accounts Chem. Res.*, **9**, 201–208 (1976).

Let us examine a model system: that of tetramethyldioxetane, which fragments in two molecules of acetone with light emission. This stems from its ground state being considerably higher in energy (by 63 kcal mol$^{-1}$) than that of the product molecules. Having gone through the transition state for cycloreversion (activation energy 27 kcal mol$^{-1}$), the system ends up in one of the first excited states, the singlet ($^1A^*$) or triplet ($^3A^*$) state of acetone A. It reverts to the ground state on emitting a quantum of energy: this is the mechanism for bioluminescence.

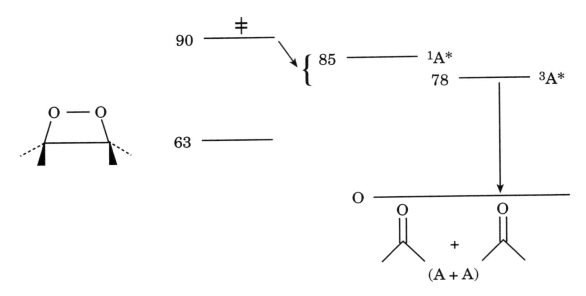

## 5   Chemiluminescence or How to Detect the Presence of Iron

Luminol (in the adjacent Scheme), when submitted to a large number of oxidizing agents (dioxygen, hypochlorite, hydrogen peroxide, permanganate etc.) gives rise to luminescence, as first described by Albrecht.

The dianion in the electronically excited state (*) emits light at 425 nm between pH 10 and 12 (acidity constants of luminol: $pK_1 = 6.2$; $pK_2 = 15.1$). The method is sufficiently sensitive to detect traces of substances with peroxidase-type activity, hemoglobin for instance (use in forensic applications: was it blood on the clothes of the suspect?). Such substances serve as catalysts for the oxidation of luminol with hydrogen peroxide.

Catalysts capable of setting-off chemiluminescence of luminol include the peroxidases, hemin, the ferricyanide $Fe(CN)_6^{3-}$ anion and transition metal cations such as $Co^{2+}$ and $Cr^{3+}$.

References: H. O. Albrecht, *Z. Physik. Chem.*, **136**, 321 (1928).
             T. Olsson *et al.*, *Bioluminescence and Chemiluminescence*, M. A. DeLuca and W. D. McElroy, eds, Academic Press, New York, 1981.

## 6   Study in Yellow

Sunglasses darkening with increased light are familiar. They are based on photochromy: absorption of a photon of light by certain molecules isomerizes them to a tautomeric form absorbing at a different wavelength and with a greater extinction coefficient.

The *anils* are such systems (see next page). When **A**, which is both a phenol and a Schiff base, absorbs a photon, it isomerizes to **B**, a ketoenamine. The **A** → **B** step involves proton transfer from oxygen to nitrogen, coupled with migration of $\pi$ electrons. This electronic redistribution involves a facile, symmetry-allowed cyclic transition state with six centers and six electrons. It is reversible: in the dark, the unstable **B** tautomer spontaneously reverts thermally to the starting molecule **A**. The latter has straw yellow color whereas **B** absorbs in the red part of the spectrum.

The two tautomers with cyclic conformations are held by the hydrogen bonds indicated: O–H . . . N for **A** and N–H . . . O for **B**. A quick thermochemical calculation is instructive: neglecting the small energy differences between the two intramolecular hydrogen bonds and using standard bond energies, **A** is more stable than **B** by about 24 kcal mol$^{-1}$, mainly due to aromatization of the benzene ring in **A**. Photochromy occurs in a variety of molecular systems, and it enjoys many applications: in particular, to digital number display.

**Bond energies**
**(kcal mol$^{-1}$)**                        **A**                              **B**

O–H 110                          N–H   93
C–O   86                          C=O 179
C=N 147                          C–N   83

aromaticity                  $\underline{\qquad 36}$                 $\underline{\qquad -}$
                                   379                              355

Reference: J. G. Calvert and J. N. Pitts, Jr., *Photochemistry*, Wiley, New York, 1966, p. 460.

## 7   Stabilization of the Final State: Silylated Enol Ethers

We have mentioned already in Chapter 4 these compounds as traps for regioselectively-formed enolates. A first example will illustrate here alkylations. The problem is to introduce an R group in either the $\alpha$ or in the $\alpha'$ position to the carbonyl group of a ketone. This is successfully carried out by formation of the enolate either under kinetic or thermodynamic control followed by capture as a silylated enol ether, which is then allowed to react with an alkyl halide RX.

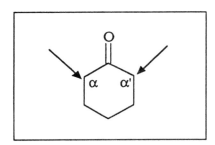

A second application of these silylated enol ethers (cleaved after use with fluoride anions, in an $S_N2$ substitution at silicon) is to the aldol condensation.

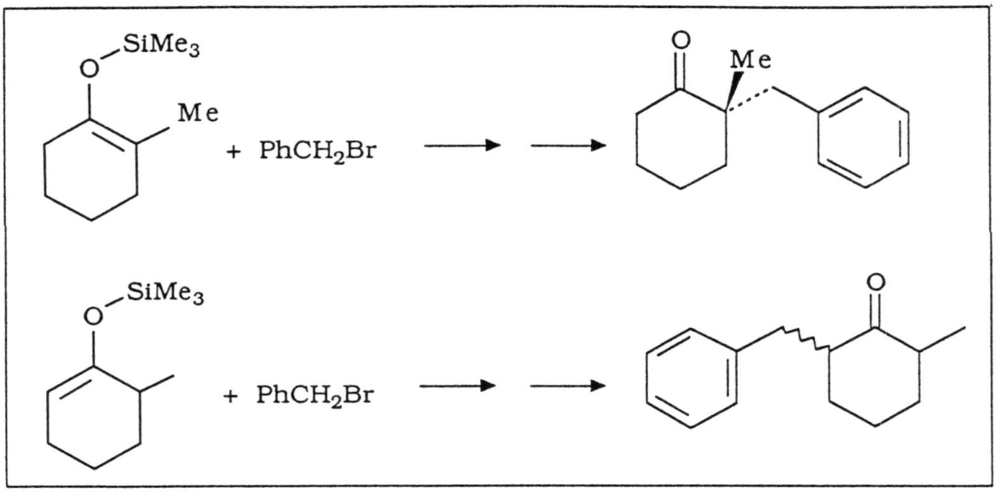

These silylated enol ethers are formed from trimethylsilyl chloride $(H_3C)_3SiCl$ or better still from $(H_3C)_3SiCH_2COOC_2H_5$ in the presence of a catalytic amount of tetra-*n*-butylammonium fluoride (0.5–3 mol%). Still another advantage of these derivatives is their improved volatility conferred by the globular trimethylsilyl group substituent, of low polarity: isolation and purification are simplified, and also all gas phase studies, such as gas chromatography and mass spectrometry.

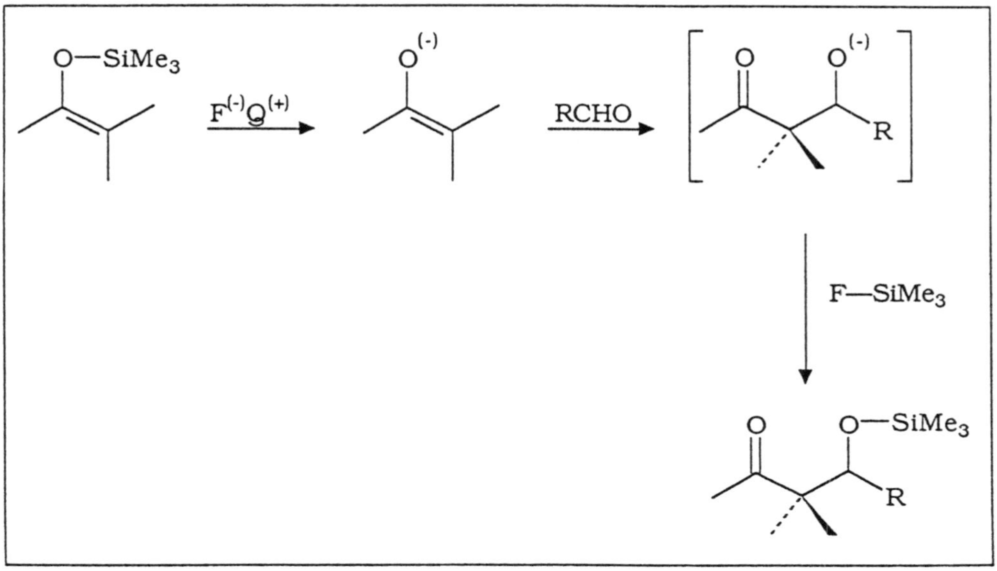

Another manifestation of the same concept: is to drive a reaction to the final state, taking advantage of the silicon–oxygen high bond energy. In an aromatic ether (such as anisole, shown in the next Scheme) the carbon substituent is replaced by a silicon group.

The reagent used is trimethylsilyl iodide. An oxonium salt is formed as the intermediate. It is characterized by a strong (*) silicon–oxygen bond. It will thus tend to fragment more toward the final state, composed of the target molecule plus methyl iodide. The reaction is quantitative (99–100% yield).

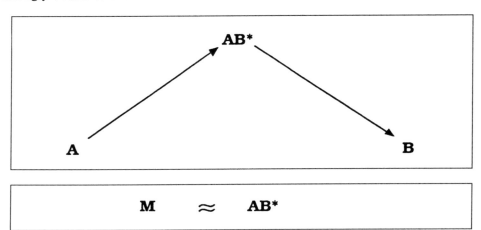

References: I. Kuwajima and E. Nakamura, *Accounts Chem. Res.*, **18**, 181–187 (1985).
          M. E. Jung and M. A. Lyster, *J. Am. Chem. Soc.*, **99**, 968 (1977).

## 8  Biological Catalysis of Organic Reactions

This recent breakthrough, an important tool for organic synthesis, is worth mentioning also for the insight into the intellectual process behind a discovery. The intuitive reasoning is here of great simplicity.

Let us first consider how an enzyme molecule (a protein) catalyzes a chemical reaction $A \rightarrow B$. Denoting as $AB^*$ the transition state for the $A \rightarrow B$ reaction, Linus Pauling's hypothesis is that binding to the enzyme at its active site stabilizes the $AB^*$ transition state. A first logical inference from this hypothesis is that any substance $M$—closely resembling in structure that of the transition state $AB^*$—will also be strongly bound to the active site.

Let us elaborate further on this last assertion, since this is the key step in the argument: if we could synthesize a protein specific for the model molecule **M** and strongly binding **M**, then there would be a high probability that this protein would catalyze the **A** → **B** reaction chosen for consideration. The idea is not total fantasy since immunology in principle provides a solution: an organism submitted to an **M**-type antigen produces antibodies to **M**. These antibodies are immunoglobulins, i.e. proteins. These immunoglobulins are specific. They are synthesized in response to the chemical structure of the antigen **M**, which is strongly bound to a receptor site on the immunoglobulin molecule.

Let us present an example of this new methodology, formation of an internal ester (or lactone) between a primary alcohol and an ester function in the substrate shown. Besides the six-membered ring lactone, a molecule of phenol is produced, since the leaving group is a phenolate anion. Accordingly, the transition state for this reaction involves a partly formed bond between the oxygen of the alcohol and the carbon of the carbonyl group and a partly broken bond between the latter and the oxygen of the phenolate group (addition–fragmentation reaction). The oxygen of the carbonyl group bears a negative partial charge. The oxygen of the alcohol, an electron donor in this nucleophilic attack at the ester carbonyl, bears a positive partial charge. These characteristics of the electronic distribution in the **AB*** transition state can be represented by a model molecule **M**, a cyclic ester named a phosphonate.

Let us now use the phosphonate as an antigen. For this purpose it is attached to a carrier protein such as serum albumin. A standard procedure of monoclonal antibody production induced by the phosphonate-protein complex provides antibodies (i.e. immunoglobulins); these antibodies have the required catalytic activity. One of these, after purification, catalyzes the $\mathbf{A} \to \mathbf{B}$ conversion, with an acceleration factor of 167 with respect to the uncatalyzed reaction.

Furthermore, with a chiral substrate $\mathbf{A'}$, the monoclonal antibody produced in response to stimulation by the phosphonate (a protein with a chiral receptor site) acts only on 50% of the substrate. A single enantiomer of $\mathbf{A'}$ is consumed whereas the other is converted to $\mathbf{B'}$: the measured enantiomeric excess is $94 \pm 8\%$.

Many chemical reactions, such as ester hydrolyses, could be catalyzed in this manner. This thus opens the perspective of biological tailor-made catalysts (*abzymes*) for all organic reactions for which we have an approximate idea of the transition state.

'It is an experience like no other experience I can describe, the best thing that can happen to a scientist, realizing that something that's happened in his or her mind exactly corresponds to something that happens in nature. It's startling every time it occurs. One is surprised that a construct of one's own mind can actually be realized in the honest-to-goodness world out there. A great shock, and a great, great joy'. Leo Kadanoff, quoted by James Gleick, *Chaos*, Viking, New York, 1987, p. 189.

References: L. Pauling, *Am. Sci.*, 51–58 (1948).
          J. Jacobs, P. G. Schultz, R. Sugasawara, and M. Powell, *J. Am. Chem. Soc.*, **109**, 2174–2176 (1987).
          A. Tramontano, K. D. Janda, and R. A. Lerner, *Proc. Nat. Acad. Sci. USA*, **83**, 6736–6740 (1986).
          A. Tramontano, K. D. Janda, and R. A. Lerner, *Science*, **234**, 1566–1573 (1986).
          A. D. Napper, S. J. Benkovic, A. Tramontano, and R. A. Lerner, *Science*, **237**. 1041–1043 (1987).
          K. M. Shokat, C. J. Leumann, R. Sugasawara, and P. G. Schultz, *Nature*, **338**, 269–271 (1989).
          S. J. Pollack and P. G. Schultz, *J. Am. Chem. Soc.*, **111**, 1929–1931 (1989).
          R. A. Lerner and S. J. Benkovic, *Chemtracts Org. Chem.*, **3**, 1–36 (1990).

# 9    Activation of Saturated Hydrocarbons

A great dream of petrochemists is to produce ethylene glycol, ethanol, acetone etc. directly from natural gas, such as that from the North Sea. At the time of writing, half-a-dozen organometallic systems capable of this direct functionalization (catalytic or stoichiometric) are known.

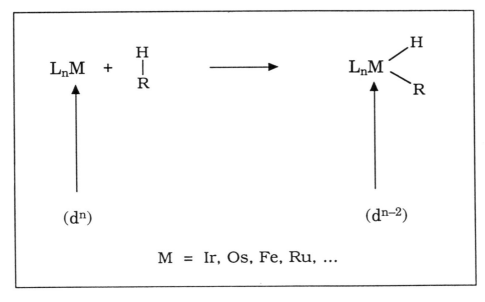

In most of the examples known, a hydrocarbon adds by way of oxidative addition to a coordinatively-unsaturated metal center **M**. The metal M is typically a transition metal.

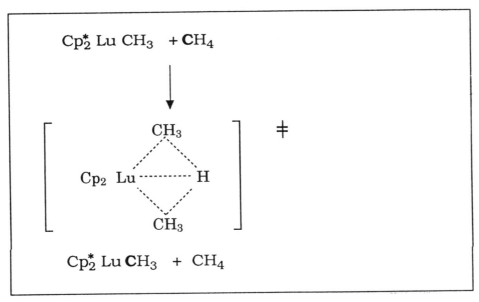

An interesting case is that studied by Watson and Parshall (Du Pont de Nemours). This lutecium complex with d⁰ electron configuration binds methane. If the methyl group and methane are distinguished by isotopic labeling of the carbons, it is found that they exchange hydrogen atoms. This exchange is proof for activation of the carbon–hydrogen bond in methane. Cp* denotes a cyclopentadienyl $C_5H_5^{(-)}$, the hydrogens of which have been replaced by methyl groups: such a ligand shields lutecium with its coordinate unsaturation.

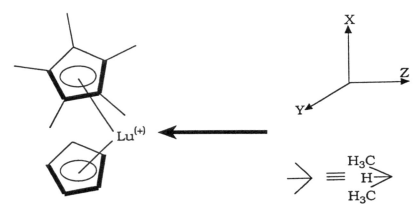

Using molecular orbital language, the stability of the system $Cp_2^*LuH(CH_3)_2$ is readily explained, i.e. stabilization of the transition state for the exchange reaction.

The pertinent orbitals of the two structural fragments are shown in the following Scheme. For lutecium $Cp_2^*$, these are levels of symmetry denoted $1a_1$, $b_2$ and $2a_1$: the latter are mainly formed from the metal atomic orbitals (s, $p_z$, $d_z2$), ($d_{yz}$ and $p_y$) and ($d_{y^2-z^2}$) respectively.

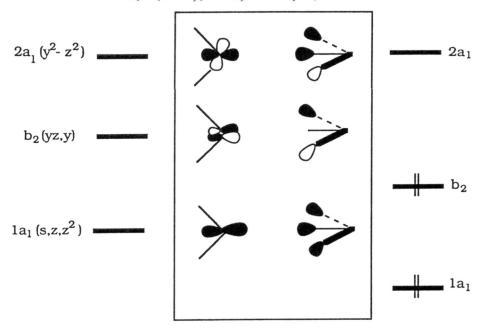

The organic part of the complex is the fragment $(H_3C,H,CH_3)^{(-)}$. Let us assume that the two carbon atoms and the hydrogen atom attached to the metal define the $(y,z)$ plane. This fragment has three molecular orbitals with the same symmetries $1a_1$, $b_2$ and $2a_1$. They resemble the three molecular orbitals (bonding, non-bonding and antibonding) of an allyl system. The two levels $1a_1$, $b_2$ are doubly occupied, the level $2a_1$ is empty. Thus the complex is formed by way of a two-electron interaction between levels of the same symmetry ($1a_1 \times 1a_1$, $b_2 \times b_2$).

References: P. J. Watson and G. W. Parshall, *Accounts Chem. Res.*, **18**, 51 (1985).
          H. Rabaâ, J. Y. Saillard, and R. Hoffmann, *J. Am. Chem. Soc.*, **108**, 4327–4333 (1986).

# 7.I
# Building of Rings

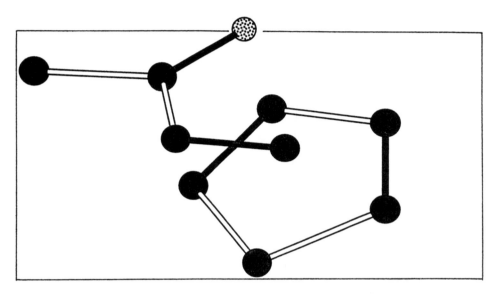

Molecules with rings allow for the very precise positioning of two groups relative to each other; this is an important determinant of biological activity. Indeed, natural products in families as numerous and as important as steroids, terpenes and alkaloids have polycyclic skeletons. For sure, there are dozens of cyclization reactions of various types. However those most frequently used can be counted on the fingers of one hand. Here we will discuss the Diels–Alder reaction, an elegant method for constructing six-membered rings with a thermal pathway. Another commonly used concerted cycloaddition is the photochemical [2 + 2] cycloaddition affording cyclobutanes. We shall also consider a third type of cycloaddition, the addition of carbenes to ethylenes leading to cyclopropanes.

# 1 Diels–Alder Reaction

**Table 14**  Diels–Alder reaction

| Diene | Dienophile | Electronic demand |
|-------|-----------|-------------------|
| rich*<br>(HOMO) | poor*<br>(LUMO) | normal |
| poor*<br>(LUMO) | rich*<br>(HOMO) | inverse |

*In electrons.

## 2  Regioselectivity

Diels–Alder reaction with normal electronic demand:

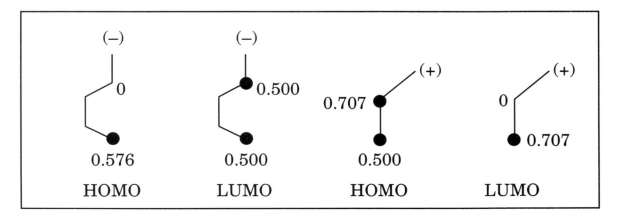

Let us consider the borderline case of the pentadienyl anion (an electron-rich diene) reacting with the allyl cation (an electron-poor dienophile) and let us use the coefficients of the Hückel molecular orbitals for both reactants assumed isolated.

The predominant interaction is $HOMO_{diene}$–$LUMO_{dienophile}$. This allows prediction of the formation of the C—C bond between the centers with the largest coefficients, i.e. an 'ortho' regioselectivity:

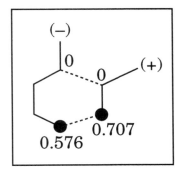

This is observed in the Diels–Alder reactions with normal electronic demand.

For example the bicyclic products of the cycloaddition:

have donor (—OMe) and acceptor (—CN, and to a lesser extent —Cl) substituents on adjacent carbon atoms.

For example, the reaction

where A = acceptor group such as —COOMe gives the following intermediate, which is not isolable.

# 3  Diels–Alder Reaction with Inverse Electronic Demand

Let us consider once again a limiting case, which will allow us to make a generalization: the case of the pentadienyl cation (an electron-poor diene) reacting with the allyl anion (an electron-rich dienophile). The coefficients of the (Hückel) molecular orbitals are:

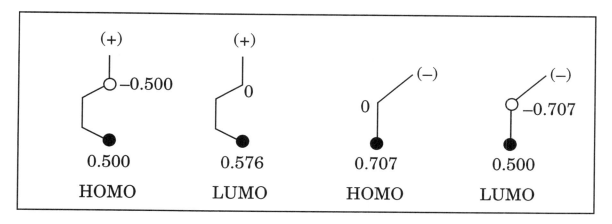

The predominant interaction, this time, is $LUMO_{diene}$—$HOMO_{dienophile}$. This predicts an 'ortho' regioselectivity, and corresponds exactly to what is observed.

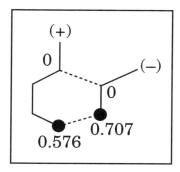

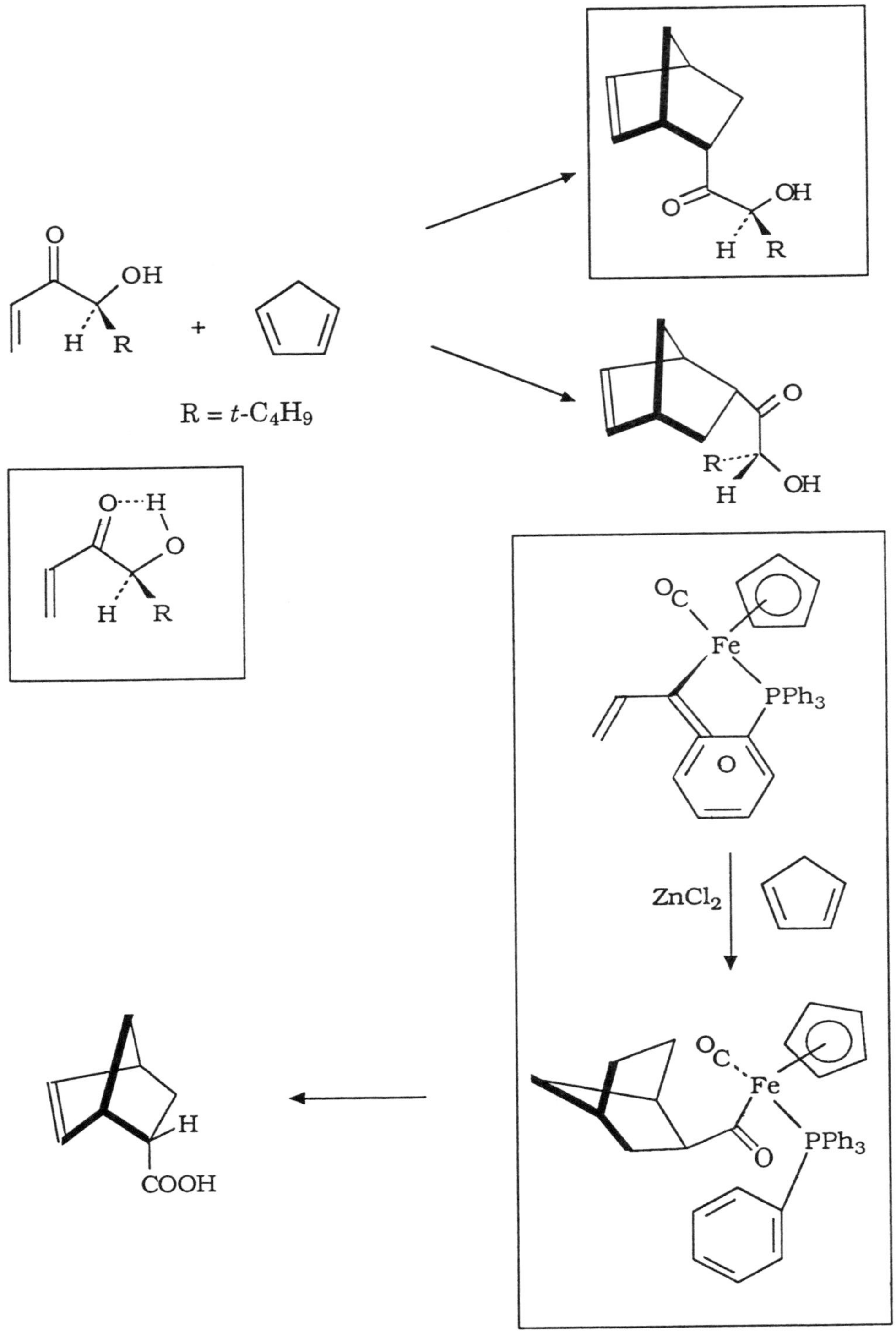

OH
H R
OH
H R
+
R = t-C4H9
O
R
H
OH
O···H
O
H R
OC
Fe
PPh3
O
ZnCl2
OC
Fe
O
PPh3
H
COOH

# 4   Stereoselectivity in the Diels–Alder Reaction

Beside its aptitude at building cyclohexenes—which opens the way to several types of six-membered rings—the Diels–Alder reaction can create *up to four chiral centers* in a single step. This is a remarkable ability, yet to be fully exploited.

The first example (see adjacent Scheme) is that of a chiral dienophile reacting with cyclopentadiene. Of the two endo cycloadducts (*exo* adducts are not present, in this case) formed, the enantiomer in the box predominates over its congener in a ratio of 100:1. This stereoselectivity can be accounted for by the locked conformation of the enone which is strongly hydrogen-bonded. Therefore, the two diastereotopic faces of the dienophile are clearly differentiated: the upper face is masked by the *t*-butyl group leaving the lower face more readily accessible. This accounts for the high enantioselectivity observed.

The second example, similar from a conceptual point of view, has an additional asset, use of a removable chiral auxiliary group; after the cycloaddition, oxidation by Ce(IV) cleaves off the organometallic moiety. The free acid with a 2-S configuration is formed in an enantiomeric excess greater than 95%.

References: W. Choy, L. A. Reed III, and S. Masamune, *J. Org. Chem.*, **48**, 1137–1139 (1983).
          S. G. Davies, and J. C. Walker, *J. Chem. Soc. Chem. Comm.*, **1986**, 609–610.

'Chemists still form a distinct part of the population, not very numerous. They have their own language, their laws, their mysteries. They are living almost in isolation amidst a lot of people who are not very curious about and do not show any great respect for their profession.

The physicist will see masses, forces, qualities; the chemist will see small bodies, proportions, and essential principles. The former will calculate rigorously whereas the theories of the latter will be vague and approximate; yet they will present clear expositions of the nature and of the chemical properties of a given substance, or of a certain principle, considered for all the combinations that can be subjected to by nature or by art ( . . . ).

The chemist does not stop till he has broken down and identified the constituents of a mixture.

The color considered within a colored body is for the physicist a certain state of the surface of the body which enables it to send back this or that light ray. For the chemist, on the other hand, the green color of a plant is inherent to a certain green resinous body that he knows how to extract from this plant; the blue color of a clay is due to a metallic material that he also knows how to separate; the color of jasper, which seems to be so perfectly attached to this fossil substance, can also be extracted from it, according to the famous experiment by Becher.

An observation that is appropriate here is that, in their two viewpoints on the phenomenon of color, the physicist and the chemist say different things, but these do not contradict one another.'

Reference: Venel, entry on 'Chemistry' in the *Encyclopédie* by Diderot and D'Alembert, 1751–1772.

# 5  Ring Strains

**Table 15**  Ring strain

| $n$ | | Strain (kcal mol$^{-1}$) |
|---|---|---|
| 3 | cyclopropane | 27.5 |
| 4 | cyclobutane | 26.5 |
| 5 | cyclopentane | 6.2 |
| 6 | cyclohexane | 0.0 |
| 7 | cycloheptane | 6.3 |
| 8 | cyclooctane | 9.7 |
| 3 | cyclopropene | 55.2 |
| 4 | cyclobutene | 28.4 |
| 5 | cyclopentene | 4.1 |
| 6 | cyclohexene* | −0.3 |
| 7 | cycloheptene* | 3.6 |
| 8 | cyclooctene | 4.2 |

*_cis_ double bond.

One can draw a general conclusion from the table of ring strains: medium rings ($C_5$—$C_8$) are built more readily than small rings ($C_3$ and $C_4$). The latter suffer from considerable angular tensions. The destabilization of cyclopropene can be easily understood: a trigonal carbon with a normal valence angle of about 120° is forced to take up a valence angle of the order of 60°. This deformation increases the ring strain by 13 kcal mol$^{-1}$ if one trigonal carbon is introduced into cyclopropane and by 28 kcal mol$^{-1}$ when two are introduced.

$$2\,H_2C{=}CH_2 \longrightarrow \begin{array}{c} H_2C{-}CH_2 \\ \mid \qquad \mid \\ H_2C{-}CH_2 \end{array}$$

$\Delta H^0 = -18.2\,\text{kcal mol}^{-1}\,(25\,°\text{C})$

$\Delta S^0 = -41.5\,\text{cal mol}^{-1}\,\text{K}^{-1}$

$\Delta G^0 = -5.8\,\text{kcal mol}^{-1}$

$K = \quad 1.8 \times 10^4$

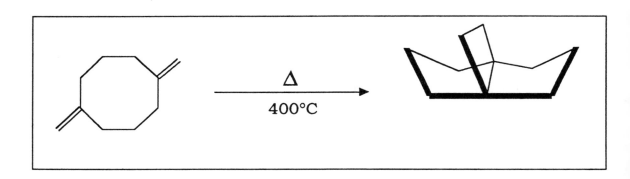

# 6  Formation of Cyclobutanes (1)

## (i) Thermal $(2\pi + 2\pi)$ Dimerization

One would expect naively that formation of cyclobutane by dimerization of two molecules of ethylene would be impaired by the ring strain of the cyclic product. The thermochemical balance is in fact as shown opposite.

The reaction is strongly exothermic at room temperature! This is because the $C-C$ $\sigma$ bond energy is about $84\,\text{kcal mol}^{-1}$, whereas the $\pi$ bond has an energy only of $64\,\text{kcal mol}^{-1}$.

The thermal dimerization of ethylene is not a practical access to cyclobutane because it requires a large activation energy (the process is forbidden by the Woodward–Hoffmann rules) and therefore a high reaction temperature: the reaction entropy is strongly negative and at higher temperature implies a positive $\Delta G^{\ddagger}$, hence an unfavorable equilibrium constant. If the entropic term is less negative, e.g. in an intramolecular dimerization, the thermal dimerization becomes viable.

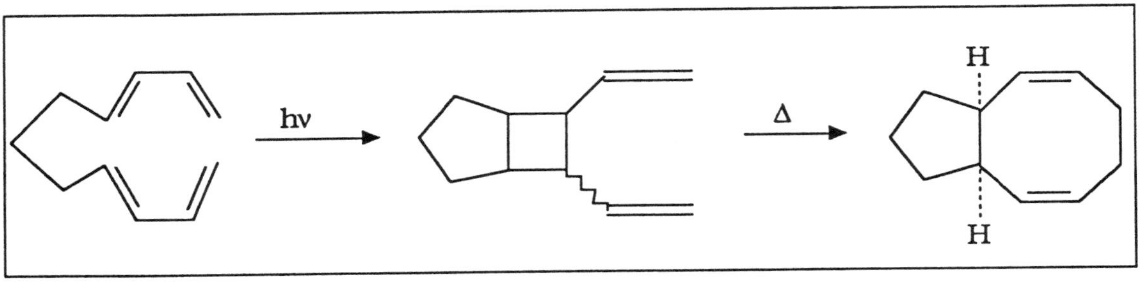
hν
Δ
H
H

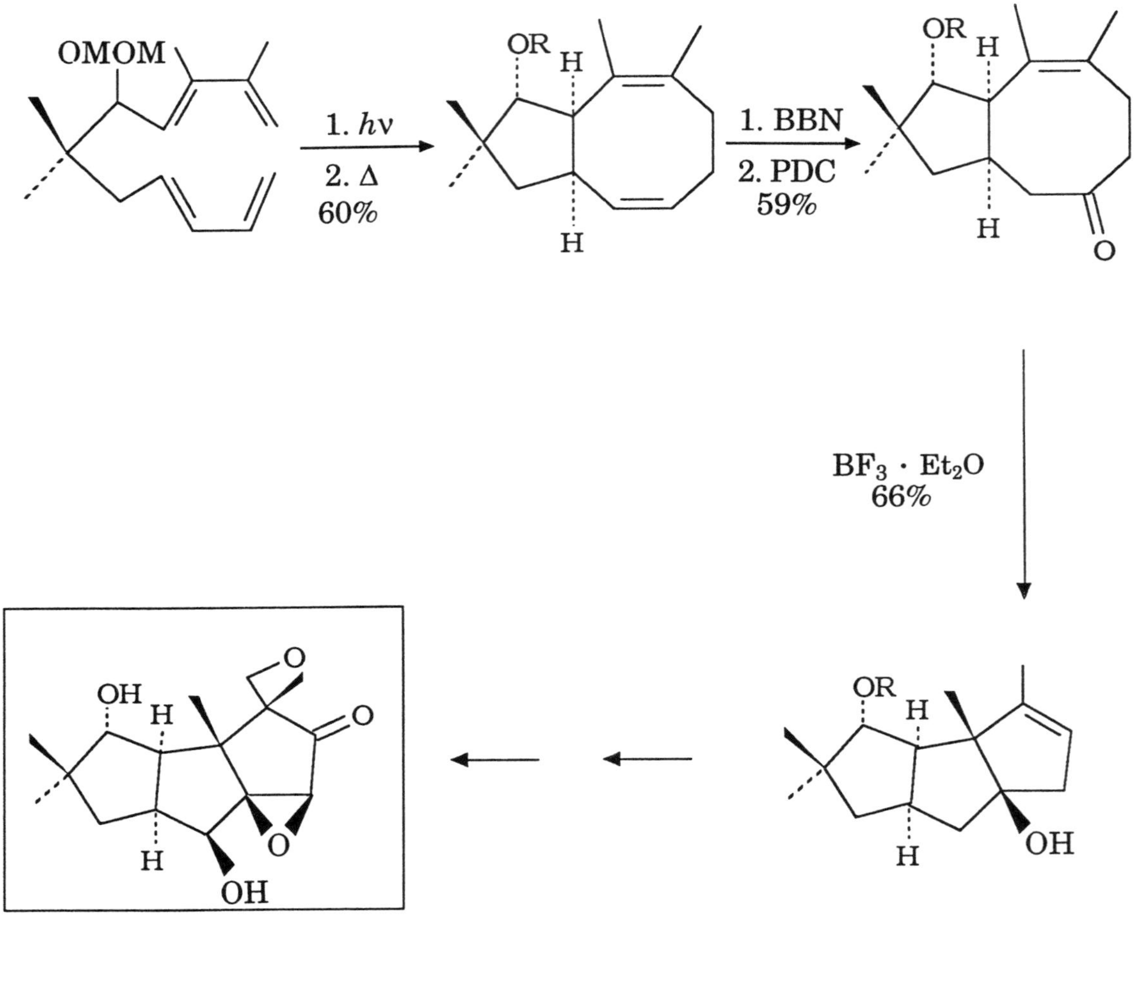
OMOM
1. hν
2. Δ
60%
OR
H
H
1. BBN
2. PDC
59%
OR
H
H
O
BF3 · Et2O
66%
OR
H
H
OH
OH
H
O
O
H
O
OH

# 7  Formation of Cyclobutanes (2)

The classical preparation of cyclobutanes is the photochemical $(2\pi + 2\pi)$ *supra–supra* cycloaddition. Hence a problem of regioselectivity arises: the two ethylenes can add head-to-head, or head-to-tail. To illustrate this way of building cyclobutane, I have chosen the synthesis of the anti-tumor agent coriolin. The key step is the intramolecular [2+2] cycloaddition of a double diene with the assistance of a third substance serving as photosensitizer. The bicyclic adduct then undergoes a sigmatropic [3,3] shift, a Cope rearrangement, which gives rise to a cyclooctadiene.

The synthesis starts from an alcohol protected as the MOM ($-CH_2OCH_3$) ether. The cyclooctadiene, resulting from a cyclization, followed by a sigmatropic shift, is hydroborated with BBN (bora-9-bicyclo[3.3.1]nonane), and the resulting alcohol is oxidized (by the $CrO_3$ (pyridine)$_2$ complex) to the corresponding ketone. The differences in stability and in accessibility of the two double bonds (tetrasubstituted and disubstituted, respectively) in the cyclooctadiene are responsible for the regioselectivity of the hydroboration. The Lewis acid $BF_3$ promotes the transannular cyclization by the addition of the carbocation to the tetrasubstituted double bond. Some further transformations of the functional groups lead to racemic coriolin.

Reference: P. A. Wender and C. R. D. Correia, *J. Am. Chem. Soc.*, **109**, 2523 (1987).

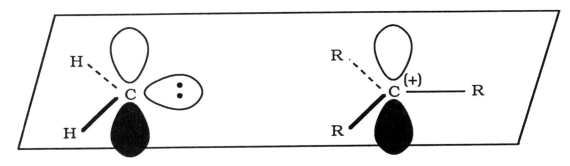

# 8 Building of Cyclopropanes: Addition of Carbenes to Alkenes

Carbenes are very reactive neutral entities with only six valence electrons. Due to this electronic deficiency carbenes are electrophiles. Their general formula is $R^1R^2C:$.

The prototype of a carbene is methylene $H_2C:$. In its singlet state (with all electrons paired), methylene is usually described as centered on a trigonal carbon. Two of the $sp^2$ hybrid orbitals are used for the two C—H bonds. Since the third $sp^2$ hybrid orbital is lower in energy than the $2p_z$ atomic orbital, it is occupied by the non-bonding electron pair, which leaves the $2p_z$ orbital vacant. There is obvious analogy to a carbocation.

One access to $:CH_2$ is through decomposition of diazomethane $H_2CN_2$, with extrusion of a neutral molecule of dinitrogen.

Another mode of production of carbenes is the $\alpha$-elimination of a molecule of hydrohalogenated acid from a precursor such as chloroform in presence of a base: from the carbanion $Cl_3C:^-$, the conjugate base of chloroform, the neutral species dichlorocarbene $:CCl_2$ is formed on elimination of the chloride anion. Overall a molecule of hydrochloric acid is abstracted from chloroform.

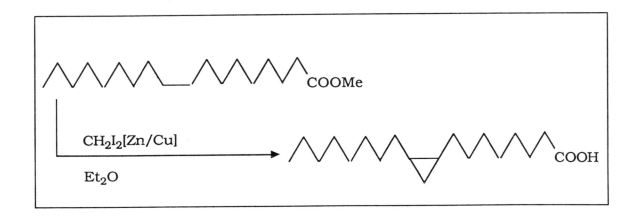

Addition of carbenes to ethylenes gives rise to cyclopropanes. The addition is stereospecific. From *cis*-2-pentene a *cis* cyclopropane is formed exclusively.

The indicated enamine with an electron rich ethylenic bond upon reaction with diazomethane in ether solution gives the corresponding cyclopropane. The yield is 70%.

If a carbene—an electron deficient species—has the choice between addition to a normal C=C bond or to an electron-rich C=C bond, the latter obviously is preferable. In this reaction a methylene dihalide $H_2CXY$ on treatment with an Zn/Ag metal couple affords carbene; the yield of cyclopropane is 75%.

It is not uncommon to use methylene iodide in the presence of a zinc/copper couple (the so-called Simmons–Smith conditions). The yield of this reaction amounts to 90%.

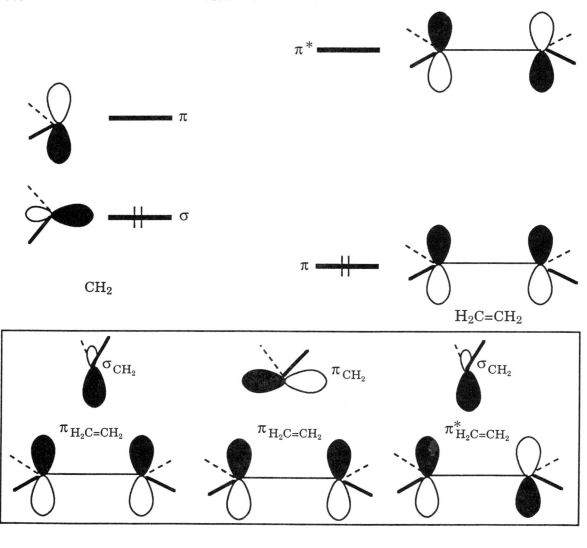

$CH_2$

$H_2C=CH_2$

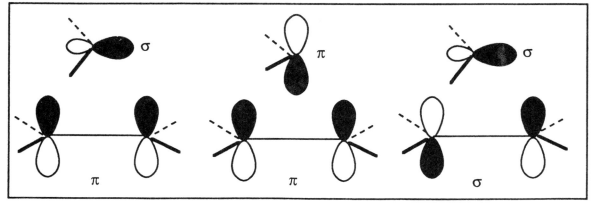

What can we say about the geometry of the transition state of a carbene addition to a C=C double bond? If we compare the frontier orbitals of methylene and ethylene, the two predominant two-electron stabilizing interactions are on the one hand that between a $\sigma$ orbital of the carbene (occupied by two electrons)—and the vacant $\pi^*$ of the ethylene; and on the other hand that between the vacant $\pi$ orbital of the carbene and the doubly occupied $\pi$ orbital of the ethylene.

Let us consider first a minimum atomic motion attack with a transition state similar in geometry to the cyclopropane product: it has the plane of the carbene molecule bisecting the $\pi$-plane of the ethylene. Both, the interaction $\sigma$ (CH$_2$) + $\pi^*$ (H$_2$C=CH$_2$) as well as the interaction $\pi$ (CH$_2$) + $\pi$ (H$_2$C=CH$_2$) give no overall contribution to the stabilization.

On the other hand, if the methylene CH$_2$ is placed so that its $\sigma$ orbital lies in the $\pi$ plane of the ethylene, there is a bonding interaction: hence a symmetric product is formed via an asymmetric transition state maximizing the stabilizing interactions and minimizing the destabilizing interactions.

Reference: R. Hoffmann, *J. Am. Chem. Soc.*, **90**, 1475–1485 (1968).

# Summary: 'Building of Rings'

The Diels–Alder cycloaddition is a privileged construction of six-membered rings, omnipresent in natural products. The D–A with a normal (inverse) electronic demand enables an electron rich (poor) diene to react with an electron poor (rich) dienophile. Coordination of the dienophile with a Lewis acid reduces the gap between the frontier orbitals and hence catalyses the normal electronic demand D–A.

This section proceeds with the concrete cases in Part II. The synthesis of forskolin, an anti-tumor agent, illustrates the use of an asymmetric acetylene as a dienophile: the regioselectivity is due to the formation of the first C–C bond between the centers with the largest coefficients. The reserpine synthesis (Paul Wender) starts with a regio- and stereoselective D–A reaction and continues with a sigmatropic [3,3] shift to elaborate the bicyclic precursor of the rings D and E of this antihypertensive alkaloid, extracted from *Rauwolfia serpentina*.

The synthesis of quadrone (Paul Wender), an antitumor metabolite from a yeast, illustrates the advantage of the D–A reaction as well as the use of a Lewis acid to lower the LUMO of the dienophile (normal electron demand).

The synthesis of occidentalol (E. J. Corey) illustrates a D–A with inverse electronic demand. Furthermore, it uses a lactone so that the $CO_2$ formed in a spontaneous D–A cycloreversion becomes the driving force for the reaction. The synthesis of giberellic acid (E. J. Corey), a growth factor for plants, incorporates a D–A with formation of three new asymmetric centers.

A metal atom (e.g. iron) guarantees an excellent stereoselectivity through simultaneous coordination of the diene and dienophile; after reaction, this auxiliary is removable by oxidative cleavage with Ce(IV) (formation of *endo* 2-norbornene carboxylic acid, due to secondary orbital interactions). Enantioselective D–A reactions can be performed with a chiral auxiliary, typically with $C_2$ symmetry such as dihydrocorrins. In this way, Corey coordinates iron(III) with this ligand and obtains an enantiomeric excess of 80% and a yield of 85% with this chiral Lewis acid. The synthesis of desoxyloganin (L. Tietze), a glucoside intermediate in the biosynthesis of numerous alkaloids, illustrates another D–A with inverse electronic demand, using an acceptor substituent to reduce the gap between the HOMO of the dienophile and the LUMO of the diene. It is examined in detail in order to present realistically the types of problems and their solutions which occur during the course of a multistep synthesis.

## 1  Regioselectivity of the Diels–Alder Reaction (1)

For the Diels–Alder reaction to proceed at practical rates, an electronic requirement is a prerequisite. It is due to the necessity of minimizing the energy gap between the frontier orbitals of the reaction partners. The Diels–Alder reaction with so-called 'normal' electron demand requires an electron-rich diene and an electron-poor dienophile.

The case shown has 1,3-butadiene obeying the first criterion and the second partner, a *p*-benzoquinone, obeying the second criterion. In the dienophile, the two ethylenic double bonds of the quinone differ by their substitution. The cycloaddition occurs only with the C=C double bond substituted with the weakest electron-donor. Clearly the methyl group (donor by hyperconjugation) is inferior in its electron-donating ability to the methoxy group, able to conjugate its non-bonding electrons. Note the stereospecificity: since this is a cycloaddition, it proceeds in *supra–supra* manner, which places the methyl group and the hydrogen *cis* to one another at the ring junction.

Reference: G. Büchi, W. Hofheinz, and J. V. Paukstelis, *J. Am. Chem. Soc.*, **88**, 4113 (1966).

## 2  Regioselectivity of the Diels–Alder Reaction (2)

Another aspect of the regioselectivity in this cycloaddition is illustrated in the next Scheme: discrimination between head-to-head and head-to-tail addition. The latter occurs. The first new bond forms between the carbon centers having the largest coefficient in the frontier orbitals.

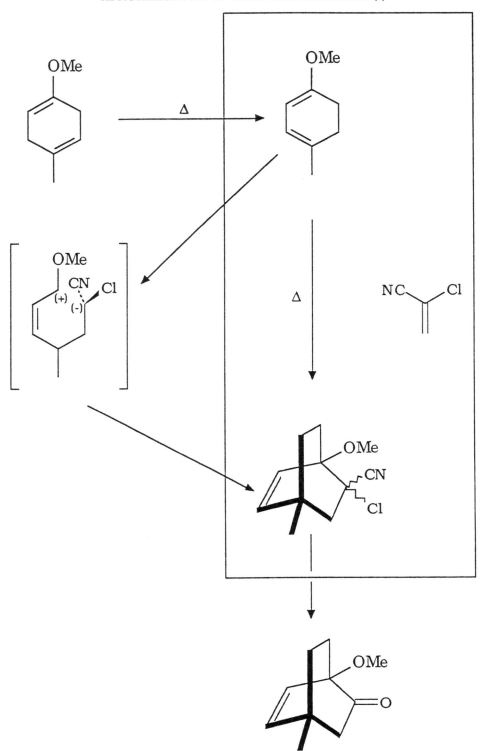

Another manner of explaining (at least in a heuristic sense) the predominance of the 'ortho' cycloadduct (with exclusion of the 'meta' product) is consideration of the zwitterionic component in this cycloaddition between a polarized diene and dienophile. The best stabilized zwitterion leads to the 'ortho' product.

The starting material here is the 1,4-cyclohexadiene resulting from the Birch reduction of the corresponding aromatic compound. The first step is thermal isomerization of the 1,4-diene to the conjugated 1,3-diene required for the cyclization. Alkaline hydrolysis ($Na_2S$ hydrate in ethanol) of the adducts affords the bicyclic ketone (69%). Less than 0.1% of the *meta* addition product is formed indicating a high selectivity in the reaction.

References: D. A. Evans, W. L. Scott, and L. K. Truesdale, *Tetrahedron Lett.*, 121 (1972).

## 3   Dienophiles Owing Their Activation to Ring Strain

The first example is addition of cyclopropene to cyclopentadiene: the reaction takes place under exceptionally mild conditions (at 0°C) with a yield of 97% (exclusive formation of the *endo* stereoisomer).

The second example is a cycloaddition to benzyne. Furans are reluctant to undergo Diels–Alder reactions which make them lose their aromaticity. One can get around this difficulty: the activation volume for a Diels–Alder reaction is negative, thus Le Chatelier's principle predicts that application of pressure will reduce the activation enthalpy. Indeed, cycloaddition proceeds readily under rather harsh reaction conditions, a pressure of 15 kbar. Benzyne is prepared by treatment of 1-bromo-2-fluorobenzene with lithium amalgam for 4 h at room temperature. The yield in tricyclic product is 76%.

Consider another application of benzyne, to formation of tryptycene (shaped like a triptych) from its addition to anthracene.

References: M. L. Deem, *Synthesis*, 675 (1972).
R. H. Levin, M. Jones, Jr., and R. A. Moss, eds, *Reactive Intermediates*, 1-1, Wiley, New York, 1978.
R. W. Hoffmann, *Dehydrobenzenes and Cycloalkynes*, Verlag Chemie, Weinheim, 1967.

## 4    Diels–Alder Reaction with Opening of the Adduct

Formation of an acetophenone (at 100 °C, 2 h, in dioxane) from a furan and methylvinylketone is not evident at first sight. To explain it to ourselves, let us consider the predictable consequence of bringing the two reactants together: a Diels–Alder cycloaddition between the diene (a furan) and the dienophile, activated by its attracting methylketone group, must first proceed. The 'ortho' orientation ought to be favored and corresponds indeed to the observed arrangement of substituents in the aromatic ring of the product. The diene reacts from its HOMO, the dienophile from its LUMO and the first bond is formed between the centers with the largest coefficients, in the frontier orbital approximation.

Reference: W. J. Nixon, Jr., J. T. Garland, and C. De Witt Blanton, Jr., *Synthesis*, 56 (1980).

Subsequent opening of the strained cycloadduct is assisted by the non-bonding electron pair on nitrogen. Simple migration of a proton from nitrogen to oxygen suffices to neutralize the resulting zwitterion. A tautomerization follows, with migration of another proton from carbon to nitrogen, increase of conjugation being the driving force. The sequence ends with elimination of a molecule of water, with aromatization as the driving force.

## 5   Synthetic Utility of Two Diels–Alder Reactions in Succession

In the example figured the reaction proceeds via an inverse Diels–Alder reaction (or cycloreversion), with elimination of a molecule of carbon dioxide, the bicyclic product formed aromatizing with loss of a molecule of neutral methanol (D. L. Boger and M. D. Mullican, *Org. Synth.*, **65**, 98–107 (1967).

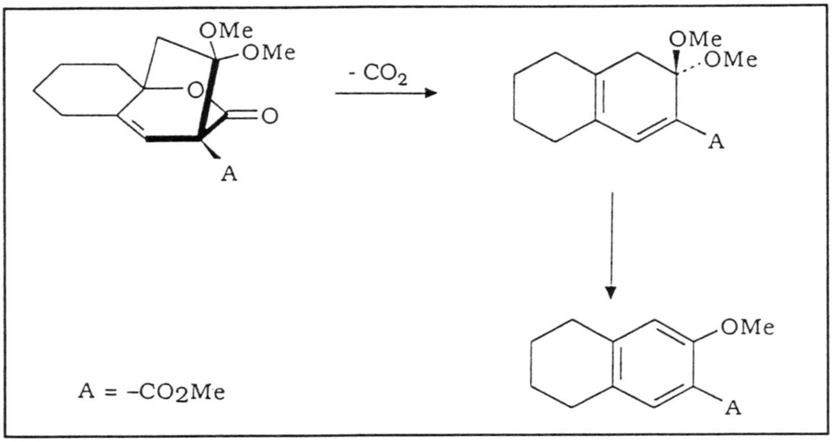

A = –CO2Me

## 6   A Step in the Synthesis of Morphine

A Diels–Alder reaction is part of the first steps of the morphine synthesis by Marshall Gates·

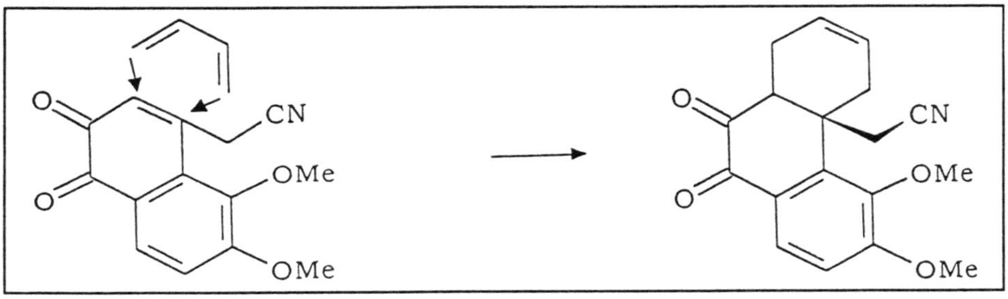

It transforms a bicyclic to a tricyclic compound which already possesses the characteristic structural elements of morphine, particularly the benzene ring with two adjacent oxygen functions:

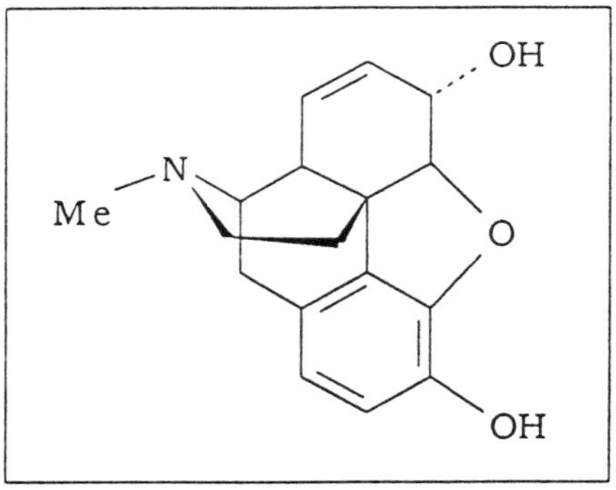

*morphine*

References: M. Gates and W. F. Newhall, *J. Am. Chem. Soc.*, **70**, 2261 (1948); *Experientia*, **5**, 285 (1949).
M. Gates, R. B. Woodward, W. F. Newhall, and R. Kunzle, *J. Am. Chem. Soc.*, **72**, 1141 (1950).

This Diels–Alder reaction allows 1,3-butadiene to react with an electron-poor double bond, conjugated with a carbonyl group.

## 7 Other Uses of the Diels–Alder Reaction in Important Syntheses

The majority of the important classical total syntheses (1940–1960) use the Diels–Alder reaction, if not at the very start of the synthesis as is often the case, or for entry into the synthetic scheme somewhere along the line. In the onocerane (tetracyclic triterpenes) series, Eschenmoser combines two bicyclic fragments into an intermediate closely related to the final product.

bis-nor-onoceranediol

References: E. Romann, A. J. Frey, P. A. Stadler, and A. Eschenmoser, *Helv. Chim. Acta*, **40**, 1900 (1957).
J. S. Scheinber, W. Leimgruber, M. Pessaro, P. Schudel, and A. Eschenmoser, *Angew. Chem.*, **71**, 637 (1959), *Helv. Chim. Acta*, **44**, 540 (1961).

Colchicine is a powerful anti-mitotic which prevents the assembly of the microtubules by strong and specific attachment to tubuline. In like manner as above, a step in Eschenmoser's colchicine synthesis is:

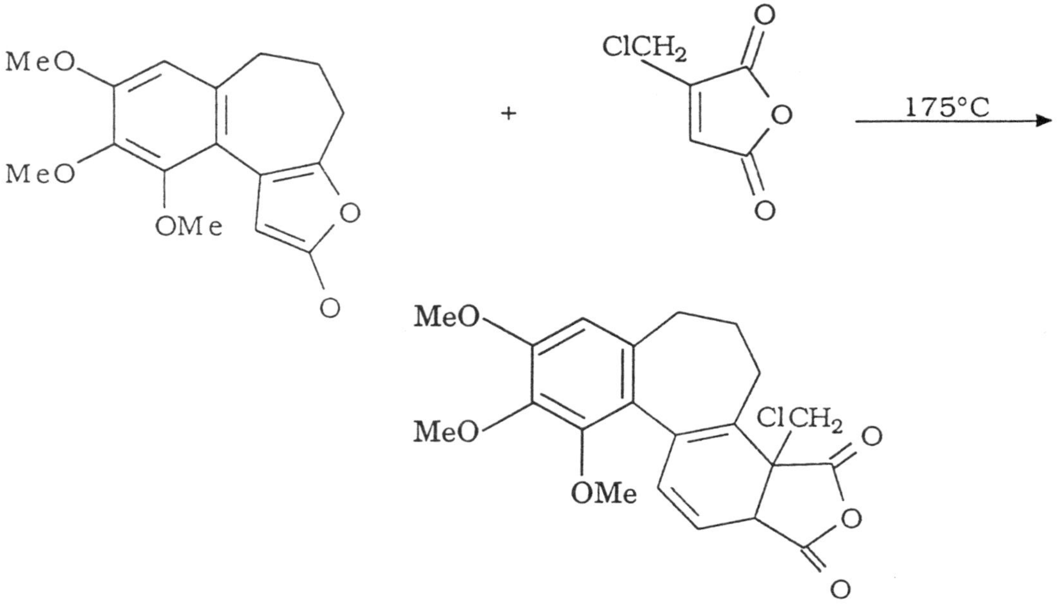

## 8   Stereochemical Control in the Diels–Alder Reaction

As a general rule only the *endo* stereoisomer is formed. The reason for such selectivity is stabilization of the *endo* transition state by secondary orbital interactions:

> 'it is also a good rule not to put too much confidence in experimental results until they have been confirmed by theory', Sir Arthur Eddington, quoted in *More Random Walks in Science*, R. I. Weber, ed., Institute of Physics, Bristol, 1982, p. 111.

The adjacent scheme exemplifies a cycloaddition between maleic anhydride (electron-poor, hence an excellent dienophile) and a diene conjugated with a carbonyl group: the latter is electron-deficient, which increases the frontier orbital gap between the diene and the dienophile, thus making the reaction difficult: temperatures of 200 °C are required, for a low yield (34%).

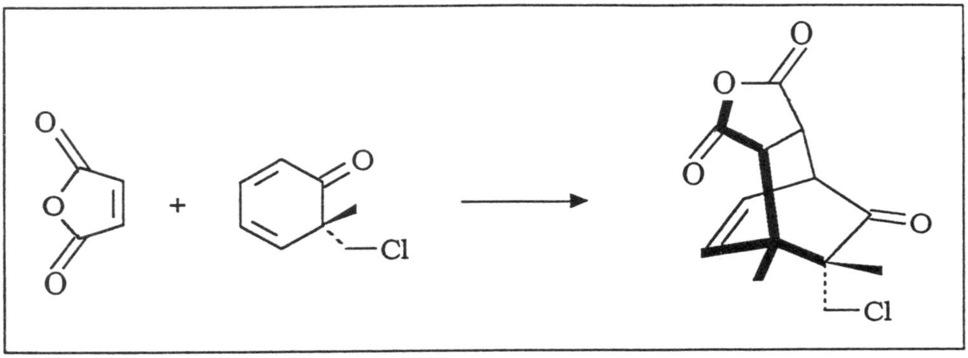

# 9 Retro Diels–Alder Reaction

The selection rules allow both the direct and the reverse reaction. The latter is observed mainly when it can lead to cleavage of a stable, neutral molecule. Addition of methyl acrylate to α-pyrone gives an adduct which undergoes a retro Diels–Alder reaction, with cleavage of a molecule of $CO_2$. The new diene formed adds a second molecule of methyl acrylate (with *endo* preference). This is one of the preparations of barrelene, thus named for its barrel shape.

**barrelene**

Reference: H. E. Zimmerman, G. L. Grunewald, R. M. Paufler, and M. A. Sherwin, *J. Am. Chem. Soc.*, **91**, 2330 (1969).

## 10　Benzocyclopropene

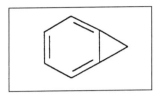

This hydrocarbon has considerable ring strain:

It was prepared in the following subtle manner:

A = - CO₂Me

Reference: E. Vogel, W. Grimme, and S. Korte, *Tetrahedron Lett.*, 3625 (1963).

## 11   The Toxic Bird

homobatrachotoxin

One might mistake this for the title of a surrealist poem! In the battle for survival, anything goes. A sparrow of New Guinea, the crested pitohui (*Pitohui dichrous*) contains the extremely toxic steroidal alkaloid homobatrachotoxin in its skin and feathers. When as little as 0.03 μg of this substance is injected into laboratory animals (mice), it causes convulsions and death within 15 mn. Each pitohui contains 15 to 20 μg of poison in its feathers and about half a μg in its skin.

The New Guineans have been aware of this toxicity for a long time. They named the pitohui 'litter bird' and knew that its skin was bitter and astringent. They distrusted it very much and always skinned it carefully before eating it.

Two other species of the same genus, *P. kirhocephalus* and *P. ferrugineus*, are poisonous. Their toxin is related to batrachotoxin, found in the skin of small Amazonian frogs, and used by the Indians to coat their arrow heads.

References: J. P. Dumbacher, B. M. Beehler, T. F. Spande, H. M. Garaffo, and J. W. Daly, *Science*, **258**, 799–801 (1992).
N. Angier, *International Herald Tribune*, 1992, 31 October–1 November.
I. S. Majnep and R. Bulmer, *Birds of My Kalam Country*, Auckland University Press, Auckland, NZ, 1977.

## 12   Synthesis of a Coffee Poison

Cyclopropanation, from addition of a carbene to an olefinic C=C double bond, is illustrated in one of the steps of the total synthesis of racemic atractyligenine. The diazoketone indicated is decomposed for 10 mn in refluxing toluene, in the presence of a copper(II) salt. The carbene resulting from cleavage of a dinitrogen molecule is captured by the double bond: note the good regio- and stereoselectivity due to improved accessibility. The yield is 67%.

References: R. Santi and S. Luciani, eds, *Atractyloside: Chemistry, Biochemistry, and Toxicology*, Piccin Medical Books, Padua, 1978.
M. Klingenberg. *TIBS*, **4**, 249–252 (1979).
A. K. Singh, R. K. Bakshi, and E. J. Corey, *J. Am. Chem. Soc.*, **109**, 6187–6189 (1987).

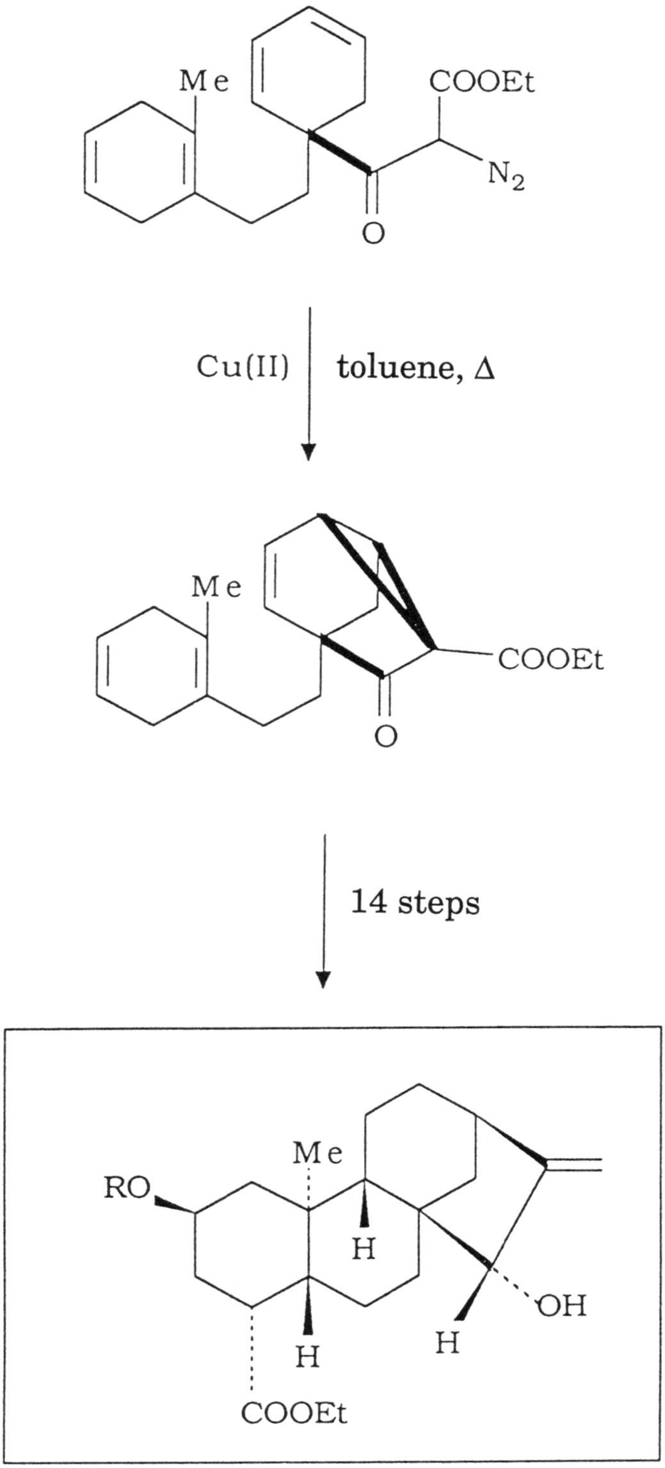
Me
COOEt
$N_2$
O
Cu(II)
toluene, Δ
Me
COOEt
O
14 steps
Me
RO
H
H
H
OH
COOEt

The target molecule atractyligenine is a diterpene, two extremely toxic derivatives of which have been isolated from the thistle *Atractylis gummifera*. These substances block ADP transport to the mitochondria which inhibits in turn ATP production. These atractyligenine glucosides (or atractylosides, R = sugar residue), are also found in the coffee plants *Coffea arabica* and *Coffea robusta*. They are consumed by coffee drinkers in considerable quantities!

## 13   Biomimetic Synthesis of Proto Daphnyphylline

In this superb synthesis which we owe to Clayton H. Heathcock (Berkeley) (only 10 linear steps with an overall yield of 42%!) a multiple cyclization is the key step. Initiated by formation of a carbocation, it forms successively five bonds and four rings! Biosynthesis of the alkaloids from *Daphniphyllum macropoda* (a tree called 'Yuzuriha' in Japan) is likely to follow a similar pathway.

Reference: S. Piettre and C. H. Heathcock, *Science*, **248**, 1532–1534 (1990).

$R = (CH_2)_3OCH_2Ph$

The conditions used (i) are addition of ammonia to a solution of the starting dialdehyde in methylene chloride, in the presence of ammonium acetate and triethylammonium chloride. After 16 h at room temperature and evaporation of the solvent, the residue, which is likely to contain 2, is dissolved in acetic acid. The plausible mechanism involves an intramolecular Michael reaction 1 → 4, followed by formation of the azadiene 2 as a result of treatment with ammonia. The following step is an acid catalyzed intramolecular Diels–Alder cycloaddition.

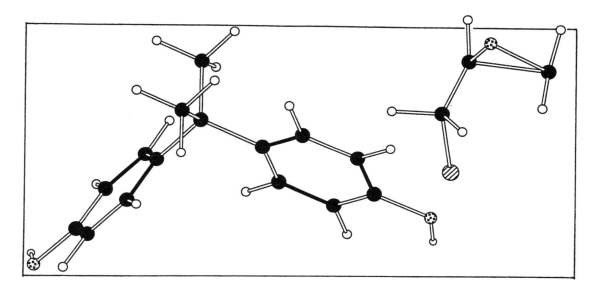

Since the stereochemical consequence of a $S_N1$ type substitution is racemization, caused by the intermediacy of a carbocation, such reactions are hardly useful in organic synthesis. One resorts to either bimolecular substitutions $S_N2$ which, as a general rule, occur with inversion of configuration, or to reactions where the substrate is a carbanion and serves as a nucleophile, attacking a reagent that bears an appropriate leaving group.

— OTs

— OMs

# 1   S$_N$2 Substitution

Let us describe the reaction in the usual way denoting the nucleophile :Nu and the leaving group Y.

The leaving group ability varies when Y = halogen in the sequence: I > Br > Cl > F in an aprotic solvent (the sequence is the inverse in a protic medium, hardly ever used for S$_N$2 reactions). The longer the bond C—Y, the weaker, and the more readily it can be cleaved. Br and I are the most frequently used leaving groups. Groups such as OH, OR and NH$_2$ are very poor leaving groups due to the strong carbon–oxygen or carbon–nitrogen bond. In order to improve the leaving ability of the group, one can stabilize the negative charge created on Y$^{(-)}$. Esters formed from alcohols and sulfonic acids are excellent leaving groups since Y$^{(-)}$ is then the conjugate base of a strong acid. The examples encountered most often are the tosylate (—OTs) and the mesylate (—OMs) groups.

The carbon center that constitutes the substitution site undergoes an inversion of its configuration, the so-called Walden inversion. The pentacoordinate transition state has the geometry of a trigonal bipyramid with an antarafacial disposition of the nucleophile :Nu and the leaving group :Y.

| RY | + | : Nu | = | R–Nu product |
|----|---|------|---|--------------|

| : Nu | R–Nu product |
|------|--------------|
| HO$^{(-)}$ | R — OH |
| R'O$^{(-)}$ | R — OR' |
| R'S$^{(-)}$ | R — SR' |
| R'CO$_2$$^{(-)}$ | R — O–C(=O)–R' |
| R'C ≡ C : $^{(-)}$ | R — C ≡ C — R' |
| : N ≡ C : $^{(-)}$ | R — C ≡ N : |
| H$_3$N : | R — NH$_2$ |
| R'$_3$N : | R'$_3$RN$^{(+)}$ Y$^{(-)}$ |

|  | CH$_3$—Br | CH$_3$—CH$_2$—Br | (CH$_3$)$_2$CH—Br | (CH$_3$)$_3$C—Br |
|--|-----------|------------------|-------------------|-------------------|
| k$_{rel}$ | 145 | 1 | 8×10$^{-3}$ | 5×10$^{-4}$ |

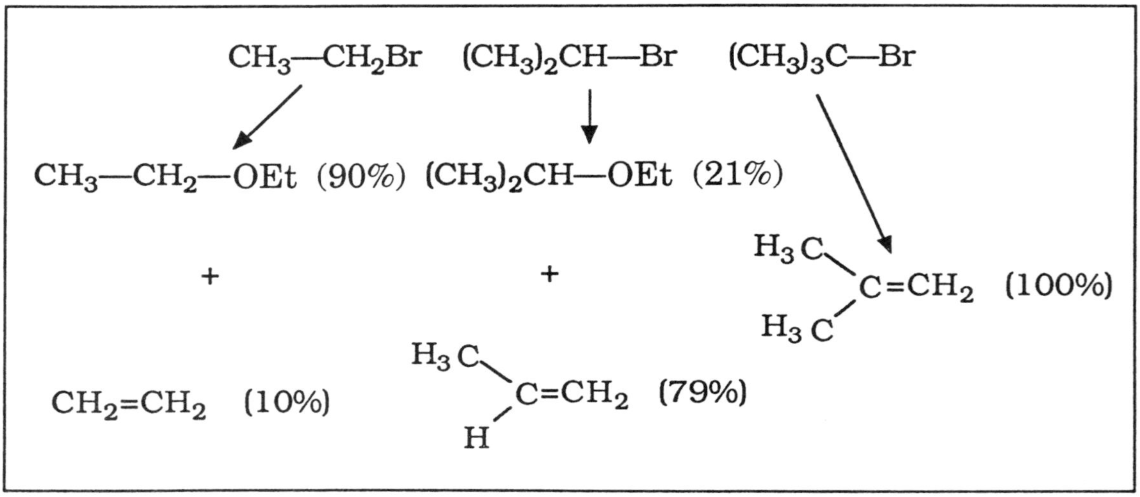

CH$_3$—CH$_2$Br     (CH$_3$)$_2$CH—Br     (CH$_3$)$_3$C—Br

CH$_3$—CH$_2$—OEt (90%)   (CH$_3$)$_2$CH—OEt (21%)

H$_3$C \
    C=CH$_2$  (100%) \
H$_3$C

+                    +

CH$_2$=CH$_2$  (10%)

H$_3$C \
    C=CH$_2$  (79%) \
H

If the nucleophile :Nu is the hydroxide anion, then alcohols are formed; ethers are obtained from alkoxide anions; thioethers from sulfides; esters from carboxylates; non-terminal acetylenes from monosubstituted acetylides; nitriles from the cyanide anion. In addition to anionic nucleophiles, neutral nucleophiles such as ammonia can be used affording primary amines; or tertiary amines affording the corresponding ammonium salts.

Since the nucleophile :Nu is also a base, a proton from the carbon adjacent to that bearing the leaving group can be readily removed, i.e. an E2 elimination occurs as a side reaction.

The carbon reactivity sequence in the $S_N2$ reaction is:

$$\text{primary} > \text{secondary} > \text{tertiary}.$$

To explain this sequence the steric effect and the accessibility of the carbon site under substitution can be invoked.

For the E2 elimination reaction the inverse obtains: the ratio of $S_N2$ substitution to E2 elimination products is largest for primary carbons. The results given in the adjacent scheme refer to reactions with ethoxide anion $C_2H_5O^{(-)}$ in ethanol at 55 °C.

$$: Nu^{(-)} + R\text{—}Y \rightarrow Nu\text{ - }R + Y^{(-)}$$

| | | |
|---|---|---|
| $H_3C\text{—}X$ | $3 \times 10^6$ | |
| $H_3C\text{—}CH_2\text{—}X$ | $10^5$ | |
| $H_3C\text{—}CH_2\text{—}CH_2\text{—}X$ | $4 \times 10^4$ | |
| $(H_3C)_2CH\text{—}X$ | $2{,}5 \times 10^3$ | **R—Y** |
| $(H_3C)_3CCH_2\text{—}X$ | $1$ | |
| $(H_3C)_3C\text{—}X$ | $0$ | |

| | | |
|---|---|---|
| $PhS^{(-)}$ | $4{,}7 \times 10^5$ | |
| $I^{(-)}$ | $3{,}7 \times 10^3$ | |
| $EtO^{(-)}$ | $10^3$ | |
| $Br^{(-)}$ | $5 \times 10^2$ | |
| $PhO^{(-)}$ | $4 \times 10^2$ | **Nu$^{(-)}$** |
| $Cl^{(-)}$ | $80$ | |
| $AcO^{(-)}$ | $20$ | |
| $NO_3^{(-)}$ | $1$ | |

| | | | |
|---|---|---|---|
| | $TsO^{(-)}$ | $6 \times 10^4$ | $(-6{,}5)$ |
| | $I^{(-)}$ | $3 \times 10^4$ | $(-9{,}5)$ |
| | $Br^{(-)}$ | $10^4$ | $(-9)$ |
| **Y$^{(-)}$** | $Cl^{(-)}$ | $2 \times 10^2$ | $(-7)$ |
| | $F^{(-)}$ | $1$ | $(\ 3{,}2)$ |
| | $HO^{(-)}$ | $0$ | $(15{,}7)$ |
| | $EtO^{(-)}$ | $0$ | $(16)$ |
| | $H_2N^{(-)}$ | $0$ | $(35)$ |

# 2 Structure–Reactivity Relationship

An earlier scheme compared the reactivity of primary, secondary and tertiary halides in $S_N2$ substitution. The adjacent page presents a fuller Table for alkyl halides. The reactivity is expressed by the rate constant $k_{rel}$.

The relative nucleophilicity $k_{rel}$ of the different reagents is given: these are indicative of 'standard' values representing the expected reactivity in an $S_N2$-type reaction. The sequence shows that for a given atom (e.g. oxygen) the nucleophilicity is directly related to the basicity: ethoxide > phenoxide > acetate > nitrate; whereas a large polarizable atom is more nucleophilic: thiophenoxide > phenoxide and iodide > bromide > chloride.

In the third Table ($k_{rel}$ vs $pK_a$) the relative leaving group abilities are displayed. The weakest conjugate bases (i.e. the conjugate bases of the strongest Brønsted acids) are the most reactive leaving groups. Only the tosylate, an excellent leaving group, does not comply with this correlation.

References: R. Breslow, *Organic Reaction Mechanisms*, W. A. Benjamin, Inc., New York, pp. 75–80 (1965).
J. McMurry, *Organic Chemistry*, Brooks-Cole, Monterey, pp. 298–302 (1984).

$k_{rel}$        1        5 000

$k_{rel}$        1        10 000

50%        +        50%

# 3 Neighboring Group Participation in the $S_N2$ Reaction

The rate constant for the $S_N2$ reaction increases when a nucleophilic atom is adjacent to the carbon atom being substituted. The substitution reaction with the nucleophilic sodium ethoxide is 5000 times faster with chlorohydrin than with ethyl chloride. Hydrolysis of the $\beta$-chlorosulfide indicated is in turn 10 000 times faster than that of the ether analog.

Participation of the heteroatom present in the same molecule reduces the activation entropy (since the entropy loss for the internal rotation is less than the entropy loss for translational degrees of freedom). The $S_N2$ reaction in this case involves the two steps: a first ionization step with participation of the heteroatom (e.g. sulfur) is followed by opening of this onium-type cyclic intermediate.

The stereochemical consequence is an apparent retention of configuration, resulting from two successive steps each involving inversion of configuration. The intermediate is a bicyclic cation, with the two oxygens from the OAc group attached to the adjacent 1 and 2 positions, both in the front half-space.

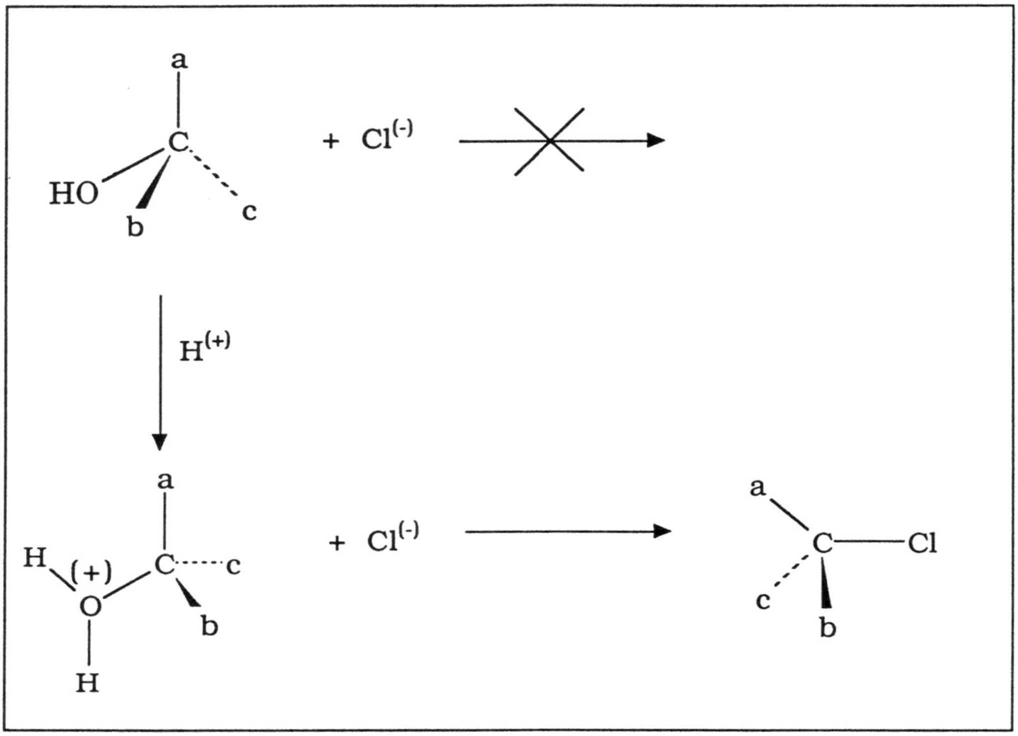
a
HO
C
b
c
+ Cl(-)
H(+)
H
(+)
O
H
+ Cl(-)
Cl

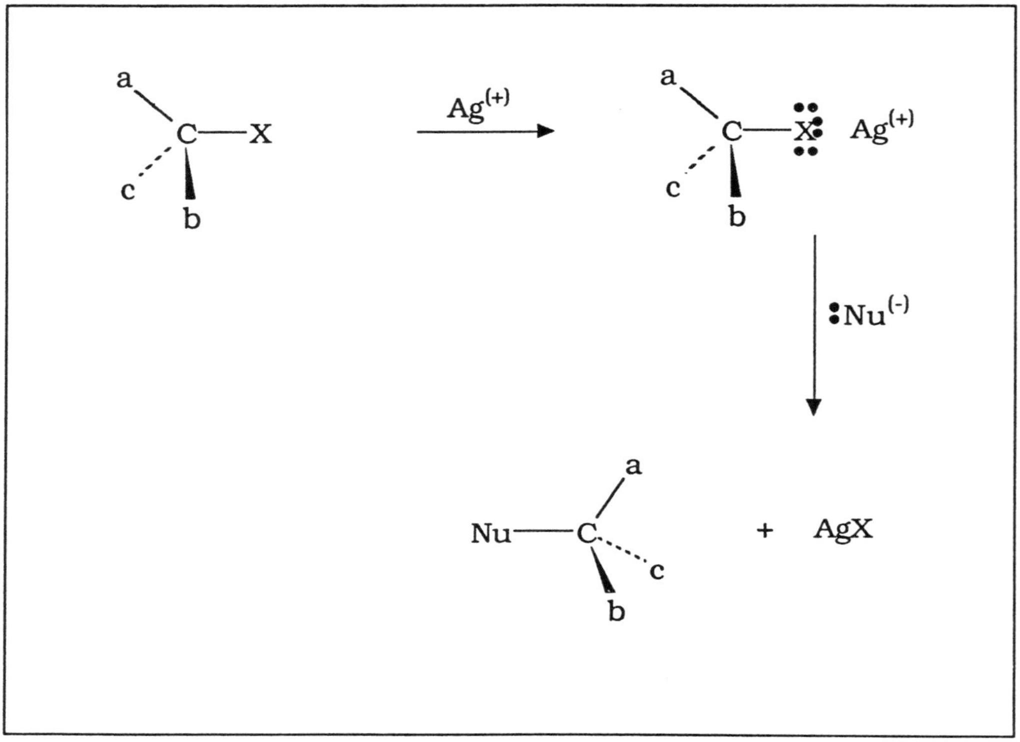
a
C
X
c
b
Ag(+)
Ag(+)
Nu(-)
Nu
+ AgX

# 4  Catalysis of the $S_N2$ Substitution by Electrophiles

Direct substitution of an alcohol with a chloride is not possible, even when a dipolar aprotic solvent which enhances the nucleophilicity of the chloride anion is used—the carbon–oxygen bond is strong and $OH^{(-)}$ is not a good leaving group.

Protonation of the alcohol in a prior step with subsequent loss of a neutral water molecule is sufficient to drive the reaction to completion. The action of hydrochloric acid thus converts the alcohol to the corresponding chloride. The same principle is used for other reagents involved in the transformation: $ROH \rightarrow RX$, $X = Br$, $Cl$ such as $PBr_3$ or $SOCl_2$, where first of all the alcohol function is converted to a better leaving group.

Catalysis of nucleophilic substitution by Lewis acids, e.g. $Ag^{(+)}$ or $Cu^{(++)}$, is similarly explained: activation by stabilization of the final state by way of a strong bond ($X^{(-)}$ . . . metal cation).

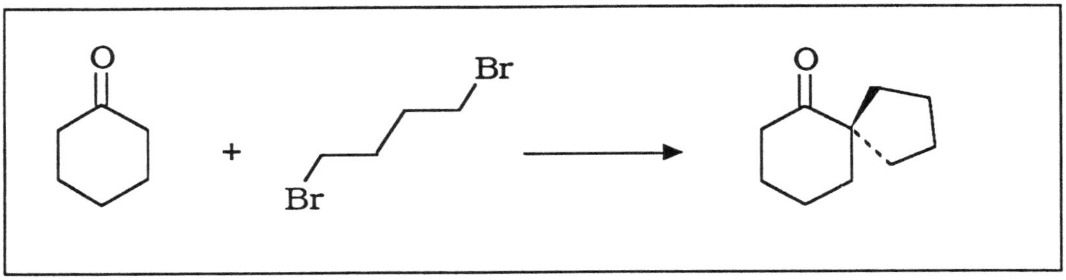
O
Br
Br
O

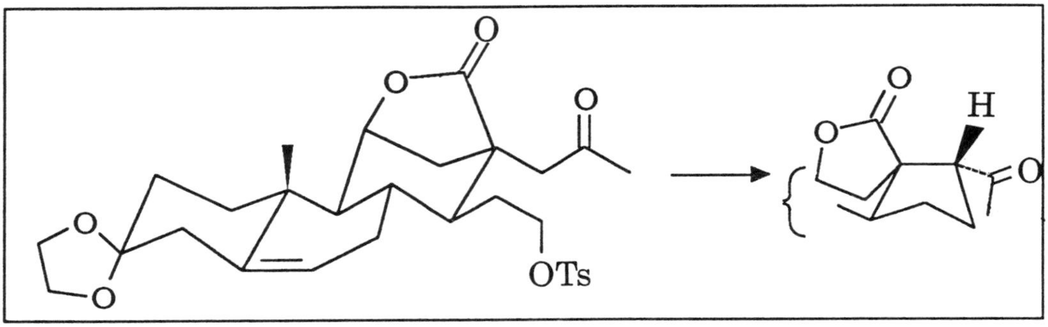
O
O
O
O
O
OTs
O
H
O

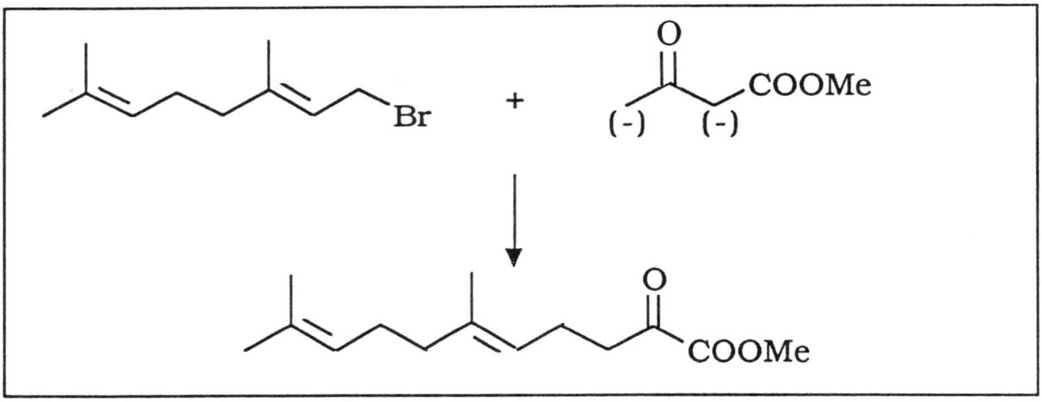
Br
O
COOMe
(-)
(-)
O
COOMe

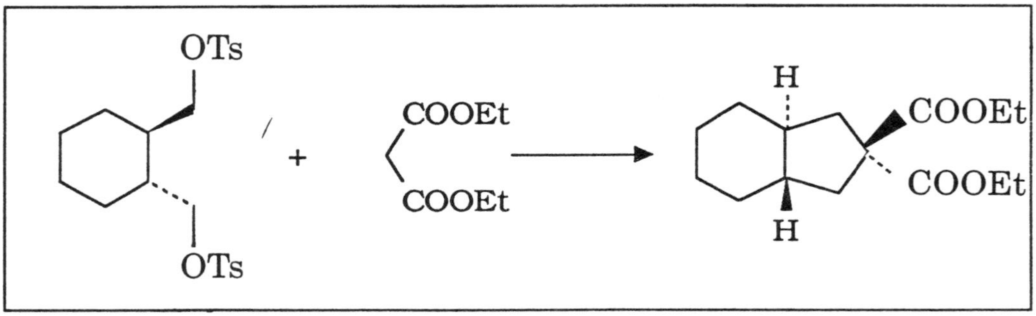
OTs
OTs
COOEt
COOEt
H
H
COOEt
COOEt

# 5 Regioselectivity in Alkylations

In cyclizations where alkylation is effected with dihalides, formation of five- and six-membered rings occurs more readily as compared to smaller or larger rings. This explains here the formation of the bicyclic spiro system as opposed to a seven-membered ring. Conditions: potassium *t*-butoxide in *t*-butanol, with benzene as solvent, heating for 5 hours; yield 70%.

The second example uses similar reaction conditions. The formation of the five-membered ring is preferred over that of a seven-membered ring (smaller activation entropy due to fewer internal rotations immobilized in the transition state); yield 54%.

The third example demonstrates regioselectivity resulting from differences in nucleophilicity of the two potential donor centers: in the dianion, reaction takes place at the least stabilized and hence most nucleophilic end (under conditions of kinetic control: THF at 0 °C); yield 95%.

Finally, when two nucleophiles react with a molecule bearing two acceptor sites (each with a leaving group), an intramolecular substitution is preferred, whenever possible (NaOEt/EtOH; yield 87%).

References: M. Mousseron, R. Jacquier, and H. Christol, *Bull. Soc. Chim. France*, 346 (1957).
W. S. Johnson, J. C. Collins, Jr., R. Pappo, M. B. Rubin, P. J. Kropp, J. E. Pike, and W. Bartmann, *J. Am. Chem. Soc.*, **85**, 1409–1430 (1963).
F. W. Sum and L. Weiler, *J. Am. Chem. Soc.*, **101**, 4401–4403 (1979).
B. C. Ayres and R. A. Raphael, *J. Chem. Soc.*, 1779 (1958).

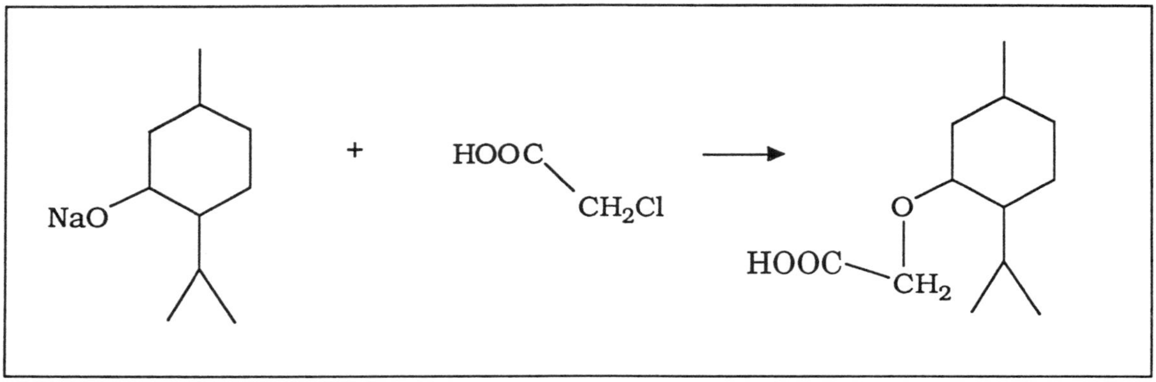
NaO
+
HOOC
CH2Cl
O
HOOC
CH2

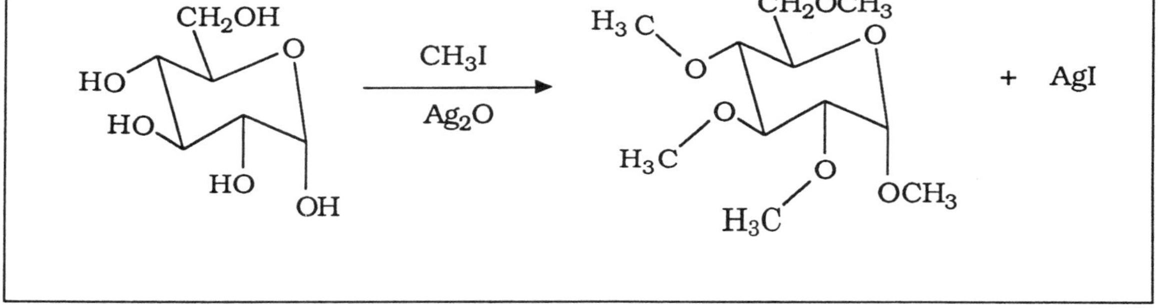
CH2OH
O
HO
HO
HO
OH
CH3I
Ag2O
CH2OCH3
H3 C
O
O
H3C
O
H3C
OCH3
+ AgI

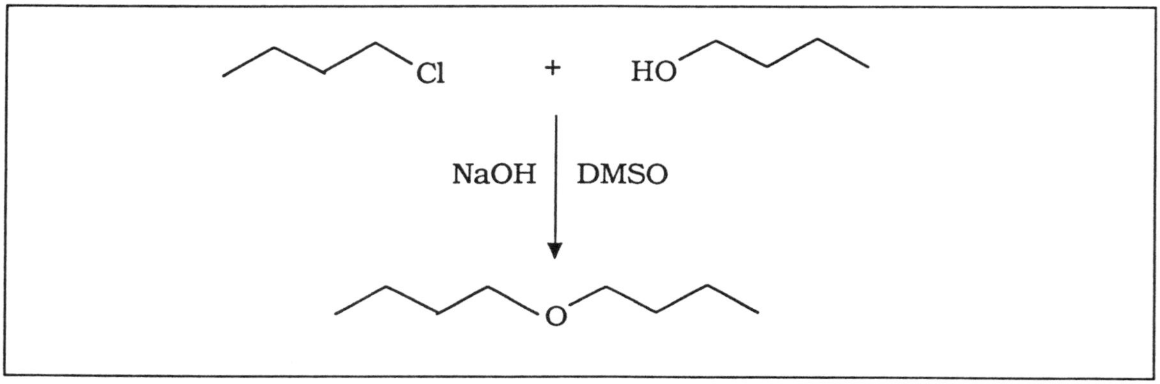
Cl
+
HO
NaOH
DMSO
O

# 6 Williamson Ether Synthesis

An ether $R-O-R'$ is formed by an $S_N2$ reaction, for instance by attack of an alkyl halide $R'X$ with an alkoxide $Na^{(+)(-)}OR$.

A useful modification which avoids the formation of the alkoxide (ROH + NaH, for instance) is when the alcohol ROH reacts directly with the halide $R'X$ in the presence of silver oxide. With glucose, the yield of the pentaether approaches 85%.

Use of a dipolar aprotic solvent improves the Williamson synthesis: the almost unsolvated alkoxide anion is very active.

Alexander W. Williamson (1824–1904) obtained his doctorate in 1846 at Giessen, under the direction of Liebig, who enjoyed a very high reputation throughout Europe for his excellent laboratory organization, for his pre-eminence in experimental organic chemistry, and his contacts with agriculture and industry. From 1849 until his death, Williamson was professor at University College in London.

References: *Org. Synth.*, Collective vol., III, 544–546.
R. G. Smith and A. Vanterpool, *Can. J. Chem.*, **47**, 2015–2019 (1969).
C. A. Russell, *Ambix*, **34**, 169–180 (1987).

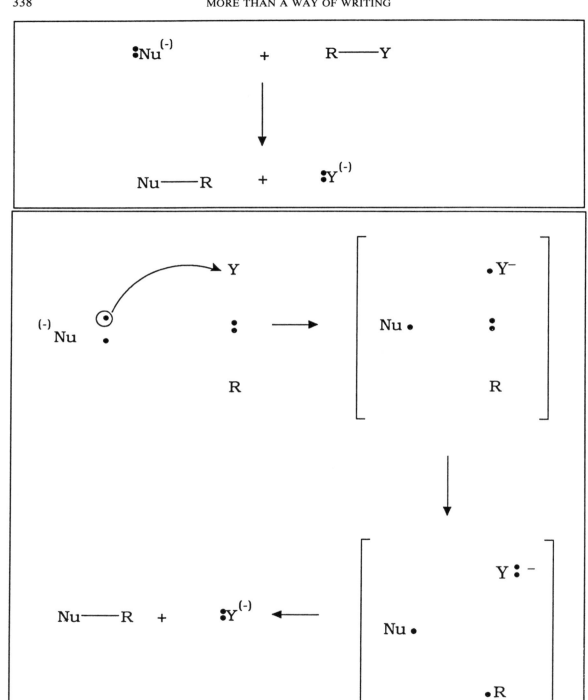

# 7  More than a Way of Writing

The accepted mechanism of the $S_N2$ reaction involves attack of the substrate R—Y by a lone pair (the nucleophile :Nu$^{(-)}$) with departure of the leaving group Y bearing a lone pair of electrons. In the course of several observations made during the past twenty years, an alternative mechanism for the $S_N2$ substitution involving a single electron transfer has been proposed.

In a first step, the nucleophile donates an electron to the atom Y (thus being transformed to the radical anion Y$^{\bullet-}$). In this way a pair of free radicals is formed. Consequently a transfer of a single electron from Y to R can occur, with localization of the electron pair of the $\sigma$-bond R—Y on Y. Finally coupling of the radicals Nu$^\bullet$ and R$^\bullet$ yields the reaction product.

The second mechanism involving a monoelectron transfer is in play if an $S_N2$ reaction is carried out between a fluorenyl anion and an alkyl halide such as $CH_3CH_2I$ (a good electron acceptor). In the majority of cases, however, the monoelectronic radical pathway appears insignificant as compared to the usual polar pathway.

References: A. Pross and S. S. Shaik, *Accounts Chem. Res.*, **16**, 363–370 (1983).
A. Pross, *Accounts Chem. Res.*, **18**, 212–219 (1985).
F. G. Bordwell and C. A. Wilson, *J. Am. Chem. Soc.*, **109**, 5470–5474 (1987).
F. G. Bordwell and J. A. Harrelson, *J. Am. Chem. Soc.*, **109**, 8112–8113 (1987).

# Summary: 'Functionalizations: Substitution'

Let us recall the distinction between the two limiting cases of nucleophilic substitution: the **dissociative $S_N1$ mechanism** with decreased coordination in the intermediate state compared to the initial and final states and hence the stereochemical consequence of **racemization**: the intermediate carbocation is planar. **The associative $S_N2$ mechanism**, on the other hand, is characterized by an increase of coordination in the transition state, hence the stereochemical consequence is **inversion of configuration**. The structure of the transition state is that of a trigonal bipyramid with the nucleophile and leaving group in apical positions, sharing the delocalized negative charge. It is important to choose the appropriate leaving group: the conjugate base of a strong acid.

Amongst the numerous applications in synthesis, noteworthy is the introduction of a polyene chain into a system by attack of a halide (e.g. a bromide) with an enolate formed by treatment of a carbonyl compound with a strong base (LDA): as in the phorbol synthesis by Paul Wender. A second standard modification is the standard sequence for chain extension: $\mathbf{RCOOMe \rightarrow RCH_2OH \rightarrow RCH_2OTs \rightarrow RCH_2CN \rightarrow RCH_2CHO \rightarrow RCH_2CH_2OH}$.

A third general use is to join the fragments in an ideally convergent synthesis; this is illustrated by Corey's atractyloside synthesis, that of a toxic ingredient of coffee.

Alkylating agents such as mustard gases are known for their toxicity. Biological methylations involve S-adenosyl methionine as the alkylating agent; bacterial chemotaxism is based on it.

Alkylations can be directed in a stereoselective manner, by a detour via a sigmatropic shift: an example for this is the juvabione synthesis by David Evans; or even more directly by means of an auxiliary ring, e.g. in numerous syntheses of prostaglandins, natural products ubiquitous in organisms, with potent physiological effects.

**Plate 1**

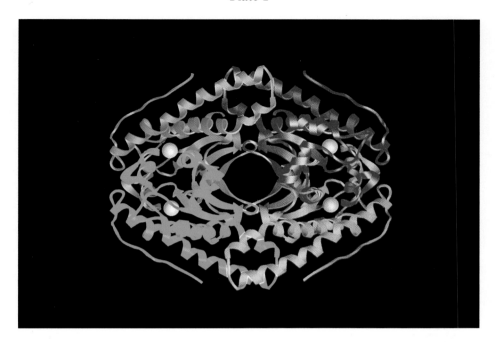

**Figure 1** Mitochondrial manganese superoxide dismutase shows four identical subunits with manganese ions in pink. The tetrameric enzyme contains two dimeric units (blue/green and red/orange) that contribute residues to each others' catalytically active sides. Enzymes of this type may offer better protection against infections. (Courtesy of Prof. J.A. Tainer, Scripps Clinic, La Jolla, USA)

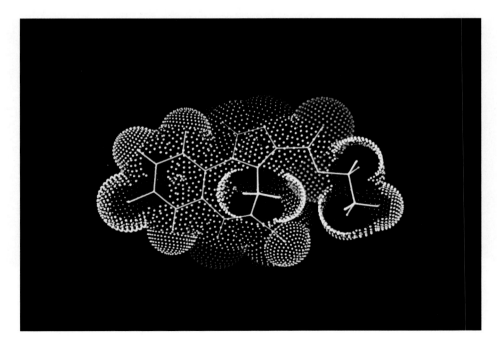

**Figure 2** Computer-graphic design, based on radiographic structure analysis of a molecule of the benzodiazepine antagonist ANEXATE. The dotted, spherical structures give some idea of the dimensions of the individual atoms of this molecule and its electron shell. (Courtesy of Hoffmann La Roche Ltd, Basel, Switzerland)

# Plate 2

**Figures 3** and **4**   The parasitic wasp *Cotesia marginiventris* approaches and stings a beet armyworm caterpillar on a maize seedling. The wasps are attracted to odors that are actively released by the plant in response to caterpillar feeding. Because these semiochemicals are beneficial to both the emitter and the receiver they are termed synomones. (Courtesy of Dr T.C. Turlings and Dr P. Greavy, United States Department of Agriculture, Gainesville, USA)

**Figure 5**   Coriander — the seeds of this plant have high levels of petroselenic acid. Genes involved in petroselenic acid biosynthesis are now being cloned in order to transfer them to rapeseed. (Courtesy of Dr D.J. Murphy, Plant Science Research Ltd, Cambridge Laboratory, Norwich, UK)

**Plate 3**

**Figure 6**  Vitamin A crystals — a deficiency leads to night-blindness, later to total blindness, in extreme cases even to death. Already with 100 g carrots per day (provitamin beta-carotene), this deficiency can be prevented. (Courtesy of Hoffmann La Roche Ltd, Basel, Switzerland)

**Figure 7**  Vitamin $B_{12}$ crystals — a deficiency leads to anemia, possibly also to pernicious anemia which can be fatal. $B_{12}$ is not present in vegetables. Very low quantities can be found in liver, meat and kidney of animals. (Courtesy of Hoffmann La Roche Ltd, Basel, Switzerland)

**Plate 4**

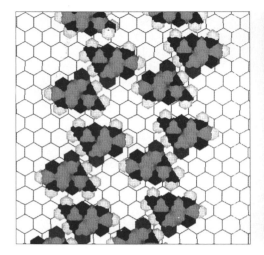

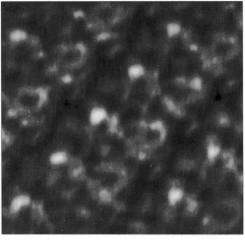

**Figure 8** Model of a layer of guanine molecules on graphite

**Figure 9** Scanning Tunnelling Spectrocopy (STM) image of guanine on graphite. (Courtesy of Prof. W. Hechl, TU Munich, Germany)

**Figure 10** Flow visualization in a coal gasification reactor. (Courtesy of Shell Research Ltd, Amsterdam, The Netherlands)

# 8.II Functionalizations: Substitutions

## 1 Extension of a Carbon Chain

xtension of a carbon chain by one carbon atom is a frequent necessity. Standard solutions have een developed. Such a routine is illustrated by a sequence in a recent (−)-grandisol synthesis. This a sex pheromone of the male of *Anthonomus grandis*, the cotton weevil:

Starting with an ester, lithium aluminum hydride reduction transforms it into the alcohol. This cohol is esterified with tosyl chloride into the tosylate —OTs, an excellent leaving group. $S_N2$ bstitution by the cyano group gives the nitrile. This is the key step for insertion of the additional rbon. Reduction of the nitrile with DIBAL (di-isobutyl aluminum hydride) affords after hydrolysis e aldehyde, then reduced to the alcohol. This sequence is not without analogy to the transformation COOMe ⟶ $-CH_2SEt$ in Stork's cantharidin synthesis, presented among the illustrations Chapter 7.

$$\mathrm{RCOOMe} \xrightarrow[\mathrm{THF}]{\mathrm{LiAlH_4}} \mathrm{RCH_2OH} \xrightarrow[\mathrm{C_5H_5N}]{\mathrm{TsCl}} \mathrm{RCH_2OTs} \xrightarrow[\mathrm{HMPA}]{\mathrm{NaCN}}$$

$$\mathrm{RCH_2CN} \xrightarrow[\mathrm{CH_2Cl_2}]{\mathrm{(iBu)_2AlH}} \mathrm{RCH_2CHO} \xrightarrow[\mathrm{THF}]{\mathrm{LiAlH_4}} \mathrm{RCH_2CH_2OH}$$

ference: A. I. Meyers and S. A. Fleming, *J. Am. Chem. Soc.*, **108**, 306 (1986).

## 2  The Initial Steps in Synthesis of a Poison

We have already mentioned in the previous chapter atractylosides, these extremely toxic ingredien[t] of coffee. This recent synthesis starts with an alkylation: (i) lithium diisopropylamide (LDA) in TH[F] at − 78 °C for 1 h deprotonates the methyl ester of 1,3-cyclohexadiene-5-carboxylic acid; (ii) the resultir[g] carbanion, stabilized by conjugation, is allowed to react with the indicated iodide in the same solven[t] to which one equivalent of hexamethylphosphoric acid triamide (HMPT) has been added. An $S_N$ reaction affords the bicyclic product in 82% yield (a single enantiomer is shown).

    Very often iodides RI are used as alkylating agents: the carbon–iodine bond is long and has relatively low bond energy; it is readily cleaved hence rendering iodide an excellent leaving grou[p].

Reference: A. K. Singh, R. K. Bakshi, and E. J. Corey, *J. Am. Chem. Soc.*, **109**, 6187–6189 (1987).

# 3 Activation of a Carbonyl Group by Substitution

Hydrolysis of the ester with aqueous hydroxide affords the acid ($RCOO_2Et \rightarrow RCO_2H$) which on treatment with oxalyl chloride ClCO.COCl leads to the acid chloride ($RCO_2H \rightarrow RCOCl$). The latter is transformed to the acyl azide by reaction with sodium azide $NaN_3$.

These functional group interchange reactions set up the acyl azide for rearrangement to an isocyanate, then hydrated to an amino acid derivative (in the so-called Curtius rearrangement). This sequence is part of the synthesis of ergotamine, an alkaloid from the ergot of rye, the drug responsible for ergotism.

Reference: A. Hofmann, A. J. Frey, and H. Ott, *Experientia*, **17**, 206 (1961).

## 4  Mustard Gas

Mustard gas with formula $ClCH_2CH_2SCH_2CH_2Cl$ can undergo a spontaneous internal $S_N2$ reaction, forming the indicated sulfonium chloride. This intermediate, with its activated strained three-membered ring, is very reactive; for instance a nucleophile such as the $\epsilon$-amino group from a lysine residue in a protein, will be directly alkylated. This is the reason for the blistering action of mustard gas on the skin and the mucous membranes. Mustard gas claimed about 400 000 victims during World War I.

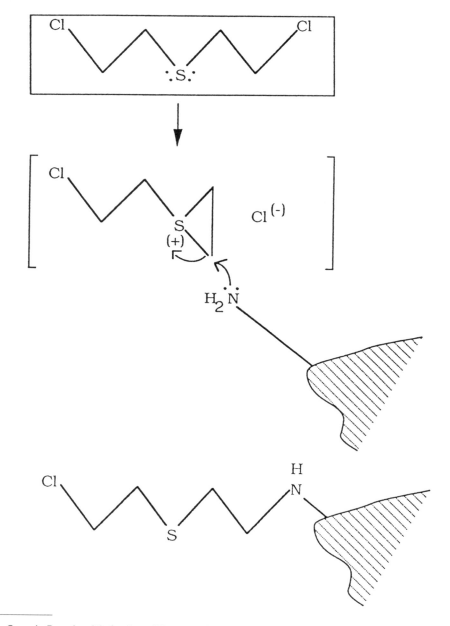

Reference: R. Breslow, *Organic Reaction Mechanisms*, W. A. Benjamin, Inc., New York, pp. 74–75, (1965).

## 5 Epoxy Adhesives

ɔoxy resins are remarkable adhesives for glass, ceramics and metals. Such glues are crosslinked resins ridges between the polymer chains), prepared in two steps:

The first step is reaction of the disodium salt of bisphenol A with epichlorohydrin: the nucleophile rO$^{(-)}$ attacks and opens the epoxide. A pre-polymer of low molecular weight results from this $S_N2$ action.

In the second crosslinking step, this pre-polymer is treated with a triamine such as $_2N(CH_2)_2NH(CH_2)_2NH_2$, the free OH groups thus formed then react with the terminal epoxy groups form crosslinks between the chains of the pre-polymer.

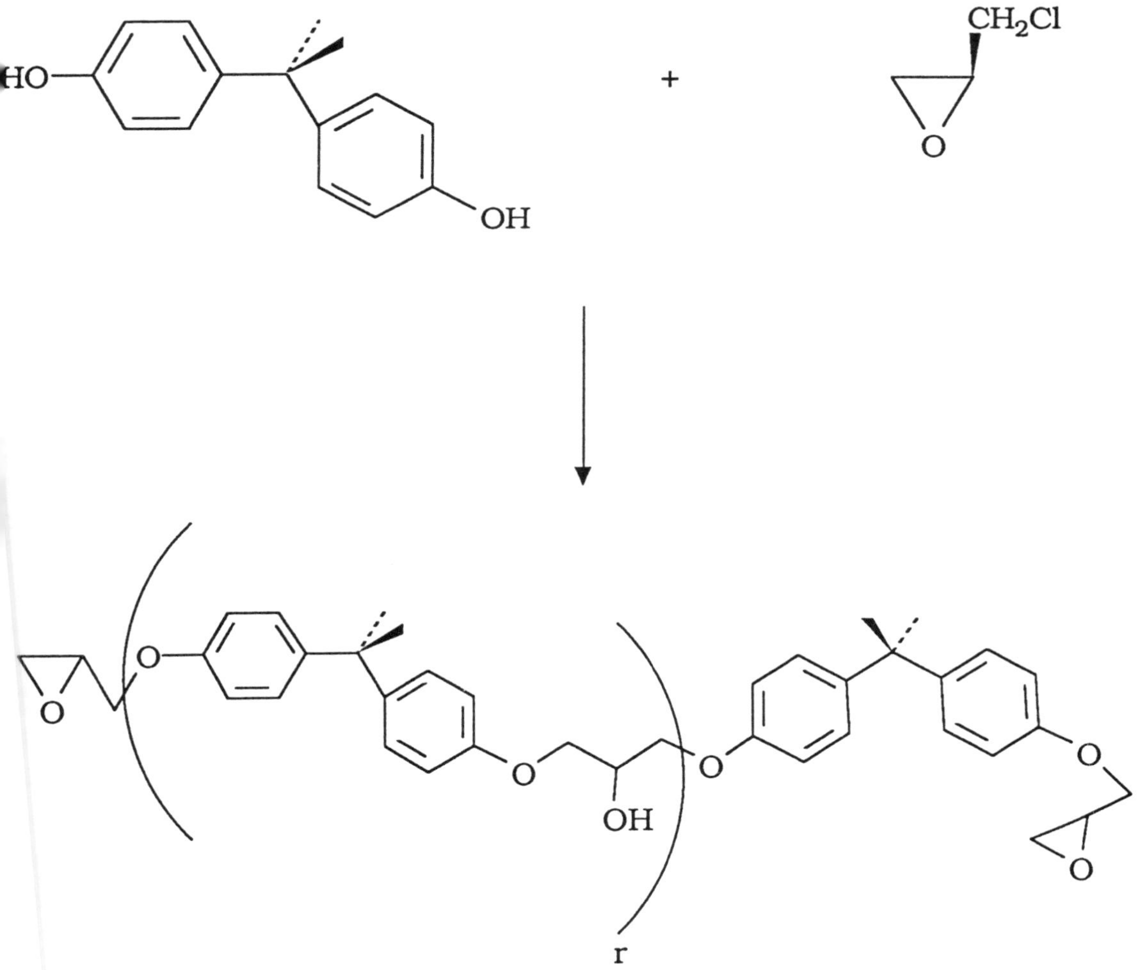


Only an artist is capable of making solution into an enigma', Karl Kraus.

## 6 A Biological Methylation Agent

For laboratory transformations $Nu^{(-)} + H_3C-X \rightarrow Nu-CH_3 + X^{(-)}$ one makes use of reagents such
methyl iodide ($X = 1$), i.e. a methyl attached to a good leaving group. Nature makes its methylations
like manner. It also uses a methyl group attached to a good leaving group, viz. a conjugate base
a strong acid.

Nature prepares this reagent by adding the sulfur atom of methionine—its most nucleophilic center—to adenosine triphosphate (ATP). This $S_N2$ substitution produces a sulfonium ion (tricoordinate sulfur, where the three attachments are methyl, R standing for the methionine molecule, and R' for the adenosine (= adenine + ribose) residue).

This biological methylating reagent is called S-adenosylmethionine and serves for carrying out numerous methylating reactions, for instance the conversion of norepinephrine to adrenaline in the medullar cortex. Let us examine in detail the following example.

References: L. Stryer, *Biochemistry*, W. H. Freeman, San Francisco, 2nd ed., p. 909 (1981).
J. McMurry, *Organic Chemistry*, Brooks-Cole, Monterey, pp. 328–329 (1984).

## 7  Chemotaxy of Bacteria, or the 100 $\mu$m Free Style

This complex and fascinating biological phenomenon was discovered separately by Pfeffer and Engelmann in the 1880s. It was neglected until the Fifties, when three laboratories established the main lines: Julius Alder (University of Wisconsin); Harold Berg (Cal Tech, then Harvard) and Donald Koshland (Berkeley). The best studied bacteria were the coliform bacillus (*Escherichia coli*) and *Salmonella typhimurium*, both of which are Gram-positive.

Positive chemotaxy is when the bacteria swim toward an attractor. Negative chemotaxy is when the bacteria withdraw from a repulsive chemical substance. One finds amongst attractive substances: N-acetylglucosamine; D-fructose; D-galactose; D-glucose and other sugars; L-aspartate; L-alanine; L-serine etc. Amongst the repelling substances are: fatty acids; simple alcohols such as methanol or ethanol; phenol; benzoate and salicylate anions; skatol; etc.

Bacteria move rapidly even though they are very light, and their motion is opposed by a highly viscous aqueous environment: they can swim 10 to 20 times their own length in a mere second! Their swim is effected by a rotating flagellum, similar to a propeller of a ship, at about 15 revolutions per second. But the motion is far from being uniform. In the absence of a chemical signal, the bacterium goes forward in any given direction for about a second, tumbles around head over heels, moves in another direction, changes direction again, and so on. Such abrupt changes of direction come from reversion of the rotation of the flagellum.

When an attracting chemical substance, galactose for instance, is present in the medium, the evolution continues to resemble Brownian motion, but the intervals during which the bacterium swims in the 'right' direction become longer than the converse; the overall displacement becomes biased and occurs towards the source of the concentration gradient detected by the bacterium. As it goes along, the bacterium monitors the concentrations from time to time, as if to climb up the concentration gradient. The bacterium thus has a chemical memory of the concentrations it has sampled already.

The receptors responsible for chemotaxy have been localized and identified. They are found between the outer cell membrane, which resembles a loose net, and the tighter internal membrane, in the so-called periplasmic space. A dozen receptors for repelling substances and twenty or so attracting ones have been identified. One can induce certain receptors to respond to a stimulating chemical species. Other receptors are permanent; for instance, the aspartic acid receptor is sensitive to concentrations as low as $10^{-8}$ M. The number of receptors of a given type in a cell varies from 1000 to 10 000. Stimulation of only one out of a thousand receptors is enough for a cellular response.

These receptors are proteins with molecular masses of the order of 30 000. Other proteins serve to transfer information from the periplasmic receptor to the cytoplasmic components that control the movement of the flagellum. Binding to (and release from) the receptors occurs in a fraction of a second.

Four homologous proteins, responsible for this transduction, have been identified in *E. coli* and are named Tsr, Tar, Trg and Tap. These proteins contain about 550 amino acid residues, for a molecular weight of about 60 000. The way in which information translates into the structure of these proteins is simply through methylation. S-adenosylmethionine methylates specific glutamate residues in these messenger proteins, in the presence of an enzyme, a methyltransferase specific to chemotaxy. The number of methylation sites is known: five for Tsr, four for Tar, five for Trg and about the same number for Tap.

*Attractive (repulsive)* substances increase (reduce) the proportion of methylated glutamates. The inverse reaction, demethylation of methyl glutamate, transfers the methyl groups to solvent molecules thus forming methanol. It is catalyzed by another enzyme, an esterase. Methylation/demethylation is slower than binding to the receptors, the reaction time being of the order of minutes.

This makes for a rather refined chemical code: binary for attraction/repulsion; and also quantitative since the level of methylation/demethylation is a function of the concentration of the chemical stimulus.

References: L. Stryer, *Biochemistry*, 2nd ed., W. H. Freeman, San Francisco, pp. 905–909 (1981).
         H. C. Berg and E. M. Purcell, *Biophys. J.*, **20**, 193–219 (1977).
         R. C. Stewart and F. W. Dahlquist, *Chem. Rev.*, **87**, 997–1025 (1987).

## 8 Substitution with Participation and Rearrangement

The 1,2-glycol is esterified with *p*-toluenesulfonyl chloride TsCl in the presence of pyridine, at 0 °C in methylene chloride. The selectivity is noteworthy: only the more accessible secondary alcohol is esterified to the tosylate. Cleavage of the TsO$^{(-)}$ leaving group is triggered off by electrophilic catalysis with Li$^{(+)}$ ClO$_4^{(-)}$ in THF. Participation of the $\pi$-electrons of the neighboring unsaturated carbon causes the latter to migrate in this pinacol rearrangement. The resulting homoallylic carbocation expels a proton, forming a $\beta$, $\gamma$-unsaturated, non-conjugated seven-membered ring ketone. Treatment with 2N HCl in ethanol at room temperature has two effects: isomerization to the conjugated cycloheptenone and regeneration of the carbonyl group earlier protected as an ethylene ketal.

This is part of the classical longifolene synthesis by Corey.

Reference: E. J. Corey, M. Ohno, P. A. Vatakencherry, and R. B. Mitra, *J. Am. Chem. Soc.*, **83**, 1251 (1961); **86**, 478 (1964).

## 9 Enantioselective Alkylation via Oxazolines

The goal is enantioselective introduction of a group R′ on a carbon adjacent to a carbonyl. The idea makes use of a heterocycle differentiating neatly between the two half-spaces which it defines: one is blocked by a bulky group, the other comprises an oxygen atom which serves as an attractor.

Such chiral oxazolines are prepared from the enantiomerically pure amino acid S-serine. The reaction of serine, first with phenyllithium, and second with sodium borohydride affords the indicated aminoalcohol.

On the other hand, addition of ethanol and hydrochloric acid to the nitrile RCH$_2$CN affords a carbocation, stabilized by two heteroatoms O and N. Reaction with the aminoalcohol ((i) NH$_4$Cl/-EtOH, (ii) NaH/-H$_2$; (iii) MeI/-NaI) leads to the enantiomerically pure chiral Δ-oxazoline of S,S configuration in high 60–90% yield.

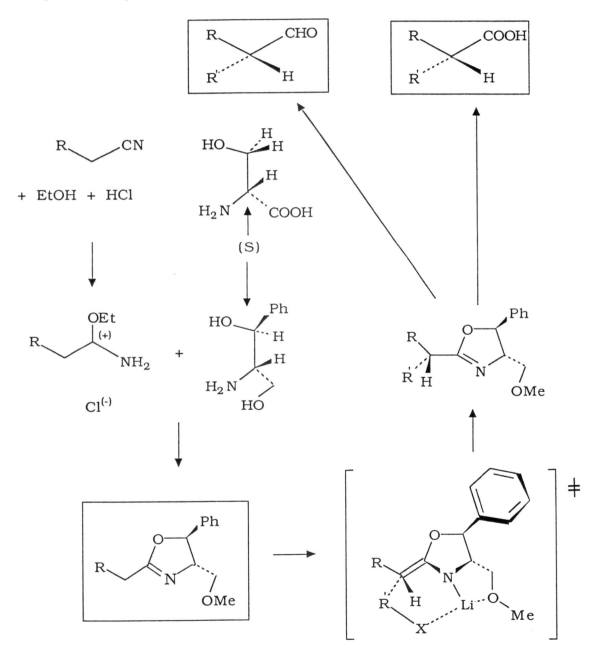

This oxazoline is metallated with *n*-butyllithium or lithium diisopropylamide (LDA) in THF at −78 °C. The R′X alkylating agent is then introduced at −98 °C in the solution, slowly allowed to warm up to room temperature. R′ reacts almost exclusively from below, with the terminal carbon of the double bond preferentially in the Z configuration. The alkylation product is hydrolyzed to the acid (3 N HCl) or reduced (NaBH$_4$) and then hydrolyzed (H$_3$O$^+$) to the aldehyde. The resulting enantiomeric excesses are typically between 75 and 85%.

Reference: A. I. Meyers, G. Knauss, K. Kamata, and M. E. Ford, *J. Am. Chem. Soc.*, **98**, 567 (1976).

## 10 Synthesis of α-Alkylated Amino Acids

The procedure illustrated overleaf was developed by the Kanegafuchi Japanese company. Step (i) is an enzymatic cyclization. At the DSM company, double addition of HCN and ammonia to a ketone affords a racemic amide R$^1$R$^2$C(NH$_2$)C(=O)NH$_2$. The latter in the presence of the amidase α-alkylaminopeptidase from *Mycobacterium neoaurum* affords the amide of the required product R$^1$R$^2$C(NH$_2$)COOH with high enantioselectivity.

References: Kanegafuchi, European patent 0.175.312 (1985).
DSM/Stamicarbon, European patent 0.231.546, 0.236.591, 0.232.562 (1986); 0.150.834 (1988).
V. H. M. Elferink, D. Breitgof, M. Kloosterman, J. Kamphuis, W. J. J. van der Tweel, and E. M. Meijer, *Recl. Trav. Chim. Pays-Bas*, **110**, 63–74 (1991).

Why synthesize such non-naturally occurring amino acids? So as to trick nature: incorporation into new proteins having enzyme inhibiting properties, or hormonal activity, slower to metabolize, etc See the text by Roald Hoffmann in Chapter 4.

Such biotechnologies (industrial use of enzymes) are now commonly used by the chemical industry

# Functionalizations: Reductions

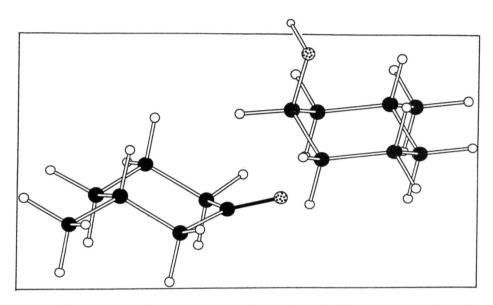

Let us now consider functionalizations with a change in oxidation state at carbon. Reductions of both carbon–carbon and carbon–oxygen multiple bonds are discussed here. The aim is of course to perform these reductions, if possible, in a regioselective and stereoselective manner.

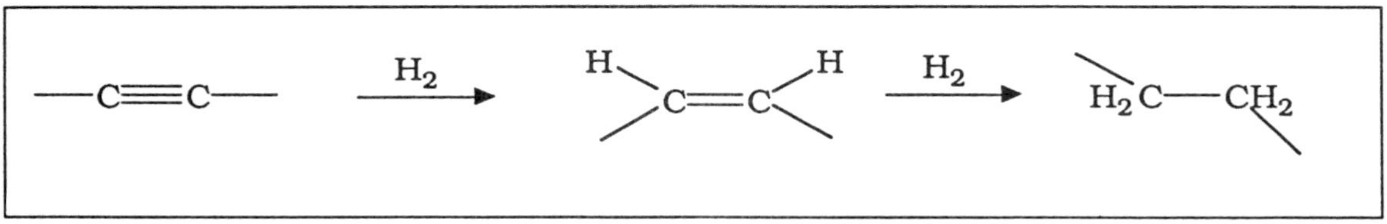

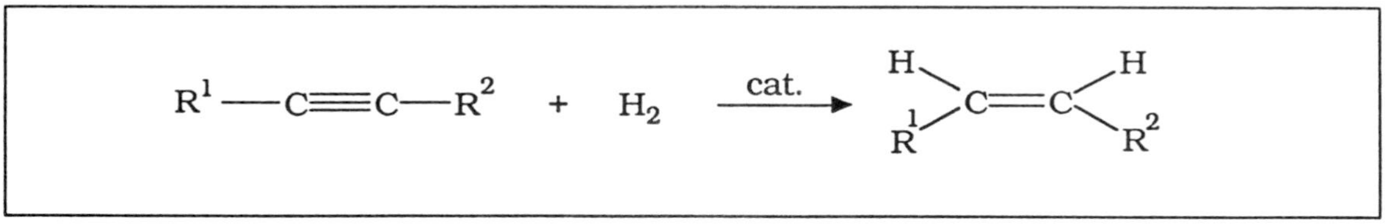

# 1  Hydrogenation of Alkynes

C≡C triple bonds can be completely hydrogenated with molecular hydrogen in the presence of a catalyst. Access to alkenes is possible in a first step and in a second step alkanes are produced. The first of these two steps is the faster. The first hydrogenation can be stopped at the alkene stage by using the Lindlar catalyst: palladium supported on calcium carbonate partially poisoned by treatment with lead acetate and quinoline, an aromatic amine.

Hydrogenation with this catalyst gives selective *cis*-hydrogenation, thus furnishing a C=C double bond with *cis* (Z)-configuration in the olefinic product. In the same way diisobutylaluminum (DIBAL) gives *cis*-hydrogenation. *trans*-Hydrogenation can be effected by way of alkali metals dissolved in liquid ammonia (or an amine; under so-called Birch conditions).

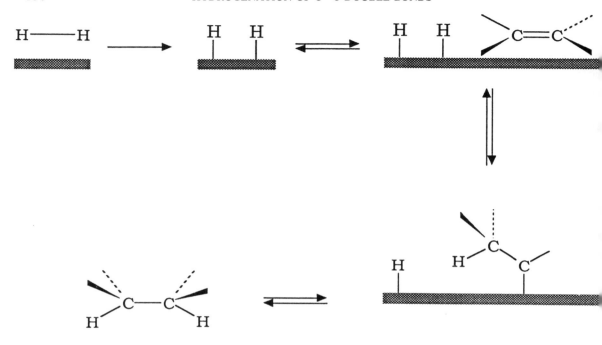

# 2   Hydrogenation of C=C Double Bonds

Carbon–carbon double bonds are most frequently hydrogenated in the presence of a heterogeneous catalyst such as a noble metal: palladium on carbon, finely dispersed; rhodium, nickel, ruthenium; platinum used in form of the oxide $PtO_2$ (Adam's catalyst).

The accepted mechanism of hydrogenation is dissociative chemisorption of $H_2$. The olefin then attaches itself to the catalyst surface, from the least hindered face. The addition of two hydrogen atoms absorbed on the catalyst surface proceeds in two successive steps (demonstrated by isotope exchange ($H_2$—$D_2$)).

When a continuous layer of silica covers the surface of the platinum catalyst, the hydrogen atoms can permeate into it and can effect an isotope exchange R—D→R—H.

The enthalpy difference -$\Delta H^o$ during the course of the hydrogenation is about $30\,kcal\,mol^{-1}$. This reflects the thermodynamic stability of olefins: *trans*>*cis*; *tetra*->tri->di->monosubstituted.

Reference: A. B. McEwen, W. F. Maier, R. H. Fleming and S. M. Baumann, *Nature*, **329**, 531–534 (1987).

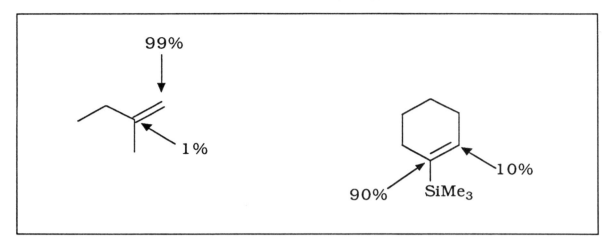

# 3  Reduction with Boranes

The electron deficient boron atom (with an empty 2p orbital) adds to ethylene and acetylene bonds, either to the sterically most accessible carbon or the carbon which most effectively stabilizes a positive charge.

The adduct is cleaved by using hydrogen peroxide in a basic medium. The regioselectivity is enhanced by use of disiamylborane instead of diborane: 91% yield of methylketone as compared to 74% (cf. adjacent scheme).

References: B. M. Mikhailov and Y. N. Bubnov, *Organoboron Compounds in Organic Synthesis*, 1984, Ch. 2.
G. Zweifel, G. M. Clark and N. L. Polston, *J. Am. Chem. Soc.*, **93**, 3395 (1971).

$$ArH \longrightarrow [ArH^{(-)}] \xrightarrow{\ H^{(+)}\ } ArH_2$$

# 4   Birch Reduction

This procedure (alkali metals dissolved in liquid ammonia or amines) is applied, above all, for hydrogenation of aromatic rings to non-conjugated 1,4-cyclohexadienes. The mechanism of this reaction involves an anion or radical anion formed in the first step; it is selectively reprotonated at the site of highest electron density. This was one of the first 'triumphs' of the MO method which can, even with the Hückel model, easily account for the regioselectivities observed in Birch-type reductions. Let us consider the case of naphthalene. The electron densities for the intermediate anion are given as being the squares of the coefficients $c_r^2$ for the lowest unoccupied molecular orbital (LUMO): hence the proton preferentially attacks position 1. In the case of phenanthrene, positions 9 and 10 have the highest electron density, where protonation occurs in the second step of the reduction.

References: A. J. Birch, *Australian J. Chem.*, **7**, 256–261 (1954); **8**, 95 (1955).
   A. J. Birch and G. Subba Rao, *Adv. Org. Chem.*, **8**, 1 (1972).
   A. A. Akhrem, I. G. Reshetova, and Yu. A. Titov, *Birch Reduction of Aromatic Compounds*, Plenum Press, New York, 1972.
   A. Streitwieser, Jr. *Molecular Orbital Theory for Organic Chemists*, Wiley, New York, 1961.

94%

70%

# 5 Reduction of Carbonyl Groups

*Reminder*: This concerns a nucleophilic addition of the hydride anion $H^{(-)}$ provided by the metal hydride $M^{(+)}H^{(-)}$. The most commonly used reagents are: the very reactive lithium aluminum hydride $LiAlH_4$ and the less reactive sodium borohydride $NaBH_4$. The latter is used for reduction of aldehydes and ketones. Due to its rapid hydrolysis in protic milieu $LiAlH_4$ must be used in well dried ether (THF). $NaBH_4$ on the other hand is hydrolyzed only very slowly and this allows for the use of water or alcohols as solvents.

$LiAlH_4$ is used for the reduction of the less reactive esters and amides, due to their conjugation. In the example given, $LiAlH_4$ is not selective: the alcohol has to be reoxidized ($CrO_3/H_2SO_4$) to the ketone. Overall yield is 88%.

Borane $BH_3$ (available in the more stable form, $H_3B.THF$) is another reducing agent for amides which does not affect the ester group; yield is 94%; however $C=C$ double bonds are also reduced (see above).

Lewis acids such as Co(II) enhance the reactivity of $NaBH_4$ and thus reduce amides to amines with yields in the order of 70%.

References: W. A. Ayer, W. R. Bowman, T. C. Joseph, and P. Smith, *J. Am. Chem. Soc.*, **90**, 1648 (1968).
H. C. Brown and P. Heim, *J. Am. Chem. Soc.*, **86**, 3566 (1964).
T. Satoh, S. Suzuki, Y. Suzuki, Y. Miyaji, and Z. Imai, *Tetrahedron Lett.*, 4555 (1969).

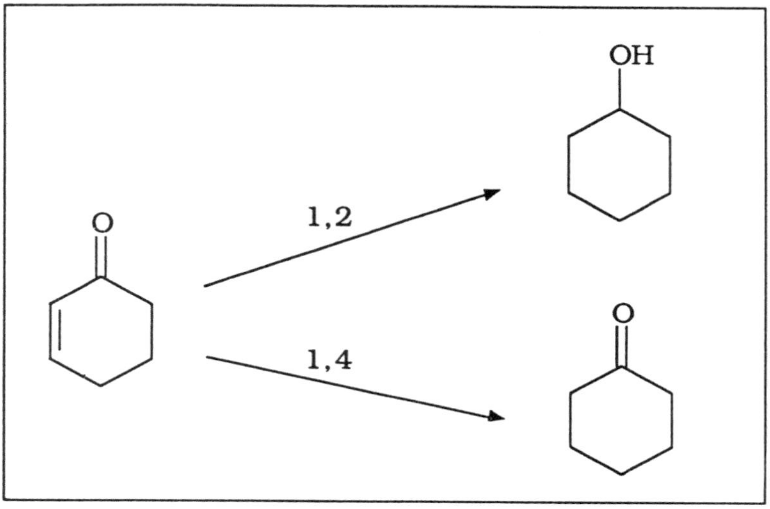
OH
1,2
O
1,4
O

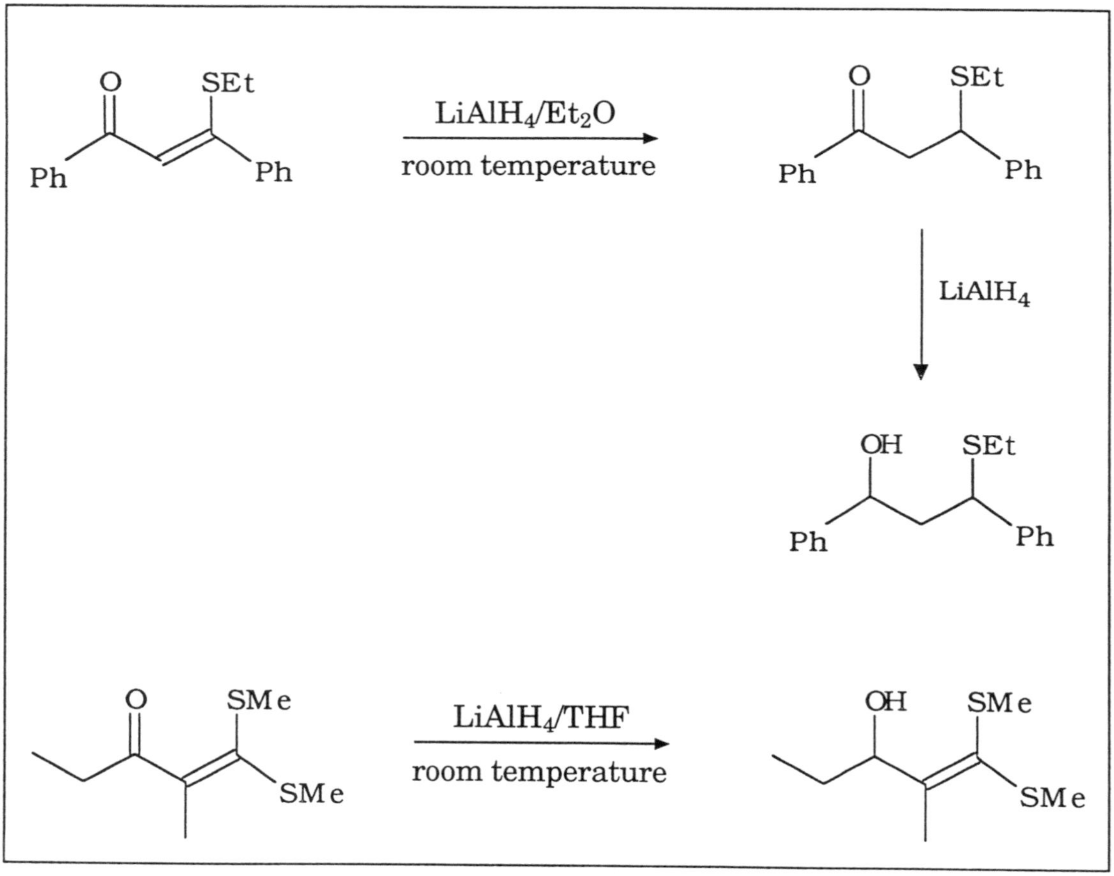
O
SEt
Ph
Ph
LiAlH4/Et2O
room temperature
O
SEt
Ph
Ph
LiAlH4
OH
SEt
Ph
Ph
O
SMe
SMe
LiAlH4/THF
room temperature
OH
SMe
SMe

# 6 Reduction of $\alpha,\beta$-Unsaturated Carbonyl Compounds

There is a competition between the two types of addition: 1,2-addition provides the alcohol; 1,4-addition leads to the saturated ketone (via a keto–enol tautomerization). Normally, 1,2-addition predominates.

An electron poor C=C double bond is more prone to 1,4-addition.

The nature of the metal hydride is important: the percentages of 1,4-addition to 2-cyclohexenone are 5, 22 and 95% with LiAlH(OMe)$_3$, LiAlH$_4$ and LiAlH(O-$t$-Bu)$_3$. **Hard (soft) metal hydrides favor 1,2-(1,4-)addition**, complying with the frontier orbital coefficients.

References: O. Eisenstein, J. M. Lefour, C. Minot, G. Soussan, and T. A. Nguyen, *Compt. Rend. Ac. Sci.*, **274**, 1310 (1972).
J. Durand, T. A. Nguyen, and J. Huet, *Tetrahedron Lett.*, 2397 (1974).
J. Bottin, O. Eisenstein, C. Minot, and T. A. Nguyen, *Tetrahedron Lett.*, 3015 (1972).

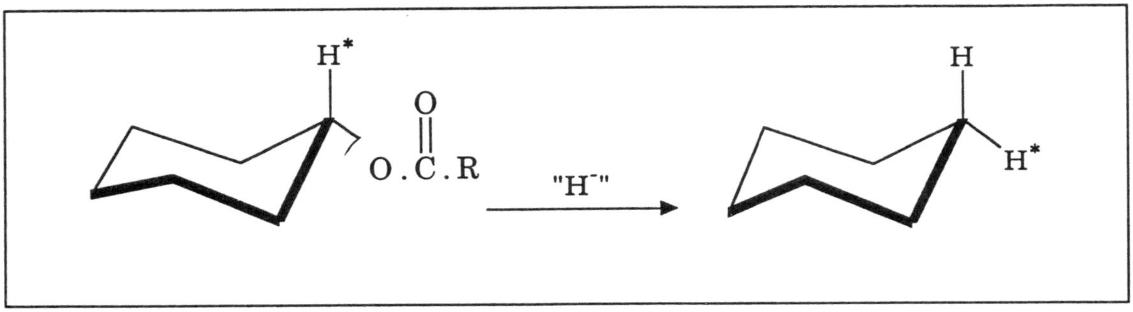
H*
O.C.R
"H⁻"
H
H*

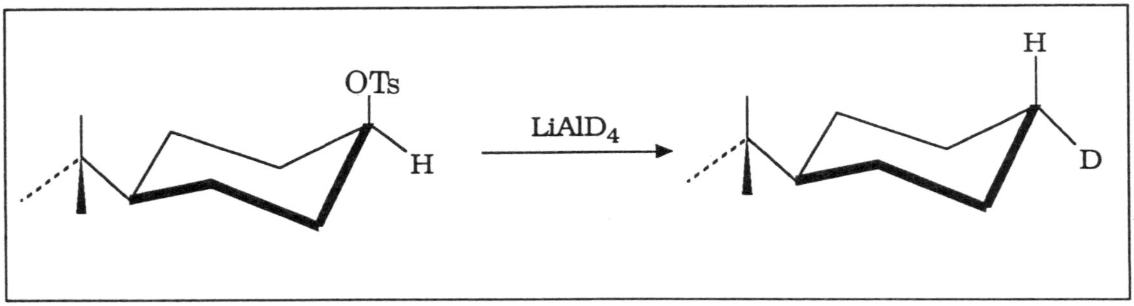
OTs
H
LiAlD4
H
D

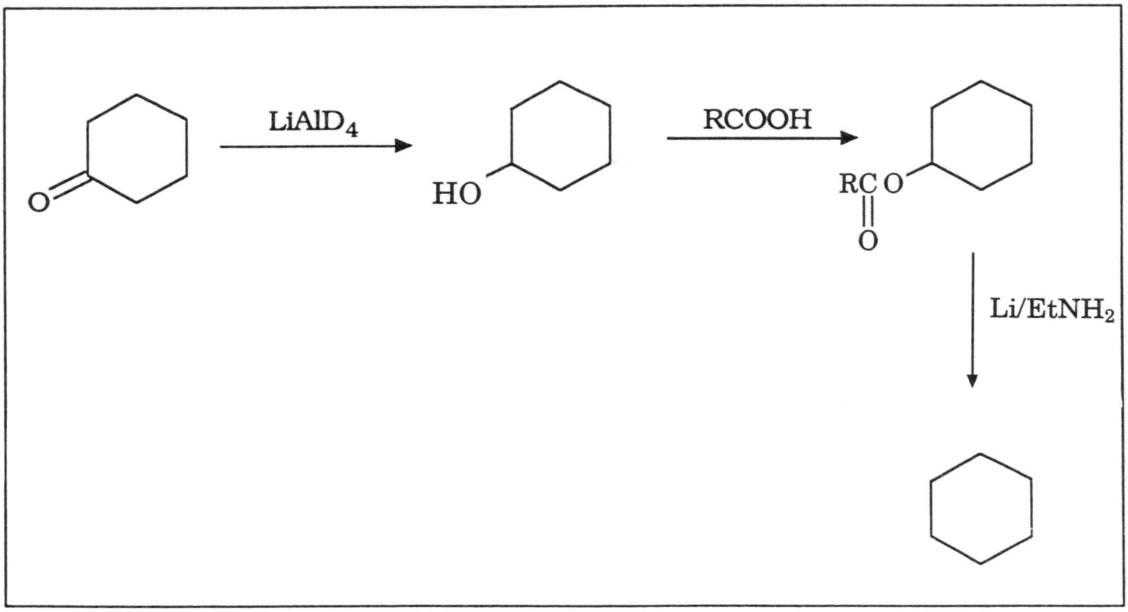
LiAlD4
HO
RCOOH
RC O
O
Li/EtNH2

# 7 Reductive Cleavage of an Ester

Since certain esters are good or even excellent leaving groups, their nucleophilic substitution by hydrides has a double effect: reduction of the carbon that bears them and departure of the ester function. Several reagents function as equivalents of hydride anion 'H$^-$'; for instance lithium aluminum hydride, commonly used with tosylates (esters of $p$-toluenesulfonic acid) OTs; or alkali metals, e.g. sodium and lithium, dissolved in liquid ammonia or an amine (so-called Birch conditions).

Since it is an S$_N$2 substitution, the reaction proceeds with inversion of configuration at the carbon. Reductive cleavage is thus a good method of introducing a labeled hydrogen stereospecifically into the molecule. The sequence ketone → alcohol → ester → hydrocarbon, even though requiring three steps, allows a smooth conversion of a carbonyl group C=O to the corresponding methylene derivative CH$_2$ under mild conditions.

Reductive opening of epoxides with attack at the least hindered carbon (LiAlH$_4$ or NaBH$_4$) also follows a S$_N$2 mechanism.

References: F. Rolla, *J. Org. Chem.*, **46**, 3909 (1981).
A. Ookawa, M. Kitade, and K. Soai, *Heterocycles*, **27**, 213 (1988).

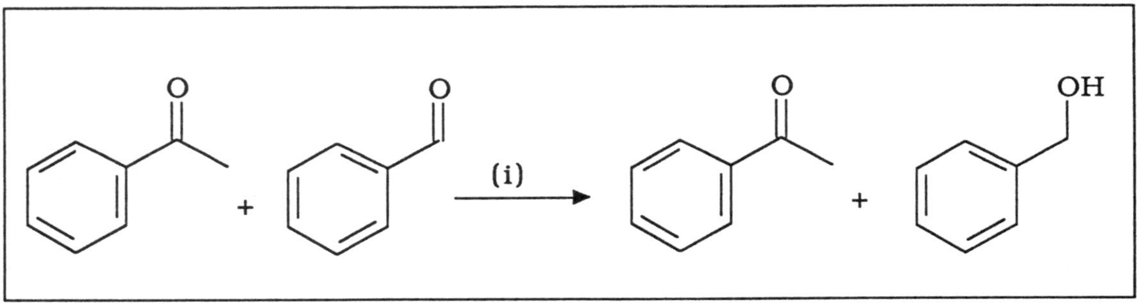
O
O
(i)
O
OH
+
+

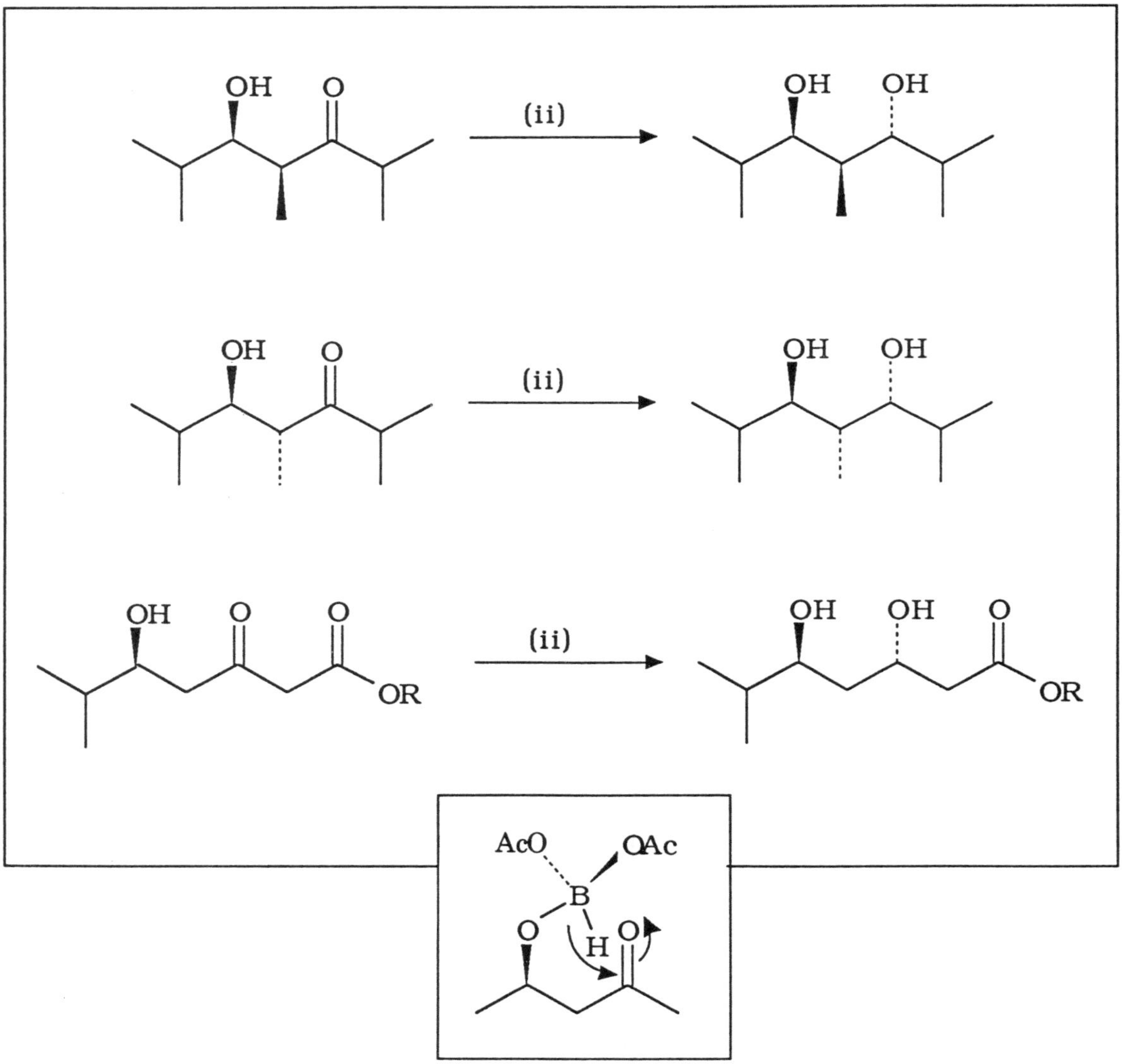
OH
O
(ii)
OH
OH
OH
O
(ii)
OH
OH
OH
O
O
OR
(ii)
OH
OH
O
OR
AcO
OAc
B
O
H
O

# 8 The Bewildering Selectivity of Triacetoxyborohydrides

Selective reduction of an aldehyde in presence of a ketone is possible using the reagent (i) $NaBH(OAc)_3$. Borohydride triacetates are thus mild and chemoselective reagents: acetophenone is recovered in about 90%; the yield of benzylalcohol is 95%.

Regarding their stereoselectivity: the hydroxy group OH of 3-hydroxyketones serves as the directing group; the use of $Me_4N^{(+)(-)}BH(OAc)_3$ in a mixture of acetic acid and acetonitrile at $-20\,°C$ (ii) provides the 1,3-diol resulting from predominant attack of the hydride ion from the upper face of the carbonyl group (the half-space where the OH group is situated). The yields of the isolated diol are in the order of 90%; the stereoselectivity varies between 95:5 and 98:2.

The reason for this high stereoselectivity is due to attachment of boron to the alcohol oxygen, in a $S_N2$ reaction, with departure of an acetate group; the B—H bond can attack the carbonyl from the upper face ($\beta$-face in a steroid type notation); noteworthy is the primary influence of the OH group, irrespective of the orientation of the methyl group at the carbon adjacent to the carbonyl group.

References: G. W. Gribble and D. C. Ferguson, *Chem. Comm.*, 535 (1975).
D. A. Evans and K. T. Chapman, *Tetrahedron Lett.*, **27**, 5939 (1986).

## Summary: 'Functionalizations: Reductions'

The transformation of a cyclohexene—resulting from a Diels–Alder reaction—to the corresponding cyclohexane or a cyclohexanone, obtained by an aldol or Claisen condensation, being reduced to the corresponding alcohol which in turn can undergo—after esterification to the tosylate—numerous nucleophilic substitution, links this chapter to the preceding ones.

Hydrogenation of C=C double bonds is a reaction commonly catalyzed by a noble-metal catalyst such as rhodium, palladium, ruthenium or platinum. This heterogeneous catalysis is initiated by dissociation of $H_2$ on the catalyst surface. The alkene with the least hindered face is attached to the catalyst, with consequently a sequential addition of two hydrogen atoms to the former. The least stable C=C bonds are the first to be hydrogenated, for example, the exocyclic C=C bond of limonene. Enantioselectivity can be achieved by homogeneous catalysis with the Wilkinson's catalyst, a chiral rhodium complex where simultaneous coordination of the olefin and the two hydrogen atoms allows the addition to take place from the most hindered side, and at the same time providing a high enantiomeric excess. An application of this is in the industrial production of L-DOPA (Monsanto process), a drug used against Parkinson disease.

The Birch reduction uses alkali metals dissolved in liquid ammonia or amines. The aromatic ring successively takes up two electrons from the metal. The resulting conjugated anion is protonated at the site of highest electron density: this is an elegant illustration of the success of the MO theory, even in its rudimentary Hückel version. Thus Birch hydrogenation *transforms benzenes to non-conjugated 1,4-cyclohexadienes.*

Finally, a discussion on the reduction of carbonyl groups with metal hydrides is presented. Aldehydes are again more reactive than ketones. The very reactive lithium aluminum hydride $LiAlH_4$ requires ethers as solvents and is not chemoselective. Sodium borohydride $NaBH_4$ can be used in protic solvents and reduces only aldehydes and ketones. $LiAlH_4$ and $NaBH_4$ activated by a Lewis acid such as $Co^{2+}$ reduce also amides. Borane $BH_3$ reduces both amides and C=C double bonds but not esters. Alane $AlH_3$, as an amine complex, reduces nitriles, amides and esters. The strong reducing agent $LiAlH_4$ also reduces esters.

Complex metal hydrides allow for flexibility, e.g. di-*i*-butylaluminum (DIBAL) transforms amides and esters to aldehydes. High stereoselectivity can be obtained by use of $NaBH(OAc)_3$ in the reduction of $\beta$-ketols (David Evans). Examples of biotechnologies (reductases) are also given.

# Functionalizations: Reductions

## 1  Access to an Aromatic Annulene

Starting from the indicated precursor, *cis*-hydrogenation of the three carbon–carbon triple bonds is carried out with the Lindlar catalyst. This leads to an intermediate with the following sequence of configurations for the olefinic C=C bonds (*c* = *cis*, *t* = *trans*):

$$(c-c-t-c-c-t-c-c-t)$$

This intermediate, destabilized by its strain, isomerizes to a hydrocarbon with the following sequence for the C=C bonds:

$$(c-t-t-c-t-t-c-t-t)$$

It is the [18] annulene with 18 $\pi$-electrons delocalized in the macrocyclic plane, thus fulfilling the Hückel rule for aromaticity ($4n + 2$; $n = 4$).

Reference: F. Sondheimer, R. Wolovsky, and Y. Amiel, *J. Am. Chem. Soc.*, **84**, 274 (1962).

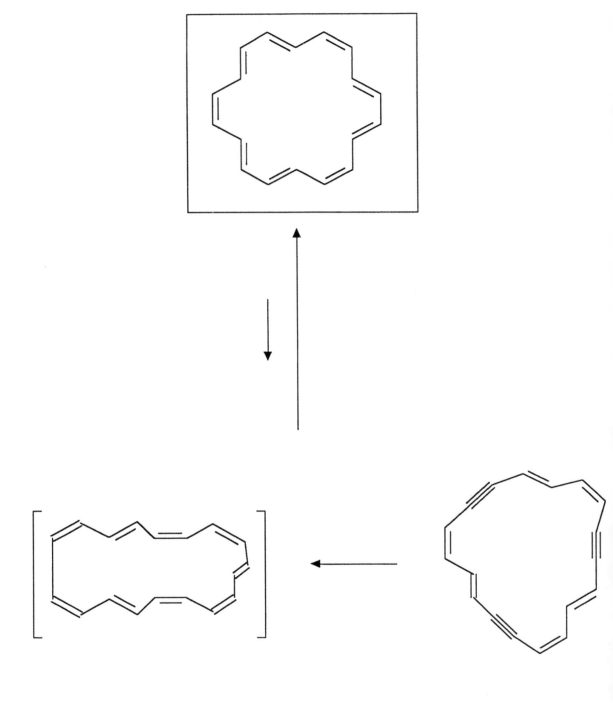

# 2 Selectivity in Catalytic Hydrogenation of Double Bonds

The substrate is limonene, the essence of lemon, which gives this fruit its characteristic odor. Catalytic hydrogenation (3–4 bar $H_2$; Pt/C) affords the monohydrogenation product in quantitative yield (98%). The regioselectivity is due to the difference in stability (reactivity) between the two olefinic double bonds present in limonene: the less stable disubstituted double bond is hydrogenated more rapidly (regioselectivity).

The second example demonstrates the stereoselectivity of catalytic hydrogenation. The two methyl groups render the superior face less accessible than the inferior face of the C=C double bond. In the presence of a Pd/C catalyst, under dihydrogen pressure of 1 bar, the hydrogenation product is formed in 84% yield. Note that this procedure does not affect the more stable aromatic ring, inert under these conditions.

Reference: G. Stork and J. W. Schulenberg, *J. Am. Chem. Soc.*, **84**, 284 (1962).

## 3   Industrial Use of the Lindlar Catalyst: Retinoids

*cis*-Hydrogenation of a triple bond with dihydrogen, in the presence of the Lindlar catalyst, is of utmost importance in retinoid chemistry carried out at Hoffmann–Roche (Basel). These substances are very efficient, particularly in a preventive manner, against epithelial cancer (skin, bladder etc.). The reaction described in the adjacent scheme affords 7-*cis*-retinol, an isomer of vitamin A; the latter has *trans*-configuration of the 7–8 double bond.

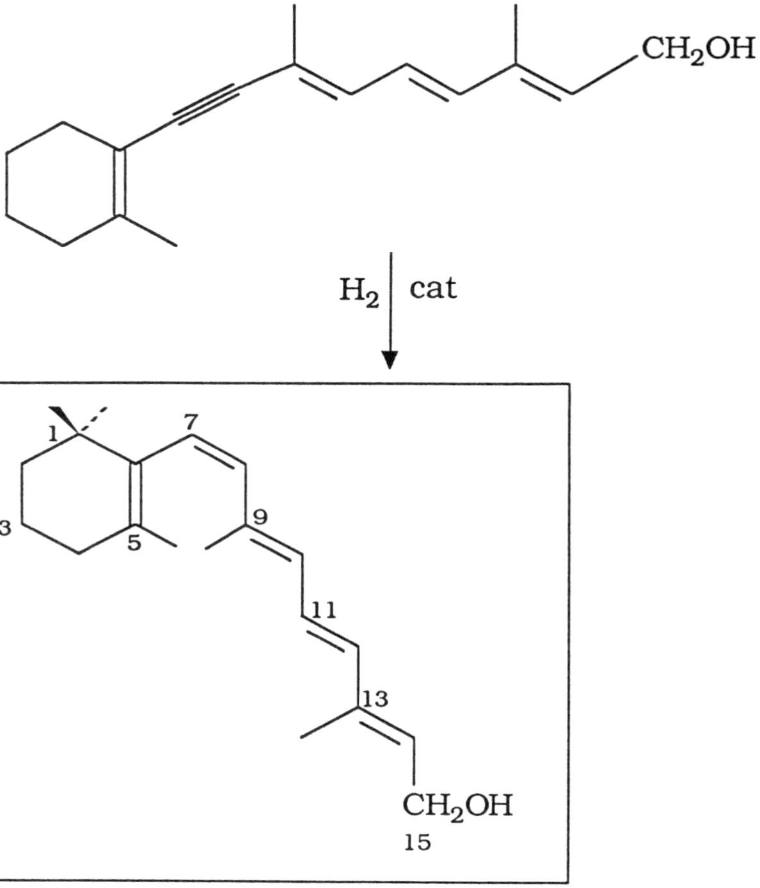

## 4   Comparison of Two Industrial Hydrogenation Procedures

Starting from the acetylenic acid in the next scheme, hydrogenation with $H_2$, using Lindlar catalyst (i) affords the *cis*-olefin quantitatively (yield $>98\%$). In contrast, if the reduction is carried out under Birch conditions (ii) ($Li/NH_3$ in THF at $-40\,°C$), the *trans*-olefin is formed in 98% yield.

References: R. E. A. Dear and F. L. M. Pattison, *J. Am. Chem. Soc.*, **85**, 622 (1963).
        J. Fried, S. Heim, S. J. Etheredge, P. Sunder-Plassmann, T. S. Santhanakrishnan, J. L. Himuzu, and C. H. Lin, *Chem. Comm.*, 634 (1968).

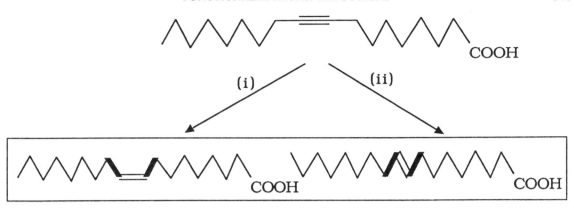

# 5 Stereoselectivity in Birch Reduction of a C=C Double Bond

Reduction of a cyclohexenone under Birch conditions is chemoselective: it affects exclusively the C=C double bond. The result is surprising: the *trans*-decalone is formed in 94% yield; less than 1% of the *cis*-decalone is produced; the former however is destabilized by 1,3-interactions between the two axial substituents (Me, OMe).

The reason is the mechanism of the reduction: the intermediate dianion is either a carbanion conjugated with the enolate (A and B) or a non-conjugated carbanion (C). B suffers from a second diaxial 1,3-repulsion on its lower face. Hence the preferred reaction pathway via A.

Reference: G. Stork and S. D. Darling, *J. Am. Chem. Soc.*, **86**, 1761 (1964).

**A**

**B**

**C**

## 6 Examples of Industrial Catalytic Hydrogenations

Benzene is hydrogenated to cyclohexane in the liquid phase between 150 and 200 °C at 20 to 40 bar $H_2$ over a nickel or platinum catalyst. This recent process requires benzene feed with less than 1 ppm sulfur lest it poisons the catalyst. Also, the exothermicity of the reaction ($\Delta H° = -5$ kcal mol$^{-1}$) requires efficient dissipation of the heat. This is the major production pathway (*ca* 85%) for cyclohexane, produced in about 1 million tons a year in the United States and in about 500 000 tons in Japan.

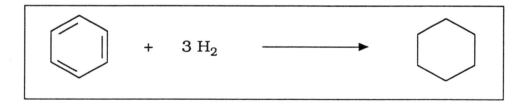

Transformation of phenol to cyclohexanol is the major production pathway for the latter: a nickel catalyst, 150 bar hydrogen pressure, at 150 to 200 °C. The cyclohexanol thus produced is oxidized to cyclohexanone, and then by oxidative cleavage ($HNO_3$ or $O_2$) to adipic acid, used for the preparation of nylon-6,6. The production of adipic acid in the United States amounted to 841 000 tons a year in 1977. It remains now of the same order.

Reference: K. Weissermel and H. J. Arpe, *Chimie organique industrielle*, Masson, Paris, 1981.

## 7  Enantioselective Catalytic Hydrogenation

Several natural products can be used as chiral auxiliary agents, for instance the alkaloid cinchonidine. Hydrogenation (10 bar $H_2$; ethanol, 17 °C) of methyl pyruvate leads to methyl lactate MeCHOH—COOMe of $R$-configuration with a high enantiomeric excess when cinchonidine is pre-adsorbed on the catalyst (platinum deposited on silica).

Cinchonidine can be represented schematically by a capital L. Adsorption on the metal catalyst surface results in almost full coverage. Nevertheless, it does leave catalytic sites open where the substrate can fit in the manner indicated, and not from its other, enantiotopic side: this accounts for the enantioselectivity. Other alkaloids (e.g. cinchonine) allow for hydrogenation of pyruvate to the enantiomeric lactate of $S$ configuration.

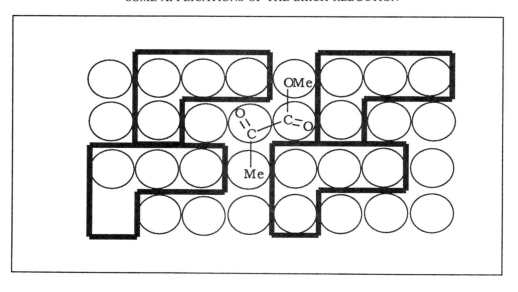

References: D. Lipkin and T. D. Stewart, *J. Am. Chem. Soc.*, **61**, 3295, 3297 (1939).
Y. Orito, S. Imai and S. Niwa, *Nippon Kagaku Kaishi*, 118 (1979); 670 (1980).
A. Ibbotson and P. Wells, *Chem. Brit.*, November, 1004–1005 (1992).

## 8 Some Applications of the Birch Reduction

Anisole affords the 1,4-cyclohexadiene in 79% yield with the donor methoxy group at C-1.

Benzoic acid is also hydrogenated to a 1,4-cyclohexadiene, however, the electron-attracting substituent is found at the $sp^3$ carbon; yield 89%.

The extensive Birch reduction of naphthalene (14 h, $-70\,°C \rightarrow -30\,°C$, Na in $EtOH/Et_2O/NH_3$ 1:1:5) leads to a bis(1,4-cyclohexadiene). The yield is quantitative ($\sim 100\%$).

Birch reduction is possible in the presence of a protected carbonyl group. Of course, if there is competition between an olefinic C=C bond and an aromatic ring, the olefinic bond is reduced selectively.

Reference: J. E. Cole, W. S. Johnson, P. A. Robins, and J. Walker, *J. Chem. Soc.*, 244 (1962).

# 9  Selectivity in Reduction with Hydrides

Cyanoborohydride $BH_3CN^{(-)}$, with the bulky counter ion $^{(+)}NBu_4$, reduces aldehydes much faster than ketones, at room temperature in hexamethylphosphoric acid triamide (HMPA), acidified with $H_2SO_4$ (a pH of about 4). The product mixture consists of 90% of the primary alcohol and only 5% of the 1,2-diol.

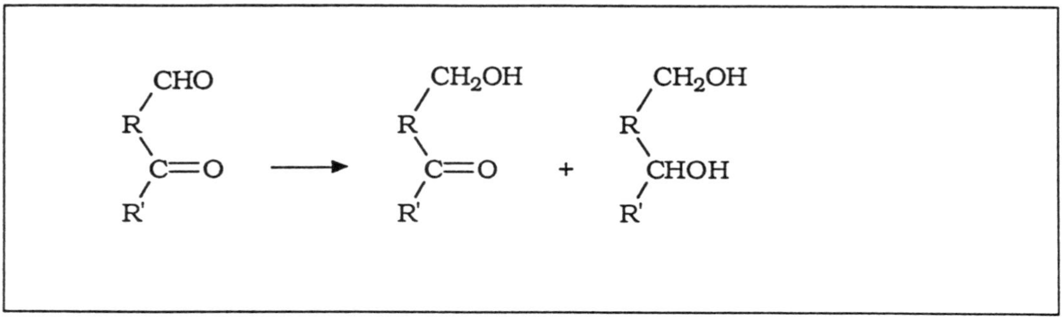

Aluminum hydrides with bulky groups at the metal atom (for instance DIBAL) reduce esters and amide to aldehydes and hemiacetals, respectively. It is interesting (and also useful) that this reduction stops at the aldehyde oxidation stage without leading to the alcohol.

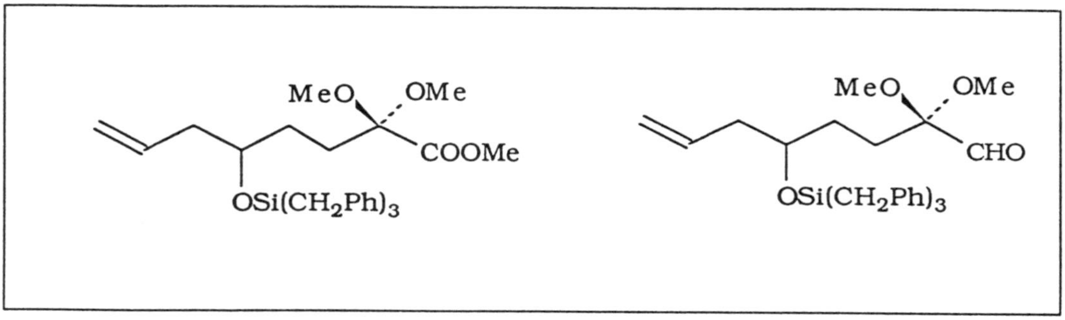

Esters can be reduced to ethers with a combinaton of sodium borohydride $NaBH_4$ and boron trifluoride $BF_3$ in ether provided that it is an ester of an $\alpha$-branched alcohol:

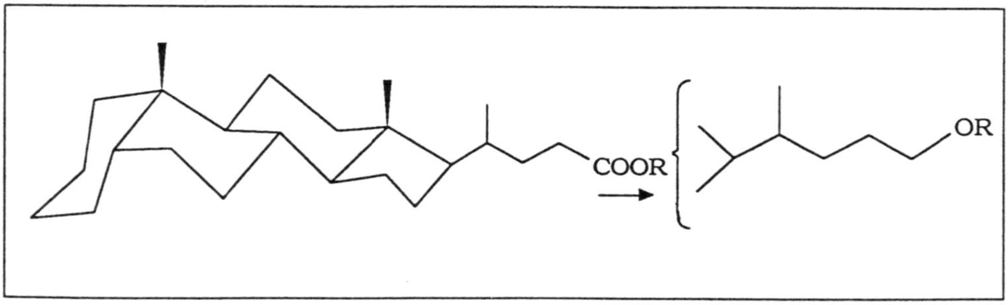

References: R. O. Hutchins and D. Kandasamy, *J. Am. Chem. Soc.*, **95**, 6131 (1973).
E. J. Corey, K. C. Nicolaou, and T. Tora, *J. Am. Chem. Soc.*, **97**, 2287 (1975).
G. R. Pettit and D. M. Piatrak, *J. Org. Chem.*, **27**, 2127 (1962).

| R | yield (%) |
|---|---|
| Ar | 0 |
| n-Bu | 7 |
| s-Bu | 76 |
| t-Bu | 41 |

We have already discussed chemoselectivity in reduction of ketones containing conjugated C=C double bonds: a C=C bond is reduced under Birch conditions without affecting the C=O bond. In contrast, can a C=O bond be reduced without affecting the conjugated C=C bond?

The solution is to use alane $AlH_3$, produced *in situ* by the reaction:

$$3 \ LiALH_4 + AlCl_3 \rightarrow 2 \ AlH_3 + 3 \ LiCl$$

in like manner to $BH_3$ (isoelectronic systems), $AlH_3$ exists in the form of the dimer with two three-center bonds Al—H—Al].

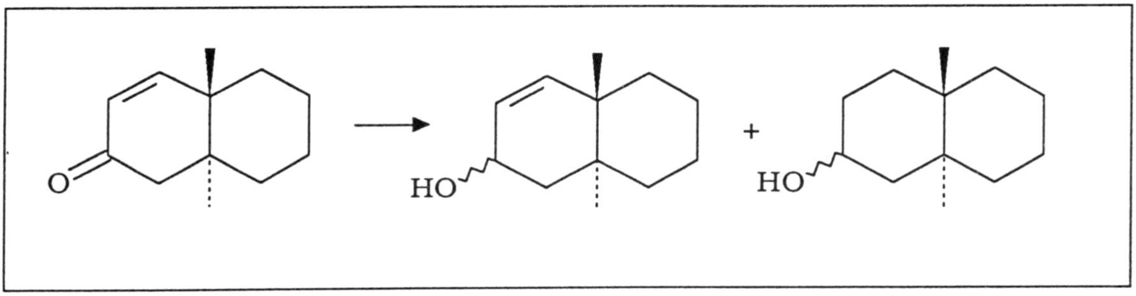

In the first case shown in the above scheme, 93% of allylic alcohols and only 2% of saturated alcohols form.

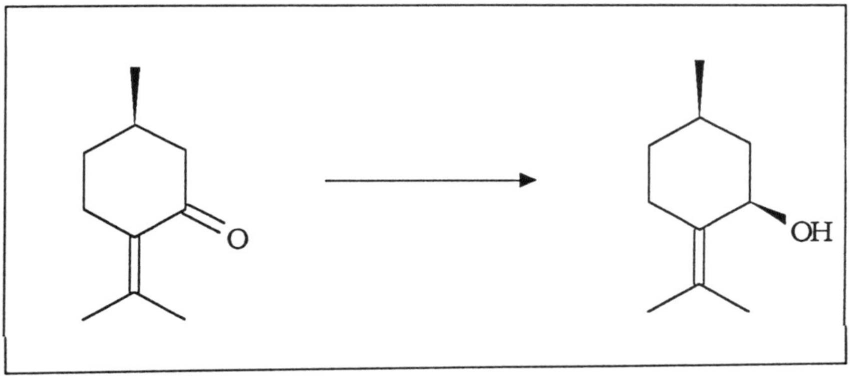

A second solution is possible by resorting to sodium borohydride, in presence of cerium (III) salts. This is the reagent used in the second case (yield 100%). The stereospecificity of the reaction should also be noted.

References: D. C. Wigfield and K. Taymaz, *Tetrahedron Lett.*, 4841 (1973).
W. G. Dauben, R. G. Williams, and R. D. McKelvey, *J. Am. Chem. Soc.*, **95**, 3932 (1973).
J. L. Luche, L. Rodriguez-Hahn, and P. Crabbé, *Chem. Comm.*, 601 (1978).

## 10　Stereoselectivity in Reduction with Metal Hydrides

Luche conditions ($NaBH_4$/Ce(III)) allow for quantitative reduction (99%) of the carbonyl group only, and the only alcohol formed results from attack of the nucleophile $H^{(-)}$ on the rear side (i.e. the $\alpha$-face of the steroid).

The bulkier the substituents at the metal hydrides $M^{(+)}H^{(-)}$, the greater is their tendency to attack from the most accessible side. For instance, lithium tri-$t$-butoxyaluminate $Li^{(+)}$ $AlH(O\text{-}t\text{-}But)_3$ gives a quantitative yield of the two epimeric alcohols (adjacent scheme), with a stereoselectivity favoring the 17$\alpha$-epimer by a factor 9:1.

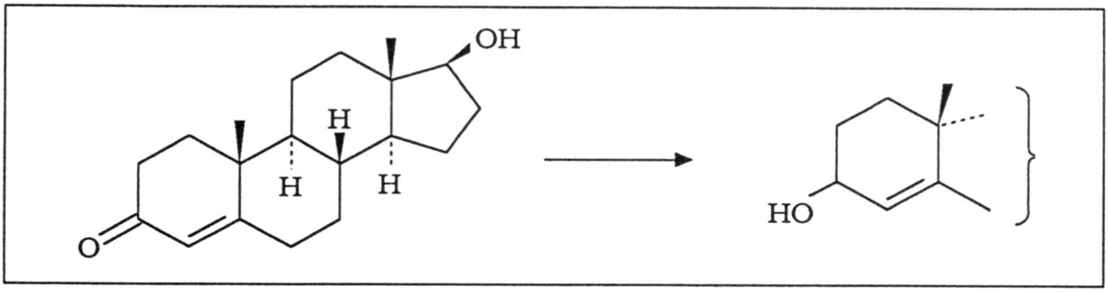

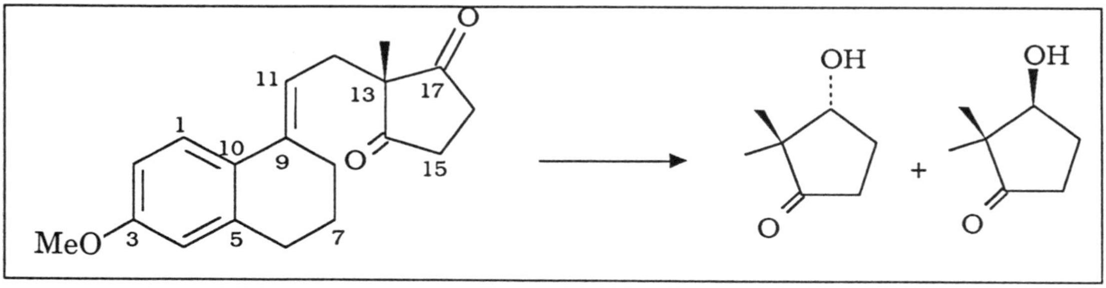

The same is true for more complex boranes such as $Li^+BHSia_3^-$ (Sia = $s$-isoamyl = 1,2-dimethylpropyl): this reagent reduces 3-methylcyclohexanone to the axial alcohol, the thermodynamically less stable product. Attack leading to the equatorial alcohol is disfavored due to steric repulsive interactions with the substituents at C-3 and C-5.

References: J. L. Luche, L. Rodriguez-Hahn, and P. Crabbé, *Chem. Comm.*, 601 (1978).
　　　　　　C. H. Kuo, D. Taub, and N. L. Wendler, *J. Org. Chem.*, **33**, 3126 (1968).
　　　　　　H. C. Brown and S. Krishnamurthy, *J. Am. Chem. Soc.*, **94**, 7159 (1972).

$$Li^{(+)}AlH(OMe)_3 \ + \ \left( \begin{array}{c} \end{array} \right)_3 \xrightarrow{\Delta} Al(OMe)_3 \ + \ Li^{(+)}BHSia_3{}^{(-)}$$

## 11 Increasing the Productivity of Silk Worms

The tetracyclic molecule indicated is 20-hydroxyecdysone. This hormone of the steroid family provokes moulting in silk worms. In the latter it is formed from the precursor α-ecdysone without an oxygen function at C-20.

When ecdysones are added to the food of silk worms, to the extent of 15 mg per 20 000 larvae at a given moment in their development, the colony synchronizes the spinning of the silk cocoons. This treatment is usefully complemented with administration of lauryl alcohol. This has the effect of repelling the larvae to their nesting sites where the spinning shows the best quality.

[Lauryl alcohol has the formula $H_3C(CH_2)_{10}CH_2OH$]

cholesterol

The silk worm *Bombyx mori* is unable to synthesize these ecdysones! Generally, insects are deprived of the ability of biosynthesizing endogenous steroids required for their growth, development and reproduction. For this purpose, they usually resort to metabolic transformations of their cholesterol. The latter is no exception; insects are unable even to carry out total biosynthesis of cholesterol. Vegetarian and omnivorous insects use plant steroids which they transform to cholesterol.

Thus *Bombyx mori* is capable of using sitosterol, a plant steroid, and transforms it to cholesterol as indicated. Other plant steroids serving as starting material for cholesterol production by insects such as *Bombyx mori* are stigmasterol and camposterol. Their structural formulas are not given here; however the metabolic pathways for their conversion to cholesterol are known in detail.

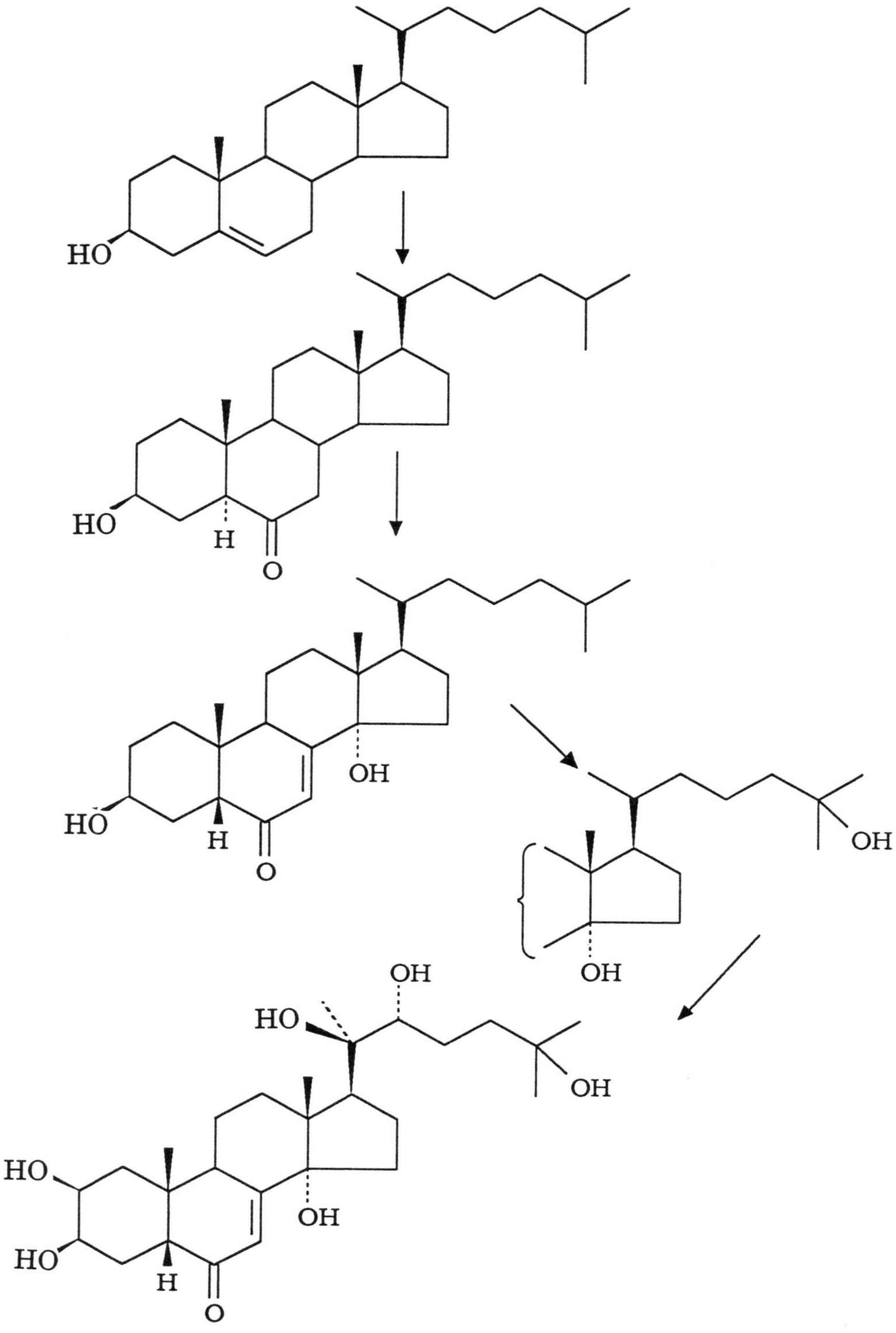
HO
HO
H
O
OH
HO
H
O
OH
OH
OH
OH
HO
OH
HO
HO
OH
H
O

The silk worms *Bombyx mori* prosper when their food contains 0.1 to 0.5% of plant sterols (or directly cholesterol). Their prothoracical glands secrete $\alpha$-ecdysone, whose biosynthesis from cholesterol is shown in the adjacent scheme. The conversion from $\alpha$-ecdysone to 20-hydroxyecdysone occurs in other organs (such as the Malpighi tubules, and the membrane tissues of the body and the intestine). The ovary is a privileged site for this transformation. It accounts for 60% of the body weight of the insect at certain stages of the development (pupa). This is a remarkable organ since ecdysone steroids from the prothorax bring it to maturity; it is enriched with ecdysteroids from the hemolymph; it metabolizes ecdysteroids to glycoconjugates, the form for storage.

Synthesis of ecdysone is of great industrial importance since (as stated above) it improves the quality of silk production and also increases the quantity by synchronization of the cocoon spinning by the larvae. In contrast to biosynthesis, one of the chemical synthesis of ecdysones uses diosgenine as starting material. Before describing the key steps of this synthesis by Nakanishi, a brief digression about this substance is in order.

At the beginning of the Thirties, chemists, in like manner as insects, were making steroid hormones from cholesterol or stigmasterol. The yields were low; accordingly, the costs were high: 1000 dollars per gram of progesterone.

Russell Marker was a chemist who worked in this area, first at the Rockefeller Institute, then from 1934 onwards at Pennsylvania State College. He was looking for cheap vegetal sources for steroids; this incited him to explore the possibilities of diosgenine. During a stay in Texas at the end of 1941, Marker learned the existence of a Mexican Dioscorea plant with tubercles of up to 250 pounds, which appeared to him as a promising source of diosgenin.

He left for Mexico. In January 1942, at the US Embassy in Mexico he became acquainted with a Mexican botanist. They went to the State of Veracruz in order to collect plants, locally referred to as *cabeza de negro*. Marker hired a truck and the botanist showed up accompanied by his girl friend, his mother as chaperone and an interpreter. He prevailed upon Marker to accept this whole crew. The first expedition was a failure: they returned to Mexico City five days later *without any cabeza de negro*. Not letting himself be discouraged, Marker returned on his own on a bus to the Mexican countryside. He came back to Penn State with a sample of this yam. In October 42, having succeeded in convincing the general director of Parke–Davies of the economic soundness of his project, Marker left again for Mexico with half of his savings.

He came back to Penn State with the extract of about 10 tons of tubercles of *cabeza de negro* (*Dioscorea composita* and *D. terpinapensis*). In a colleague's laboratory, he transformed it to progesterone, of which he got about 3 kg. This colleague bought a third of the total amount from him and was able to resell it for 80 dollars a gram!

During his stay in Mexico in 1942, Marker concluded an agreement with the owner of Hormona Laboratories to set up a new company Syntex to manufacture steroids from *cabeza de negro* (also referred to as *barbasco* in the Latin American world). In 1944, Marker was producing already 30 kg of progesterone and 10 kg of dehydroxy-isoandrosterone yearly by this method.

The follow-up to the story was less rosy: Marker became cheated of his 40% share in Syntex. He attempted to fight back and founded another, competitive company, but was forced to close it. In 1949, he gave up chemistry. Today Marker's methods are still used, and take up about 60 000 tons of Mexican Dioscoreas a year.

References: N. Ikekawa, *Experientia*, **39**, 466–473 (1983).
R. E. Marker, R. B. Wagner, P. R. Ulshafer, E. L. Wittbecker, D. P. J. Goldsmith, and C. H. Ruof, *J. Am. Chem. Soc.*, **69**, 2167–2230 (1947).

The key step in Nakanishi's ecdysone synthesis is the reductive cleavage of the five-membered tetrahydrofuran ring with zinc powder in acetic acid.

The simplest way to understand this reaction is by analogy with the nucleophilic substitution by a hydride anion of an oxygen leaving group, which goes with inversion of configuration. In the present case, the oxygen-bearing carbon is not directly attacked. Attack of the nucleophile from the reducing agent occurs at the δ-carbon, at the end of the conjugated system: it is a vinylogous reaction, i.e. transmitted by π electrons.

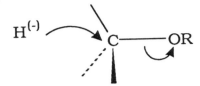

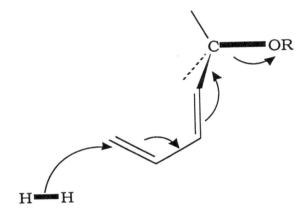

Reference: E. Lee, Y.-T. Liu, P. H. Solomon, and K. Nakanishi, *J. Am. Chem. Soc.*, **98**, 1634–1635 (1976); *Chem. Eng. News*, 13 July, p. 48 (1987).

# 10.I
# Functionalizations: Oxidations

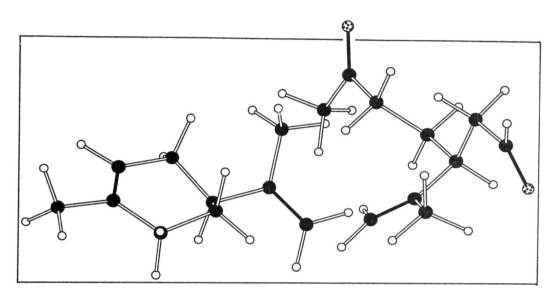

Oxidative functionalizations cover a wide range. Some of the most interesting oxidations affect simultaneously two adjacent carbons; for instance dihydroxylation and epoxidation or ozonolysis of a C=C double bond. These reactions together with the allylic oxidation and the oxidation of alcohols to carbonyl compounds are the topics of this chapter.

In multistep synthesis, oxidation affects in general carbons already partly oxidized, i.e. functionalized carbons; whereas direct functionalization of saturated hydrocarbons (natural gas from the North Sea) is an important goal for the petrochemical industry. As an example: oxidation of cyclohexane to cyclohexanol (using oxygen and cobalt(II) as catalyst) on a scale of a million tons per year leads to nylon 6.

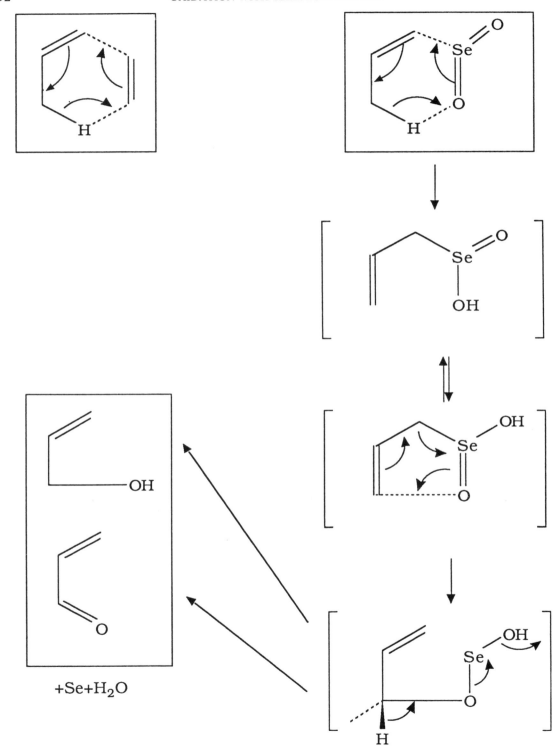

+Se+H$_2$O

# 1 Oxidation with Selenium Dioxide

(A) Molecules with allylic C—H bonds constitute a first type of substrate. This oxidation resembles the ene-reaction: it shares with it a thermally-allowed six-electron transition state, obeying the Dewar–Zimmerman rule.

The first intermediate formed still possesses an Se=O double bond. A simple internal rotation around the C—Se bond provides again a cyclic six-electron transition state.

This sigmatropic [2,3]-shift affords a second intermediate having two Se—O single bonds; the latter either forms the corresponding allylic alcohol on solvolysis or, provided the initial olefin has at least two allylic hydrogens, undergoes an oxidative cleavage at higher temperature with elimination of a selenium atom and a molecule of water. The mechanism of this fragmentation is shown with arrows to indicate electronic migration, in the usual manner.

References: K. B. Sharpless and R. F. Lauer, *J. Am. Chem. Soc.*, **94**, 7154 (1972).
D. Arigoni, A. Vasella, K. B. Sharpless, and H. P. Jensen, *J. Am. Chem.*, **95**, 7917 (1973).
K. B. Sharpless, A. Y. Teranishi, and J. E. Bäckvall, *J. Am. Chem. Soc.*, **99**, 3120 (1977).

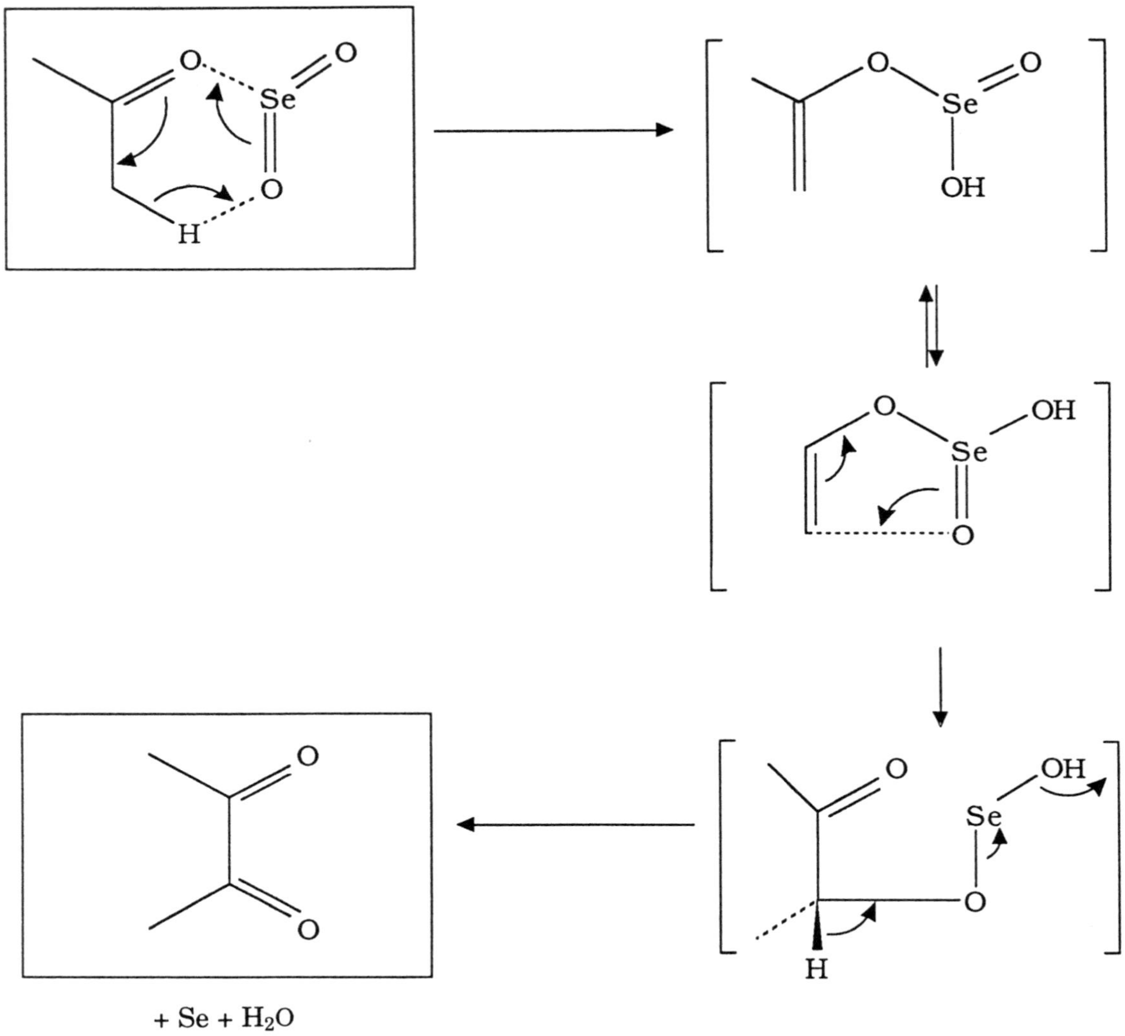
O
Se
O
O
H
O
Se
O
OH
O
Se
OH
O
O
O
Se
OH
O
H
O
O
+ Se + $H_2O$

(B) In a similar way molecules with enolizable carbon–hydrogen bonds $\alpha$ to the carbonyl are also oxidized with $SeO_2$ and transformed to 1,2-diketones. The latter are rather reactive molecules since they incorporate the Coulomb repulsion of two electric dipoles associated with two adjacent carbonyl groups. The driving force of this process is the fragmentation to a molecule of water and a selenium atom.

References: N. Rabjohn, *Org. React.*, **5**, 331 (1949); **24**, 261 (1976).
E. N. Trachtenberg, in R. L. Augustine, Ed., *Oxidation*, M. Dekker, New York, 1969, Vol. 1, Ch. 3, p. 125.

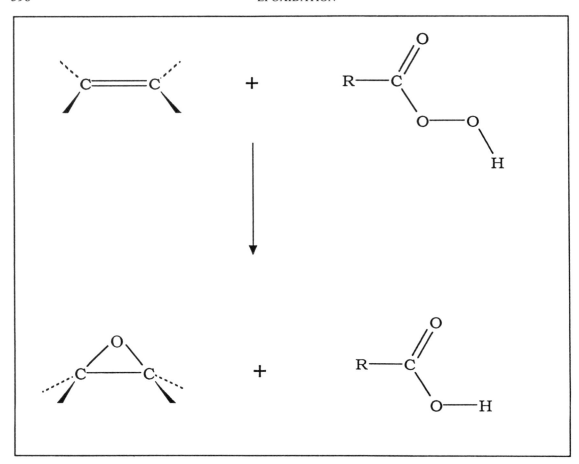

## 2 Epoxidation

Epoxidation of C=C double bonds is often used for introducing an oxygen atom into an organic molecule. This reaction, carried out with peracids RCOOOH, can be considered as addition of an electrophile (the oxygen atom) to a double bond. The reactivity of the peracids increases with their increasing acidity:

| R: | $CH_3$ | $C_6H_5$ | $m\text{-}Cl-C_6H_4$ | H | $p\text{-}NO_2-C_6H_4$ | $CF_3$ |
|---|---|---|---|---|---|---|
| $pK_a$: | 4.8 | 4.2 | 3.9 | 3.8 | 3.4 | <0 |

The peracid most commonly used in practice used to be *m*-chloro-perbenzoic acid. As a hazardous chemical (storage; transport) it is no longer used in the laboratory. Numerous possible replacements have been suggested but none of these has yet emerged as a true alternative.

Since epoxidation is an electrophilic addition to a double bond, it occurs on the least hindered face (stereoselectivity); the oxygen atom affixes itself onto the most electron-rich double bonds, thus the most substituted ethylenic bonds are preferentially epoxidized (regioselectivity).

References: B. Plesnicar, in W. S. Trahanovsky, Ed., *Oxidation in Organic Chemistry*, Academic Press, New York, Part C, Ch. III, p. 211, 1978; H. C. Brown, J. H. Kawakami, and S. Ikegami, *J. Am. Chem. Soc.*, **92**, 6914 (1970).
R. B. Woodward, F. E. Bader, H. Bickel, A. J. Frey, and R. W. Kierstead, *Tetrahedron*, **2**, 1 (1958).
W. Knöll and Ch. Tamm, *Helv. Chim. Acta*, **58**, 1162 (1975).

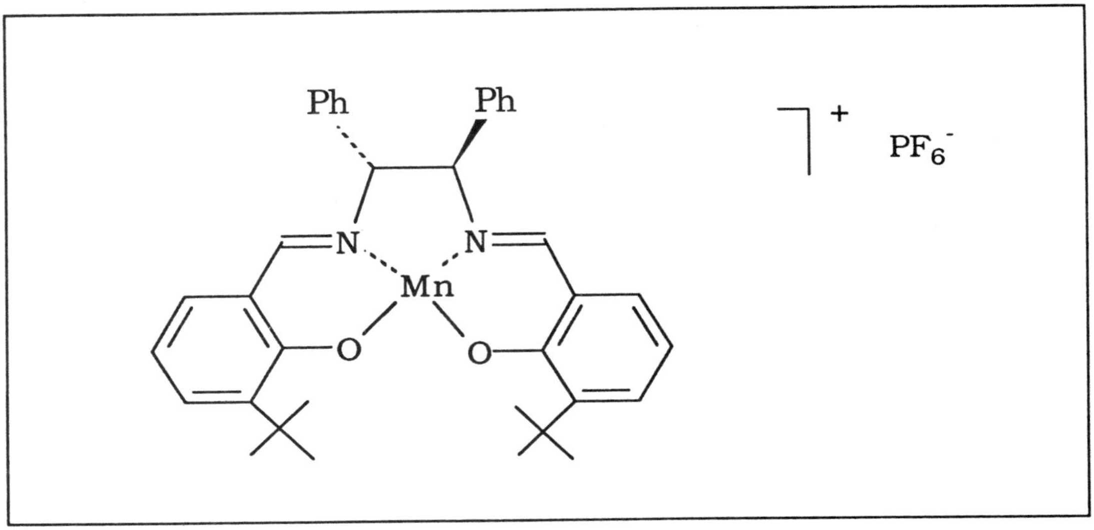
Ph
Ph
N
N
Mn
O
O
+
PF6-

# 3 Enantioselective Epoxidation

There are several procedures available. The Jacobsen procedure uses iodosomesitylene $C_6H_3Me_3IO$ or, better yet, sodium hypochlorite NaOCl (Javel water) as the oxidizing agent. A two-phase system methylene chloride/water at pH 9.5 is necessary. A manganese catalyst (1–8 mol%; 200 catalytic cycles) with the metal forced into a chiral complex of $C_2$ symmetry was developed by Corey. With substrates such as 1-methylcyclohexene an enantiomeric excess of between 20 and 93% is obtained, with the Jacobsen procedure.

References: E. J. Corey, R. Imwinkelried, S. Pikul, and Y. B. Xiang, *J. Am. Chem. Soc.*, **111**, 5493 (1989).
W. Zhang, J. L. Loebach, S. R. Wilson, and E. N. Jacobsen, *J. Am. Chem. Soc.*, **112**, 2801 (1990).

| K | Ca | Sc | **Ti** | **V** | **Cr** | **Mn** | **Fe** | **Co** | **Ni** | Cu | Zn |
|---|----|----|----|----|----|----|----|----|----|----|----|
| Rb | Sr | Y | **Zr** | **Nb** | **Mo** | **Tc** | **Ru** | **Rh** | **Pd** | Ag | Cd |
| Cs | Ba | La | **Hf** | **Ta** | **W** | **Re** | **Os** | **Ir** | **Pt** | Au | Hg |

$$MnO_4^- + 3e^- \rightarrow MnO_2 \qquad 1{,}695V$$

$$Ce^{4+} + e^- \rightarrow Ce^{3+} \qquad 1{,}61$$

$$MnO_4^- + 8H^+ + 5e^- \rightarrow Mn^{2+} \qquad 1{,}51$$

$$Mn^{3+} + e^- \rightarrow Mn^{2+} \qquad 1{,}51$$

$$Tl^{3+} + 2e^- \rightarrow Tl^+ \qquad 1{,}25$$

$$MnO_2 + 4H^+ + 2e^- \rightarrow Mn^{2+} \qquad 1{,}23$$

$$OsO_4 + 8H^+ + 8e^- \rightarrow Os \qquad 0{,}85$$

$$Fe^{3+} + e^- \rightarrow Fe^{2+} \qquad 0{,}771$$

# 4 Correct Choice of an Oxidizing Agent

The most common transition metals used in oxidations are indicated in bold faced letters.

The correct choice of a reagent depends on its redox potential. This explains the frequent use of permanganate and of osmium tetroxide in oxidations.

References: P. Laszlo, in *Comprehensive Organic Synthesis*, Vol. 7, *Oxidation*, S. V. Ley, Ed. Pergamon Press, Oxford, 1991, pp. 839–848.

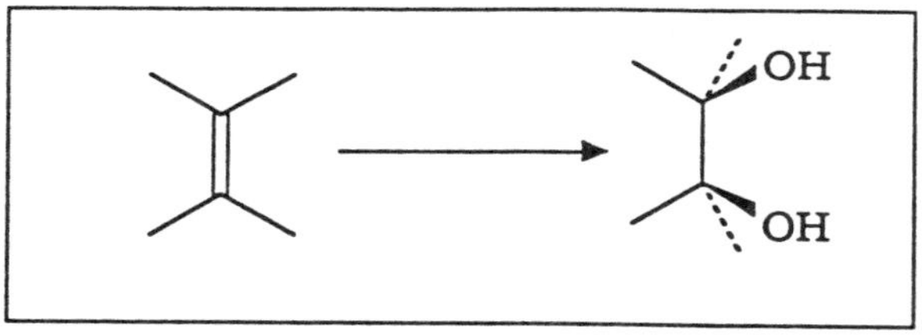
OH
OH

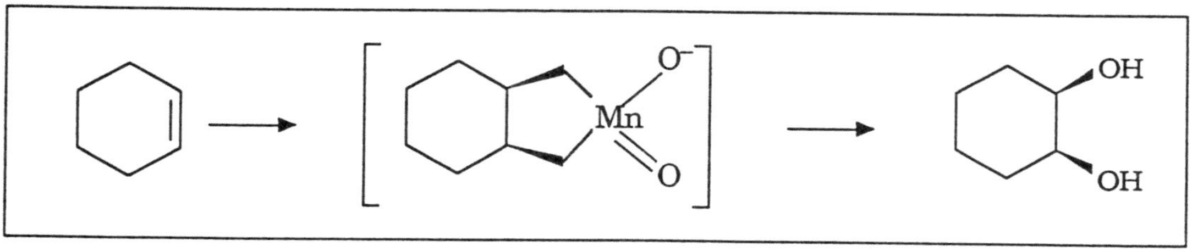
O⁻
Mn
O
OH
OH

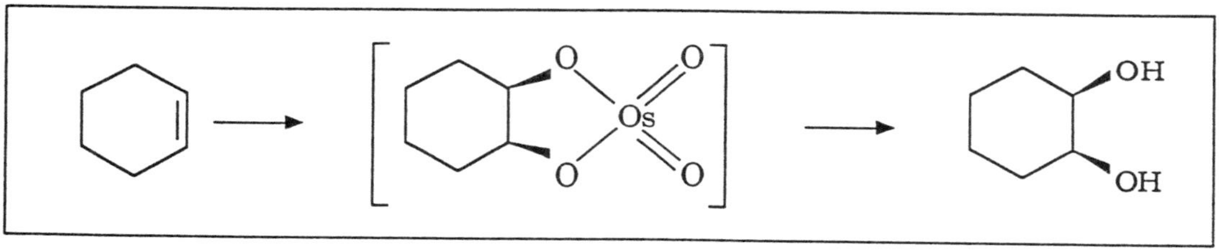
O
O
Os
O
O
OH
OH

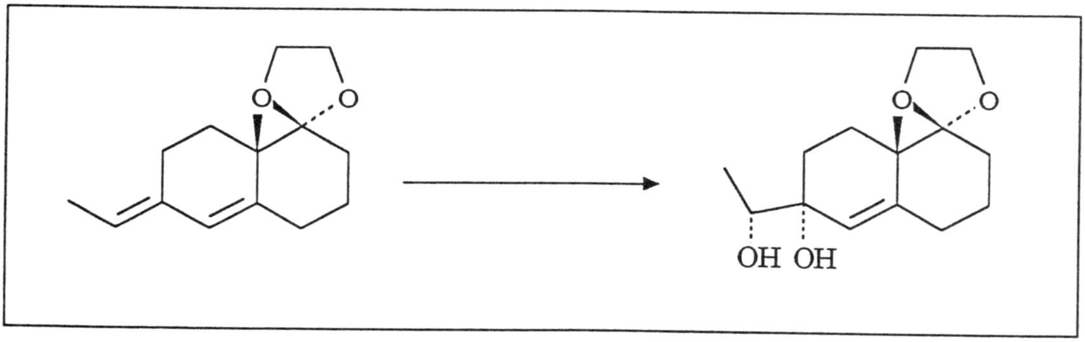
O
O
O
O
OH OH

# 5  *cis*-Hydroxylation of Alkenes

The two most common reagents used for this transformation are potassium permanganate $KMnO_4$ and osmium tetroxide $OsO_4$. The latter, despite being both costly and toxic, affords better yields than the former; hence its common use in fine chemistry.

Permanganate oxidation of an olefin is carried out invariably with $KMnO_4$ in aqueous base (NaOH). The 1,2-diol, corresponding to *syn*-dihydroxylation, results from hydrolysis of the cyclic intermediate (yields of about 40%).

The mechanism of the oxidation with osmium tetroxide is similar; however, due to the absence of side reactions, yields are in the range of 90%. High selectivity is obtained, especially with the pyridine complex $OsO_4.py_2$; this bulky reagent is both regio- and stereoselective, attacking preferentially electron-rich and readily accessible C=C double bonds, as shown in the example indicated in the adjacent scheme; the yield is quantitative (100%).

References: S. Bernstein, R. H. Lenhard, W. S. Allen, M. Heller, R. Littell, S. M. Stolar, L. I. Feldmann, and R. H. Blank, *J. Am. Chem. Soc.*, **78**, 5693 (1956).
E. J. Corey, M. Ohno, P. A. Vatakencherry, and R. B. Mitra, *J. Am. Chem. Soc.*, **86**, 478 (1964).

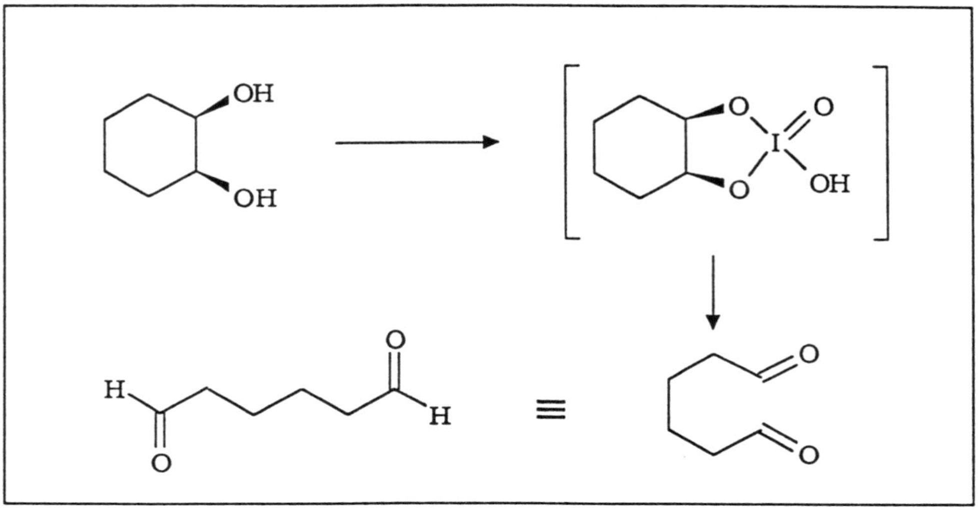
OH
OH
O
O
I
O
OH
H
O
O
H
O
O

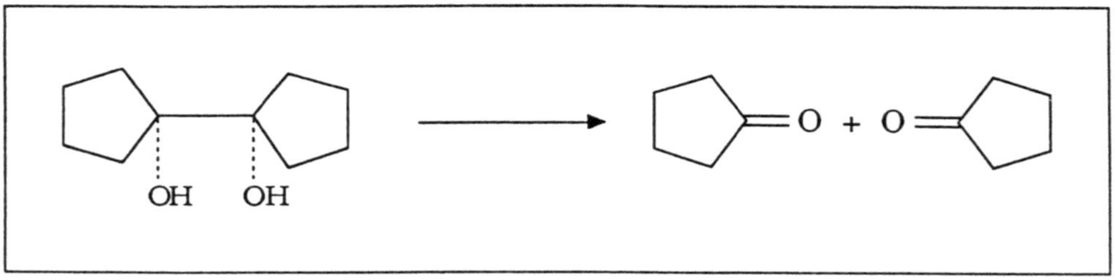
OH OH
O + O

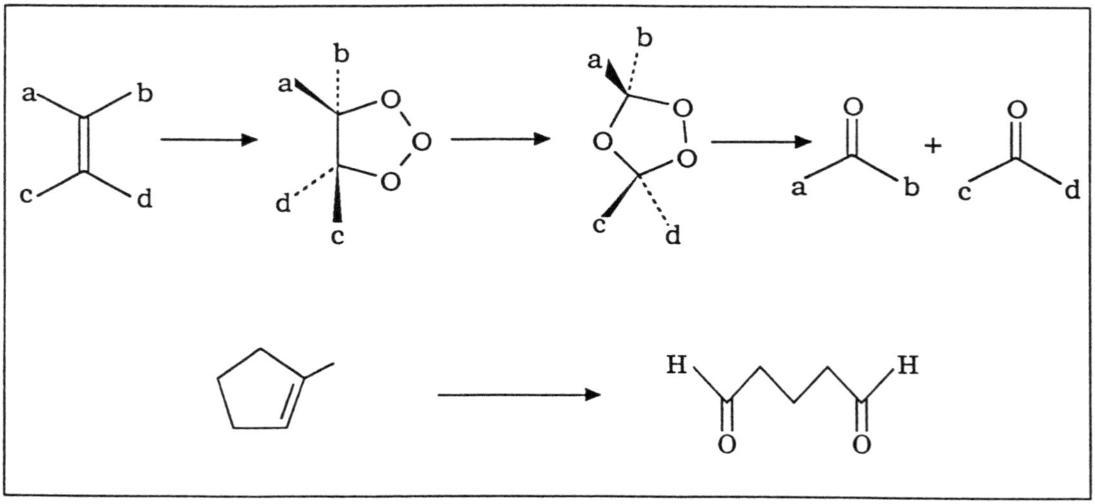
a b c d
O O O
a b c d
H H
O O

# 6   Cleavage of 1,2-Diols

Often the oxidation carried out with $OsO_4$ precedes an oxidative cleavage with periodic acid $HIO_4$. Again a cyclic intermediate is formed. This periodate is hydrolyzed to (a) product(s) characterized by an opening of the C(1)—C(2) bond of the 1,2-diol; yields are in the range of 80%.

# 7   Ozonolysis of Alkenes

Another method for oxidative cleavage of alkenes is the addition of ozone $O_3$. The cycloadduct formed (typical conditions: $CH_2Cl_2$, $-78°$) rearranges to an ozonide with a peroxy bond O—O. Cleavage (Zn, $CH_3COOH/H_2O$) of this ozonide gives rise to the two corresponding carbonyl compounds.

The difficulty which arises in respect to internal relations is to explain how any particular truth is possible. In so far as there are internal relations, everything must depend on everything else. But if this be the case, we cannot know about anything till we equally know about everything else. Apparently, therefore, we are under the necessity of saying everything at once. This supposed necessity is palpably untrue. Accordingly it is incumbent on us to explain how there can be internal relations, seeing that we admit finite truths. A. N. Whitehead, *Science and the Modern World*, Macmillan, New York, 1967, p. 163.

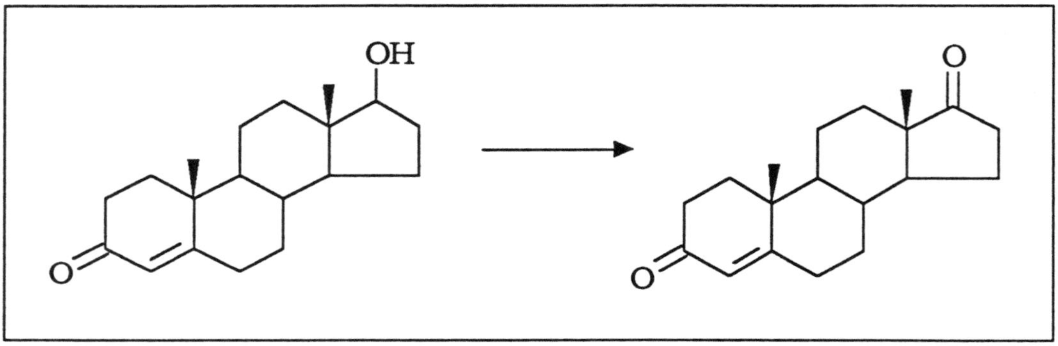
OH
O
O
O

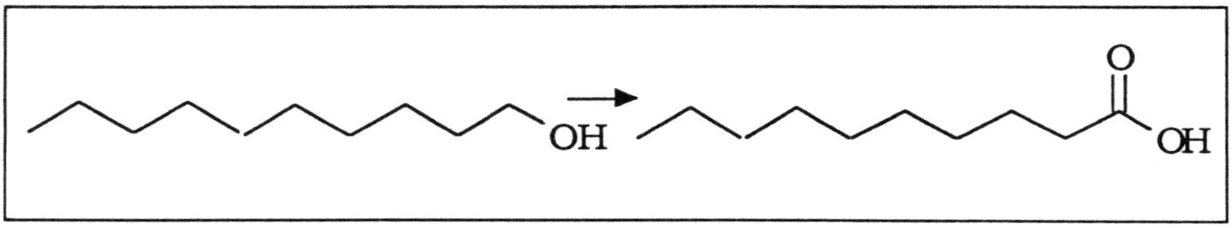
OH
O
OH

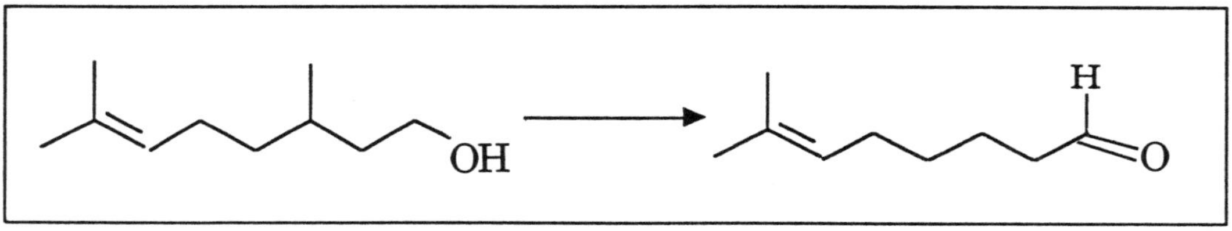
OH
H
O

# 8  Oxidation of Alcohols

Only primary and secondary alcohols can be oxidized under mild conditions, into aldehydes and ketones, respectively. A large number (this is an understatement!) of reagents effect this oxidation.

For secondary alcohols, the Jones reagent (chromium trioxide $CrO_3$ in aqueous sulfur acid) or better pyridine chlorochromate ($C_5H_5N.CrO_3Cl$) are used. The latter for instance oxidizes testosterone to 4-androstene-3,17-dione ($CH_2Cl_2$, 25 °C, yield 82%).

With primary alcohols as substrates, there is a problem of overoxidation: the aldehyde formed can become further oxidized into the carboxylic acid. This happens with the Jones reagent, transformation to the acid is quantitative.

If the aldehyde is the target, a choice reagent is pyridine chlorochromate in methylene chloride: for example the transformation of citronellol (from the rose essence) to citronellal is illustrated in the adjacent scheme.

Reference: H. G. Bosch, in Houben-Weyl, *Methoden der Organische Chemie*, Vol. IV/1b: *Oxidation II*, Thieme, Stuttgart, 1975, p. 429.

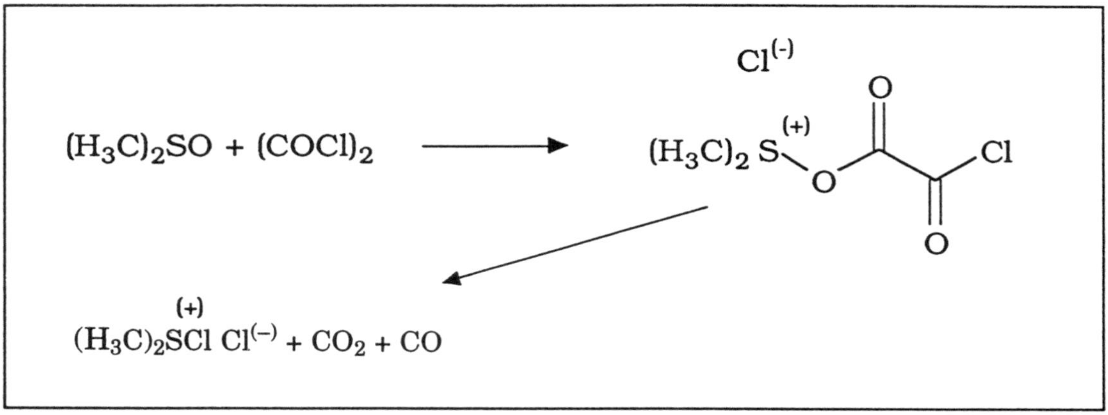

$(H_3C)_2SO + (COCl)_2 \longrightarrow$

$Cl^{(-)}$

$(H_3C)_2 \overset{(+)}{S} \overset{O}{\underset{O}{\overset{\parallel}{C}}} \overset{\parallel}{C} Cl$

$\overset{(+)}{(H_3C)_2SCl}\ Cl^{(-)} + CO_2 + CO$

---

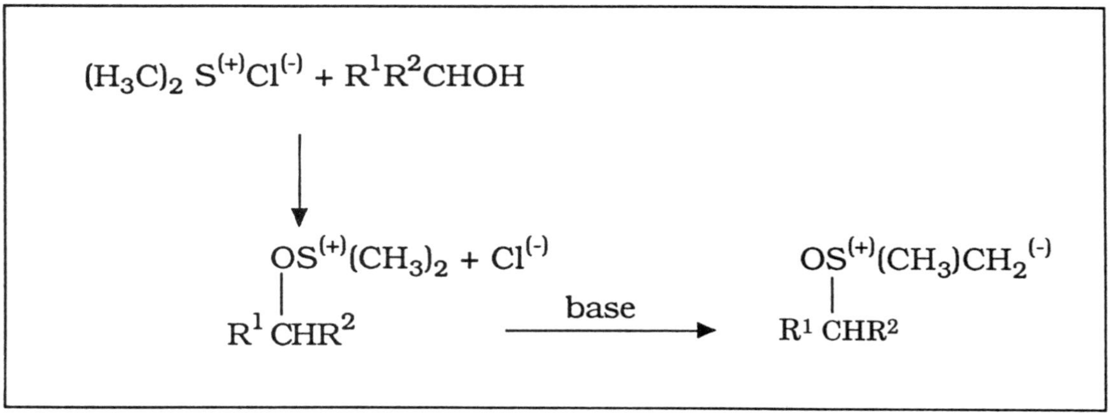

$(H_3C)_2 S^{(+)}Cl^{(-)} + R^1R^2CHOH$

$\overset{OS^{(+)}(CH_3)_2}{\underset{R^1 CHR^2}{|}} + Cl^{(-)} \quad \xrightarrow{\text{base}} \quad \overset{OS^{(+)}(CH_3)CH_2^{(-)}}{\underset{R^1 CHR^2}{|}}$

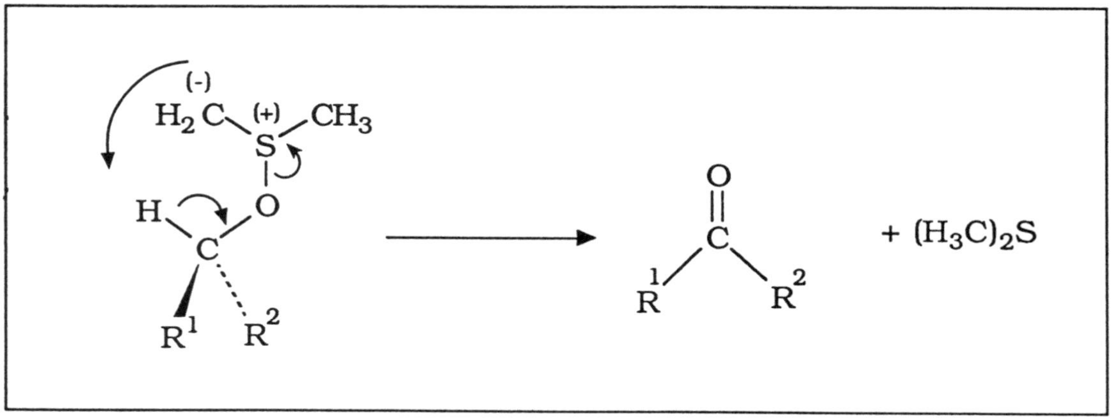

---

$H_2\overset{(-)}{C}\diagdown\overset{(+)}{S}\diagup CH_3$

$H\diagdown\underset{R^1\ \ R^2}{\overset{O}{\underset{|}{C}}}$

$\longrightarrow$

$\overset{O}{\underset{R^1\ \ \ R^2}{\overset{\parallel}{C}}}$

$+ (H_3C)_2S$

# 9 Oxidation of Alcohols by Activated Dimethyl Sulfoxide (Swern)

In the first step, dimethyl sulfoxide (DMSO) is activated by various reagents. The example indicated here is that of oxalyl chloride $(COCl)_2$. The oldest activating agent used for this purpose is dicyclohexyl-carbodiimide (Pfitzner and Moffatt).

The resulting 'activated' sulfonium ion adds the alcohol to be oxidized; and a base (frequently triethylamine) forms a sulfur ylide.

Cleavage of this sulfur ylide affords the ketone and dimethyl sulfide (driving force of the overall transformation); the reagent $DMSO/(COCl)_2$ cleaves also silyl ethers except for protecting groups such as $-SiMe_2t$-Bu.

References: K. E. Pfitzner and J. G. Moffatt, *J. Am. Chem. Soc.*, **85**, 3027 (1963).
   A. J. Mancuso, D. Swern, *Synthesis*, 165 (1981).
   T. T. Tidwell, *Synthesis*, 857–870 (1990).

## Summary: 'Functionalizations: Oxidations'

Allylic oxidation can be effected by selenium dioxide $SeO_2$. The initial addition step is isoelectronic with the ene reaction, which is closely related to the Diels–Alder reaction. The resulting product is either an *allylic alcohol* or an $\alpha$, $\beta$-unsaturated ketone. Stereoselectivity stems from abstraction of the allylic hydrogen from the most accessible half-space.

Epoxidation has the merit of building-up considerable activation in a cheap way. This explains why ethylene oxide—produced from ethylene—is number ten worldwide of the most important organic chemicals, with an annual production in the range of 5 million tons.

Peracids approach an ethylenic double bond on its most accessible face. Their terminal oxygen undergoes electrophilic addition to the double bond with the highest electron density. *m*-Chloro-perbenzoic acid (now classified as a hazardous chemical) used to be the reagent of choice. The reaction, being very sensitive to steric hindrance is highly *selective*. The enantioselective Sharpless epoxidation requires allylic alcohols. The oxidant is *t*-butyl hydroperoxide. With a titanium catalyst and $(+)$- or $(-)$-diethyl tartrate as the chiral auxiliary agent, excellent enantiomeric excesses occur. Enantioselective epoxidation of olefins, more generally, can be effected with reagents such as iodosomesitylene or Javel water (or dioxygen, ideal in practice), with transition metal (Mn, Ni) complexes with $C_2$ symmetry as chiral auxiliary agents.

The more generally used transition metals, the derivatives of which serve as oxidizing agents, include also vanadium, chromium, manganese, iron, molybdenum and osmium. The choice of an oxidizing agent is based on the pertinent redox potentials. Permanganate $MnO_4^-$ affords *cis*-hydroxylation of alkenes whereas osmium tetroxide $OsO_4$ is a very selective reagent for the same goal. Selective oxidations of alcohols use reagents such as chromium trioxide or pyridine chlorochromate.

Oxidative cleavage of a $C=C$ bond is most frequently carried out with ozonolysis; the reaction type is oxidative bond breaking.

Synthesis of geosmine, with the odor of freshly dug soil, is depicted as a succession of reductions and oxidations, which summarizes the contents of chapters 9 and 10.

## 1   Oxidation of Activated C—H Bonds
## with SeO₂

Oxidation of the bicyclic olefin shown with selenium dioxide affords the allylic alcohol. Regio-selectivity as well as the *exo* stereoselectivity are both explained by better accessibility of the allylic C—H bond converted to a C—O bond, toward formation of the first cyclic, six-center transition state.

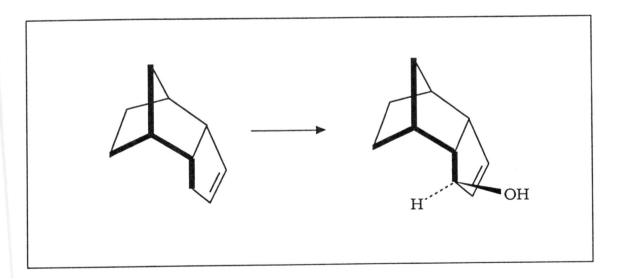

References: N. Rabjohn, *Org. React.*, **24**, 261 (1976).
      H. J. Reich, in W. S. Trahanovsky, Ed., *Oxidation in Organic Chemistry*, Academic Press, New York, Part C, 1978, p. 1.

Oxidation occurs in a regioselective manner, *trans* to the bulkier substituent on the C=C double bond. In the example indicated, the respective yields of the major and minor product are: allylic alcohol of *E*-configuration 70%, alcohol of *Z*-configuration <2%.

Oxidation at the carbon $\alpha$ to the carbonyl group in the cyclopentenone leads quantitatively to the diketone (97% yield).

## 2   Improvement of the Allylic Oxidation
## by SeO$_2$

Oxidations with SeO$_2$ have two disadvantages: they require a stoichiometric amount of SeO$_2$. The selenium by-product is in a colloidal form, difficult to remove. The idea is to re-oxidize the selenium formed, using for this purpose t-butylhydroperoxide $(H_3C)_3C-O-O-H$ (TBHP).

The first example is that of a terminal olefin: transformation to an allylic alcohol (60% yield) occurs with half an equivalent of SeO$_2$, in the presence of two equivalents of TBHP.

In the second example, that of geraniol acetate, the conditions are: 0.1 mol of substrate, 2 mol% SeO$_2$, 0.4 mol of TBHP, affording the product (note the regio- and stereoselectivity) in 55% yield.

## 3   Oxidation of Propene to Acrolein

This reaction is carried out in the gas phase with air dioxygen as the oxidant, with molybdenum(VI) oxygen complexes as catalysts, at temperatures between 350 and 450 °C. The conversion is about 20%, the selectivity with respect to acrolein is ca. 80%. The reaction is strongly exothermic $(-88\,\text{kcal mol}^{-1})$; its driving force is the fragmentation into two neutral molecules, both quite stable.

The largest producer worldwide (24 000 tons a year) is the French company ATOCHEM; they succeeded to Péchiney–Ugine–Kuhlmann in this production.

## 4   Oxidation of Ethylene to Ethylene Oxide

This moderately exothermic reaction ($-25\,\text{kcal mol}^{-1}$) is made feasible by heterogeneous silver catalysts. In fact, the reaction is accompanied by combustion of part of the ethylene and ethylene oxide. The industrial process attains a 70% selectivity for ethylene, with a heat release of about $120\,\text{kcal mol}^{-1}$. Worldwide, production of ethylene oxide—ranking 24th in bulk chemicals and 10th in organic chemicals—is about 5 megatons per year: 2.7 in the US, 0.7 in Japan and 2 in Western Europe. Ethylene oxide owes its importance to its chemical activity, the ring strain makes it a very reactive molecule.

## 5   Selectivities in Epoxidation

### A. Stereoselectivity

The double bond of norbornene (= bicyclo[2.2.1]heptene) is more accessible from the convex *exo* face than from the concave *endo* face. The latter is somewhat masked by the *endo* C(5)-H and C(6)-H bonds. Epoxidation affords 99% of the *exo* epoxide (**X**) and less than 1% of its *endo* isomer (**N**). If however the *exo* face is blocked by a *syn* methyl group at C-7, the stereoselectivity reverses: 90% **N** and less than 10% **X**.

References: H. C. Brown, J. H. Kawakami, and S. Ikegami, *J. Am. Chem. Soc.*, **92**, 6914 (1970).
         W. Knöll and Ch. Tamm, *Helv. Chim. Acta*, **58**, 1162 (1975).
         E. Vogel, M. Biskup, W. Pretzer, and W. A. Böll, *Angew. Chem. Int. Ed. Engl.*, **3**, 642 (1964).

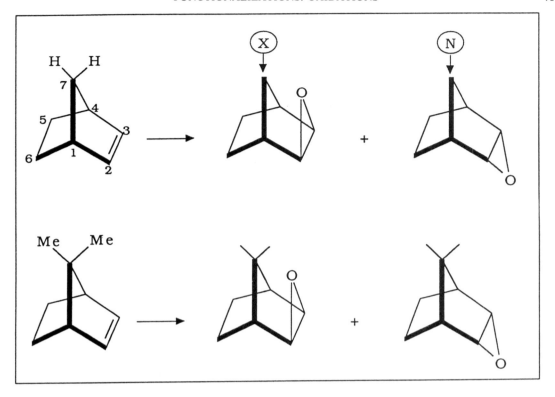

## B. Regioselectivity

Substituted olefinic double bonds are electron-rich due to hyperconjugation of carbon–carbon bonds. Double bonds with the highest electron density are those with the highest number of substituents. Limonene, with an exocyclic disubstituted and an endocyclic trisubstituted double bond, is epoxidized selectively at the latter.

1,4,5,8-Tetrahydronaphthalene, produced by Birch reduction of naphthalene, is epoxidized selectively at the tetrasubstituted double bond.

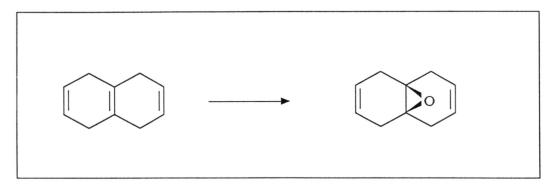

## 6  The Enantioselective Sharpless Epoxidation

*My own beliefs are that the road to a scientific discovery is seldom direct, and that it does not necessarily require great expertise. In fact, I am convinced that often a newcomer to a field has a great advantage because he is ignorant and does not know all the complicated reasons why a particular experiment should not be attempted.—Ivar Glaever, Nobel Prize in physics (1973)*

The selective Sharpless epoxidation, a novel reaction, is a relatively recent acquisition of organic chemistry. We shall sketch out the slow and somewhat tortuous path which led to it. This will provide a representative example of the pathway to discovery: a mix of painstaking patience, plausible inferences, Edinsonian empirism, quest for external suggestions, luck and opportunism, and, above all, a solid dose of intuition.

Let us define the problem: depending on whether the electrophilic oxygen atom attacks one or the other face of *trans*-butene, either the epoxide of *R,R* configuration or that of *S,S* configuration is formed. The desired enantioselectivity is obtaining as high as possible an excess of either enantiomer.

The initial idea was to use an OH group at the allylic carbon as an anchoring group; it will preferably direct the epoxidation to one or the other face. The resulting epoxides will thus be diastereoisomers. Indeed use of TBHP, in the presence of $V^{5+}$ salts as catalyst, affords an epoxidation product (98%) having the *two oxygen functions in the same half-space*.

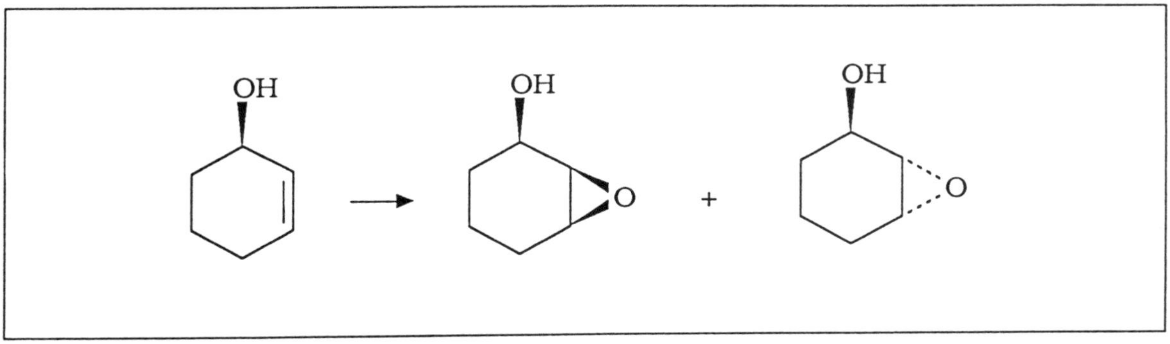

As a working hypothesis, catalysis by the metal can be represented as in the adjacent scheme (denoting the metal by M). An intermediate metal complex with a covalent O—M bond between the allylic alcohol and the metal would form. It would account for the observation that only allylic alcohols ionizable into alkoxides are eligible for such diastereoselectivity. This is not the case for their methyl ethers, for instance, as tested in control experiments. R in the Scheme stands for the TBHP *t*-butyl group.

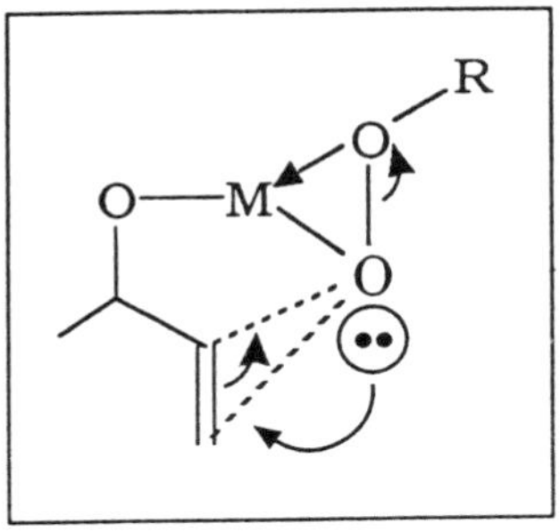

Generalization is forthcoming: when open-chain allylic alcohols are submitted to the same epoxidation conditions (TBHP/$V^{5+}$/CH$_2$Cl$_2$), there is a strong predominance (at least 80%) for the *erythro* isomers. This can be explained by the same scheme with a covalent intermediate involving the metal. How could one increase even more this *erythro* preference?

Let us consider a plausible structure for the catalyst: the —OR functions at vanadium (V) (as well as at molybdenum, which also functions as catalyst) correspond, in the above hypothesis, to catalytic sites. The oxo groups (of the type M=O) are not usable; on the contrary, they block potential sites for attachment of a chiral coordinating group L, which could induce enantioselectivity. This analysis led Sharpless to titanium complexes.

K. Barry Sharpless therefore attempted to epoxidize geraniol with TBHP, using Ti(O-*i*Pr)$_4$ as a catalyst. He found—this is the key observation—that epoxidation works better (faster and with higher yields) in the presence of a chelating ligand L, as indicated in the following Scheme: a pyridine-based dicarboxylic acid.

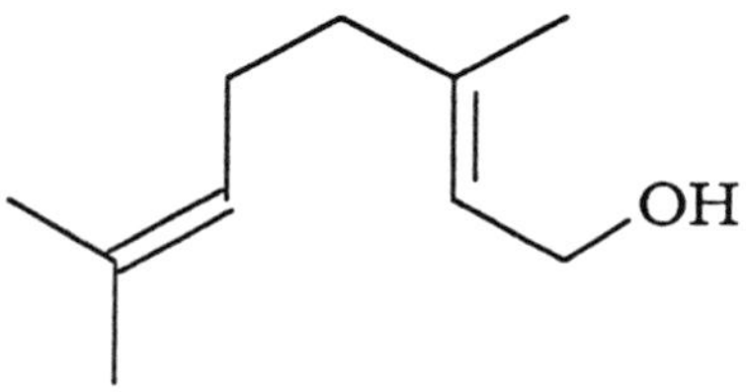

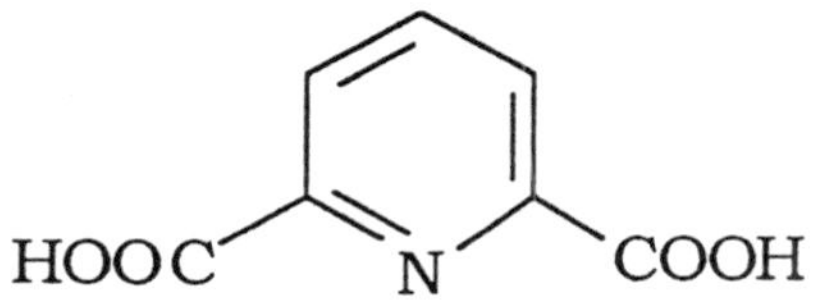

A plausible transition state, responsible for this catalysis, is indicated in the following Scheme.

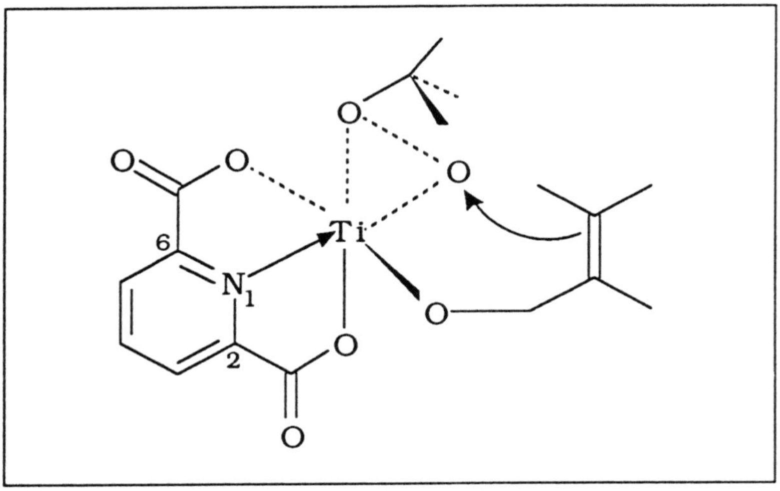

Sharpless and his team then tested a great number of pyridine ligands of this type, bearing potentially coordinating groups at carbons C-2 and C-6, and having at most C$_2$ symmetry (i.e. having a binary C$_2$ axis), consistent with chirality: none of these were satisfactory.

A proposal formulated by his colleague Harry Mosher, a specialist in asymmetric induction, finally cleared the way. When Sharpless carried out epoxidation using Ti(O-*i*Pr)$_4$ + TBHP, in the presence of diethyl tartrate, the ratio *erythro/threo* is 97:3, and the reaction stops exactly at 50% conversion.

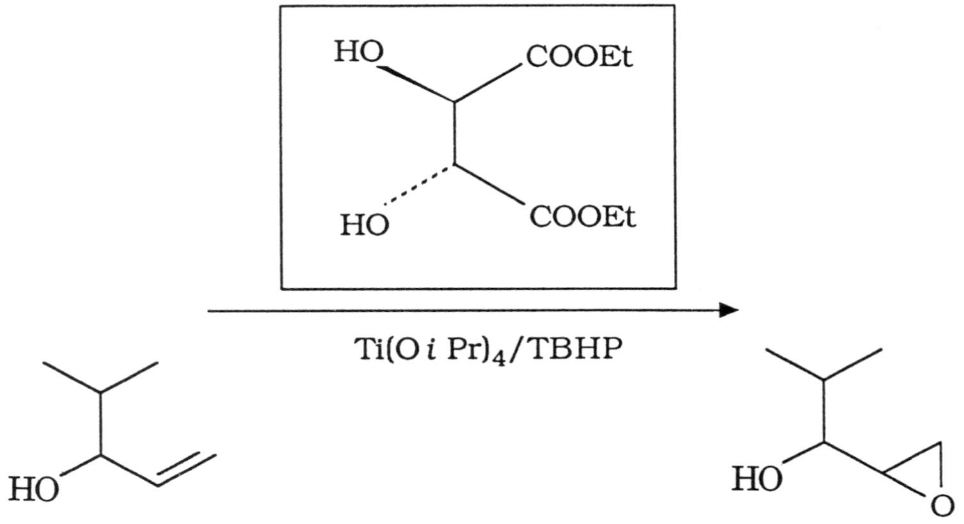

HO.          COOEt

HO          COOEt

$Ti(O\, i\, Pr)_4/TBHP$

HO

HO          O

The explanation is self-evident: a single enantiomer of the allylic alcohol substrate is being epoxidized. Enantioselective epoxidation had been discovered!

The procedure established in this way is summarized in the following Scheme. The conditions are: $TBHP/Ti(O\text{-}iPr)_4/CH_2Cl_2/-20\,°C/\text{diethyl tartrate}$.

Natural dextrorotatory ($+$) diethyl tartrate is part of the complex which performs epoxidation almost exclusively from the inferior face (in the following Scheme), whereas use of non-natural (synthetic) laevorotatory ($-$) diethyl tartrate leads to attack only from the superior face. The yields are good (70–90%), and the enantiomeric excess ($>90\%$) is particularly good.

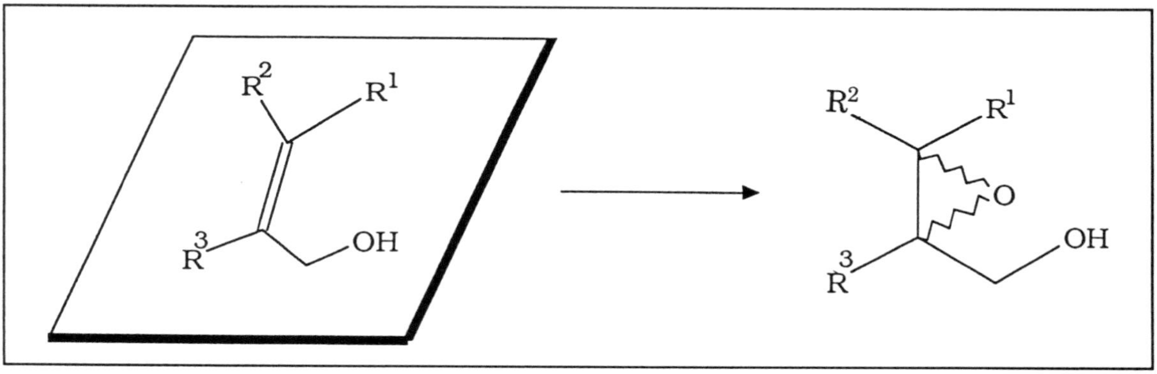

$R^2$     $R^1$

$R^3$         OH

$R^2$     $R^1$

                    O

$R^3$          OH

This Sharpless epoxidation was discovered in 1980. The first commercial application (1981) was the synthesis of disparlure, a sex pheromone of a moth, the *gipsy moth*. In 1982, it enabled the total synthesis of eight L-hexoses. Since 1985, the pharmaceutical company Upjohn has used this procedure for preparing a C-8 epoxy-alcohol on a scale of some ten kilograms in a reactor of more than 2000 L.

The obstinacy of Barry Sharpless had its returns: 'I thought and thought and thought for months and years. Ninety-nine times, the conclusions were false, the hundredth time I was right' (Albert Einstein).

References: T. Katsuki and K. B. Sharpless, *J. Am. Chem. Soc.*, 102, 5974 (1980).
             S. S. Woodard, M. G. Finn, and K. B. Sharpless, *J. Am. Chem. Soc.*, 113, 106–113 (1991).
             M. G. Finn and K. B. Sharpless, *J. Am. Chem. Soc.*, 113, 113–126 (1991).

## 7 Enantioselective Dihydroxylation of Olefins

Chiral diamine auxiliaries, A*, able to complex with osmium, confer a great enantioselectivity on formation of *cis*-glycols. They lead to good yields, in the range of 60 to 95%, and to enantiomeric excesses between 62 and 95%.

The enantioselectivity is explainable within frontier orbital formalism: The olefin adds to the diamine–osmium tetroxide complex, forming a bond with two oxygens, one apical and the other equatorial: the stereochemical preference is due to orbital control.

References: K. Tomioka, M. Nakajima, and K. Koga, *Tetrahedron Lett.*, **31**, 1741–1742 (1990).
K. A. Jørgensen, *Tetrahedron Lett.*, **31**, 6417–6420 (1990).

Ar = phenyl, 3,5-xylyl

## 8 Selectivities in Ozonolysis

Limonene will serve again as the example: upon ozonolysis, the more electron-rich endocyclic double bond is preferably attacked. The yield of the dicarbonyl compound exceeds 90%.

Is it possible, in spite of such regioselectivity, to attack exclusively the other, exocyclic double bond? The path to be followed is obvious: protection of the more reactive double bond by way of epoxidation (with *m*-chloroperbenzoic acid). The epoxide formed is submitted to ozonolysis, and finally the cyclic double bond is regenerated with Zn/NaI (in NaOAc(AcOH) in 36% yield.

Reference: J. A. Edwards, J. Sundeen, W. Salmond, T. Iwadare, and J. H. Fried, *Tetrahedron Lett.*, 791 (1972).

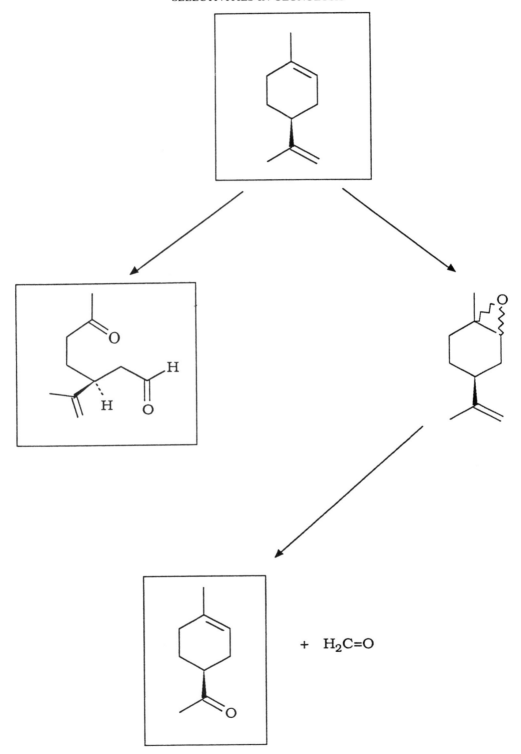

## 9 Selective Oxidation of Alcohols

The following Scheme shows a spectacular example of the selectivity and mildness of Collin's reagent $CrO_3.py_2$, related to pyridinium chlorochromate. The substrate is a steroid having no less than three alcohol functions, primary, secondary and tertiary. Since their reactivity sequence is: tertiary << primary < secondary, only the secondary alcohol is oxidized to the corresponding ketone.

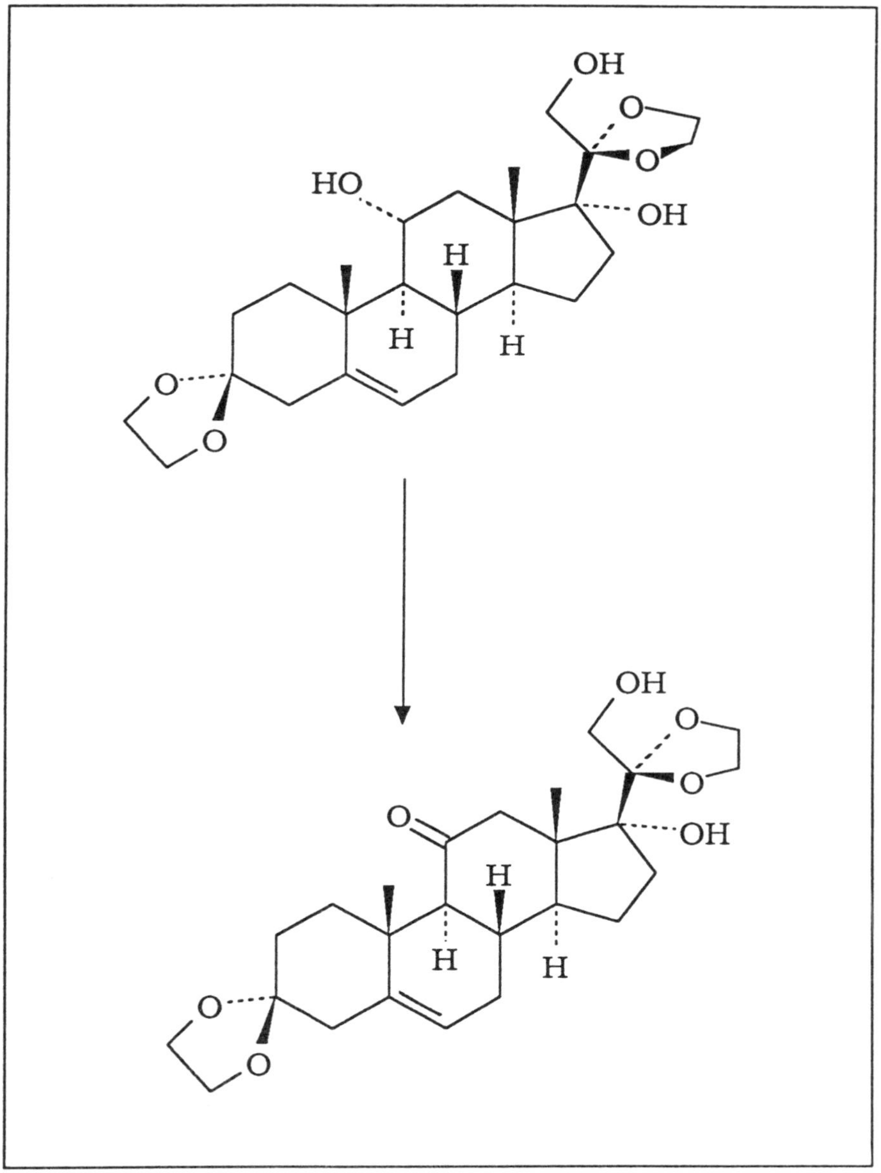

'In these days, a man who says a thing cannot be done is quite apt to be interrupted by some idiot doing it', Elbert Hubbard, *More Random Walks in Science*, R. L. Weber (ed.), Vol. 2, Institute of Physics, Bristol, 1982.

Another mild oxidizing agent, capable of selectively oxidizing a secondary alcohol without affecting a primary alcohol present in the same molecule, is silver carbonate supported on celite (powdered skeletons of diatoms). This reagent was devised by Fétizon and Golfier, and is the prototype of a series of reagents on solid supports, efficient and easy to handle, which have become nowadays very popular.

References: W. S. Allen, S. Bernstein, and R. Littell, *J. Am. Chem. Soc.*, **76**, 6116 (1954).
C. Djerassi, R. R. Engle, and A. Bowers, *J. Org. Chem.*, **21**, 1457 (1956).
M. Fétizon and M. Golfier, *C. R. Ac. Sci. Paris*, **267B**, 900 (1968).
P. Laszlo, ed., *Preparative Chemistry Using Supported Reagents*, Academic Press, San Diego, 1987.

## 10    Biological Oxidation of Alcohols

Alcohol dehydrogenase enzymes oxidize alcohols to the corresponding carbonyl compounds:

ethanol     $(R^1 = CH_3,\ R^2 = H) \rightarrow$ acetaldehyde

lactic acid   $(R^1 = CH_3,\ R^2 = CO_2H) \rightarrow$ pyruvic acid

The redox system $NAD^{(+)}/NADH$ is the co-enzyme. The reduced nicotinamide adenine dinucleotide (NADH) has a ring tetrahedral carbon and is therefore not aromatic. The NADH sets free a hydride $H^{(-)}$ which converts it into the aromatic nicotinamide adenine dinucleotide $NAD^{(+)}$.

The $NADH \rightarrow NAD^{(+)}$ conversion increases the stability of the system and is exothermic. Oxidation of an alcohol in the organism thus requires a base :B to deprotonate the alcohol. If water is sufficiently basic, it forms $H_3O^{(+)}$; otherwise the conjugate base of water $(OH^{(-)})$ is used, affording $H_2O$.

'Essential phenomena are not to be interpreted at random', Heraclitus (fragment 52).

Reference: L. Stryer, *Biochemistry*, W. H. Freeman, San Francisco, 1981.

## 11 And the Enzyme Solved the Problem

Metabolic oxidation of aldehydes to acids is worth mentioning, since it provides an example for improvement of the direct oxidation process, usually either difficult or impossible. When oxidizing an aldehyde to the acid, the negatively-charged hydrogen atom (the hydride $H^{(-)}$ ion) has to be replaced by an $OH^{(-)}$ group. A dissociative mechanism would imply a separation $H^{(-)} O{=}C^{(+)}{-}R$, difficult to achieve because of the coulombic attraction between the fragments. The trick the organism utilizes is addition of a nucleophile $X^{(-)}$ followed by formation of an acylated intermediate having hydride as leaving group. The reaction is concluded by addition of a reagent serving as an $OH^{(-)}$ donor: orthophosphate and regeneration of the nucleophile $X^{(-)}$ by elimination. Cleavage of the oxygen–phosphorus bond leads to the orthophosphate anion, highly stabilized by resonance, and to the carboxylic acid.

Such an *oxidative phosphorylation* is realized by glyceraldehyde-*D*-phosphate dehydrogenase. The nucleophilic group $X^{(-)}$ is a negatively charged sulfur atom of a cysteine residue. The hydride ion is removed from the tetrahedral intermediate by the coenzyme $NAD^{(+)}$, attached to the enzyme.

## 12  Production of Acetaldehyde by Oxidation of Ethylene

The first observation is almost a century old: platinum salts allow for selective oxidation of ethylene to acetaldehyde. The reaction is stoichiometric since the platinum salts are reduced to platinum(0) (F. C. Philips, 1894). The Wacker–Hoechst process, established at the end of the fifties, is catalytic.

The catalyst is a mixture of $PdCl_2$ and $CuCl_2$. The gaseous components (ethylene and oxygen or air) react with the aqueous solution of the catalyst, at 3–10 bar above 100 °C. The selectivities are about 95%.

Reference: K. Weissermel and H. J. Arpe, *Chimie organique industrielle*, Masson, Paris, 1981.

The mechanism is formation of a π-complex between ethylene and palladium with a square planar geometry. The complex is transformed to a σ-complex with formation of a bond between the two ligands ethylene and OH. The σ-complex is cleaved as indicated. The palladium is then re-oxidized by copper(II) which in turn is re-oxidized by oxygen. The annual production of acetaldehyde by this process amounted to 2.6 million metric tons in 1978. Half of the annual production of acetaldehyde serves to manufacture acetic acid by an oxidation process.

An analogous Wacker process exists for conversion of propylene to acetone, the annual production of which is about 2 million metric tons.

## 13   A Taste of Soil in the Mouth

The natural product responsible for the odor of freshly dug soil or for the muddy taste of certain fish has been recently identified. It is a metabolite from actinomycete molds, known as 'geosmin'. Geosmin has a *trans*-decalin skeleton, with a methyl group and a tertiary alcohol function at the ring junctions, and a second methyl group at the carbon adjacent to that bearing the alcohol function. This second methyl group is *cis* to the OH hydroxy group.

A synthesis of geosmin starts from the cyclohexenone shown. Treatment with LiAlH$_4$ (i) affords a mixture of alcohols, esterified with acetic anhydride Ac$_2$O (ii); reductive cleavage of the resulting acetates with lithium in ethylamine solution (iii) forms the bicyclic olefin.

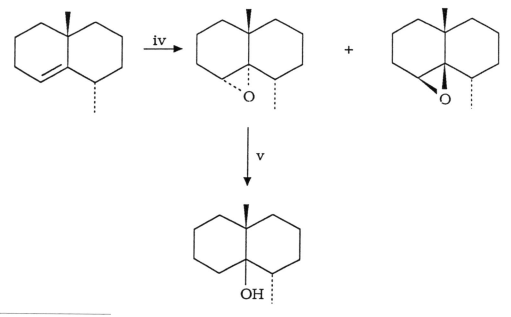

The latter is converted to the epoxide with *meta*-chloroperbenzoic acid $m$-Cl—C$_6$H$_4$—CO$_3$H (iv). The prevailing stereoisomer (57%) results from attack of the C=C double bond from the most accessible face, i.e. the one devoid of a methyl group at the ring junction. After separation from the mixture, it only remains to open this major epoxide with LiAlH$_4$ (v); a reductive cleavage similar to step (iii) thus leads to geosmin.

References: N. N. Gerber and H. A. Lechalier, *Appl. Microbiol.*, **13**, 935 (1965).
          J. A. Marshall and A. R. Hochsteler, *J. Org. Chem.*, **33**, 2593–2595 (1968).

# 11.I
# Addition–Elimination Reactions

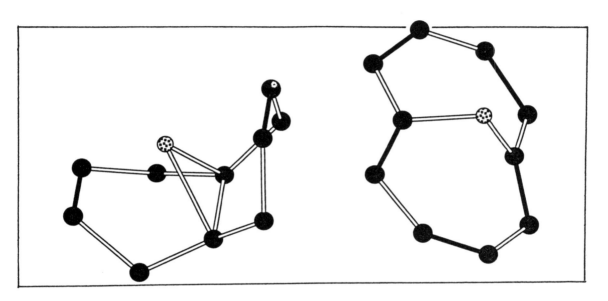

Addition of electrophiles to alkenes is an easy and useful way to introduce heteroatoms. Conversely, formation of alkenes is typically brought about by elimination reactions. This chapter has both a mechanistic and utilitarian perspective.

|       |     |         |     |
|-------|-----|---------|-----|
| C—Cl  | 79  | C = C   | 55  |
| C—H   | 99  | H—Cl    | 103 |
|       | 178 |         | 158 |

# 1   Study of a Simple Case

First, let us examine in some depth a prototypic example. We shall consider, for this purpose, the unimolecular elimination reaction E1 of a molecule of hydrochloric acid and of one of isobutene from *t*-butyl chloride in the vapor phase. The overall reaction scheme, breaking the carbon−chlorine and the carbon−hydrogen bonds and forming a $\pi$ carbon−carbon and a hydrogen−chlorine bond, is endothermic (by about $20 \, \text{kcal mol}^{-1}$).

Conversely, the entropy difference is favorable since a fragmentation of one initial particle into two in the final state increases the entropy of the system.

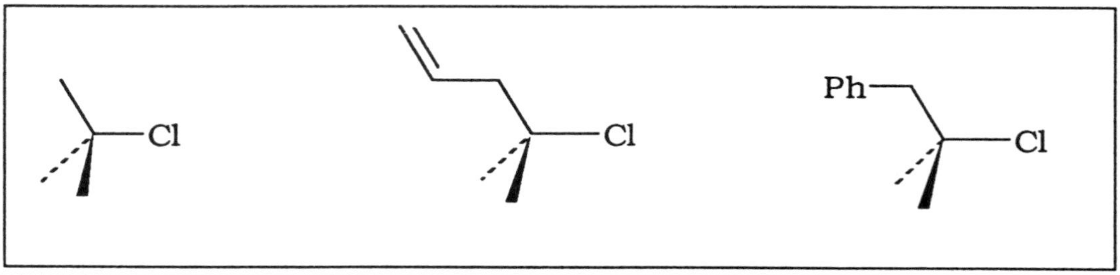
Cl
Cl
Ph
Cl

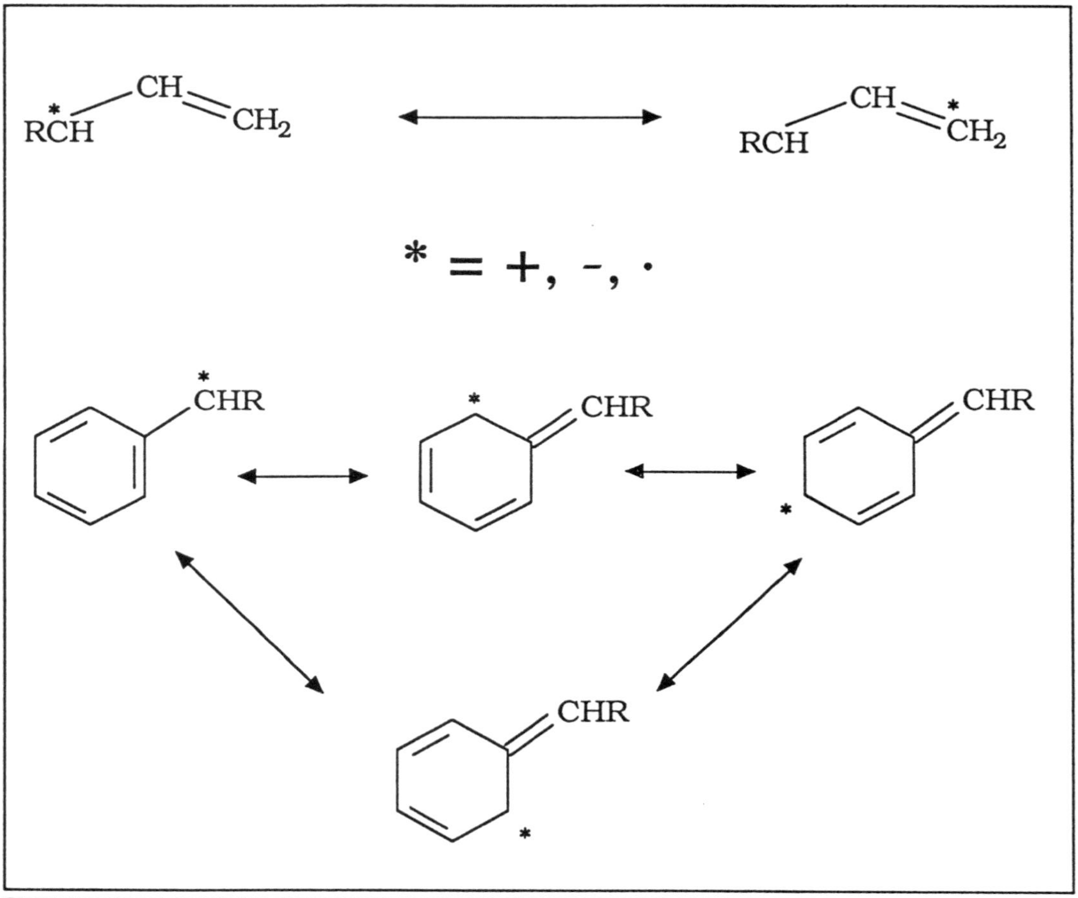
RCH
CH
CH2
RCH
CH
CH2
* = +, −, ·
CHR
CHR
CHR
CHR

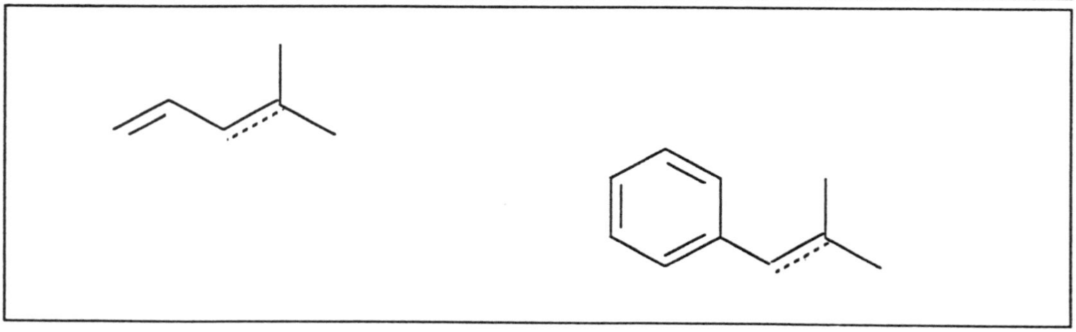

Kinetic measurements admit of a first conclusion: under the same conditions, the decomposition rates of the three tertiary chlorides indicated are practically the same. Recall that incorporation into an allylic or benzylic system renders a carbon–hydrogen bond considerably more labile. Irrespective of the nature of the carbon–chlorine bond breaking mechanism—whether an anionic, cationic or radical center is formed—allylic or benzylic resonance would stabilize it.

In the same way, if the double bond C=C made its incipient presence felt in the transition state, partial conjugation with the pre-existing ethylene or benzene system would lower the energy of the transition state. Thus one would expect an acceleration on passing from *t*-butyl chloride to the corresponding unsaturated chlorides. However this is not observed.

Such a comparison therefore implies the absence of any weakening of the carbon–hydrogen bond in the transition state. The C=C double bond present in the product is definitely not formed in the transition state.

| | $E_a$ | $D(R^{(+)} + Cl^{(-)})$ | $D(R^{.} + Cl^{.})$ kJ/mol$^{-1}$ |
|---|---|---|---|
| | 250 | 805 | 335 |
| | 210 | 690 | 343 |
| | 190 | 625 | 328 |

Kinetic measurements (in the gas phase, let us emphasize once again) give an Arrhenius activation energy $E_a$ of $188 \pm 8 \, \mathrm{kJ \, mol^{-1}}$ and a pre-exponential factor A of $1.9 \times 10^{14}$. In the series ethyl chloride, isopropyl chloride, $t$-butyl chloride, the activation energy decreases in a significant and regular manner.

On comparing these activation energies with the homolytic ($R-Cl \rightarrow R^{\cdot} + Cl^{\cdot}$) and heterolytic ($R-Cl \rightarrow R^{(+)} + Cl^{(-)}$) cleavage reactions, it is obvious that the sequence of activation energies observed evokes and correlates with the *heterolytic* dissociation energies.

Accordingly, the mechanism appears to consist of an initial slow (rate-determining) step during which $t$-butyl chloride dissociates to the $t$-butyl cation and chloride anion; the two subsequent steps are fast: the carbocation expels a proton and forms isobutene; the proton attaches to the chloride ion, affording hydrochloric acid.

Since the step that determines the global rate is formation of the tertiary cation, there is good reason to believe that the transition state strongly resembles the intermediate state, an ion pair $Me_3C^+ + Cl^-$.

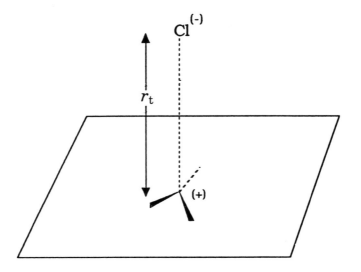

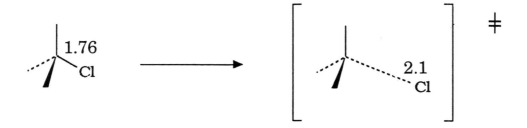

Let us now assume that the transition state identifies with such an ion pair. We denote $r_t$ the distance between the carbocation and chloride centers. Then the activation energy can be written:

$$E_a = D(R^{(+)} + Cl^{(-)}) - e^2/4\pi\epsilon_0 r_t$$

Upon replacing $E_a$ and $D$ in the above equation with the experimental values ($E_a = 188$; $D = 625$), one comes up with a value of about 2.1 Å for $r_t$.

This very simple calculation, assuming complete ionization, neglecting the contributions from other forces, especially the repulsive, equating the permittivity of vacuum with its value of one to the effective dielectric constant—nevertheless leads to quite a realistic result: *the carbon—chlorine bond is stretched by 15–20% in the transition state*, since its original length in *t*-butyl chloride is 1.76 Å.

Numerous other observations support the proposed mechanism: in particular the considerably lower value of the $E_a$ Arrhenius activation energy ($90 \pm 3 \, kJ \, mol^{-1}$) in 50% aqueous ethanol. This decrease of the activation energy, as compared to the gas phase, stems from greater solvation of the ionic transition state than the initial state, unionized neutral *t*-butyl chloride.

---

'colored chlorine colored the hydrogen'—Raymond Queneau, *Petite cosmogonie portative*, 1950.

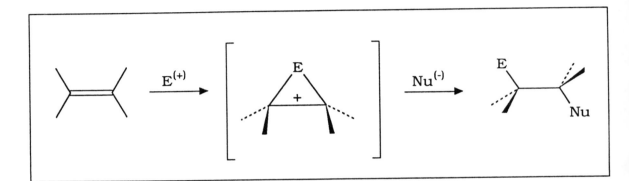

And what about the mechanism of the reverse reaction? Reading the above reaction backwards, one witnesses addition of hydrochloric acid to isobutene, to form *t*-butyl chloride. The principle of microscopic reversibility would not have surprised the legendary Frenchman La Palisse ('one hour before he died, he was still alive', according to the epitaph on his tombstone; which really means that he fought like a devil before he was killed; a statement that has remained exemplary of stating the obvious). It is defined thus: the forward and the reverse reactions share identical mechanisms provided that the two reactions (direct and reverse reaction) follow the same pathway.

# 2 Classical Mechanism of Electrophilic Addition (Ad$_E$) and Markovnikov's Rule

In this very generally accepted mechanism, more than 50 years old, an electrophile $E^{(+)}$ adds to a C=C double bond, affording an intermediate 'onium ion' (e.g. a halonium ion if addition of a positive halogen ion is involved). This intermediate is a three-membered ring with delocalization of the positive charge over three centers, the atom E and the two carbon atoms. In the second step of this electrophilic addition (Ad$_E$) mechanism, a nucleophile $Nu^{(-)}$ attacks the onium ion. This attack is stereoselective. It takes place preferentially in the half-space opposite to E. Thus a product of anti(periplanar) stereochemistry with a a torsion angle (ECC, CCNu) of 180° is formed.

References: I. Roberts and G. E. Kimball, *J. Am. Chem. Soc.*, **59**, 947 (1937); P.B.D. De la Mare, *Electrophilic Halogenation*, Cambridge University Press, Cambridge, 1976.

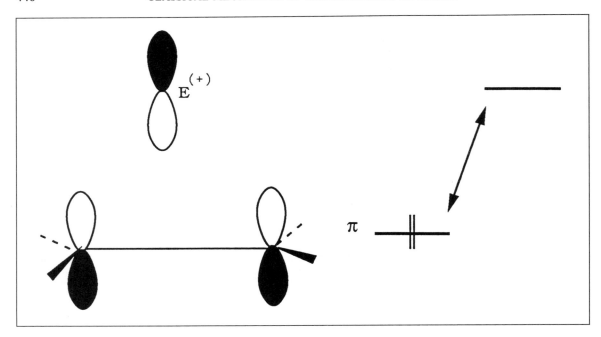

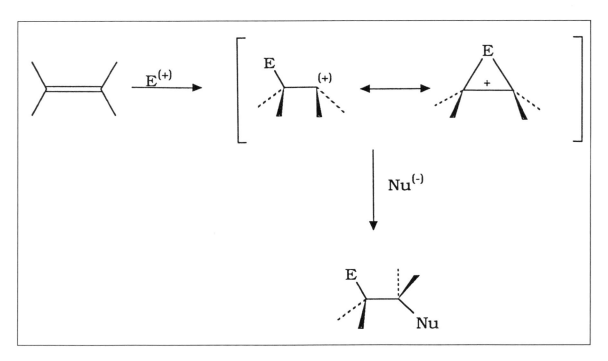

It is easy to link this classical $Ad_E$ mechanism to orbital control (whereas in the previous section the mechanism obeyed exclusive charge control). In the frontier orbitals (Fukui) approximation, the stabilizing two-electron interaction connects the LUMO of the electrophile $E^{(+)}$ and the HOMO of the olefin. The latter is therefore a doubly occupied $\pi$-orbital. In order to ensure maximum overlap between the LUMO of the electrophile—for simplification let us consider a 1 s atomic orbital for $H^+$ and a 3 p orbital for $Cl^+$—and the HOMO of the olefin, the triangle ECC in the transition state is isosceles. This MO explanation agrees totally with the classical description of the intermediate 'onium ion'.

The everyday reactions are carried out in solution, the dielectric constant of which, higher than unity (typically 5–20) leads to a competition between charge and orbital control. The pure $Ad_E$ transition state is a hybrid with carbocation and onium ion as limiting forms. The adjacent Scheme shows the full standard mechanism for the $Ad_E$ electrophilic addition.

'Once upon a time there was a twelve-foot line
Feeling alone it sought a companion
It did not go further than the tip of its nose
And right away found the buddy it was looking for'—Raymond Queneau, *Le chien à la mandoline*, 1965.

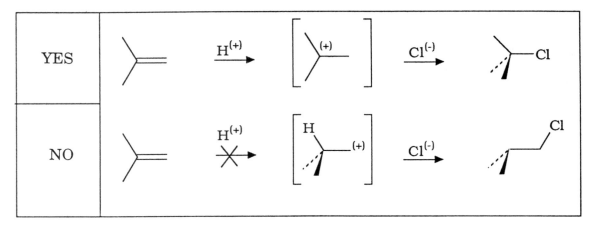

Markovnikov's rule (1833–1904) in its modern form, accounts for the regioselectivity observed in electrophilic additions where the most substituted carbocation (hence the most stable) is preferred. We could have already noted this from microscopic reversibility when addition of hydrochloric acid to isobutene leads to *t*-butyl chloride via the *t*-butyl cation, tertiary and hence more stable; not even a slight trace of 1-chloro-2-methylpropane is found that would indicate intermediate formation of the *i*-butyl cation.

In the same way hydrogen iodine HI adds regiospecifically to 1-methylcyclohexene, affording 1-iodo-1-methylcyclohexane via the tertiary cation; no 1-iodo-2-methylcyclohexane—which would imply intermediate formation of a secondary cation—is formed.

'What if life were
only this argument
between two principles
both a little soft'—Raymond Queneau, *Oeuvres Complètes*, Gallimard-La Pléiade, Paris, 1989, Vol. 1, p. 921.

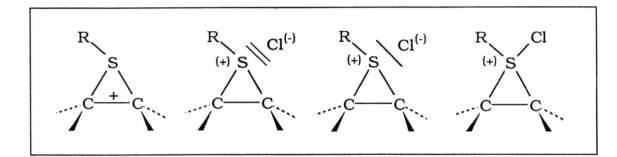

The term 'onium ion' remains to be explained: In a general way, an 'onium' ion bears a positively charged atom with a valence greater than the normal. Ammonium ions $R_4N^{(+)}$ are familiar. Similarly there are oxonium ions $R_3O^{(+)}$, sulfonium ions $R_3S^{(+)}$ etc. The cyclic cation formed in $Ad_E$ reaction complies well with this definition.

Another question arises: which structure is correct? There is not a single answer. There are families of species. For example all the structures depicted in the adjacent Scheme can be used to describe the intermediate formed in the addition of a sulfenyl chloride RSCl to a C=C double bond. Since $RS^{(+)}$ is a very strong electrophile, the best structural representation is probably that of a three-membered ring with delocalization of the positive charge on sulfur *and* on the carbon atoms.

With decreasing electron attracting character of the $RS^{(+)}$ group, the positive charge is more localized on sulfur and hence there is a greater tendency to form an ion pair, first weakly and then more strongly bonded. Finally, there is a continuum between the latter (where the two ions are separated by a simple hyphen) and the neutral molecule with a true sigma bond between sulfur and chlorine. Please note that in these four structural representations the two $\pi$-electrons of the C=C bond have been used to form the two carbon—sulfur bonds which are thus 'one electron bonds'.

---

'Particles are a kind of verbal shorthand; it is entirely possible that we won't ever do away with it. An example is that of the valence bars in chemistry; while knowing how much such a symbolism is simplified, one continues to use it.'—E. Schrödinger, *L'homme devant la science*, La Baconnière, Neuchâtel, 1953, p. 36.

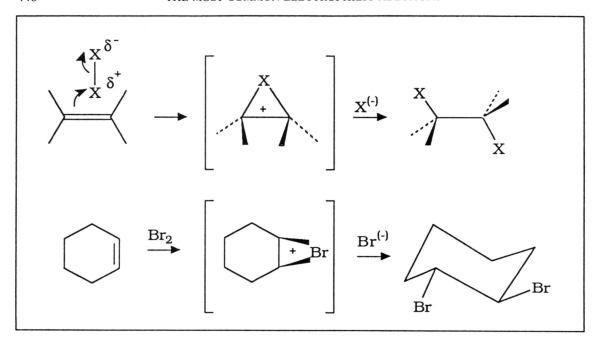

# 3 The Most Common Electrophilic Additions

## 3.1 Halogens

A dihalogen molecule $X_2$ is polarized by an ethylenic double bond and is then cleaved to the ions $X^{(+)}$ and $X^{(-)}$. Addition of the electrophile to the double bond affords the halonium ion; the nucleophile then attacks the halonium ion in an *antiperiplanar* fashion with formation of the dihalogenated addition product. In this way, cyclohexene adds bromine, affording trans-1,2-dibromocyclohexane.

## 3.2 Hydrogen Halides

Dissociation of the acid HX forms a proton, which on electrophilic addition to the ethylenic double bond leads to an intermediate carbocation according to Markovnikov's rule. Attack of this intermediate by the nucleophilic anion $X^{(-)}$ affords the reaction product. Once again we are confronted with the by now familiar example of the addition of hydrochloric to isobutene.

Addition of deuterated hydrochloric acid DCl to 1-methylcyclohexene is regiospecific (exclusive formation of 1-chloro-1-methylcyclohexane) and stereoselective (preferential formation of the product with a *cis*-relationship between D and Cl; only one enantiomer is indicated).

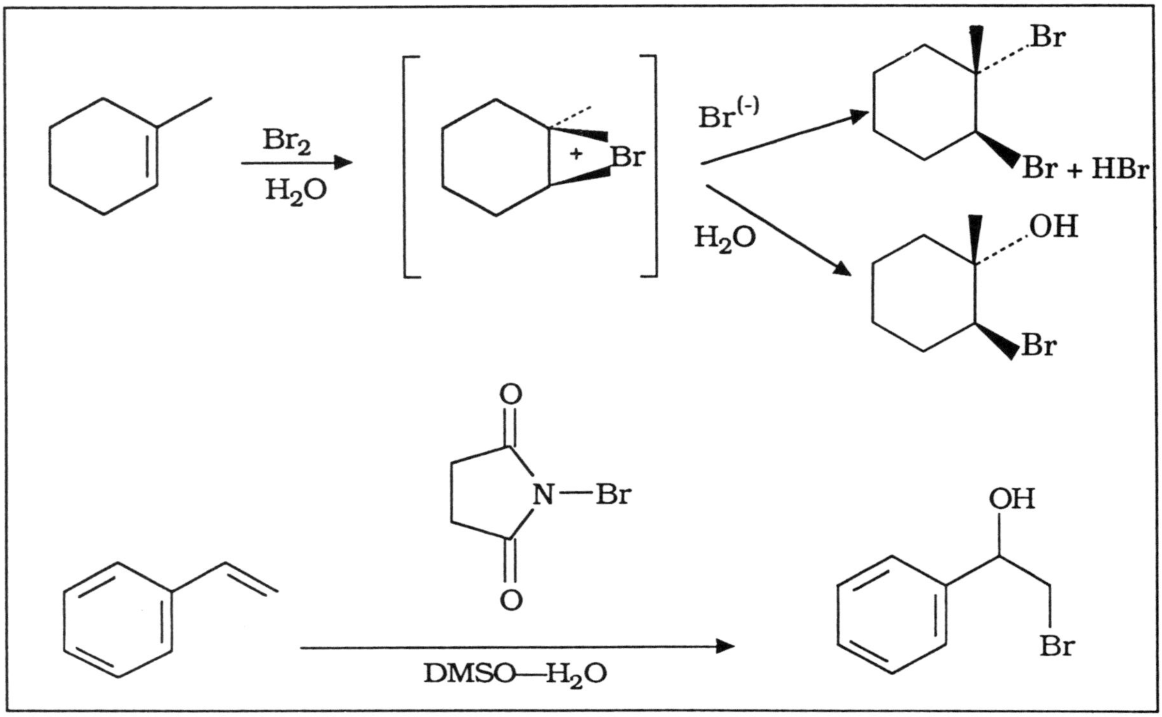
Br2
H2O
+ Br
Br(-)
Br
Br + HBr
H2O
OH
Br
O
N—Br
O
OH
Br
DMSO—H2O

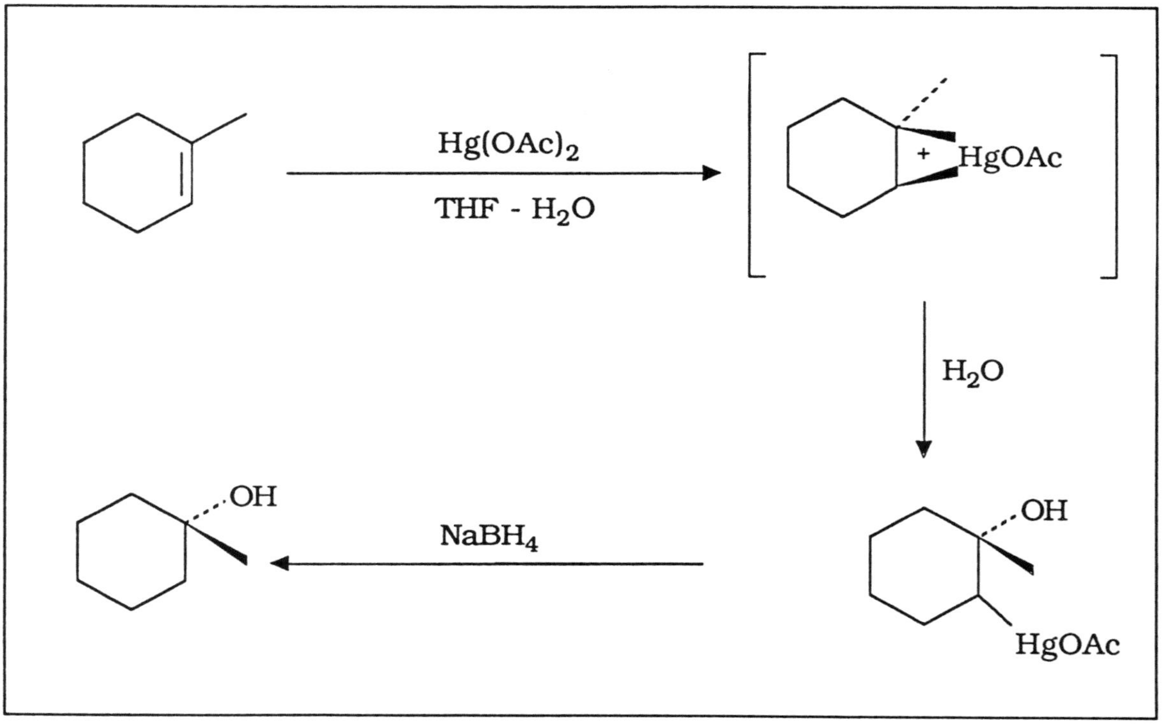
Hg(OAc)2
THF - H2O
+ HgOAc
H2O
OH
HgOAc
NaBH4
OH

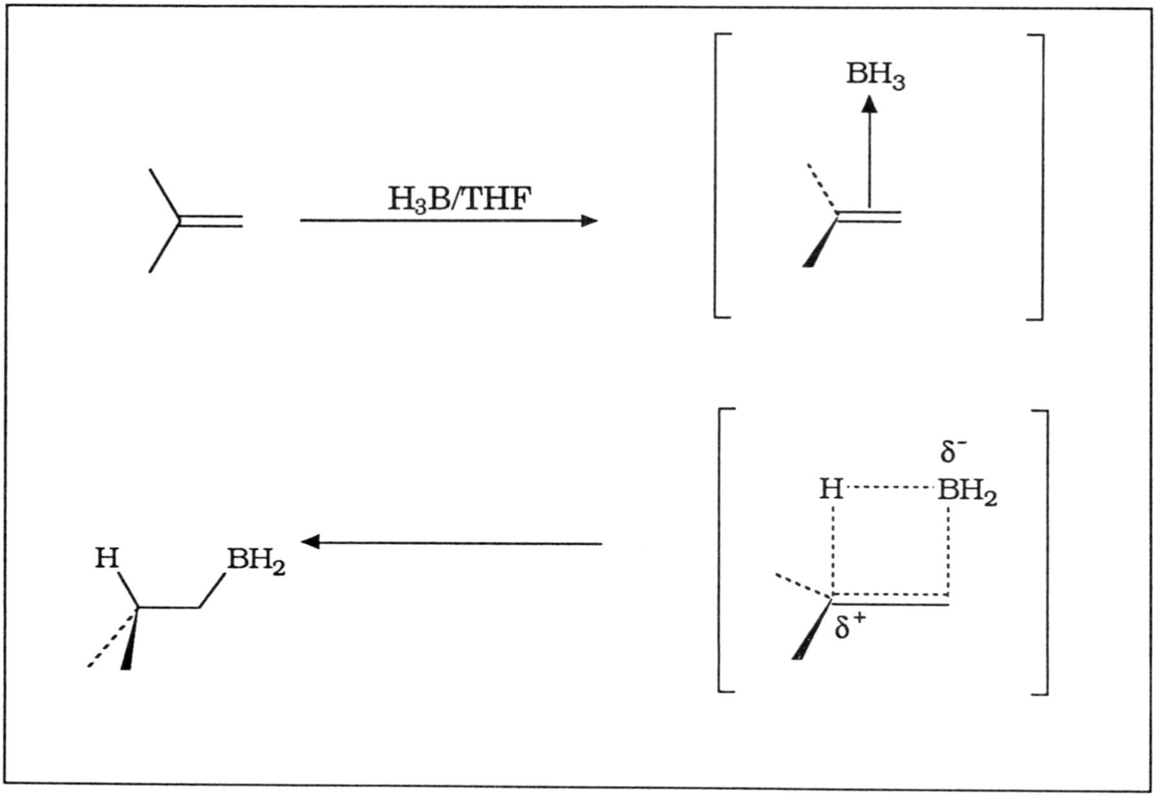
H3B/THF
BH3
δ-
H
BH2
δ+
H
BH2

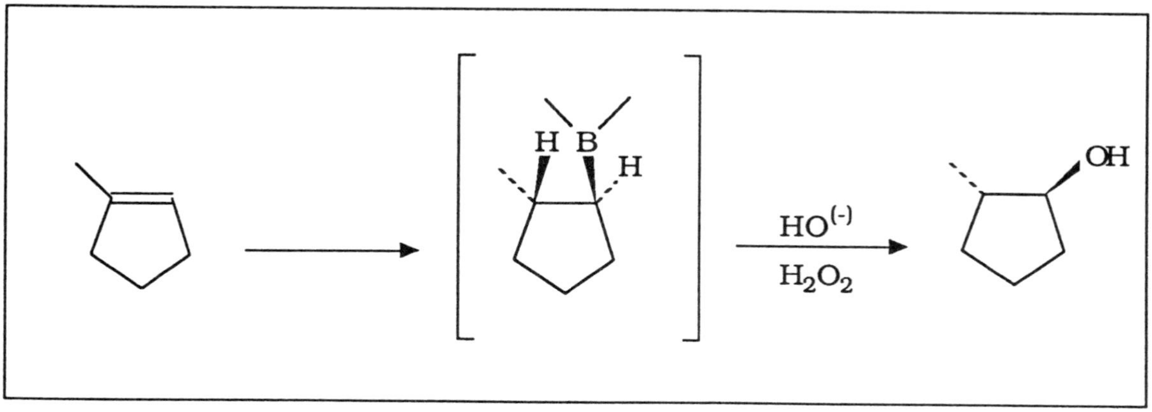
H B
H
HO(-)
H2O2
OH

## 3.5 Hydroboration

Hydroboration is a $[2+2]$ cycloaddition of a boron—hydrogen bond to a $\pi$ C=C double bond. Therefore thermally it cannot proceed in a concerted manner. In the first step, borane $BH_3$ (present as its Lewis acid-base complex with THF), a rather weak electrophile, forms a $\pi$ complex with the C=C double bond. The ethylenic C=C bond is the electron-donor, $BH_3$ the electron-acceptor. The next is the key step: boron has attached itself to the least substituted carbon, sterically the most accessible; a hydrogen atom is then transferred from boron to the most substituted carbon, that can best accommodate a partial positive charge (in the hypothesis of charge control). This transfer involves a four-center transition state, depicted here for the addition of borane to isobutene. This *syn* and *anti*-Markovnikov addition results in an organoborane of the general formula $H_2BR$. Since it still has two boron—hydrogen bonds, the process can be repeated till the trialkylborane $BR_3$ is produced. These trialkylboranes undergo oxidative cleavage with hydrogen peroxide in basic medium which affords the alcohol with retention of configuration (only one enantiomer is indicated for the hydroboration of 1-methylcyclopentene).

Herbert Charles Brown (1912–   ) was awarded the Nobel prize in chemistry for his work on hydroboration. He optimized this reaction and found numerous applications for it. Brown, who obtained his PhD at the University of Chicago with Schlessinger, has been at Purdue University since 1947.

---

'Tinkal [sodium borate] inflates like drooling albumin
refrigerated won't dent the pure spirit from coal
From this four-hooked brother
will architecture themselves the drooling albumin
and the petrol lamp and comforting aspirin
and sugar and alcohol, indole and indigo . . .'—Raymond Queneau, *Petite cosmogonie portative*, 1950.

R,R' = H,Ac

65 %     11 %     7 %

8 %

### 3.6  Oxythallation

Thallium(III) acetate adds to ethylenic double bonds from the least hindered side. the onium cation formed in this $Ad_E$ reaction is hence attacked by the acetate nucleophile from the most hindered face. From this second intermediate a cyclic *cis*-dioxolenium cation results, from an intramolecular $S_N2$ substitution reaction, of course also on the most hindered face of the olefin. Hydrolysis affords a mixture of *cis*-glycols and the corresponding acetates.

In the steroid series for instance, this procedure gives *cis*-dihydroxylation with good selectivity.

'Why do I welcome within myself a thousand arguments
with which my mind is already obsessed?—Maurice Scève, *Délie*, 1544.

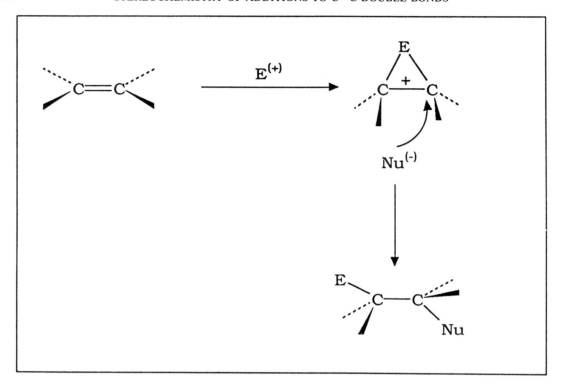

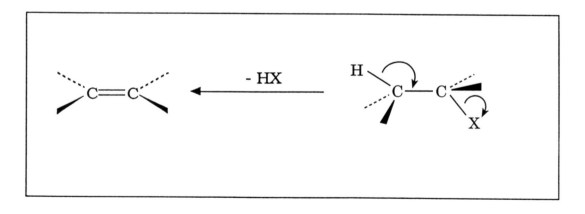

# 4  Stereochemistry of Additions to C=C Double Bonds

The addition proceeds preferentially in a *trans* (antiperiplanar) fashion. Indeed, the onium ion is more susceptible to attack by the nucleophile $Nu^{(-)}$ from the face opposite to E.

According to the principle of microscopic reversibility, one would expect and indeed one finds that the antiperiplanar disposition of the C—H and C—X bond is more favorable to elimination $E_1$ of a molecule HX (the proton $H^+$ corresponds to the electrophile $E^{(+)}$; the leaving group $X^{(-)}$ corresponds to the nucleophile $Nu^{(-)}$).

---

'And that is not all, it is enough, as we just found, to shift the molecule before the collision by an infinitely small amount for it to deviate, after the collision, by a finite amount'—Henri Poincaré, 'Le hasard', *Revue du mois*, **3**, 257–276 (1907).

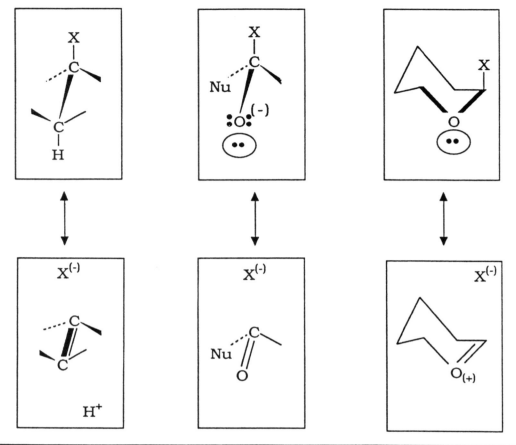

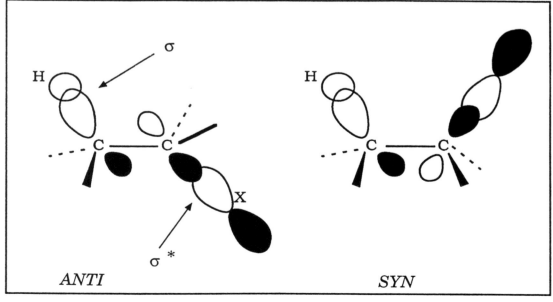

*ANTI*                                    *SYN*

Why is the antiperiplanar arrangement more favorable? Notice the analogy between this stereoelectronic effect and, for example, the anomeric effect (lengthening and stabilization of an axial C—X bond adjacent to oxygen, when X is an electron-attracting atom or group); and, as another example, the lengthening of the C—X bond in the tetrahedral intermediate of an addition–elimination reaction at a carbonyl group (X = attracting atom or group, $X^{(-)}$ = leaving group).

In the resonance formalism, each of these three manifestations of basically the same stereoelectronic effect receives also very similar descriptions. Namely, resonance occurs between a limiting form having the two atoms C and X bonded, and a limiting form without a covalent bond between C and X.

Finally, we provide a description of the same effect in the MO language. We do so for its heuristic value. A two-electron interaction between the doubly occupied C—H $\sigma$-level and the empty antibonding C—X $\sigma^*$-level lengthens the C—X bond and promotes cleavage. This interaction occurs by lateral $\pi$-type overlap between the rear lobe of the C—H and the adjacent antibonding orbital of the C—X bond, both in phase (i.e. having the same sign of their wave functions). Were the geometry synperiplanar instead of antiperiplanar, the same overlap would be disfavored.

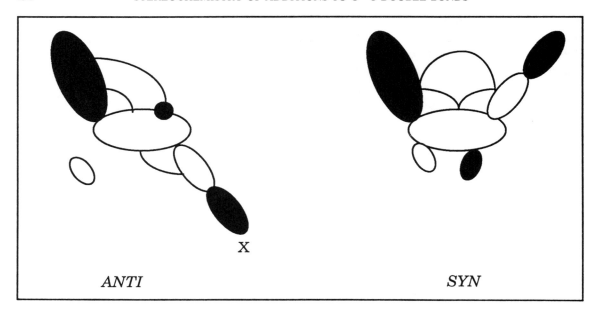

ANTI                                                    SYN

This analysis of the stereoelectronic effect remains somewhat rudimentary. Besides the doubly occupied electron donating C—H MO and the electron accepting empty antibonding C—X $\sigma^*$-orbital we should also consider the $\sigma$ MO of the median C—C bond which is also doubly occupied in the ground state and is also electron donating.

What really differentiates between the *anti* and *syn* geometries under these conditions becomes apparent from the following analysis of the two criteria: (i) donating orbitals should have opposite phases (true for both geometries) and (ii) donor and acceptor orbitals should have the same phase (true only for the *anti* geometry).

References: S. David, O. Eisenstein, W. J. Hehre, L. Salem, and R. Hoffmann, *J. Am. Chem. Soc.*, **95**, 3806 (1973).
S. Inagaki, K. Iwase, and Y. Mori, *Chem. Lett.*, **1986**, 417–420.

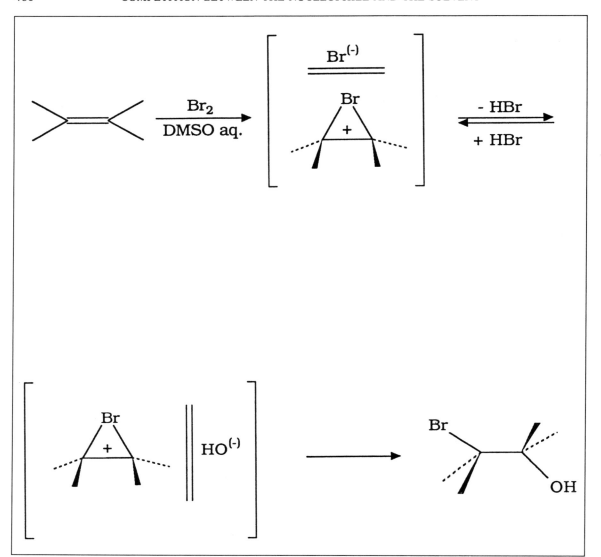

# 5 Competition between the Nucleophile and the Solvent

In the description of electrophilic addition given so far, we have failed to consider the presence of a counter ion in the first addition step. The formulation as an 'onium ion' assumes involvement of a dissociated ion pair $A^{(-)} + C^{(+)}$ where the anion $A^{(-)}$ remains at an infinite distance from the onium ion $C^{(+)}$.

Two other types of ion pairs can exist; they predominate in most organic solvents, which commonly have dielectric constants $D$ in the range 2–25, i.e. insufficient for full dissociation of the ion pairs. They are 'loose' and 'tight' ion pairs. In 'tight' ion pairs (also named by various authors 'intimate' or 'contact' ion pairs), anions and cations are nearest neighbors; the notation used is $A^{(-)}, C^{(+)}$. In 'loose' ion pairs denoted $A^{(-)} \| C^{(+)}$ the ions are separated by one or several solvent molecules.

The intuitive postulate is that only 'loose' ion pairs can exchange nucleophiles. The addition of bromine to a double bond *in the presence of water*, in dimethyl sulfoxide can be seen as competition between $Br^{(-)}$ and $H_2O$ for attack on the intermediate bromonium cation, present as a loose ion pair.

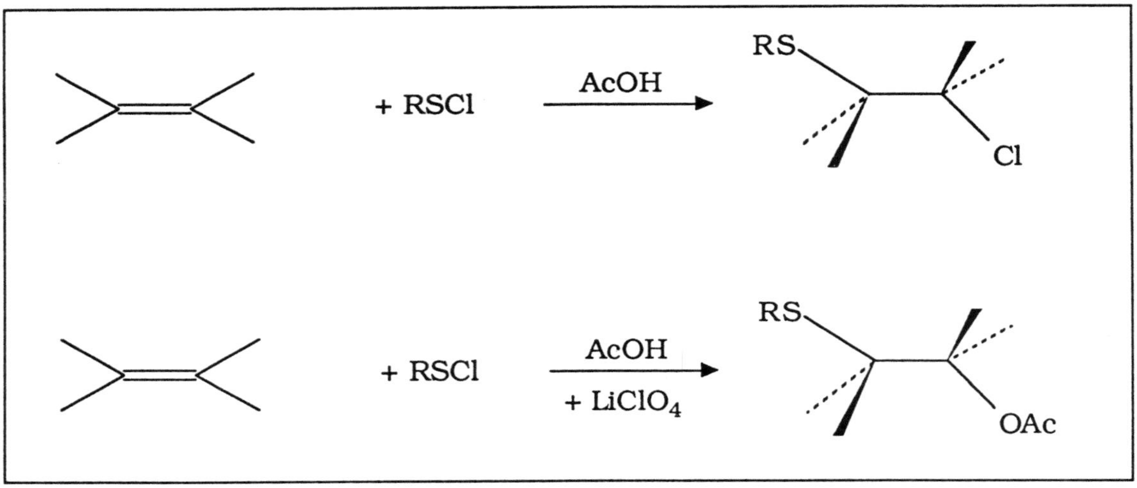
+ RSCl
AcOH
RS
Cl
+ RSCl
AcOH
+ LiClO4
RS
OAc

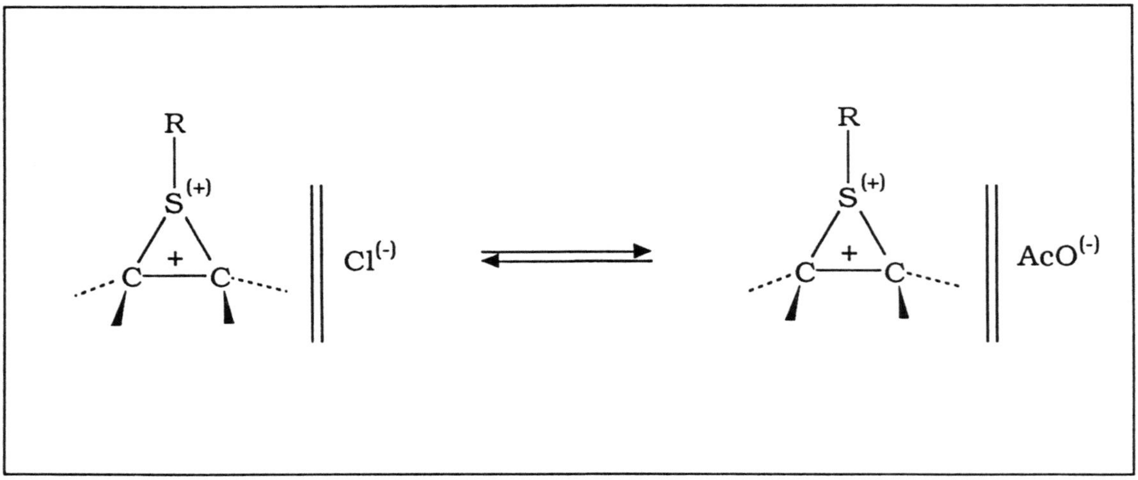
R
S(+)
+
C
C
Cl(-)
R
S(+)
+
C
C
AcO(-)

An experiment in support is reaction of an olefin with chlorosulfenyl chloride RSCl, in acetic acid as solvent. It affords the 'normal' addition product; addition of the electrophile $RS^{(+)}$ to the nucleophile $Cl^{(-)}$ has occurred in *anti* manner.

If the solvent is 'doped' with a salt such as lithium perchlorate which enhances the ionic strength, the equilibrium between the 'tight' ion pair $A^{(-)}, C^{(+)}$ and the 'loose' ion pair $A^{(-)} \| C^{(+)}$ is shifted toward the latter; the acetate forms instead of the chloride.

In the presence of lithium perchlorate, the predominant loose ion pairs exchange their chloride and acetate counterions (the latter being of course the conjugate bases of the solvent molecules).

Reference: W. A. Smit, N. S. Zefirov, I. V. Bodrikov, and M. Z. Krimer, *Accounts Chem. Res.*, **12**, 282–288 (1979).

$$R = H, CH_3, \overset{\overset{\displaystyle O}{\|}}{C}CH_3$$

Obviously, oxymercuration is another reaction with an exchange of nucleophiles since in aqueous tetrahydrofuran water, and not the acetate counterion, attacks the intermediate onium ion.

Accordingly, one will be able to change the chemoselectivity of oxymercuration simply by playing with the nature of the solvent. Use of methanol and demercuration with sodium borohydride leads to methyl ethers. Use of acetic acid under the same conditions affords acetates. It is quite convenient to be able to modulate the nature of a product just by choosing the appropriate solvent.

|              |       |     |     |
| ------------ | ----- | --- | --- |
| $Hg(OAc)_2$  | trans | cis | cis |
| DCl          | trans | cis | cis |

|                | + | cis exo | trans | rearrangement |
| -------------- | - | ------- | ----- | ------------- |
| $CH_3COOD$     |   | 95      | 0     | 5             |
| DCL            |   | 60      | 0     | 40            |
| $DBr(D_2O)$    |   | 48      | 0     | 52            |
| $Hg(OAc)_2$    |   | 100     | 0     | 0             |
| $TI(OAC)_2$    |   | 100     | 0     | 0             |
| NOCl           |   | 100     | 0     | 0             |

# 6 Addition to Strained Olefins/*Cis*-Additions

The usual *trans*-addition is pushed aside when the substrate is a strained olefin. *Cis*-addition tends to predominate. This is illustrated by the behavior of cyclohexene, norbornene and bicyclo[2.1.1]hexene towards addition of mercuric acetate and $d_l$-hydrochloric acid.

With respect to stereochemistry, focusing on electrophilic addition to norbornene, most reagents give a large preponderance of *cis* addition products. With some reagents, (Wagner–Meerwein) rearrangement products are formed from carbocationic intermediates, following opening of the initially formed onium ion.

Whenever *cis* and *trans*-additions to a *strained* olefin are compared, destabilization of the initial state translates into an acceleration of the electrophilic addition, whether to the *cis*- or to the *trans*-product, provided that the angular tension increases more in the initial state than in the transition state (La Palisse, once again!).

Preference for the *cis*- rather than the *trans*-product when starting from a strained olefin thus indicates a greater destabilization of the *trans* as compared to the *cis* transition state.

Reference: T. G. Traylor, *Accounts Chem. Res.*, **2**, 152–160 (1969).

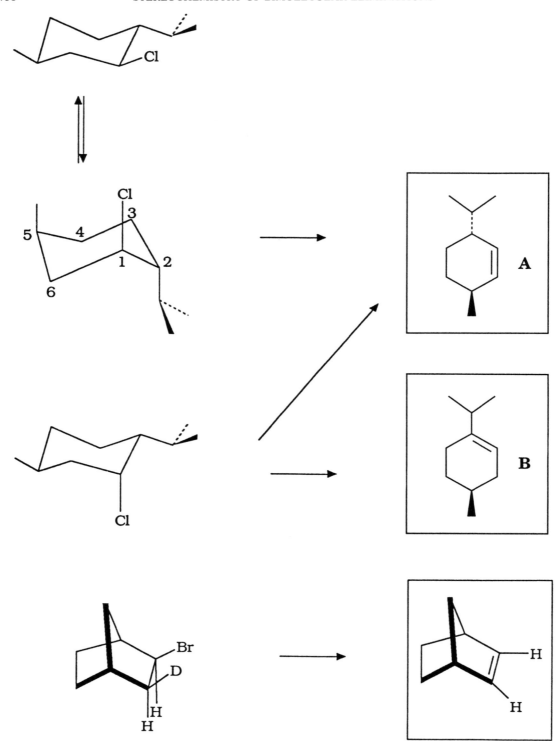

# 7   Stereochemistry of Bimolecular Eliminations

The preferential mode of the $E_2$ elimination is antiperiplanar. Menthyl chloride in the energetically lowest conformation has all three substituents (Me, $i$Pr, Cl) in equatorial positions which precludes an antiperiplanar elimination. Such an *anti*-elimination can only take place in a triaxial conformer. In contrast to this, in neomenthyl chloride the most stable conformation has chlorine in an axial position. With neomenthyl chloride the $E_2$ reaction takes place much faster than with menthyl chloride. The olefin preferentially produced from neomenthyl chloride is **B** (B/A = 3), in agreement with the Zaitsev rule: the most substituted olefin is preferentially formed. The $E_2$ elimination from menthyl chloride is thus slower and forms exclusively the *anti*-Zaitsev olefin, since in the conformation with the axial chlorine atom, only the carbon atom C-6 bears an axial hydrogen with a torsion angle H—C—C—Cl of 180°, as required by the antiperiplanar arrangement.

In a certain number of cases, $E_2$ elimations with *syn* rather than *anti*-stereochemistry are observed. The former occurs when *anti*-eliminations are made impossible by molecular construction. This is the case for the bromide in the norbornane series indicated here: deuterium is absent in 94% of the norbornene formed.

References: E. D. Hughes, C. K. Ingold, and J. B. Rose, *J. Chem. Soc.,* **1953,** 3839; H. Kwart, T. Takeshita, and J. L. Nyce, *J. Am. Chem. Soc.,* **86,** 2606 (1964).

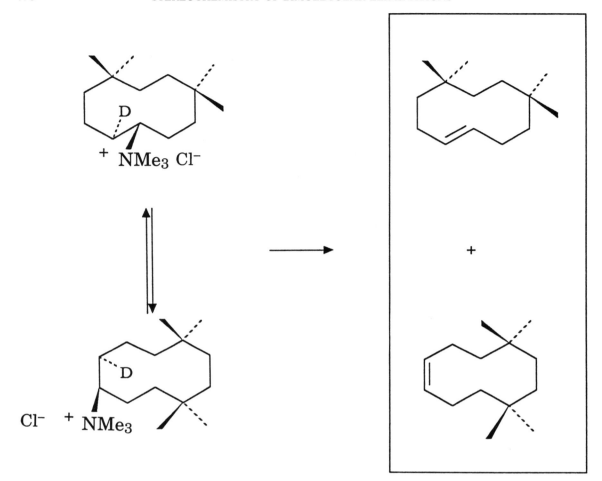

In medium and large rings, *anti* and *syn* E$_2$ eliminations coexist. When the cyclodecane, indicated in the adjacent scheme, undergoes a Hofmann elimination, a mixture of *cis*- and *trans*-cyclodecenes is obtained. In the presence of deuterium, primary isotopic effects of deuterium affect formation of both olefinic bonds. In contrast, when the substrate is deuterated on the adjacent carbon *cis* to the trimethylamino—NMe$_3$ leaving group, no isotopic effect is observed in any of the olefins formed. Since an E$_2$ mechanism implies an isotopic effect, these results mean that only the hydrogen *trans* to the leaving group is eliminated in both cases (*cis* or *trans* product). In other words:

|  |  |
|---|---|
| (olefin) | *cis*-isomer → *anti*-elimination |
|  | *trans*-isomer → *syn*-elimination |

A third case of a switch to *syn*-elimination is when ion pairs are present: we won't consider these further, they would lead us to byzantine discussions.

References: J. Sicher, J Zavada, and J. Krupicka, *Tetrahedron Lett.*, **1966**, 1619.
        J. Sicher and J. Zavada, *Coll. Czech. Chem. Comm.*, **32**, 2122 (1967).
        J. Sicher and J. Zavada, *Coll. Czech. Chem. Comm.*, **32**, 3701 (1967).

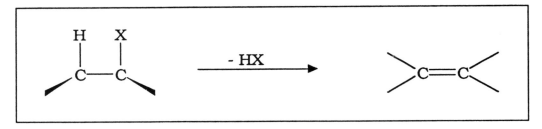

X = OH $\longrightarrow$ OH$_2^{(+)}$                                        AH

X = Cl, Br, I

= NMe$_3^{(+)}$, N$^{(+)}$Me$_2$, SMe$_2^{(+)}$
        |
        O$^{(-)}$

= OH $\longrightarrow$ O–S–O–S–CH$_3$                    : B

X = OCR, OCR                                        Δ

# 8  Modes of $\beta$-Eliminations

Cleavage of a neutral molecule HX can be carried out under *acidic* or *basic conditions* or by 'brutal' heating (*pyrolysis*), affording an olefinic double bond:

 (i) under acidic conditions: acid-catalyzed dehydration of an alcohol;
 (ii) under basic conditions: if X = Cl, Br, I, dehydrohalogenation can be effected either with strong bases (alcoholic potassium hydroxide, or potassium *t*-butoxide) or by weak bases (LiX/DMF); if X = NMe$_3^{(+)}$, SMe$_2^{(+)}$ or (+)N($-$O)Me$_2$ with strong bases (e.g. Hofmann elimination); if X = OSO$_2$Mes (mesyl sulfite prepared from the corresponding alcohol) with CH$_3$SO$_2$Cl/SO$_2$ and a weak base;
(iii) by pyrolysis of esters or thionoesters.

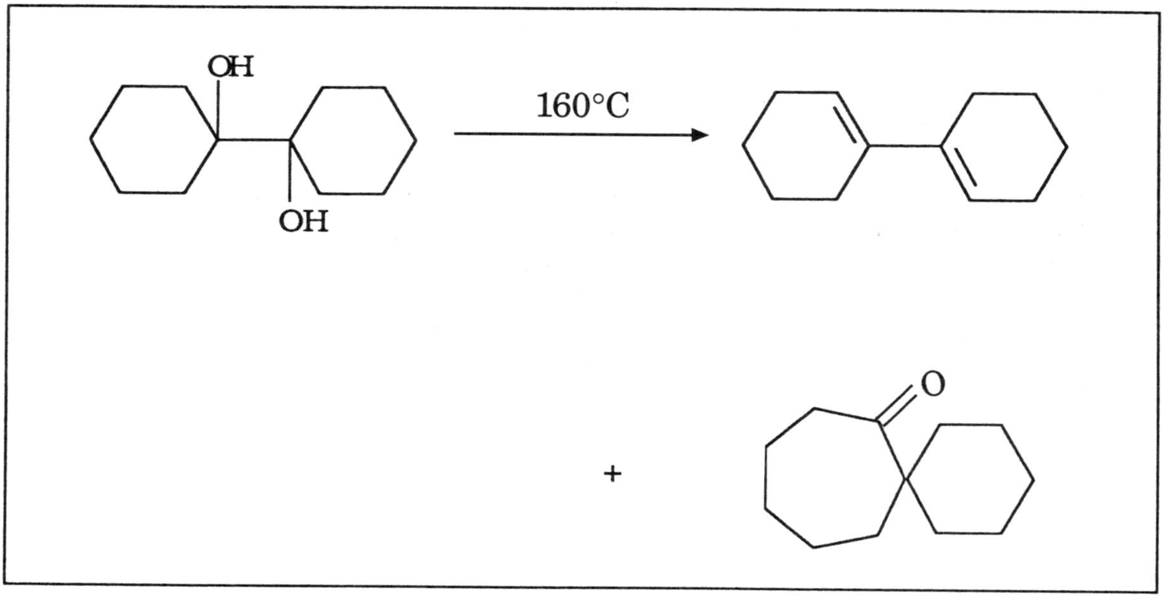
OH
OH
160°C
+
O

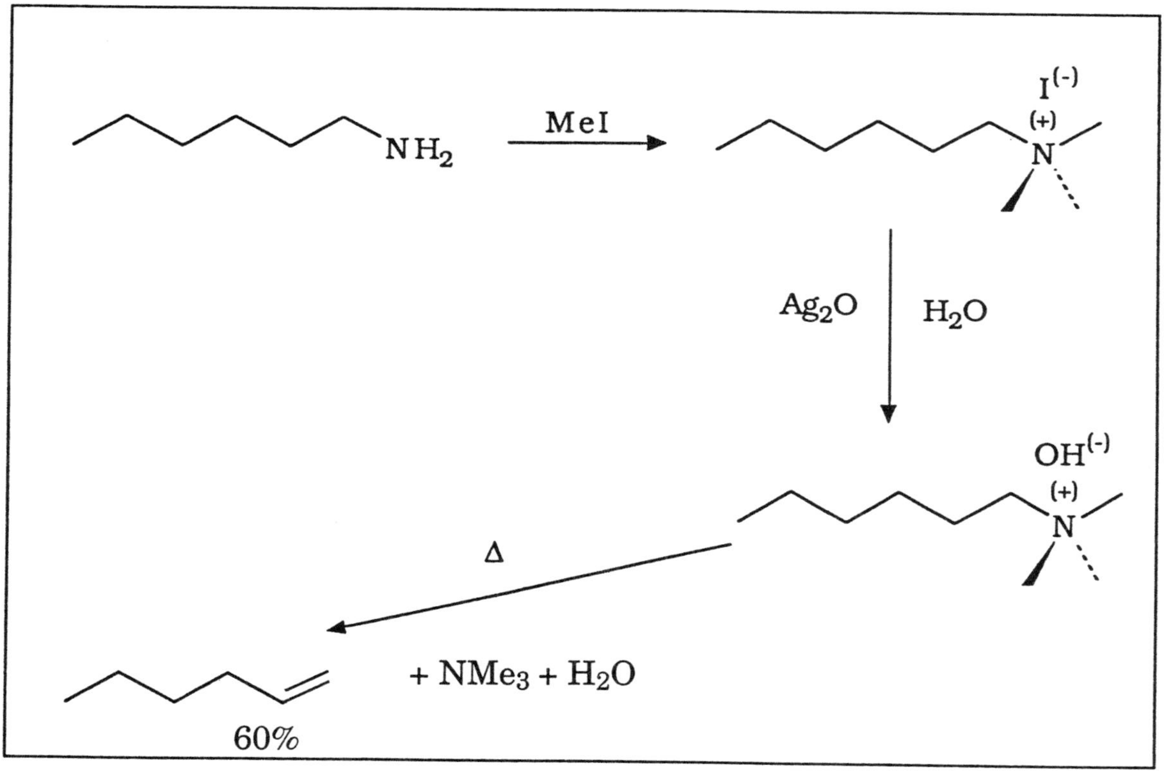
NH2
MeI
I(-)
(+)
N
Ag2O
H2O
OH(-)
(+)
N
Δ
+ NMe3 + H2O
60%

## (i) Avoiding carbocationic rearrangements under acidic conditions

In cases where the carbocation formed by cleavage of the group $C—OH_2^{(+)}$ could induce migration of an adjacent $C—C$ bond, dimethyl sulfoxide under drastic conditions (typically 16 h at 160°C) is used. In the case indicated only 4% of the corresponding rearrangement product pinacolone is formed; the expected diolefin is produced in 85% yield.

## (ii) Hofmann elimination

One of the advantages of this reaction is that it starts from the readily accessible amines. Firstly a quaternary ammonium salt is formed on methylation with an excess of methyl iodide. The iodide thus obtained is then heated in the presence of silver oxide, the role of which is to effect an exchange of the counter ion $(I^{(-)} \rightarrow HO^{(-)})$. It is the hydroxide anion which is responsible for the elimination. Thus starting from *n*-hexylamine, *n*-hexene is formed in moderately good yield (60%)

Reference: V. J. Traynelis, W. L. Hergenrother *et al*, *J. Org. Chem.*, **27**, 2377 (1962); **29**, 123 (1964).

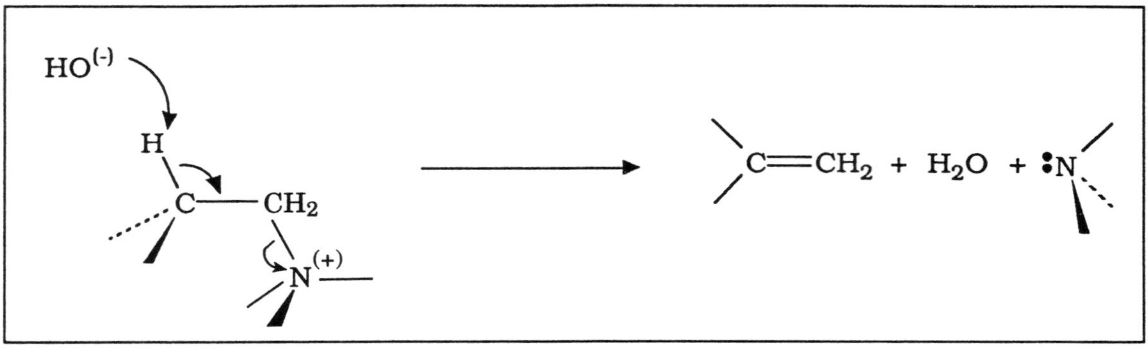

HO(-)
H
C
CH2
N(+)
C=CH2 + H2O + :N


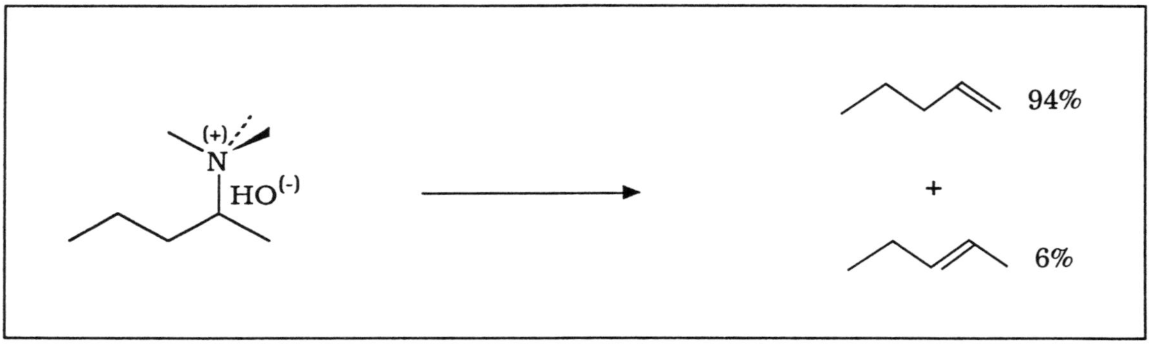

(+)
N
HO(-)
94%
+
6%


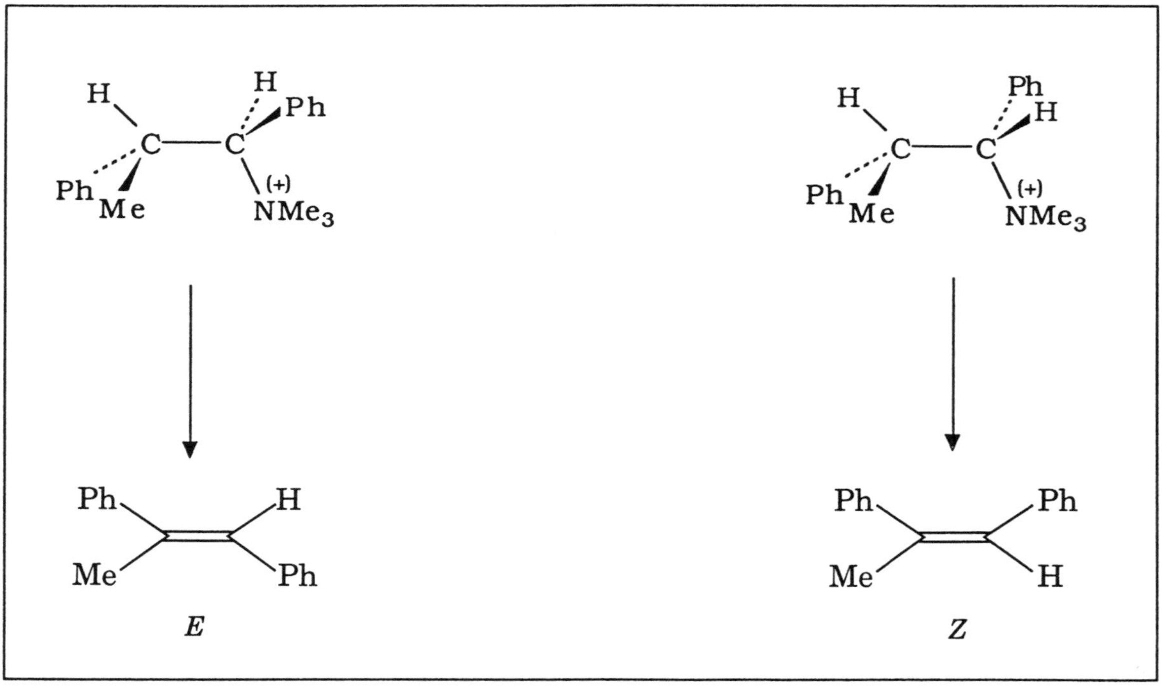

H
Ph
Me
C
C
H
Ph
(+)
NMe3
Ph
H
Me
Ph
E
H
Ph
Me
C
C
Ph
H
(+)
NMe3
Ph
Ph
Me
H
Z

The Hofmann elimination is probably a bimolecular reaction $E_2$: removal of a proton by means of a base and expulsion of a neutral molecule of trimethylamine (leaving group) from the ammonium cation.

One of the advantages of the Hofmann elimination is its regioselectivity: the least substituted olefin (hence the least thermodynamically stable) prevails in the reaction mixture. For instance, the indicated ammonium salt leads to 1-pentene and 2-pentene in a ratio of 94:6.

The Hofmann elimination is also stereoselective favoring the antiperiplanar disposition of the proton and the leaving group. In the adjacent scheme, the *threo*-isomer affords the *E*-olefin whereas the *X*-olefin is produced from the *erythro*-isomer, although 57 times slower: the indicated conformation, prone to antiperiplanar elimination, suffers from a destabilizing *gauche* Ph–Ph interaction in the *erythro*-isomer.

Reference: D. J. Cram, F. D. Greene, and C. H. De Puy, *J. Am. Chem. Soc.*, **78**, 790 (1956).

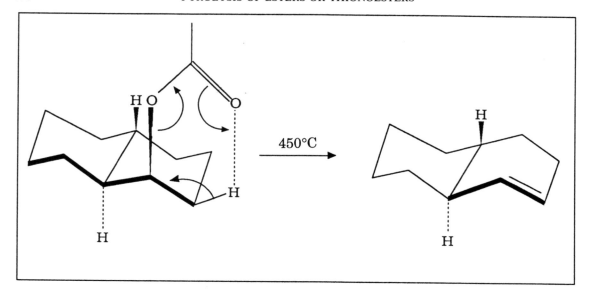

As mentioned earlier, esterification of an alcohol with $H_3C.SO_3-SOCl$ creates a leaving group $X = OSO_2Mes$ which can result in a Hofmann type elimination; however only a weak base such as 2,4,6-trimethylpyridine (collidine) is then required.

### (iii) Pyrolysis of esters or thionoesters

The driving force for such reactions is the cleavage of a neutral molecule ($RCOOH$, $CO_2$, $O=C=S$). A hexatomic transition state formally satisfying the Dewar–Zimmerman rule can be formulated. This mechanism (the so-called $E_i$ reaction) is that of a *syn*-periplanar elimination.

The reactions indicated can both be carried out on glass beads in the vapor phase at 450 and 230°C, respectively, with yields of 65%. In the second case, the target molecule was synthesized because of the possibility of an aromatic type stabilization due to interactions of the three $C=C$ double bond; these however turned out to be very weak.

References: W. Hückel and D. Rücker, *Liebigs Ann.*, **666**, 30 (1963).
L. A. Paquette, P. B. Lavrik, and R. H. Sommerville, *J. Org. Chem.*, **42**, 2659 (1977).

## Summary: 'Addition and Elimination Reactions'

The ionization of *t*-butyl chloride affording *i*-butene is a good example of the reasoning by which a reaction mechanism is derived from experimental data. Markovnikov's rule—reaction via the most stable carbocation—accounts for the regioselective additions of electrophiles to ethylenic double bonds. In contrast, hydroboration is a general procedure leading to anti-Markovnikov products.

The *stereoelectronic* effect, in the elimination of a molecule of a hydracid HX from adjacent C—H and C—X bonds dictates its antiperiplanar disposition with 180° torsion angle.

$\beta$-Eliminations of $E_2$ type, Hofmann eliminations especially, provide a good access to alkenes.

# 11.II
# Addition–Elimination Reactions

## 1   Polyethylenes

Almost half the ethylene produced goes into polymerization (figures for 1977): US 41%, Japan 37%, West Europe 53%, Federal Republic of Germany 48%. The yearly American production of polyethylene is about 4 M metric tons.

Fawcett and Gibson discovered this plastic serendipitously in the early Thirties. In the course of attempts at condensing ethylene with benzaldehyde under high pressure (2500 atm), a solid formed. In fact, traces of oxygen had served as polymerization initiator for ethylene. Polyethylene found its first industrial application as insulation material for coaxial cables used in radar, manufactured by ICI during World War II.

Linear high density polyethylene results from polymerization of ethylene, catalyzed either by $TiCl_3$ and $Me_2AlCl$ (Ziegler–Natta process) or by $CrO_3$ deposited on silica (Hogan and Banks process).

These thermoplastic polymers enjoy steadily increasing production, especially high density polyethylene: in the US low (high) density polyethylene production amounted to 4.7 (3.8) MT (metric tons) in 1988, with a rate of increase of 4% (7%) between 1978 and 1988.

*Note*: With the hope that we are presently going through only a temporary recession, I choose to quote the 1988 figures as more representative.

---

References: K. Weissermel and H. J. Arpe, *Chimie Organique Industrielle*, Masson, Paris, 1981.
R. B. Seymour, *J. Chem. Ed.*, 1987, **64**, 63–68 (1987).

## 2   Some Industrial Additions to Ethylene

Addition of water to ethylene forms *ethanol* in an exothermic reaction ($-11$ kcal mol$^{-1}$). It is carried out in the gas phase in the presence of a heterogeneous Brønsted acid catalyst: the system $H_3PO_4$ is often used. The reaction takes place at temperatures around 300°C and under pressures of 70 bar. The molar ratio (water/ethylene) should be less than 0.6 since water vapor deactivates the catalyst by washing out the phosphoric acid. As a consequence, the conversion remains below 4%, hence ethylene has to be recycled, which in turn imposes high purity on the ethylene feed. Extraction of

---

'That scientific problems as often have the form of finding auxiliary hypotheses as they do of finding and checking predictions is something that has been too much neglected in philosophy of science; this neglect is largely the result of the acceptance of the positivist model and its uncritical application to actual physical theories'. H. Putnam, *Mind Language and Reality* (Philosophical Papers 2), Cambridge University Press, 1975, p. 214.

**Table 16** Production of ethyl acetate (kT)

|  | 1974 | 1975 | 1976 | 1977 |
|---|---|---|---|---|
| United States | 65 | 66 | 61 | 62 |
| Japan | 70 | 78 | 91 | * |
| Germany | 82 | 60 | 72 | 66 |

*unpublished.

the alcohol from the aqueous solution is done by azeotropic distillation (the azeotrope contains 95% ethanol). Complete elimination of the water is brought about by addition of an agent such as benzene. All these difficulties notwithstanding, the Shell process dating back to 1947, is used in numerous plants all over the world: UCC in the US for about 380 kT per year; the VEBA process, similar to the Shell process, amounts to a total capacity of 710 kT per year.

Acid-catalyzed addition of acetic acid to ethylene forms ethyl acetate, a solvent used mainly in the paint industry. This is a promising process, not yet feasible on an industrial scale, but which could replace the Tischtschenko process, a reaction involving two molecules of acetaldehyde.

Addition of chlorine to ethylene affords 1,2-dichloroethane: this is one of the major products of the chemical industry, ranking 12th in 1986, but 15th in 1988, and ranking 3rd (after ethylene and propylene) among organics. One of the major uses of 1,2-dichloroethane is cracking to vinyl chloride. The old process entailed direct chlorination of ethylene in the liquid phase. The modern process is oxychlorination of ethylene in the gas phase, the reaction ($C_2H_4 + 2CuCl_2 \rightarrow ClCH_2CH_2Cl + 2CuCl$) being coupled with regeneration of the copper catalyst according to:

$$2CuCl + \tfrac{1}{2}2O_2 \rightarrow CuO + CuCl_2$$

$$CuO + 2HCl \rightarrow CuCl_2 + H_2O$$

Whereas direct chlorination of ethylene is exothermic ($-43$ kal mol$^{-1}$), the oxychlorination reaction is more exothermic ($-57$ kcal mol$^{-1}$) (for the above stoichiometry). The increase in exothermicity is due to the enthalpy of formation of water. Heat dissipation is essential to avoid overheating, over-chlorination and total combustion to $CO_2$ and $H_2O$. The process takes place either in fixed-bed or in fluid-bed reactors. The gaseous mixture $2HCl + \tfrac{1}{2}2O_2 + C_2H_4$ is heated at 220–240°C under a pressure of 2–4 bar.

Profile of 1,2-dichloroethane: 'a giant waking up'; and an index for the present recession:

Production:
14.53 $10^9$ lb in 1986
13.65 $10^9$ lb in 1988

| Annual Average Growth Rate (%) | |
|---|---|
| 1976–1986 | 6.1 |
| 1981–1986 | 7.8 |
| 1984–1985 | 13.0 |
| 1983–1986 | 20.1 |
| 1978–1988 | 2.2 |

Reference: *Chemical and Engineering News*, 13 April 1987, p. 21; 10 April 1989, p. 12.

One of the accesses to *vinyl acetate* is by acetoxylation of ethylene. The stoichiometric reaction was discovered by J. J. Moiseev and coworkers. The overall reaction is:

$$H_2C=CHOAc + Pd^0 + 2Cl^- + H^+$$

$$H_2C=CH_2 + PdCl_2 + 2CH_3COONa \rightarrow H_2C=CHO.CO.CH_3 + 2NaCl + Pd^0 + CH_3COOH$$

The mechanism is most probably formation of a $\pi$ complex (see the Scheme) with *syn* attack of the nucleophilic acetate ligand, followed by cleavage of the metal, whose oxidation state is reduced from 2 to 0.

A catalytic version of this reaction, in the liquid phase, was then developed: cuprous salts oxidize palladium metal to palladium(II). Thus $CuCl_2$ is reduced to $CuCl$ which is re-oxidized by atmospheric oxygen to copper(II). The overall process is described by the following equations:

$$H_2C=CH_2 + CH_3COOH + \tfrac{1}{2}O_2 \xrightarrow{\text{cat.}} H_2C=CH.O.CO.CH_3 + H_2O$$

The reaction is exothermic ($-42\,\text{kcal mol}^{-1}$). It proceeds mainly in the gas phase, at 175–200°C, at 5–10 bar (Bayer process: heterogeneous palladium catalyst charged with alkaline acetates, continually replaced).

In 1978, the total capacity of the plants using this process was over 1 MT a year. The annual production of vinyl acetate in fact exceeded 1 MT in the US in 1988, the average annual growth rate being between 3 and 6.5% from 1976 to 1988.

Direct manufacture of *ethylene glycol* from ethylene is a pathway with a great future. This is the oxidative addition:

$$H_2C=CH_2 + H_2O + \tfrac{1}{2}O_2 \rightarrow HOCH_2CH_2OH$$

The most efficient catalysts are $TiO_2/HCL$ (Teijin), $CuBr_2 + CuBr + HBr$ (Teijin), $CuI_2$ (Halcon) and $Pd(NO_3)_2$ (Kuraray).

Ethylene glycol ranked 29th among bulk chemicals manufactured in the US in 1988 (15th among the organics). It is mainly used for manufacturing polyesters and also for antifreeze (figures below, % apply to the US in 1982):

polyester fibers              43
polyester films and resins     7
antifreeze                    44
other applications             6

The US production in 1988 was 2.2 Mt, 4.6% less than in the preceding year (1985 saw a negative growth rate of $-13.4\%$). Smoothed on a two-decade average, the annual growth rate was 2–3%.

Reference: K. Weissermel and H. J. Arpe, *Chimie organique industrielle*, Masson, 1981.

## 3   The Market For Some Plastics

*Low-density polyethylene* (density lower than 0.94 g mL$^{-1}$) had a total annual production in the United States of 4.7 MT in 1988. Half of this production goes into the manufacturing of films, half of which are used for packages in different forms (50% for industrial and commercial use, 50% for food packaging and litter bags).

*High-density polyethylene* (density higher than 0.94 g mL$^{-1}$) was produced in the range of 3.4 MT in 1987 (US). Its major use is to manufacture containers: bottles for milk, orange juice, non-gaseous drinks, antifreeze, motor oils etc.

*Polypropylene* (total US production in 1988: 3.8 MT) is used mainly for packaging: mold injection stoppers, capsules etc., meant for containers made from other plastics; for the manufacture of fibres and filaments. The latter are used in textiles and rugs.

*Polyvinyl chloride* (PVC: total US production in 1988: 3.8 MT) is used mainly as a construction material, siding, pipes, etc. But there are numerous other uses, from records to toys. Its production has stagnated for several reasons: carcinogenic monomer; overcapacity; recession of the building industry; difficulty in recycling as compared to other polymers such as polyesters.

**Table 17**   The plastics market in the USA*

|                                  | 1970 (%) | 1985 (%) | 2000 (projection) (%) |
|----------------------------------|----------|----------|-----------------------|
| packaging                        | 24       | 28       | 30                    |
| building                         | 21       | 22       | 20                    |
| consumer goods                   | 11       | 8        | 7                     |
| electrical, electronic           | 10       | 6        | 7                     |
| adhesives, inks, fabric finishes | 8        | 5        | 5                     |
| furniture, decoration            | 7        | 5        | 5                     |
| automotive                       | 5        | 5        | 6                     |
| industrial equipment             | 1        | 1        | 1                     |
| miscellaneous                    | 7        | 12       | 12                    |
| production (MT)                  | 8.7      | 21.7     | 34.3                  |

* According to *Chem. Eng. News*, August 24, 1987, 27–65.

*Polystyrene* (annual US production in 1988: 2.3 MT) has applications similar to polyethylene. Polystyrene and copolymers are used in the packaging, building and furniture industries.

Let us quote here Raymond Queneau's description of the manufacture of several objects made from polystyrene:

O time, suspend your bowl, o plastic material
Where do you come from? Who are you? And what explains
Your rare qualities? What exactly are you made from?
Where indeed do you stem from? Let us go back from the object

To its distant forebears! Thus let's unwind in reverse
Its exemplary history. In the first place, the mold
Including the matrix, a mysterious being
It engenders the bowl, or whatever you wish.
But the mold itself is included in a press.

That injects the paste and conforms the cast piece,
Which thus presents the very great advantage
Of providing the finished product without any more shaping.
The mold is expensive; this is a problem.
It can be leased though; even from one's competitors.

Shaping under a vacuum is another manner
To obtain objects: a mere aspiration.
At the preceding step, carefully put aside,
The lukewarm material is extruded into a plate.
To enter the nozzle a piston was needed.

And the heating mantle—or the mantling heater
To whom one was providing—What? Polystyrene
Vivacious and unruly that runs along and forms beads.
And the granulated Swarm on the vibrating sieve.
Teemed from happiness at such a beautiful dye.

*Note*: Besides the rhyming, the original loses a lot in translation. The opening phrase puns on that of Lamartine's '*Le lac*'. There are also assonances such as between '*vivace et turbulent*' and '*vibrant*', for which I have not attempted to find equivalents.

## 4   Aging of Wines

The aroma of a wine results from the combination of flavors (non-volatile constituents: sweet-bitter-salty-acidic) and odors (volatile constituents). The latter are perceived at much lower thresholds ($10^{-4}$ to $10^{-12}$ g $L^{-1}$) due to the high sensitivity of the olfactory receptors. *Hundreds* of different molecules are therefore responsible for the taste of a wine. Some of these substances are present in grapes. (e.g. aldehydes), others are produced by fermentation (*n*-propanol, *n*-butanol, 2-methylbutanol, 3-butanol, phenylethanol, ethyl acetate, ethyl lactate).

A          B          C          D

A'          B'          C'          D'

Certain monoterpene alcohols from grapes are responsible for the characteristic aroma of several moscatels and rieslings: citronellol **A**, geraniol **B**, nerol **C**, linalool **D** amongst others.

During aging of a riesling in the bottle, a decrease in the concentration of these four molecules **A–D** is observed. On the other hand, the concentration of their homologs **A'–D'** (formed by addition of a molecule of water to the double bond with a terminal *gem*-dimethyl group), i.e. hydroxycitronellol, -geraniol, -nerol and -linalool respectively, increases with time.

The interconversion of the molecules **A–D**, catalyzed by protic acids, is readily explained by the intermediacy of carbocations.

Reference: A. Rapp and H. Mandery, *Experientia*, **42**, 873–885 (1986).

In the same way, Markovnikov addition of a molecule of water, also catalyzed by acids, can account for the slow transformation of the **A–D** manifold to the **A′–D′** manifold.

## 5   Diatomic Molecules with Multiple Physiological Functions

Structural simplicity is not incompatible with sophisticated chemical messages. The first molecular signal can be followed by others with more elaborate architectures. Three ubiquitous examples are the calcium ion, carbon monoxide CO, and nitric oxide NO.

In the past years, the surprising discovery of the role of nitric oxide in cellular communication has finally explained why nitroglycerin helps cardiacs. The smooth muscle cells in the walls of the blood vessels set free NO and this softens the blood vessel and decreases blood pressure. Nitric oxide is also a neurotransmitter as well as a regulator ot the intracellular concentration of cyclic GMP (guanosine mono-phosphate). In the brain, besides intracellular communication, nitric oxide at higher levels is a cell killer. After a stroke for instance, certain cells produce NO in concentrations sufficient to eliminate other cells. The immune system does likewise: certain cells, specialized in 'cleaning', set free nitric oxide which kills bacteria.

Carbon monoxide appears also to be an ubiquitous chemical messenger. A CO-producing enzyme is localized in some areas of the brain where cyclic GMP is also produced. The inhibition of such CO-producing enzymes in olfactive cells diminishes production of cyclic GMP.

Reference: S. I. Snyder *et al.*, *Science*, **259**, 381 (1993).

## 6   The Fragrance of Roses

Historical references to the discovery of the essence of roses are known; it is most often attributed to the Mogul Empress of India, Nur Jehan. In his memoirs, her husband the Emporor Jahangir (they were married in 1611) wrote:

> This *itr* was discovered during my reign by the mother of the begum Nur-Jehan. when preparing rose water, a foam formed on the surface whenever she poured rose water into saucers. She started gradually isolating this foam. After she had accumulated enough of it, she noticed the strength of this perfume: a drop spread in the palm of the hand has such a smell that one fancies that numerous red roses have flowered simultaneously. There is no perfume to compare with it. It renews the strength of tired hearts, it brings back to life withered souls. To retribute her for the invention, I offered a pearl necklace to the inventor. This oil then became known as '*itr-i-Jahangiri*'.

This *attar* of rose, of high commercial value, is still prepared nowadays. In Bulgaria, the rose *R. damascena triginipetala* is cultivated for this purpose, in the area of Kazanlik, the Valley of the Roses. France (in the region of Nice and Grasse), Morocco, Turkey and Russia are the major producers of rose essence.

The principal ingredient of the rose fragrance is β-damascenone. This molecule is found also in grapes, apples and other fruits where it constitutes an important flavor component. Our nose can detect doses as low as 500 femtograms ($5 \times 10^{-13}$ g) in the output of a gas chromatograph.

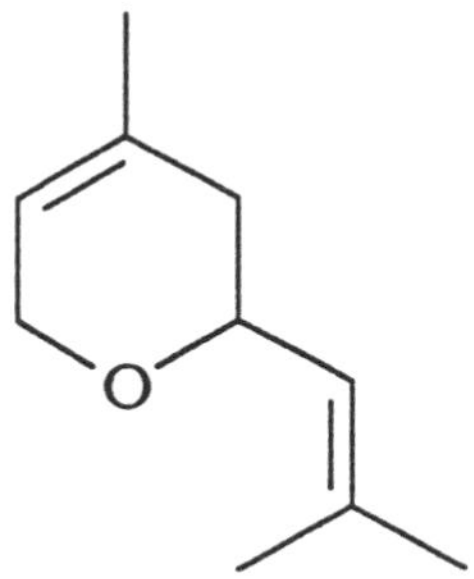

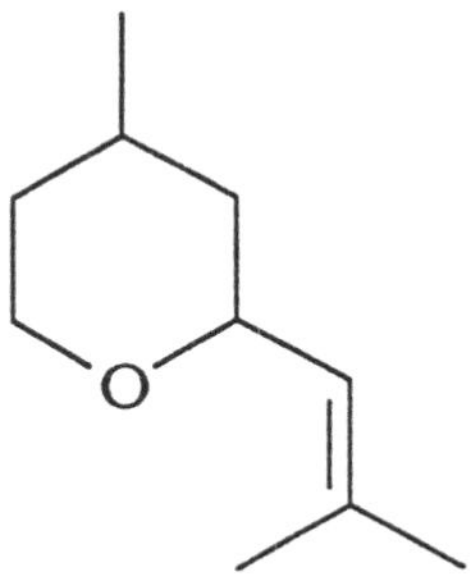

Besides damascenone, the essential rose oil contains other terpene derivatives, such as nerol oxide (a dihydropyran) and rose oxide (a tetrahydropyran). These three molecules coexist with numerous others in the aromas of grapes and wines. The essential rose oil is traditionally used in perfumery. It was one of the major assets of the perfumers in Grasse, who did not foresee the industrial and economic impact of synthetic molecules.

The fragrance of roses continues to be much prized. A recent example is the perfume *Obsession* which combines the smell of roses (damascenone) with that of jasmine (jasmone) and orange (limonene): we merely state here the probable main constituents. In the US the annual sales reached $30 M in the first year (the total annual sales of perfumes in the US reach $4 billion). The annual sales of a new perfume gradually plateau out and can then drop: the perfume dies or becomes a classic (e.g. Chanel 5). *Obsession* like its rival *Giorgio* is seductive and sensual. Presently they are both in fashion. Perfumers believe that the discreetness, or conversely the aggressivity, of a perfume bears an inverse relation to the interest of males towards their mates.

References: B. P. Pal, *Interdicipl. Sci. Rev.*, **12**, 77–88 (1987).
P. S. Zurer, *Chem. Eng. News*, **1987**, 28 September, pp. 21–22.
A. Rapp and H. Mandery, *Experientia*, **42**, 873–884 (1986).

## 7  Use of Halohydrins in Synthesis

A clever modification to addition of hypobromous acid allows for stereoselective cyclizations. When an unsaturated alcohol, such as the precursor of six-membered rings shown here, is subjected to a positive bromine ion-donor, the electrophile attaches to one face of the C=C double bond. The alcohol function serves as a nucleophile, all the more easily so that it belongs to the same molecule as the bromonium ion formed. Hence it attacks from the other face.

The Br$^+$ donor is *N*-bromosuccinimide or TBCO (2,4,4,6-tetrabromo-2,5-cyclohexadienone); the latter is cleaved to a positive bromine ion and the aromatic 2,4,6-tribromophenoxide anion.

R = CH$_2$SiMe$_3$

Reference: P. C. Ting and P. A. Bartlett, *J. Am. Chem. Soc.*, **106**, 2668–2671 (1984). (For TBCO): T. Kato, I. Ichinose, and Y. Kitahara, *Chem. Lett.*, **1976**, 1187.

## 8   A Step in Woodward's Quinine Synthesis

Quinine, an alkaloid extracted from the bark of the South American tree *Cinchona*, is the antimalarial drug of choice. The step described here forms the vinyl group substituting the 1-azabicyclo[2,2,2]heptane system of quinine.

A primary amine is treated in ethanol with methyl iodine in the presence of potassium carbonate. Mild heating affords the quaternary ammonium salt, thus set for Hofmann elimination.

The normal elimination procedure, heating on silver oxide (i), fails, affording a zwitterion as the ammonium salt, which is stabilized by as ionic bond with the carboxylate group. Drastic conditions are required (ii = 60% potassium hydroxide, at 140°C) to bring about the expected elimination. The least substituted olefin is formed according to Hofmann's Rule.

Reference: R. B. Woodward and W. von E. Doering, *J. Am. Chem. Soc.*, **67**, 860 (1945).

# 9 The Style of R. B. Woodward (1917–1979)

His presentations were memorable. He did not hesitate to speak for hours in front of a fascinated audience. His lectures were prepared in minute detail. He was meticulous in drawing chemical structures. He would take out a beautiful leather case in which he kept his own white and colored chalks. He would write on the blackboard, slowly and with much care. He cultivated his effects; his description of a synthesis would become a sequence of dramatic episodes. Each disaster, examined carefully, contained a lesson which he taught with total mastery. Such catastrophies could be avoided by a new manoeuver which the magician had thought out, and pulled out of his hat at the right time in the lecture.

After one-and-a-half or two hours of lecturing, Woodward would start being in need and he would start sipping martinis or going through a bottle of whisky; an impressive spectacle which personally I always found very sad. In character Woodward was the stereotype of the Yankee: little talkative, very attentive, witty-eyed behind his thick glasses, very reserved, somewhat timid which he overcompensated for, especially in public, by what could have been mistaken for arrogance or show-off. He was a night bird; his Harvard seminars started in the early evening and ended rarely before one or two in the morning.

Robert Burns Woodward was born in Boston on April 10, 1919. Since his early youth, he had a passion for chemistry and thought-up syntheses of natural products. According to him, he imagined his first synthesis of quinine at the age of 12! At the time when he completed his secondary education, his knowledge in chemistry was already that of a professional. When he entered MIT, this unusual student created quite a stir. The administration, recognizing his genius, was flexible enough to allow him to follow a tailor-made study program so that he could simultaneously obtain his BS and PhD degrees at 20. His subsequent academic career was rapid: instructor at the University of Illinois (1937), then at Harvard (1941), where he then became in turn an assistant professor (1944), an associate professor (1946) and a full professor (1950).

We owe to R. B. Woodward the great classic total syntheses of numerous natural products: quinine (1944), patuline (1950), cholesterol (1951), cortisone (1951), lanosterol (1954), strychnine (1954), reserpine (1956), chlorophyll (1960), colchicine (1963), cephalosporine C (1965), vitamin $B_{12}$ (together with Albert Eschenmoser, 1972), and prostaglandine $F_{2a}$ (1973). These achievements were crowned with the award in 1965 of the Nobel Prize for chemistry At the time of his death—a heart attack on July 8, 1979, after having lived for a while alone and in bad health—he had been working on the synthesis of erythromycin, an antibiotic.

The intellectual force of Woodward was that of a genius; an enormous memory together with rigorous logic and a great respect for physico-chemical methods. In the Forties, when only UV/visible and IR spectroscopies were available, i.e. before the era of nmr and mass spectroscopy, he elucidated structures which were considerable achievements at that time. During World War II, he also established the correct $\beta$-lactam structure for penicillin. His deep intuition for structure/reactivity relationship allowed him also to discover—in collaboration with Roald Hoffmann—the symmetry rules that apply to concerted reactions: a major contribution to chemistry.

Finally, Woodward—although he had little taste for teaching in the usual sense—was an outstanding teacher. His students, Americans and foreigners, hold chairs in organic chemistry at numerous universities: R. B. Woodward thus became scientific father to numerous renowned chemists.

Reference: D. H. R. Barton and H. H. Wasserman, *Tetrahedron*, **37** Suppl. 1, vii (1981).

## 10 Halogenation Followed by Dehydrohalogenation

1,6-Oxidocyclodecapentaene, the target molecule, is an aromatic system with $10\,\pi$ electrons. Cyclopentaene itself is unknown, as it would be strongly destabilized by steric repulsion of the two inner hydrogens in positions 1 and 6 of the monocycle. An oxygen bridge removes this steric hindrance and lets the $C_{10}$ macrocycle achieve coplanarity, thus providing a highly stabilized annulenic-type system.

Synthesis starts with epoxidation of the hydrocarbon resulting from double Birch reduction of naphthalene with perbenzoic acid; the central (tetrasubstituted and therefore more electron-rich) double bond is more reactive. The second step is addition of two molecules of bromine to the diene in chloroform as solvent (as in the first step) at a temperature of $-75°C$. A mixture of stereoisomers, left unseparated, forms. It is then treated with potassium *t*-butoxide in ether at $-10°C$. Dehydro-bromination ($-4HBr$) affords the target molecule in the form of the aromatic tautomer.

This (addition of halogen/dehydrohalogenation) sequence is standard.

Reference: E. Vogel, M. Biskup, W. Pretzer, and W. A. Böll, *Angew, Chem. Int. Ed. Eng.*, **3**, 642 (1964).

## 11  Dehydrohalogenation Avoiding Epimerization

Formation of a C=C double bond by elimination of a molecule of hydrohalic acid HX (X=Cl, Br, I) requires normally a strong base, a potassium alkoxide, such as potassium *t*-butoxide.

There are cases prohibiting use of such a reagent, because it could lead to enolization and therefore to epimerization of an asymmetric center. Thus a weak base is used, typically LiCl or the combination $LiBr-Li_2CO_3$ in DMF: this dipolar aprotic solvent scarcely solvates anions $X^{(-)}$ the reactivity of which is thus boosted.

It can be presumed therefore that the reaction shown is initiated by removal of the proton from the carbon adjacent to that bearing the bromine. The carbanion thus formed expels the $Br^{(-)}$ anion as leaving group in a mechanism denoted $E_1cB$ (CB stands for conjugate base).

Do note that an E1 reaction, initiated by departure of a bromide ion, would not be very favorable due to the Coulomb repulsion between two positive charges at two adjacent carbons. Noteworthy is also retention of the 'stereochemistry' at the positions $\alpha$ to the carbonyl groups during this reaction.

References: E. J. Corey and A. G. Hortmann, *J. Am. Chem. Soc.* **87**, 5736 (1965).
L. F. Fieser and M. Fieser, *Reagents for Organic Syntheses*, Vol. 1, Wiley, New York, 1967, pp. 606–609.

## 12 Selectivity in Hydroboration

It is a very mild reaction: the activation enthalpy is $2 \pm 3$ kcal mol$^{-1}$ for addition of borane $BH_3$ to ethylene in the gas phase. The transition state for transformation of the $\pi$ complex between borane and ethylene, to the organoborane $H_2B-C_2H_5$ includes a molecule of ethylene with little deformation.

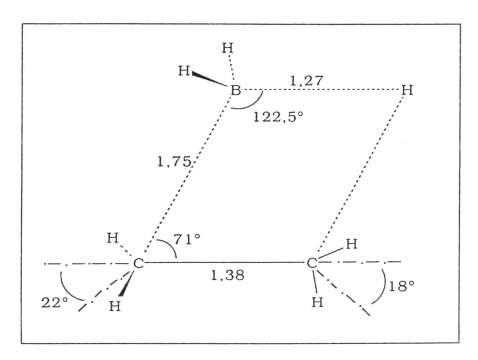

Let us ruthlessly apply a perturbational analysis to the problem of regioselectivity, when ethylenes with different substituents at carbons 1 and 2 are hydroborated. The regioselectivity is the ratio of the percentages of the products having boron attached to either carbon 1 or 2. To make a rough estimate of the interaction energy between borane $BH_3$ (LUMO) and an ethylene (HOMO) and of its positional dependence, it will hinge upon the ratio of the square of the coefficients $c_1^2/c_2^2$ for carbons 1 and 2 in the HOMO of the olefin. The calculated and experimentally determined values are summarized in the Table: the agreement is quite good and favors orbital control.

| $(C_1/C_2)^2$ | calc. | obs. |
|---|---|---|
| 0.44/0.41 | 88 : 12 | 98 : 2 |
| 0.41/0.45 | < 1 : 99 | 1 : 99 |
| 0.46/0.42 | 96 : 4 | 99 : 1 |
| 0.37/0.38 | < 1 : 99 | 1 : 99 |
| 0.47/0.45 | 93 : 7 | 94 : 6 |
| 0.41/0.42 | 34 : 66 | 2 : 98 |

'We do all have a little advance knowledge of the result, and a knowledge of what leads to the result. We are result-minded. We like to find results; but we really lack the spirit of concluding our lines of thought.' Gaston Bachelard, in *Rencontres internationales de Genève*, 1952.

Stereoselectivity of hydroboration is also of interest, especially if a borane derived from chiral terpenes, such as isopinocamphenylborane, is used as shown in the adjacent Scheme.

This leads to reactions of high enantioselectivity: starting from 2-butene of *E*-configuration, the S:R ratio of the 2-butanol product is 87:13; from *Z*-2-butene, the S:R ratio is 1:99. These experimental results are in agreement with computations of the energies of the corresponding transition states (MO theory).

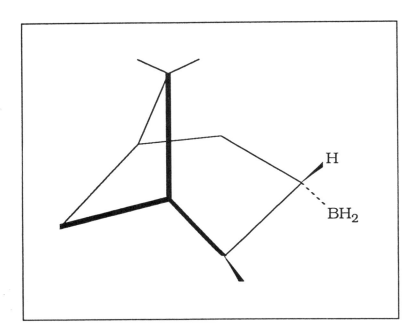

Reference: (Experimental) S. Masamune, B. Moon, K. J. S. Petersen, and T. Imai, *J. Am. Chem., Soc.*, **107**, 4549 (1985). (Theory) M. Egger and R. Keese, *Helv. Chim. Acta.* **70**, 1843–1854 (1987).

## 13   The First Step in a Synthesis of Elvirol

Elvirol is an interesting sesquiterpene ($C_{15}$); it does not conform to the isoprene head-to-tail rule.

Its synthesis involves first a thermal Claisen rearrangement (sigmatropic [3,3] rearrangement), followed by protection of the phenol group with dihydropyran; hydroboration affords the primary terminal alcohol. The last steps consist of two successive Wittig reactions required for chain extension with introduction of the terminal olefin group.

Reference: F. Bohlmann and D. Körnig, *Chem. Ber.*, **107**, 177 (1974).

## 14   Novel Useful Electrophiles

### Addition of Sulfur and Nitrogen Groups

Use of the Meerwein salt $Me_3O^{(+)}BF_4^{(-)}$ activates the reagent $MeS-NMe_2$, probably by methylation of the nitrogen and formation of a quaternary ammonium group; the electrophile thus produced can then initiate an $Ad_E$ addition. At the end of the reaction (formation of the product with the usual antiperiplanar sterochemistry), the ammonium leaving group has to be removed to avoid the inverse reaction, a Hoffmann elimination. This is done by contacting this with trimethylamine which removes the activating methyl group.

Addition of $MeS-SMe_2^{(+)}$ in the form of the tetrafluoroborate, for instance, also affords the *anti*-product. This requires only elimination of a neutral molecule of dimethylsulfide ($SMe_2$ is an excellent leaving group), either in an internal manner, with formation of the episulfonium ion (three-membered ring) and the presence of a nucleophile :Nu.

The episulfonium ion thus obtained reacts with trimethylamine, affording the same product as above. It can also be opened with dimethylsulfide which in the presence of trimethylamine leads to the same

Here is an application of this methodology: addition of $MeS-SMe_2^{(+)}$ to cyclohexene affords an episulfonium ion. This intermediate, not isolated, reacts directly with dimethyl sulfoxide (DMSO) serving both as solvent and nucleophile. Treatment of this adduct with a strongly hindered base (i.e. a poor nucleophile) gives access to a cyclohexanone bearing an $SCH_3$ group at the position adjacent to the carbonyl group: an interesting double functionalization, readily accessible by this route.

References: B. M. Trost, T. Shibata, and S. J. Martin, *J. Am. Chem. Soc.*, **104**, 3228–3230 (1982).
M. C. Caserio and J. K. Kim, *J. Am. Chem. Soc.*, **104**, 3231–3233 (1982).

## 15  Addition of Tellurium Tetrachloride

Addition of *one* equivalent of TeCl$_4$ to an olefin leads to the Markovnikov product with *syn* stereochemistry. The reaction is carried out in the presence of free radical inhibitors.

When the C=C double bond bears an ester function in the allylic position, addition of TeCl$_4$ occurs with migration of the ester group, either by a concerted mechanism (see below, top) or by a two-step ionic mechanism (see below, bottom).

References: J. E. Bäckvall *et al.*, *J. Org. Chem.*, **48**, 3918 (1983).
        L. Engman, *J. Am. Chem. Soc.* **106**, 3977–3984 (1984).

Such products—after removal of tellurium with Raney nickel—allow for formation of the epoxide by an $S_N2$ reaction (Darzens). The application illustrated here is a brief and efficient synthesis of (2S,3S)-*trans*-2,3-epoxybutene with good enantiomeric purity.

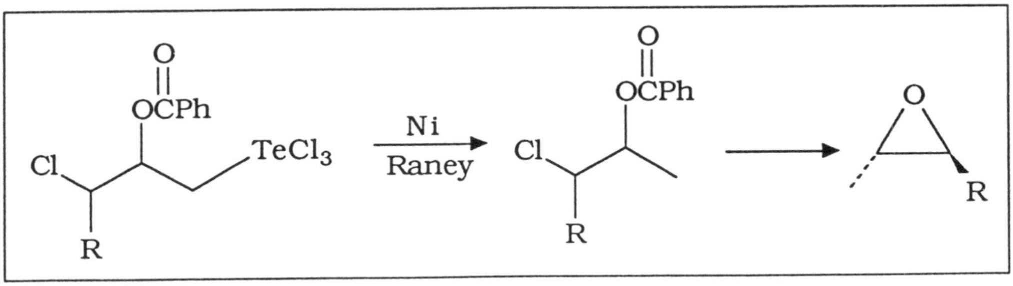

# 12.I
# Functionalizations: Wittig Conversion of Carbonyl Compounds to Olefins

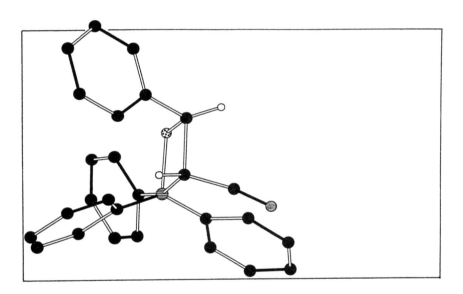

The overall yield in a multi-step synthesis is the product of the individual yields for each step. It follows from this self-evident truth that a linear synthesis should be avoided because conducive to poor overall yields. It is preferable to conceive a convergent synthesis: the target molecule is made up of a few (three to four, at most) fragments, joined together in the course of the synthesis. The Wittig reaction is one of the most convenient assembly of molecular fragments. It is also one of the very few examples of an important reaction invented during the last four decades. It is to be placed on the same high level as its antecedent, the Darzens reaction.

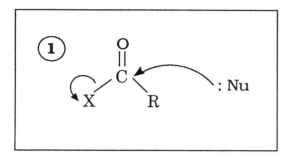

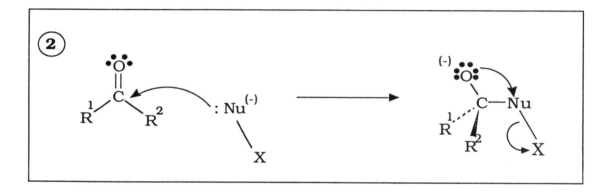

# 1 A Second Type of Addition–Elimination

At the beginning of this course we dealt with a first type of addition–elimination: attack by a nucleophile on a carbonyl group bearing a leaving group X. The addition step was accompanied or followed by cleavage of the leaving group $X^-$.

A second type of addition–elimination is observed when the potential leaving group X is attached to the nucleophile. In the first step, the nucleophile adds to the electrophilic carbon of the carbonyl group, which accumulates negative charge on the terminal oxygen. In the second step, this negatively charged oxygen attacks the center bearing the leaving group X.

---

There occurs always a slow build-up of small effects, whose sum total in the long run tips the scales, shifting to a new provisional equilibrium, ushering-in a new phase—R. Ruyer, *Esquisse d'une philosophie de la structure*, 1930.

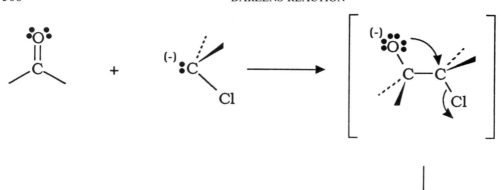

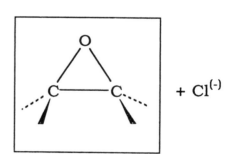

## 2  Darzens Reaction

In the present state of progress in chemistry, it is highly unlikely that a discovery can be made by pure chance.—Darzens, 1912

This reaction is historically the first example of this second type of addition–elimination reaction; it paved the way for the invention of other new processes, such as the Wittig reaction.

The nucleophile is a carbanion bearing a chlorine atom, the future leaving group. The addition step is followed by an elimination: the negatively charged oxygen attacks the carbon bearing the chlorine in an internal $S_N2$ reaction, affording an epoxide, a small strained ring. Of course, there is inversion of configuration at the chlorine-bearing carbon, as befits an $S_N2$ mechanism.

Reference: G. Darzens, *C. R. Ac. Sci. Paris*, **139**, 1214 (1904); **141**, 766 (1905); **142**, 214 (1906); **144**, 123 (1907); **145**, 1342 (1907).

## 3   Darzens, a Great Polytechnician (1867–1954)

Georges Auguste Darzens was born July 12, 1867, in Moscow. His family originated from Aude. His father Rodolphe (1823–1886) had settled in Moscow where he joined his cousins who had settled there themselves during the Napoleonic wars. R. Darzens was involved in trading between France and Russia. Having become at 13 a boarder at the Collège Sainte-Barbe in Paris, Georges Darzens proceeded to prepare the entrance examination to École polytechnique where he first became a student (Class of 1886) and where he was subsequently appointed a professor of chemistry (1913–1937). During his studies at École polytechnique, Darzens was fascinated by astronomy: however this vocation was discouraged due to his having to wear glasses.

Darzens was then seduced by chemistry: Grimaux, his professor at the École polytechnique (1886–1888), was at that time one of the rare organic chemists in France to dare to disobey the influential Marcelin Berthelot, who forbade atomic theory to be taught. Darzens wrote: 'The [atomic] theories brought me to the way of research in organic chemistry'. From 1888 to 1897, Darzens was a lab assistant in Grimaux's laboratory. He shared Grimaux's attitude in the Dreyfus affair. Both were ardent *dreyfusards*— which took considerable moral fiber and courage in the military environment which was theirs.

During the last decade of the 19th century, Darzens had still not found his way. After getting degrees of *licence* (B.S.) in mathematics, and in physics, he became *Agrégé des Sciences Physiques* in 1895. In parallel, he entered medical school, became *externe des hopitaux de Paris* in 1890, and obtained his M.D. in 1899.

At the end of the century, Darzens's activity is mind-boggling. In 1897, he had become director of the scientific laboratory of L. T. Piver, a company with a worldwide reputation for its perfumes. He maintained this position till 1920. Another line of activity was the physiology of vision: as he wrote himself, 'at the beginning of his career, when he had not yet found his call, he published (1895) a physical theory of color perception by the eye'. Darzens threw himself with passion into the beginnings of the automobile. Whereas his brother Rodolphe organized the first car rallies and actively participated in them, Georges Darzens conceived, designed, and built three or four prototypes during the period 1890–1910, one of which was constructed in a series of six to ten models. Innovations we owe to Darzens were lengthening of the pistons as compared to the 'square' version that was the earlier, prevailing

(*continued*)

dogma; introduction of ball bearings in the wheel hubs, an innovation not to the liking of his good friend Louis Renault ('You are wrong. Roller bearings are worthless, nothing beats a smooth axle'); and equalization of the diameters of all four wheels.

Darzens was not only an inventor in his leisure time, but also in science. Scientists can be classified (V. Prelog) into four classes: those who perfect the state of the art, those who find a general explanation for a collection of facts, authors of discoveries, and inventors. Darzens was clearly one of the latter.

Amongst his numerous inventions in chemistry, a general substitution method $-OH \rightarrow -Cl$, described by Darzens himself as follows:

'To prepare the $\alpha$-choropropionic ester, he has instituted a novel procedure for substituting chlorine to a hydroxy group in a molecule. This procedure uses the action of thionyl chloride in the presence of a tertiary base and it is presently used universally to effect such a substitution in fragile molecules such as terpene alcohols or sterols'.

His talents as an inventor allowed him to contribute also to the discovery of new perfumes. At Piver, he devised for example, the big successes 'Floramye' and 'Trèfle Incarnat'. Similarly with the perfume makers Grenoville (1921–1924) and Dior (1926–1931) he was involved in the fabrication of various nitrogen-containing synthetic musks. The inventive skills of Darzens also made a contribution to World War I. Affected to the gunpowder office on October 31, 1914, Darzens devised a procedure for making large quantities of picric acid from aniline. This manufacture was brought into full operation at the end of 1914. It contributed positively to the victory at the Marne by providing a sorely needed supply of explosives. During the whole war, Darzens wrote numerous confidential memoirs for the Scientific Committee on Powders and Explosives.

Darzens's major contribution to chemistry is the reaction which bears his name. He was probably the best French organic chemist in the first half of our century. The Darzens reaction anticipates the Wittig and the Corey–Chaykovsky reactions. It has also the merit of giving easy access to aldehydes and ketones. He invested himself with gusto in these studies till the end of the Forties. As early as 1904, he set up the handsome method of glycidic synthesis to provide these two classes of compounds. This method, today a classic, universally used, bears his name. It consists in the condensation of $\alpha$-chloro esters with ketones or aldehydes under the action of ethoxide or of sodium amide. In this manner, glycidic esters are produced, and they are easily saponified into the corresponding acids. The latter are decomposed simply upon distillation, providing, according to the case, novel, homologous aldehydes and ketones. Of extraordinarily general scope, this glycidic method of synthesis has given

*(continued)*

*(continued)*

access to a wide variety of aldehydes and ketones in all the areas of organic chemistry.

I shall close with a few sentences from the writings of Darzens. They show a philosophy of science from a generalist and a generalizer. His education at École polytechnique had imbued him with a unitarian spirit and with broad views on nature:

'Convinced that chemistry had a lot to gain by giving itself general methods to solve its numerous problems, that differ more in their aspects than in their deep nature, I had the desire to devote myself to such studies. As one will see, many of my efforts made their goal to devise such general methods which are for the chemist the equivalent of equations to the mathematician.'

References: 1. *Recherches et travaux scientifiques de M. Georges Darzens*, J. Dumoulin, Paris, 1912. (Note to support his candidature for the post of professor at the École polytechnique).

2. *Note sur les travaux scientifiques de M. G. Darzens*, manuscript without date, ca. 1949, by courtesy of Dr. Jean Jacques. Letter of M. Claude Darzens of November 24, 1987, and several documents, for instance, a biographical note by M. Georges Darzens (1900–1978), another son of the great scientist.

# 4  Generalization and Modernization

The generalization consists in using a carbanion $^{(-)}CR^1R^2X$ where X stands for a leaving group.

The modernization consists in taking a positively charged $X^{(+)}$ as leaving group so that the entity X expelled in the elimination step be *a neutral molecule*.

The best examples are those of dimethyl sulfide $SMe_2$ and of dimethyl sulfoxide $OSMe_2$ as leaving group: they define the Corey–Chaykovsky reaction.

Reference: E. J. Corey and M. Chaykovsky, *J. Am. Chem. Soc.*, **87**, 1353 (1965).

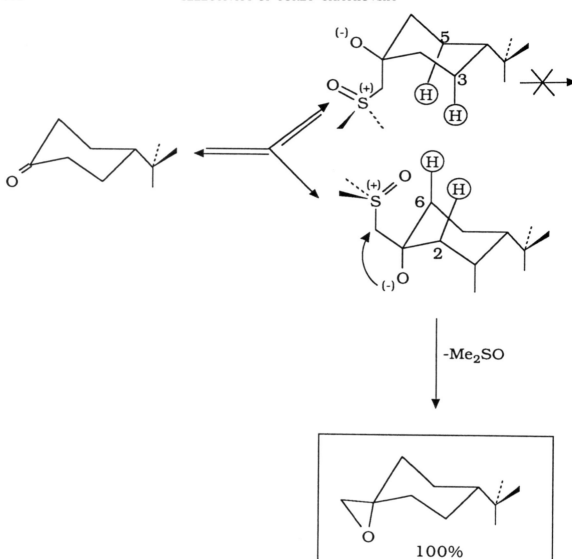

# 5 Advantage of the Corey–Chaykovsky Reaction: Its Selectivity

Consider addition of the Corey–Chaykovsky reagent $Me_2S(O)CH_2^{(-)}$ to 4-*t*-butylcyclo-hexanone. Since the *t*-butyl group is confined to the equatorial position, this substrate is locked into a single chair conformation.

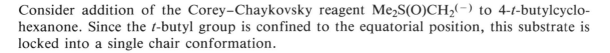

The addition step is reversible and the reagent can add a priori from either face of the carbonyl double bond. Axial addition leads to a dead end. In fact, the presence of the C(3)—H and C(5)—H axial bond prevents the $Me_2SO$- leaving group from rotation into a position placing the C—S bond—to be broken—and the C—O bond antiperiplanar to one another; this is the stereoelectronic criterion imposed by an $S_N2$ reaction.

Equatorial addition, on the other hand, is such that the steric effect favors the disposition prone to the elimination of a neutral molecule of dimethyl sulfoxide (see the adjacent Scheme). Thus, in a stereospecific manner, the single product is the epoxide with an equatorial carbon (and an axial oxygen). The dipolar reagent (a sulfur ylide) is prepared by methylation of dimethyl sulfoxide with methyl iodide followed by removal of one of the nine protons with a strong base, sodium hydride for instance.

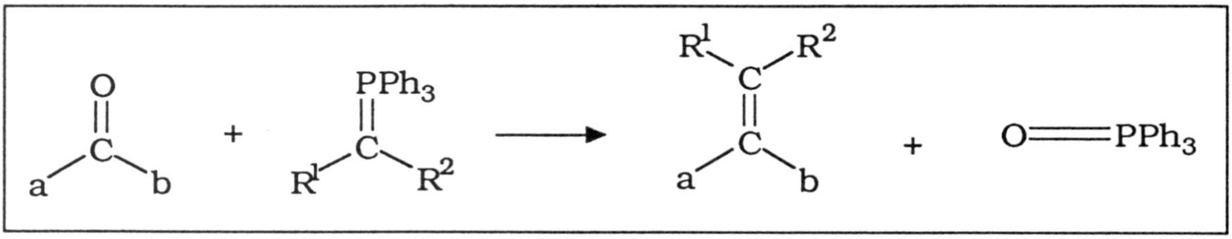

O
C
a
b
+
PPh3
C
R1
R2
R1
R2
C
C
a
b
+
O=PPh3


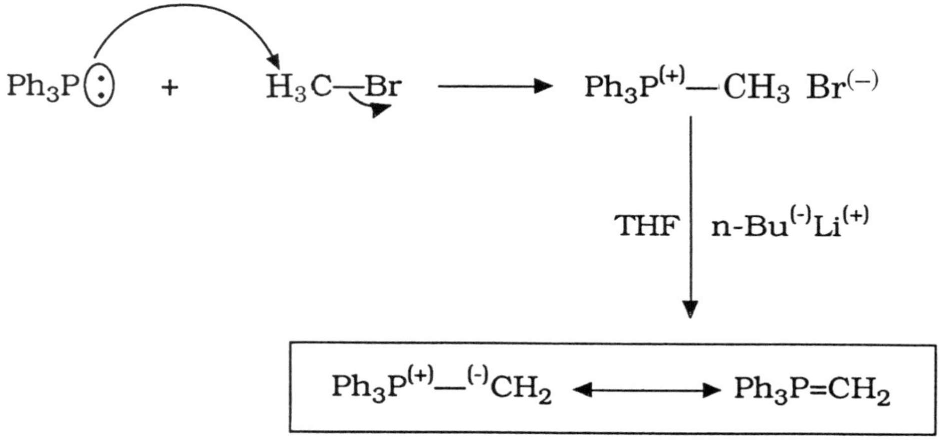

Ph3P
+
H3C—Br
Ph3P(+)—CH3 Br(−)
THF
n-Bu(-)Li(+)
Ph3P(+)—(-)CH2
Ph3P=CH2

# 6  Wittig Reaction

In the Wittig reaction a carbonyl compound O=Cab is coupled to an $R^1R^2C$ fragment and turned into the corresponding olefinic compound $R^1R^2C$=Cab. Formation of a phosphine oxide O=$PR_3$ is the driving force for this reaction, the P=O bond having a large bond energy (108 kcal mol$^{-1}$).

The Wittig reagent is readily prepared. Triphenylphosphine is frequently used as starting material. An $S_N2$ reaction, for instance with methyl bromide, leads to the phosphonium salt (keep in mind the analogy of the phosphonium with the ammonium cation). A strong base such as $n$-butyllithium removes one of the methyl protons, made more acidic by the electron-attracting $Ph_3P^{(+)}$ substituent.

In this way the simplest Wittig reagent is obtained ($R^1 = R^2 = H$). The latter is stabilized by resonance between a P=C form without charge separation (the 'ylene' form) and a dipolar $P^{(+)}$–$C^{(-)}$ form (the 'ylide' form).

Reference: G. Wittig, *Angew. Chem.*, **92**, 671 (1980).

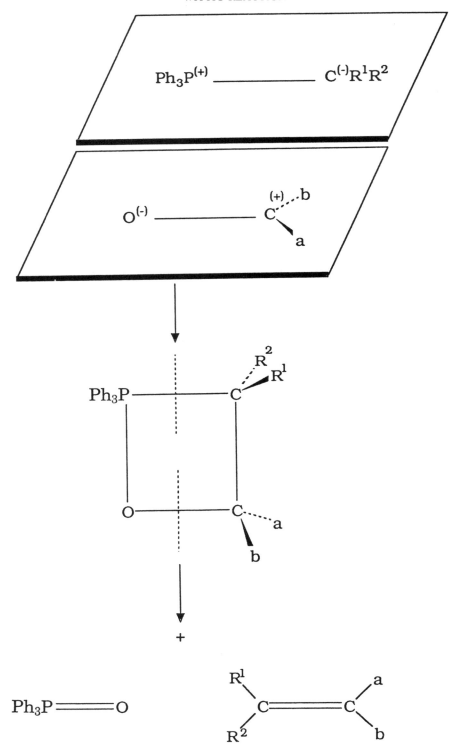

Let us view the Wittig reaction from two different angles: charge control, and orbital control. A first electrostatic description consists in disposing the two opposing dipoles in two parallel planes: the dipole of the carbonyl group and that of the Wittig reagent (i.e. both are considered in the *ylide* form).

Two new bonds are formed between the centers subjected to Coulomb attraction: O—P and C—C. The four-membered ring created is cleaved into the olefin and triphenylphosphine oxide, the formation of which renders the overall process exothermic (see above).

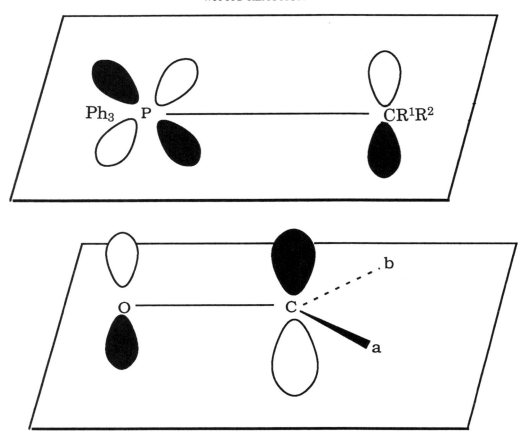

Let us discuss now the orbital description of the Wittig reaction. For this purpose, the ylene forms of the Wittig reagent and of the carbonyl group are stacked in two parallel planes. The ylene form of the Wittig reagent incorporates a $\pi$ phosphorus—carbon bond with lateral overlap of a $d$ orbital of the phosphorus and a $p$ orbital of the carbon. The predominant interaction (quantum calculations) is the two electron interaction between the frontier orbitals, the HOMO of the Wittig reagent and the LUMO of the carbonyl compound. The overlap integral $S$ for the new carbon—carbon bond is approximately 0.2 in the transition state.

Hence cycloaddition of the two partners (formally a $[2\pi_s + 2\pi_s]$ process) as well as the inverse reaction (a cycloreversion) become allowed concerted reactions.

Reference: F. Volatron and O. Eisenstein, *J. Am. Chem. Soc.*, **109**, 1–14 (1987).

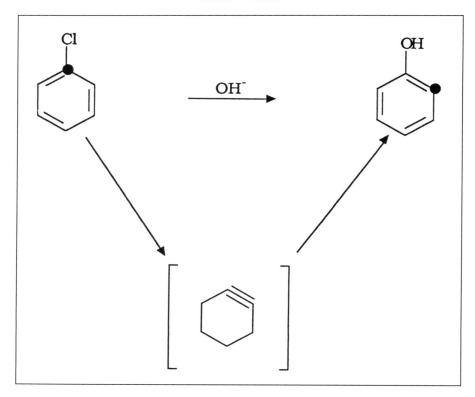

# 7   Georg Wittig (1897–1987)

Born in Berlin in 1897, he obtained his doctorate at the University of Marburg, under the direction of von Auwers (1923). He was successively professor at the Universities of Brunswick (1932), Freiburg (1937), Tübingen (1944) and Heidelberg (1956).

He was one of the founding fathers of organometallic chemistry. He had used phenyllithium (he called it his magic wand) since 1931. He was responsible for the discovery of benzyne in 1942: if chlorobenzene is treated with a strong base, phenol is obtained; the reaction seems to be a straightforward nucleophilic substitution. However, if the chlorine-bearing carbon is labeled, then part of the phenol formed has the OH group on the carbon adjacent to the labeled carbon! This can be explained by formation of an acetylenic intermediate, the so-called benzyne, prototype of reactive species known as arynes.

Wittig was awarded the Nobel Prize in chemistry in 1979 for the reaction which bears his name (1953). Wittig was an excellent pianist as well as an experienced alpinist. Georg Wittig died on August 26, 1987, in Heidelberg.

Reference: R. A. Shaw, *Chem. in Britain*, **24**, 63 (1988).

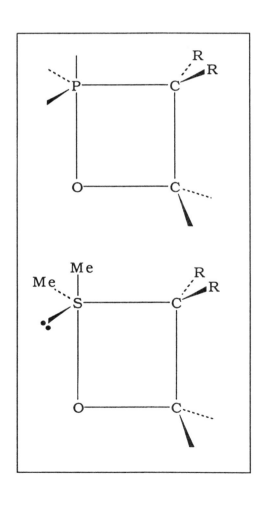

# 8   Reasons for the Difference in Behavior

You have certainly noted the contrast between the Wittig reaction with an olefinic product and the Corey–Chaykovsky reaction which affords an epoxide. What differentiates the reactivity of a sulfur ylide (C—C) from that of a phosphorus ylide (W)?

There are two answers: thermodynamic; and kinetic factors. Both reactions favor products with high thermodynamic stability: phosphine oxide in the Wittig reaction, dimethyl sulfide (or dimethyl sulfoxide) in the Corey–Chaykovsky reaction. Both reactions form a rectangular cyclic intermediate. When phosphorus occupies one of the corners, the required permutation, oxygen → carbon in the apical position of the trigonal bipyramid on phosphorus is easy (low energy barrier); hence cleavage to the olefin. When sulfur—surrounded by five electron pairs, one of them non-bonding—replaces phosphorus, a similar pseudo-rotation is impeded by a high energy barrier. The cyclic intermediate (a trigonal bipyramid around sulfur) is inert towards decomposition with carbon as the only leaving group. Prerequisite for reaction is cleavage of the sulfur–oxygen bond. The negatively charged oxygen can then attack the carbon in an $S_N2$ fashion with final breaking of the carbon–sulfur bond.

Reference: F. Volatron and O. Eisenstein, *J. Am. Chem. Soc.*, **109**, 1–14 (1987).

$$Si(CH_3)_4 \xrightarrow{\quad RLi \quad} (CH_3)_3Si-CH_2{}^{(-)}Li^{(+)}$$

# 9   A Modification of the Wittig Reaction: Peterson Olefination

The Wittig reaction suffers from the inconvenience of a phosphine oxide side-product difficult to eliminate during separation and purification of the products. Peterson has therefore recommended replacement of the phosphorus ylide with a silylated (and therefore stabilized) carbanion. The silylated carbanion obtained by means of a strong base, for instance RLi, RMgCl etc. adds to the carbonyl group. Exchange of the metal cation with a proton during isolation affords an alcohol. Elimination of silanol is readily effected, it is frequently carried out with aqueous HF (acidic conditions) or with KH or NaH or NaF under basic conditions, thus affording the required olefin. The by-product is a volatile silanol (due to the presence of the globular trimethylsilyl group) readily removed.

Sometimes it is useful to provide protection against the strong basicity and non-chemoselectivity of the organolithium agent which serves to form the silylated carbanion, by combining RLi with $CeCl_3$. The resulting organocerium reagents '$RCeCl_2$' have remarkable properties as nucleophiles for additions to carbonyl groups.

Reference: D. J. Peterson, *J. Org. Chem.*, **33**, 780 (1968); *Organometal. Chem. Rev. A.*, **7**, 295 (1972).
T. H. Chan, *Accounts Chem. Res.*, **10**, 442 (1977).
C. R. Johnson and B. D. Tait, *J. Org. Chem.*, **52**, 281–283 (1987).

β-gorgonene

Thus, Wittig elimination of the sterically hindered carbonyl group of the cyclic ketone as indicated fails. In contrast to this, Peterson olefination, less sensitive to steric hindrance, is successful. In this manner Boekman could synthesize $\beta$-gorgonene in 15% yield. This is a natural product isolated from *Pseudopteragorgia americana*, a jelly fish found on the Bermuda islands and along the coasts of Florida.

Reference: R. K. Boekman, Jr. and S. M. Silver, *Tetrahedron Lett.*, **1973**, 3497.

'If all things were to become smoke, we would know through the nostrils'—Heraclitus, fragment 8.

# Summary: 'Darzens, Wittig and Some Others . . .'

Addition–elimination of a first type is observed upon addition of a nucleophile to a carbonyl group which bears a leaving group (Claisen condensation). A second type of addition–elimination is observed when a nucleophile, bearing a leaving group, adds to a carbonyl: an epoxide results. The Darzens reaction (1904) uses a carbanion bearing a chlorine atom in this way.

A modern modification of this reaction is the Corey–Chaykovsky reaction. It starts (like the Swern oxidation discussed in Chapter 10) by activation of dimethyl sulfoxide (DMSO). This consists in sulfur methylation with methyl iodide, followed by treatment with a strong base, such as sodium hydride, to form the ylide (dipole $S^+$–$CH_2^-$). Addition–elimination (cleavage of DMSO) and reaction with the carbonyl group affords the epoxide. The innovation by Corey and Chaykovsky, as compared to the Darzens reaction, is the improved stereoselectivity of the process: starting from 4-$t$-butylcyclohexanone only one diastereoisomer with an axial C—O bond is formed.

The Wittig reaction makes use of phosphorus ylides. A phosphine is alkylated and the resulting phosphonium cation (with tetracoordinated P) is treated with a strong base, for instance $n$-butyllithium. The resulting ylides are stabilized by the ylide–ylene resonance. Addition to a carbonyl affords a four-membered ring as the intermediate which, on cleavage, affords a very stable phosphine oxide (thermodynamically highly stable) and an olefinic compound.

This addition step forms a pentacoordinated phosphorus atom, at the center of a trigonal bipyramid. Oxygen (an electron attractor) enters from an apical position. Permutational isomerism causes the oxygen to move to an equatorial position and the carbon to an apical position; this allows for breaking of the phosphorus—carbon bond and initiates the cleavage of a molecule of phosphine oxide.

The Corey–Chaykovsky reaction stands in contrast: the product is now an epoxide, not an olefinic compound as in the Wittig reaction. This difference in behavior can be explained by the easy/*difficult* pseudorotation of the cyclic intermediate in the Wittig/*Corey–Chaykovsky* reaction.

Certain reactions in industrial chemistry, especially the $\beta$-carotene (vitamin A) synthesis, make use of the Wittig reaction. And so do numerous syntheses of insect pheromones. This chapter ends with the phosphonate modification (Wittig–Horner–Emmons) and with the use of chiral Wittig–Horner reagents for synthesis of molecules with pharmacological activity.

# 12.II
# Functionalizations:
# Wittig Conversion of Carbonyl
# Compounds to Olefins

## 1 Stereochemical Control of the Product

We select for discussion one of the mechanisms proposed, by Bestmann (University of Erlangen). Besides its plausibility, it draws attention to the main factors involved in the selective formation of either the *cis(Z)* or *trans(E)* olefin.

$$R_3P{=}C\diagup^{R^1}_{\diagdown H} \quad\longleftrightarrow\quad R_3\overset{(+)}{P}{-}\overset{(-)}{C}\diagup^{R^1}_{\diagdown H}$$

+

$$O{=}C\diagup^{R^2}_{\diagdown H} \quad\longleftrightarrow\quad \overset{(-)}{O}{-}\overset{(+)}{C}\diagup^{R^2}_{\diagdown H}$$

$$\Big[\ \ \underset{(-)}{O}{-}\overset{107°}{C}\cdots\overset{R^2}{\underset{H}{}}\ \ \cdots\ C\diagup^{R^1\ (+)\ R_3P}_{\diagdown H}\ \Big]^{\ddagger} \quad\longleftrightarrow\quad$$

Both reaction partners, the carbonyl compound and the Wittig reagent, admit of a hybrid representation, combining a limiting form, having a double bond and no charge separation, and a dipolar, charge-separated form. In the initial state, Coulomb attraction favors a head-to-tail arrangement of the reactants, bringing the negative (O) end of the carbonyl group in proximity to the positive (P) end of the Wittig ylide and, conversely, the positive (C) end of the carbonyl next to the negative (C) end of the Wittig ylide.

The reaction is triggered by nucleophilic attack of the carbanionic end of the Wittig ylide on the carbonyl. As is the rule for interaction of a nucleophile :Nu with a $C{=}O$ carbonyl group, the angle of attack is 107° (Bürgi–Dunitz). The carbanionic carbon bears three groups: the trialkyl- or triphenylphosphine group (large), $R^1$ (medium) and H (small). Since the new carbon—carbon bond is already partly formed in the transition state, the system adopts a staggered conformation placing the large group ($R_3P$) at the carbanionic carbon and the large substituent on the carbonyl carbon in an antiperiplanar arrangement (180° dihedral angle).

The resulting intermediate, with positive charge on phosphorus and negative charge on oxygen (a so-called betaïne), could not be isolated, even at $-78\,°C$. It cyclizes immediately: the Coulomb attraction $P^+,O^-$ induces formation of a phosphorus—oxygen bond, whose large bond energy stabilizes the system.

This intermediate is an oxaphosphetane. The four-membered ring bears the $R^1$ and $R^2$ groups in the same half-space, in a *cis* disposition, a key point for the following events.

These take place at phosphorus. The trigonal bipyramid (TBP), as is well known, is the usual geometry for pentacoordinated compounds, such as pentacoordinate carbon in the transition state of an $S_N2$ reaction (Walden inversion). The TBP conformation of lowest energy has the two most electron-attracting groups A in the apical position (a). The two apical ligands have the longest bonds with the central atom, phosphorus in the present case. Furthermore, and for the very same reason (longest and thus weakest bonds), the bonds to apical substituents are cleaved (or formed) most readily. One witnesses apical entry of a nucleophile and apical departure of a leaving group, for instance in the $S_N2$ reaction. In the following Scheme, the TBP for the transition state for an $S_N2$ reaction between $HO^-$ and $CH_3Cl$ is depicted.

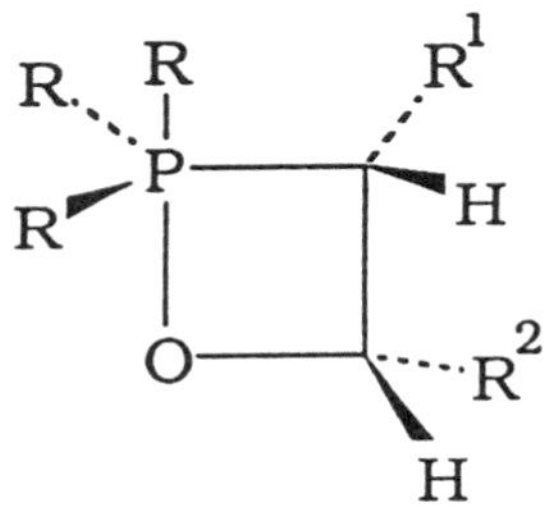

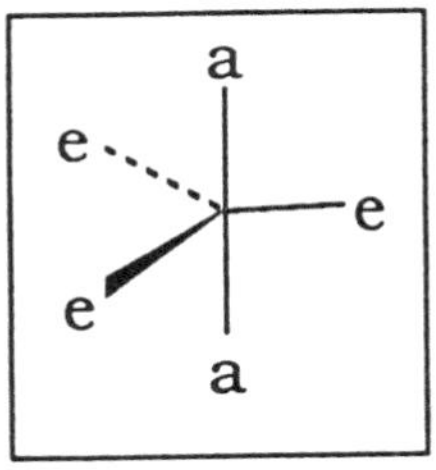

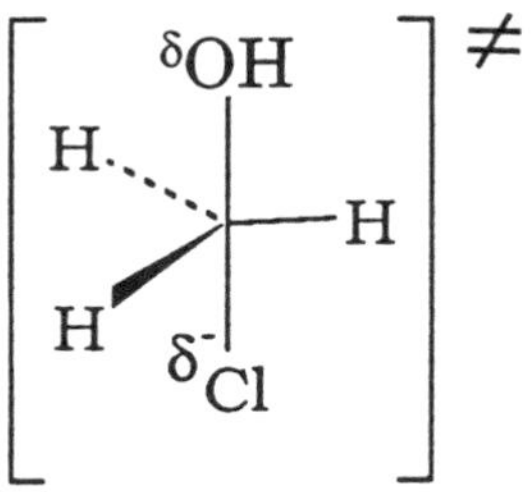

We now return to the Wittig reaction: the oxaphosphetane can open into the olefin, only after the TBP has undergone a *pseudorotation* $\Psi$ placing the carbon bearing the $R^1$ group in an apical position. This pseudo-rotation (or permutational isomerization, exchanging two apical groups with two equatorial groups) occurs probably with the third equatorial group fixed as the pivot. The angle between the two other equatorial groups opens up from 120° to 180° while the angle between the two apical groups simultaneously closes from 180° to 120°.

The new betaïne formed by opening of the oxaphosphetane, destabilized by the pseudorotation, maintains a *cis* relationship between $R^1$ and $R^2$. It gives rise to the olefin by expelling a very stable trialkyl- or triarylphosphine oxide molecule, with migration of the non-bonding electron pair from the carbanion to the new $\pi$ carbon—carbon bond. Therefore, the *cis* olefin forms selectively: at least if the reaction is carried out under *kinetic control* (fast reaction, at low temperature).

In general, the *trans* olefin is more stable. Hence, under conditions of *thermodynamic* control (long reaction times, elevated temperature) formation of the *trans* isomer is favored.

More generally, all the factors that increase the lifetime (stability) of the *cis* betaine allow its equilibration with the *trans* betaine:

1. Protic solvents, stabilizing the carbanion by hydrogen bonding.

2. Attracting $R^1$ substituents that stabilize the carbanion: $R^1 = CN$, COOR, COR, aryl, vinyl.
3. Apolar solvents with low dielectric constant in which the betaïne tends to stabilize itself by internal Coulomb attraction.
4. Addition of salts that increase the ionic strength, e.g. $Li^+$ $BPh_4^-$.
5. Donor substituents that stabilize the positive charge on phosphorus: $PR_3^+$ (R = alkyl); $PO(OR)_2$.
6. An excess of the Brønsted base which had been used to form the Wittig ylide.

On the other hand, all factors that reduce the lifetime (stability) of the *cis* betaïne favor the selective formation of the *cis* olefin:

7. Aprotic solvents.
8. Donor substituents $R^1 = $ alkyl, alkoxy.
9. Dipolar aprotic solvents with large dielectric constants which decrease the intramolecular stabilization of the betaïne.
10. Salt free solutions.
11. Electrophilic phosphorus atoms: $P(p\text{-}C_6H_4Cl)_3^+$, $PPh_3^+$.
12. Solutions containing soft Lewis bases ($I^-$, amines) interacting with the phosphonium cation.

Reference: H. J. Bestmann, *Pure Appl. Chem.*, **51**, 515 (1979).

## 2   Other Ways of Access to Phosphonium Salts

A Wittig reagent is normally obtained by treatment of a phosphonium salt with a strong base such as *n*-butyllithium, sodium amide $NaNH_2$ or potassium *t*-butoxide. The phosphonium salt is formed generally by alkylation of triphenylphosphine with a halide RX.

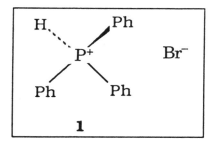

This is however not always possible. For instance, triphenylphosphine cannot be alkylated with 1-chlorotetralin.

A useful variant is reaction of triphenylphosphonium bromide with a primary, secondary or tertiary alcohol. It gives rise to the phosphonium salt **1** and to a molecule of water.

$$ \overset{+}{H}PPh_3, Br^- \;+\; R_1R_2CHOH \;\longrightarrow\; R^1R^2\overset{+}{C}HPPh_3,\ Br^- \;+\; H_2O $$

**1**

References: K. Sasse, in Houben Weyl, *Methoden der Organischen Chemie XII/1*, Georg Thieme, Stuttgart, 1963, p. 79.
H. Pommer, *Angew. Chem.*, **89**, 437 (1977).
D. Taub, R. D. Hoffsommer, and C. H. Kuo, *Tetrahedron*, **29**, 1477 (1973).

## 3   A Non-Enolizing Alternative to the Wittig Reaction

Aldehydes are substrates of choice for the Wittig reaction, they undergo fast reaction even at $-78\,°C$ (the temperature of the dry ice/acetone thermostat). Ketones react more slowly. Thus enolizable centers adjacent to a carbonyl group are vulnerable: the carbon center in a Wittig ylide is a base which can bring about inversions of configuration (epimerization) at one of these $\alpha$-carbons. This possible epimerization is one of the drawbacks of the Wittig reaction.

A solution has been proposed for methylenation of carbonyl compounds: a procedure using the reagent $Zn/CH_2Br_2/TiCl_4$. The example indicated is the conversion of (+)-isomenthone to the corresponding hydrocarbon (room temperature, 89% yield), with retention of the configuration at C-2 (the carbon bearing the isopropyl group).

References: R. L. Sowerby and R. M. Coates, *J. Am. Chem. Soc.*, **94**, 4758 (1972).
L. Lombardo, *Tetrahedron Lett.*, **23**, 4293 (1982); *Org. Synth.*, **65**, 81–87 (1987).

## 4  Industrial Use of the Wittig Reaction

The Wittig reaction is used in the pharmaceutical industry for the manufacture of carotenoids, used especially for the prevention and treatment of epithelial cancers, e.g. cancer of the bladder.

The reaction scheme as indicated (Hoffmann–La Roche) involves combination of a molecule of retinal (in the form of the aldehyde) with another molecule of retinal previously transformed to a phosphorus ylide. The Wittig reaction (retinal + retinylidenetriphenylphosphorane) affords β-carotene and triphenylphosphine oxide (in greater than 80% yield).

This large-scale preparation of β-carotene in fact uses two molecules of retinylidenetriphenyl-phosphorane. A previous reaction transforms one of these to retinal, by oxidative cleavage of the phosphorane with hydrogen peroxide. The retinal produced in this way then reacts *in situ* with the Wittig ylide.

A similar reaction produces vitamin A acetate by a Wittig reaction between a $C_{15}$ phosphorane and an $\alpha,\beta$-unsaturated $C_5$ aldehyde (BASF, 600 T of vitamin A per year)!

Reference: H. Pommer and P. C. Thieme, *Topics in Current Chem.*, **109**, 165–188 (1983).

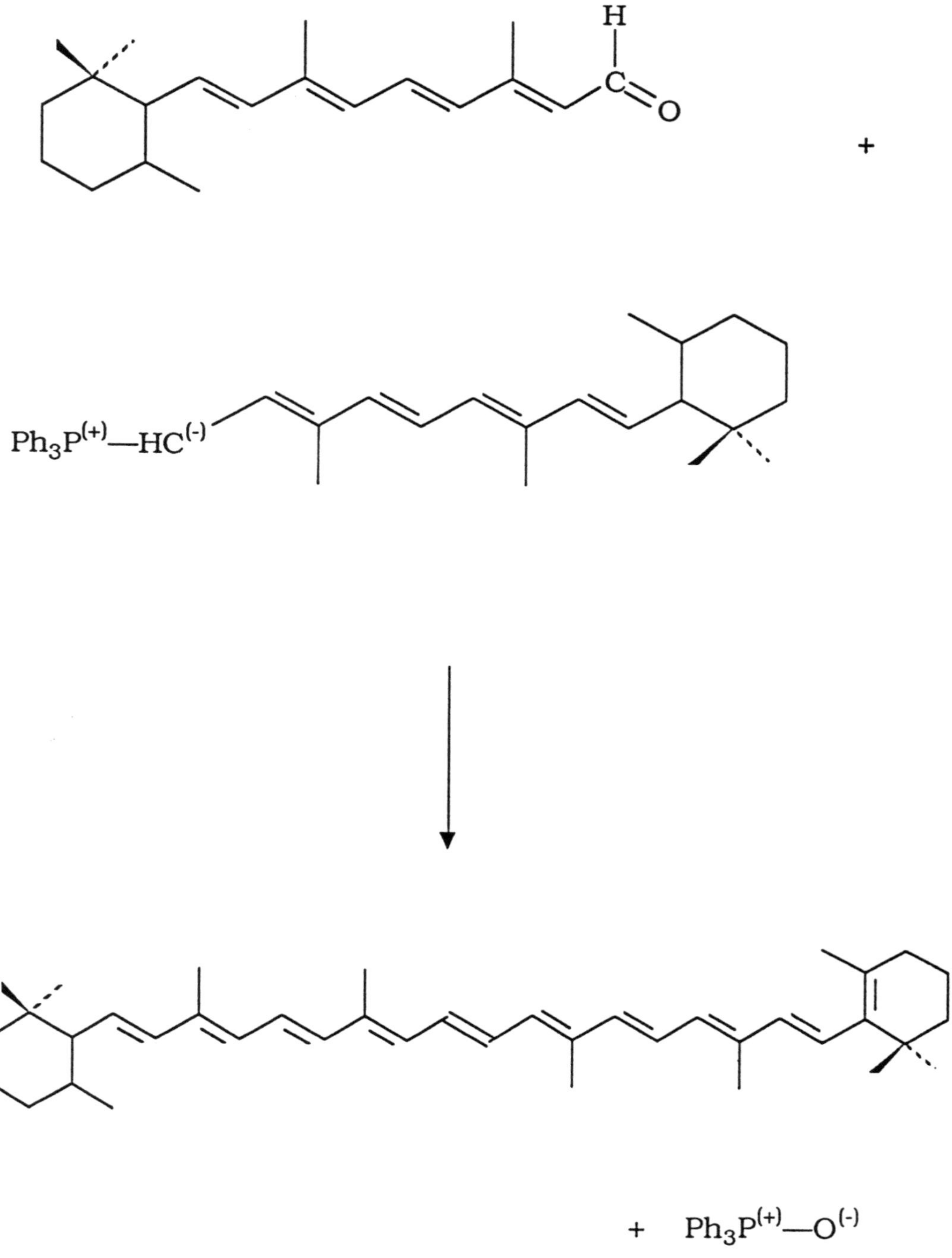
H
C
O
+
$Ph_3P^{(+)}$—$HC^{(-)}$
+ $Ph_3P^{(+)}$—$O^{(-)}$

## 5 Another Industrial Application of the Wittig Reaction

The condensation of the dialdehyde shown with two molecules of the phosphonate Ar–CH$_2$–P(OCH$_3$)$_2$ in the presence of a base (required for the formation of the ylide) affords the cyclic dinitrile. This *trans*-stilbene, known under the 'Pallanil Brilliant Weiss RS' commercial name is fluorescent and serves as an optical brightener for laundry. It is manufactured by BASF on a scale of about 50 kT per year.

## 6 Synthesis of Pheromones

The Wittig reaction has been used very often for synthesis of pheromones. The preparation of *multifidene* is an example. This triene is the sexual attractant for several species of algae. Thus the male gametes of *Cutleria multifida* can spend up to twenty hours swimming to the female gamete, that emits among other substances multifidene.

A partial synthesis involves the initial addition of one equivalent of dichloroketene to dicyclopentadiene. The four-membered ring is opened by a base.

A sequence of several steps leads to the lactone (internal ester). Reduction of the carbonyl group with DIBAL affords the tricyclic aldehyde. The unsaturated $C_4$ side chain is introduced into the aldehyde by means of a Wittig reaction (*). The alcohol is then cleaved and the required enantiomer isolated. It is then submitted to pyrolysis (retro Diels–Alder reaction), affording a 3-cyclopentenylmethanol. Oxidation with DCC (dicyclohexylcarbodiimide) leads to the corresponding aldehyde. A second Wittig reaction (**) affords multifidene.

This reaction sequence is an illustration of stereochemical control: the *cis* arrangement of the side chains of multifidene is accounted for by the $(2\pi + 2\pi)$ cycloaddition of dichloroketene to one of the double bonds of dicyclopentadiene (the Diels–Alder dimer of cyclopentadiene). The biological activity of multifidene is impressive: at a concentration of $6.5 \times 10^{-12}$ M it induces motion of the *Cutleria* gametes. Such a threshold translates into only one to ten *individual* molecules of the pheromone being required for a stimulus!

$$(CH_3)_2CH—(CH_2)_4 \overset{O}{\underset{H \quad H}{\triangle}} (CH_2)_9CH_3$$

$$(CH_3)_2CH—(CH_2)_4—CH_2—PPh_3 \quad + \quad \overset{O}{\underset{H}{\overset{\|}{C}}}—(CH_2)_9——CH_3$$

$$(CH_3)_2CH—(CH_2)_4—CH_6Br \qquad \overset{O}{\underset{H}{\overset{\|}{C}}}—OEt \; + \; Li^{(+)(-)}CH_2—(CH_2)_8—CH_3$$

$$BrCH_2—(CH_2)_8—CH_3$$

References: W. Boland and L. Jaenicke, *J. Org. Chem.*, **44**, 4819 (1979).
W. Boland, K. Jakoby, L. Jaenicke, and D. G. Müller, *Z. Naturforsch.*, **C36**, 262 (1981).
L. Jaenicke and W. Boland, *Angew. Chem. Int. Ed. Engl.*, **21**, 643–710 (1982).

## 7  Another Pheromone Synthesis via the Wittig Reaction

The target molecule (see box) is disparlure, the pheromone of *Lymantria dispar*. This insect, a moth (the disparate bombyx) has a devastating influence on forests and orchards: the larvae live on the leaves of trees. Emitted by the female, the pheromone, (7R,8S)-(+)-epoxy-7,8-methyl-2-octadecane, is detected by the male in concentrations of $10^{-10} \, g \, mL^{-1}$. The other enantiomer is inactive, even in solutions $10^6$ times more concentrated.

A racemic disparlure can be readily synthesized starting from commercially available precursors having up to ten carbons. The key step is a Wittig reaction affording a *cis*-olefin which on epoxidation with *m*-chloroperbenzoic acid produces disparlure.

The triphenylphosphonium salt is formed from the corresponding bromide. The $C_{11}$ aldehyde is obtained by condensation of the lithium reagent with ethyl formate. The lithium reagent results from metallation of *n*-decyl bromide.

## 8    Enantiospecific Synthesis of Frontalin

The starting lactone **1** is obtained from a decomposition product of lactose (next page). It is converted to the iodide **2** which on reductive cleavage affords **3**. The latter is then reduced to the primary alcohol **4** with the correct configuration.

In a first pathway, tosylation of **4**, followed by ozonolysis, affords the aldehyde **5** (64% yield) (the direct reaction **1** → **5** by way of reduction with LiAlH$_4$ and cleavage with sodium periodate is not appropriate). Wittig reaction with **5**, followed by catalytic hydrogenation leads to the methylketone **6**. Acid catalyzed regeneration of the protected carbonyl group induces cyclization to **7**. Reductive cleavage of the latter with LiEt$_3$BH (a 'super-hydride') leads to ( − )-(1*S*,5*R*)-frontaline **8**, the pheromone of the parasite *Dendroctonus frontalis Zimm* of maritime pine trees.

In a second pathway, protection by benzylation **4** → **9** is followed by hydrolysis of the ether **9**; tosylation (selectively, at the primary alcohol function) with subsequent hydrogenolysis produces the tertiary alcohol **10**. Ozonolysis to the aldehyde followed by first a Wittig reaction and then catalytic hydrogenation affords the other enantiomer ( + )-(1*R*,5*S*)-frontaline.

References: (Reductive cleavage) B. Bernet and A. Vasella, *Helv. Chim. Acta*, **67**, 1328 (1984).
(Frontaline) M. C. Trinh, J-C. Florent, and C. Monneret, *J. Chem. Soc. Chem. Comm.*, **1987**, 615; *Tetrahedron*, **44**, 6633 (1988).

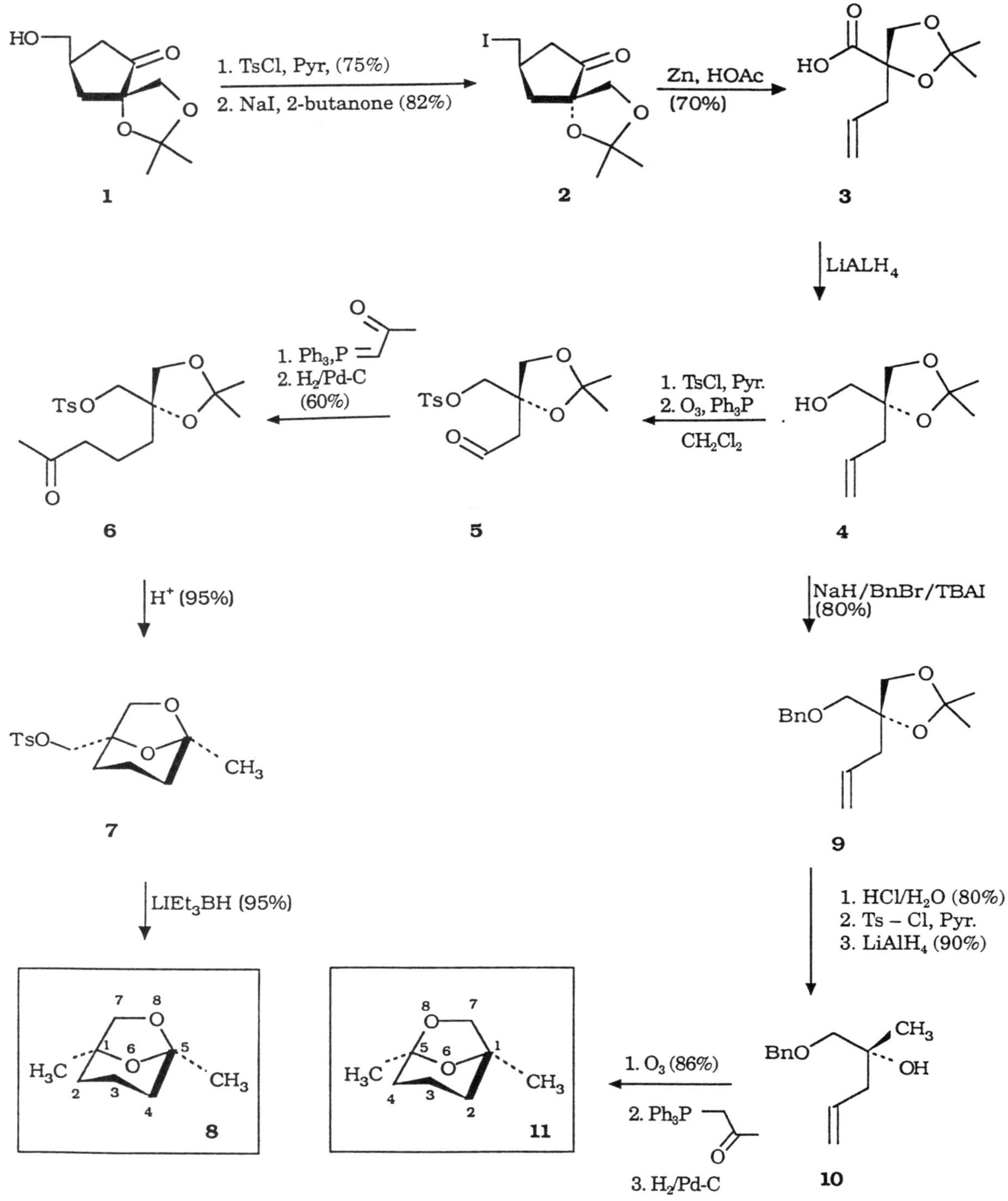
1. TsCl, Pyr, (75%)
2. NaI, 2-butanone (82%)
Zn, HOAc
(70%)
LiALH$_4$
1. TsCl, Pyr.
2. O$_3$, Ph$_3$P
CH$_2$Cl$_2$
1. Ph$_3$P
2. H$_2$/Pd-C
(60%)
H$^+$ (95%)
LIEt$_3$BH (95%)
NaH/BnBr/TBAI
(80%)
1. HCl/H$_2$O (80%)
2. Ts – Cl, Pyr.
3. LiAlH$_4$ (90%)
1. O$_3$ (86%)
2. Ph$_3$P
3. H$_2$/Pd-C
1
2
3
4
5
6
7
8
9
10
11

# 13.I
# Rearrangements

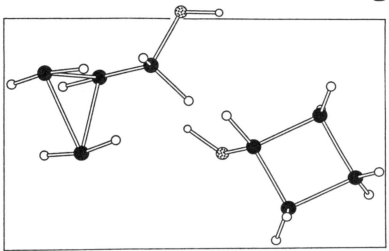

The elegance of a synthesis is acclaimed when quite a few atoms move in the course of the key step, and a deep reorganization of the molecular structure takes place. Also, at least on paper, the succession of steps appears startling from the very inception. This goes by the name of rearrangement. The creative imagination, i.e. the genius of the synthetic chemist, often finds expression in rearrangements.

In this chapter we shall confine ourselves to some very simple and common rearrangements, selected for their usefulness.

## 1  Definitions

A concise definition, even an approximate one is sufficient: a rearrangement is a reaction changing the connectivity and hence the carbon skeleton. The official definition is more detailed but not necessarily more precise:

'This term is traditionally applied to any reaction that violates the principle of minimal structural change. According to this simplistic hypothesis the chemical species would not isomerize during a rearrangement, i.e. a substitution reaction or the modification of a functional group into another would occur without breaking bonds different from those strictly required for the transformation itself. For example, every new substituent should strictly take the place of the leaving group. etc.'

Reference: *Glossary of terms for Physical Organic Chemistry (IUPAC); l'Actualité Chimique*, suppl. June–July, 1985, French Society of Chemistry

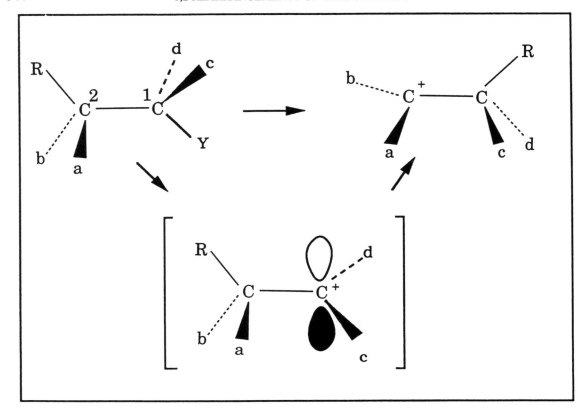

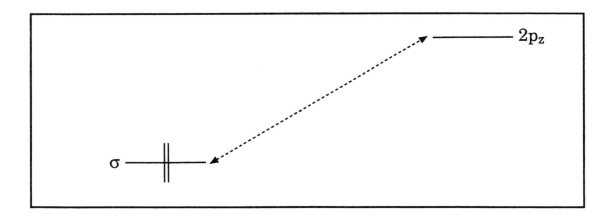

# 2   1,2-Rearrangements of Carbocations

Numerous rearrangements are triggered off by the appearance of a site of electron deficiency, then compensated by a neighboring migrating group. The prototype of these structural reorganizations is the 1,2-rearrangement of the Wagner–Meerwein type in carbocations. Heterolytic cleavage of a C—Y bond where Y is the leaving group leaves a positive charge at carbon 1; a $\sigma$ C—R bond at carbon 2 migrates either in a concerted way or in a two-step process. A transfer of the group R ($=$ hydrogen, alkyl or aryl) from carbon 2 to carbon 1 occurs. The new carbocation formed has to lose its charge either by loss of a proton $H^+$, or by trapping a nucleophile :$Nu^-$. As a general rule the bond C(2)—R that migrates is antiperiplanar to the bond C(1)—Y with the leaving group. When the reaction occurs in two steps rather than in a concerted fashion, the geometric constraint is reduced: it is sufficient for the bond C(2)—R to be parallel to the $z$-axis of the empty $2p_z$ orbital at C(1); this complies with the principle of maximum overlap between the occupied $\sigma$ orbital and the empty $2p_z$ orbital.

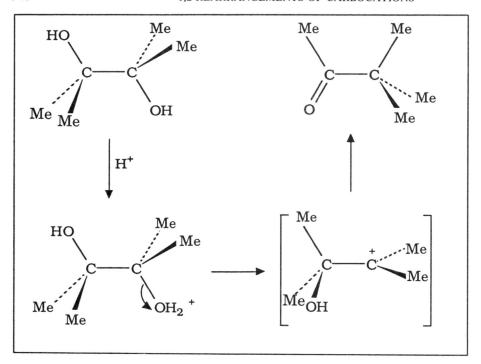

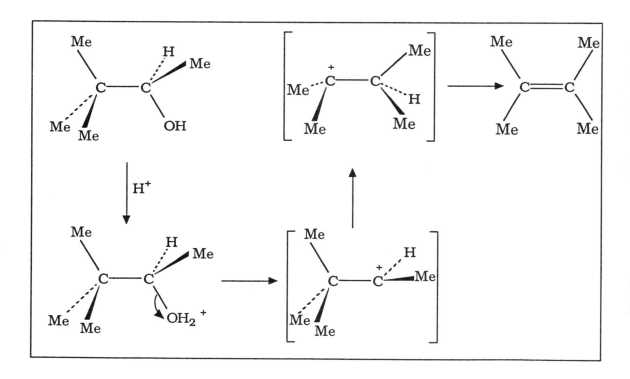

A first example is the pinacol rearrangement. The glycol (named a pinacol) with four methyl groups indicated is monoprotonated by an acid. The carbocation formed at carbon 1 is stabilized by migration of a methyl rather than an OH group (cf. bond energies for $C=C$ and $C=O$, respectively 85 and 90 kcal mol$^{-1}$). This yields the conjugate acid of the methyl ketone, i.e. pinacolone.

The second example is similar in all respects ($C=C$: 85; $C-H$: 100 kcal mol$^{-1}$). The tertiary cation, more stable than the secondary cation from which it is formed by 1,2-shift, can lose its charge by expulsion of a proton giving rise to tetramethylethylene.

Consideration of the relative stabilities of carbocations before and after the 1,2-migration explains why it is not possible to set up a scale of migration tendency for different groups. The major reason for the migration of the methyl group rather than the hydrogen atom in the above example is the higher stability of the tertiary cation. In general aryl substituents migrate preferably relative to alkyl substituents R. Within the aryl substituents, the following scale of migrative ability has been established:

| | |
|---|---|
| $p$-C$_6$H$_4$—OMe | 500 |
| $p$-C$_6$H$_4$—Me | 15.7 |
| $m$-C$_6$H$_4$—Me | 1.95 |
| C$_6$H$_5$ | 1.00 |
| $p$-C$_6$H$_4$—Cl | 0.70 |
| $o$-C$_6$H$_4$—OMe | 12.30 |

References: W. E. Bachmann and J. W. Ferguson, *J. Am. Chem. Soc.*, **56**, 2081 (1934).
R. M. Acheson, *Accounts Chem. Res.*, **4**, 171–186 (1977).

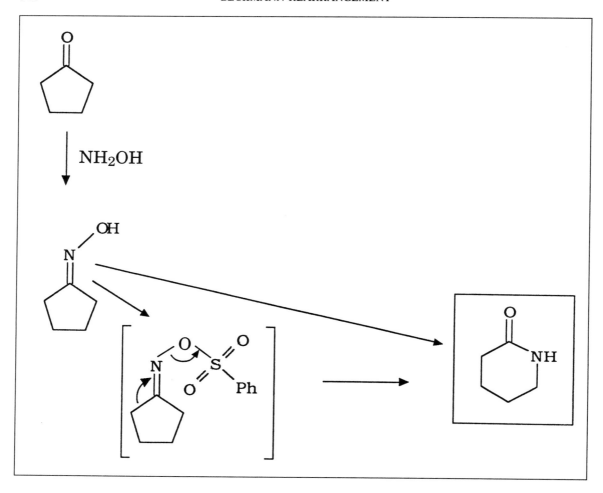

# 3   Beckmann Rearrangement

This reaction has a great synthetic potential. Starting from a ketone, it gives ready access to an amide or to a lactam (cyclic amide). Since its discovery by E. Beckmann this reaction has been extensively studied. The best operating reaction conditions are as follows:

(i) Concentrated sulfuric acid at elevated temperatures for cyclic oximes.
(ii) Either benzene sulfonyl chloride in the presence of sodium hydroxide or p-toluene sulfonyl chloride in the presence of pyridine with moderate heating are the general reagents which provide for basic conditions.
(iii) Phosphorous pentachloride $PCl_5$ is currently the most common reagent; it is used at 0°C in diethyl ether, i.e. under very mild conditions.

The oxime is formed from the ketone with hydroxylamine. Then according to the conditions (i), the rearrangement occurs directly from the protonated oxime whereas under the conditions (ii) or (iii) the leaving oxygen function is transformed into a better leaving group (e.g. $-OH \rightarrow -OTs$) before the rearrangement takes place.

From a stereochemical point of view the group antiperiplanar to the nitrogen—oxygen bond migrates with the leaving group. In our example $PhSO_2Cl$ was selected for the esterification of the oxime with the yield amounting to 95%.

Reference: L. G. Donaruma and W. Z. Herdt, *Org. React.*, **11**, 1–156 (1960).

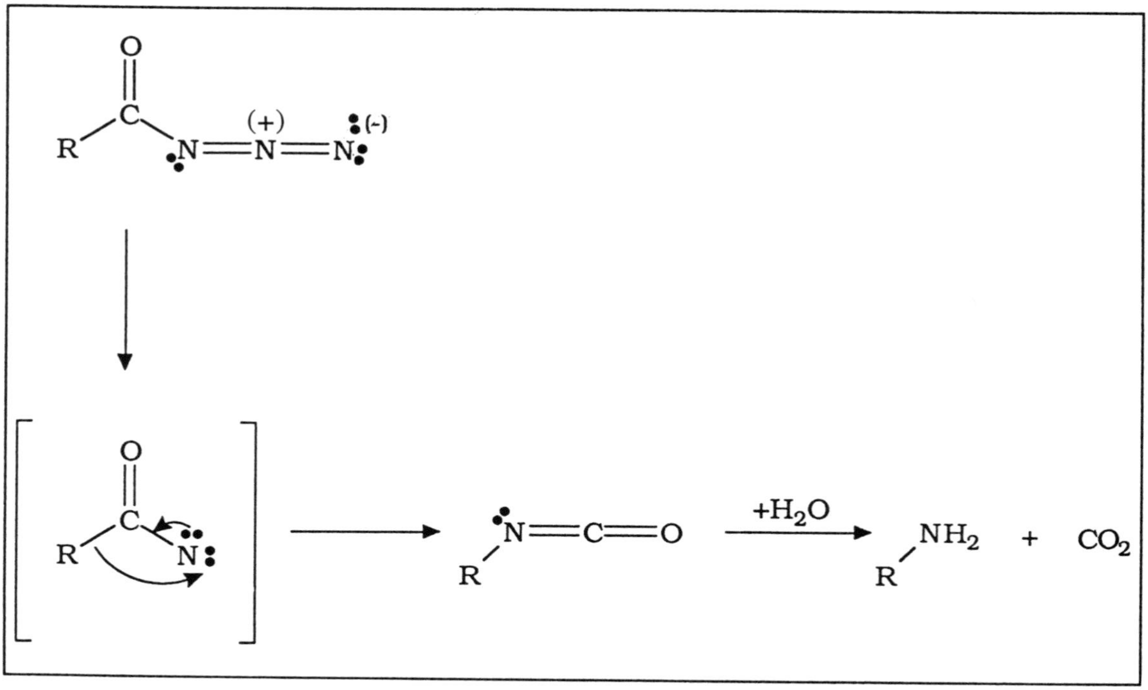

O
C
R
N
N
N
(+)
(-)
O
C
R
N
N
C
O
R
+H2O
NH2
R
+
CO2


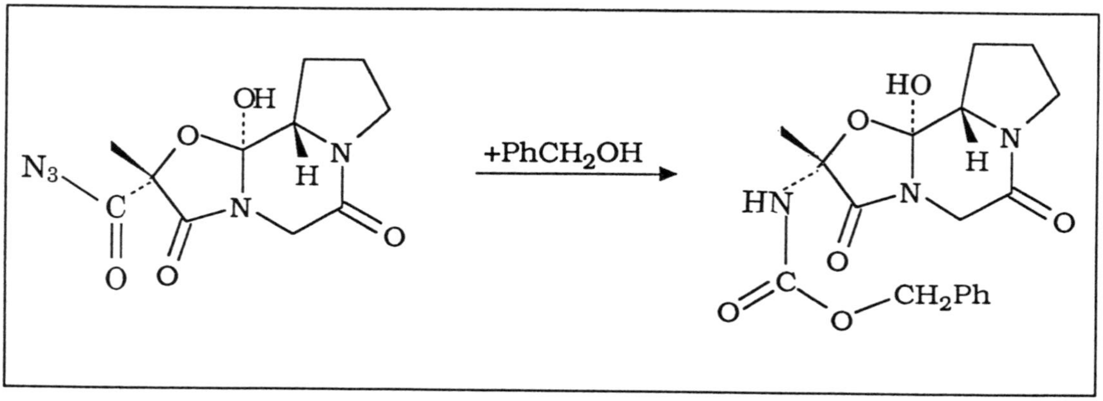

N3
C
O
O
OH
O
N
H
N
O
+PhCH2OH
HO
O
N
H
HN
N
O
O
O
C
O
CH2Ph

# 4 Curtius Rearrangement

The starting compound for this reaction is an acyl azide which can expel a molecule of dinitrogen, thus creating a nitrene—the nitrogen analog of a carbene. This intermediate—like the carbenes—has an electron deficiency: its electronic configuration features an empty 2p orbital at the nitrogen. The rearrangement occurs in response to this electron deficiency and as indicated leads to an isocyanate $R-N=C=O$, which can be converted by water into the corresponding amine, after loss of $CO_2$.

The reaction illustrated in this example is an intermediate step in a synthesis of ergotamine. It is the follow-up reaction of a transformation sequence of a functional group ($RCOO_2Et \rightarrow RCO_2H \rightarrow RCOCl \rightarrow RCON_3$) earlier featured in Chapter 8.

References: J. March, *Advanced Organic Chemistry*, 2nd Ed., McGraw-Hill, New York, p. 1005.
A. Hofmann, A. J. Frey, and H. Ott, *Experientia*, **17**, 206 (1961).

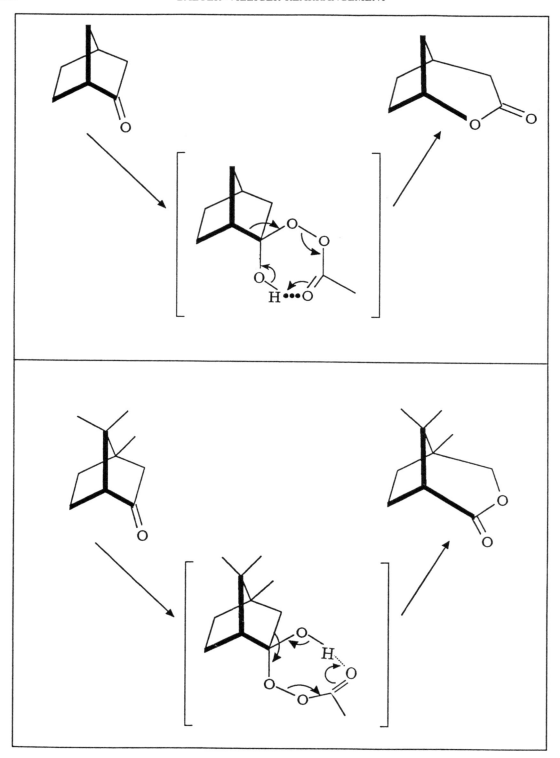

# 5   Baeyer–Villiger Rearrangement

This oxidative rearrangement carried out with peracids transforms a ketone $R^1.CO.R^2$ into an ester (or a lactone) $R^1.CO.OR^2$. The peracid first adds as a nucleophile to the carbonyl group; then one of the substituents ($R^1$ or $R^2$) migrates to the closest oxygen in the peroxy group which is positively charged. In the absence of other effects, especially steric, the aptitude at migration decreases in the following sequence: tertiary alkyl > secondary alkyl > benzyl > phenyl > $n$-alkyl > methyl. The overriding advantage of this rearrangement in the case of polycyclic ketones is the total retention of configuration in the resulting lactone.

In the first example the *exo* side of the carbonyl group is less hindered; the peracetic acid therefore adds from this side. On cleavage a molecule of acetic acid is produced. The tertiary carbon 1 migrates preferentially to the secondary carbon 2.

The second example manifests the opposite case, due to the steric effect: now it is the *endo* side of the carbonyl that is less hindered. Therefore the migration of C(2) rather than C(1) occurs; it is another manifestation of a steric constraint.

References: W. D. Emmons and G. B. Lucas. *J. Am. Chem. Soc.*, **77**, 2287 (1955).
C. H. Hassall, *Org. React.*, **9**, 73 (1957).
J. Meinwald and E. Frauenglass, *J. Am. Chem. Soc.*, **82**, 5235 (1960).
R. R. Sauers and G. P. Ahearn, *J. Am. Chem. Soc.*, **83**, 2759 (1961).

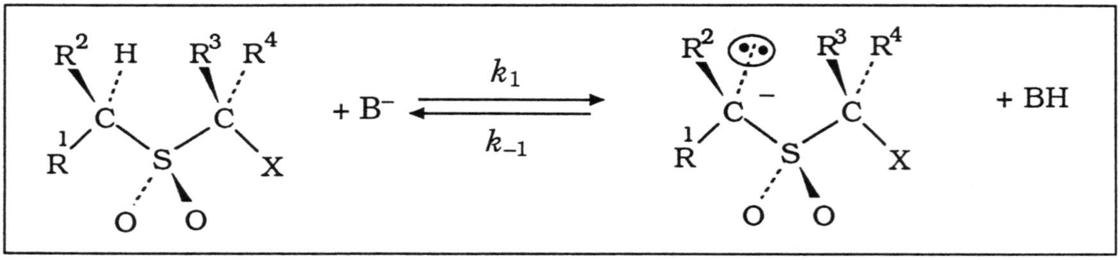
$R^2$
H
$R^3$
$R^4$
C
C
$R^1$
S
X
O
O
+ $B^-$
$k_1$
$k_{-1}$
+ BH

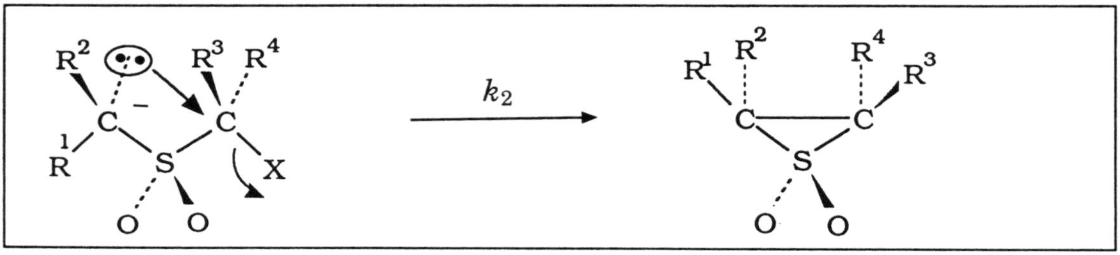
$R^2$
$R^3$
$R^4$
C
C
$R^1$
S
X
O
O
$k_2$

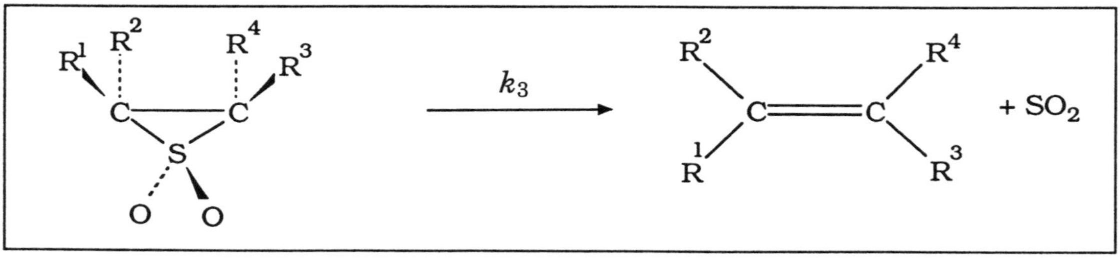
$R^1$
$R^2$
$R^4$
$R^3$
C
C
S
O
O
$k_3$
+ $SO_2$

# 6   Ramberg–Bäcklund Rearrangement

Discovered about 50 years ago, this is an olefin synthesis starting from halosulfones under basic conditions, depicted thus:

$$H_3C. CH_2. SO_2. CHX.CH_3 \xrightarrow{KOH/H_2O} H_3C.CH = CH.CH_3 + KBr + K_2SO_3$$

The accepted mechanism of this reaction comprises three major steps:

(i) Deprotonation equilibrium of the sulfone to form the conjugate base with the equilibrium constant

$$K = k_1/k_{-1}$$

(ii) Formation of a three-membered ring, i.e. an 'episulfone' by attack of the carbanion at the halogen-bearing carbon with halogen as leaving group; this step has an anology to the Darzens reaction (formation of epoxides from $\alpha$-chloroesters);

(iii) Expulsion of a neutral uncharged molecule of sulfur dioxide. The key-step of the Ramberg–Bäcklund reaction is therefore a rearrangement of a carbanion; this affords an unstable episulfone the fragmentation of which leads to the end product of the reaction, an ethylene derivative.

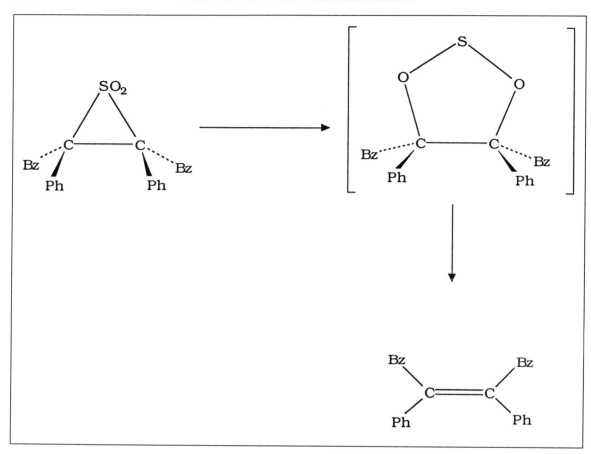

Isotope exchange experiments indicate that $k_{-1} >> k_2$ since the rate of exchange is much greater than the rate of halide ($X^-$) formation. There is a strong influence of the leaving group ($k_{Br}/k_{Cl} = 620$ at $0°C$). Episulfones prepared separately under the reaction conditions lose $SO_2$ at a rate by far superior to that of halide formation from $\alpha$-halosulfones. Therefore the third step is faster; the global rate constant can thus be expressed as:

$$k_{obs} = K.k_2$$

As a general rule the Ramberg–Bäcklund reaction leads preferentially to the *cis* olefin (kinetic control). The thermodynamic product (the *trans* olefin) in formed if a strong base in a solvent of low polarity (e.g. *t*-BuOK in *t*-BuOH) is used.

The loss of sulfur dioxide $SO_2$ probably does not occur in a concerted process which would violate the Woodward–Hoffmann rules. The intermediacy of the five-membered cycle 1,3,2-dioxathiolane has been postulated ($Bz = CO.C_6H_5$).

Reference: L. Ramberg and B. Bäcklund. *Arkiv. Kemi. Mineral Geol.*, **13a**, No. 27 (1940).

## Summary: 'Rearrangements'

A non-trivial modification of the carbon skeleton is called a rearrangement. Rearrangements often induce 1,2-migrations, so-called Wagner–Meerwein migrations. The latter are triggered off by the formation of a carbocation, then stabilized by migration of a bond (in most cases a C—C bond) from the adjacent carbon. The rearrangement ends by loss of a proton from the resulting carbocation or by trapping of a nucleophile present in the medium.

The Beckmann rearrangement is a ring expansion of a cyclic ketoxime providing a cyclic amid (or lactam). In this way, $\epsilon$-caprolactam can be prepared. It is a key intermediate in nylon synthesis.

The Curtius rearrangement also goes through an electron deficient intermediate (a nitrene) yielding an isocyanate after migration of the adjacent C—C bond. Isocyanates can react readily with water ($\rightarrow$ amines) or alcohols ($\rightarrow$ amides).

The Baeyer–Villiger rearrangement is initiated by nucleophilic addition of a peracid $RCO_3H$ to a carbonyl group. The elimination of a molecule of the corresponding acid $RCO_2H$ occurs with insertion of an oxygen atom between the carbonyl carbon and, as a general rule, the highest substituted adjacent carbon. This reaction is extremely useful to synthesis.

# 13.II
# Rearrangements

## 1   Cyclobutene Synthesis

Commercially available and inexpensive cyclopropylcarbinol is the starting material for this straight-forward synthesis. In a first step (aqueous hydrochloric acid at 100°C) it is tranformed to cyclobutanol. This transformation is initiated by protonation of the alcohol function. The conjugate acid of cyclopropylcarbinol thus formed splits off a molecule of water; a carbon—carbon bond in the ring, antiparallel to the broken carbon—oxygen bond, migrates: ring expansion occurs, with formation of the secondary cyclobutyl carbocation. This carbocation is trapped by a molecule of water and thus affords, after deprotonation, cyclobutanol.

Reference: J. Salaün and A. Fadel. *Org. Synth.*, **64**, 50–55 (1965).

The second reaction—esterification of cyclobutanol (with tosyl chloride in pyridine solution)—endows the molecule with a much improved leaving group. The tosylate thus formed is submitted to a third reaction, with base (potassium *t*-butoxide in dimethyl sulfoxide solution at 70°C): elimination of TsOH affords cyclobutene. The overall yield of this reaction sequence is about 45%.

## 2   1,2-Migrations

The same reaction can see coexistence of normal products and rearrangement products. Thus two Russian chemists, in 1903, on treatment of cyclobutylcarbinylamine with nitrous acid HNO$_2$, obtained the four products indicated. In addition, the identical product distribution is obtained when starting from cyclopentylamine, which implies involvement of the same carbocationic intermediates. This reaction is known as the Demyakov rearrangement. The first step is diazotation of the amine RNH$_2 \rightarrow$ RN$_2{}^{(+)}$.

Reference: N. Demjakov and M. Luschnikov, *J. Russ. Phys. Chem. Soc.*, **35**, 26 (1903).

Treatment of the cyclohexadienone below with aqueous sulfuric acid affords the trisubstituted phenol as indicated (50% yield). The mechanism probably implies protonation on the carbonyl oxygen, carbons 3 and 6 becoming electron-deficient. The ester group migrates, leading to a tertiary carbocation which is doubly allylic. The migration occurs regioselectively to carbon 5, sterically more accessible than carbon 3. Finally, the resulting cation discharges by deprotonation.

Reference: H. Plieninger, L. Arnold and W. Hofmann, *Chem. Ber.*, **101**, 981 (1968).

## 3   Ring Contractions by Treatment of Olefins with Thallium(III) Nitrate

The reaction is carried out under mild conditions with striking ease. If, for instance, cyclohexene is treated for one minute with thalium(III) nitrate in methanol at room temperature, and the reaction mixture is hydrolyzed with aqueous 1 M sulfuric acid for 5 minutes, cyclopentanecarboxaldehyde is obtained in good yield (85%). The mechanistic explanation is addition of the $Tl^{3+}$ electrophile to the double bond, followed by nucleophilic (*anti*) attack of the solvent on this onium ion. The metal serves as a leaving group and a secondary carbocation is formed. It stabilizes itself by way of 1,2-migration of a C—C bond (Wagner–Meerwein rearrangement), which creates yet another secondary carbocation, again stabilized, by delocalization of lone pairs from the heteroatom. This is the driving force for this ring contraction. The new cationic intermediate is attacked by a second solvent molecule, which after deprotonation affords the acetal. Acidic hydrolysis converts this acetal to the corresponding aldehyde.

In the same way methylcyclopropylketone is obtained from 1-methylcyclobutene in better than 90% yield.

References: E. C. Taylor and A. McKillop, *Accounts Chem. Res.*, **3**, 338–346 (1970).
E. C. Taylor, R. L. Robey, K. T. Liu, B. Favre, H. T. Bozino, R. A. Conley, C. S. Chiang, *J. Am. Chem. Soc.*, **98**, 3037 (1976).
R. J. Ouellette, *Oxidation in Organic Chemistry*, W. S. Trahanowsky, Ed., Academic Press, New York, 1973, Part B, p. 135.

## 4   The Aroma of Virginia Tobacco

Damascenone has already been discussed in this text: the main constituent in the fragrance of roses, it is also found in wine as well as in tobacco. In the latter case, steam distillation of Virginia tobacco affords two additional bicyclic constituents called bicyclodamascenone A and B.

**A**                                 **B**

In fact, these two molecules can be prepared by acid-catalyzed cyclization of damascenone. The mechanism of this cyclization involves first formation of a carbon–carbon bond between the hydroxyallyl cation of $E$ configuration and the ring, the latter formed by a sigmatropic [1,5] shift; this new bond joins C-2 to C-9.

Thus a bicyclic intermediate with two unsaturated six-membered rings is formed. The positive charge of the allyl cation attracts the $\pi$ electrons of the enol: a second bond between C-5 and C-8 forms. This creates a tricyclic compound having a four-membered ring and hence a strong overall angular strain.

When this cyclobutane ring opens to relieve the strain and cleaves the C-5/C-8 bond bicyclodamascenone A results. Cleavage of the C-5/C-6 bond gives rise to its isomer bicyclodamascenone B.

References: E. Demole and P. Enggist, *Helv. Chim. Acta*, **59**, 1938–1943 (1976).
C. Enzell, *Pure Appl. Chem.*, **57**, 693–700 (1985).

## 5   Small Frogs with a Giant Effect: or, What is the Beckmann Rearrangement Good For?

With tosylate and mesylate as leaving groups, fragmentation is best induced with an organoaluminum compound such as tripropylaluminum. This organometallic reagent has also the asset of introducing an alkyl group (here the propyl group) by nucleophilic attack at the carbon−nitrogen bond C=N. This allows for one-pot tranformations as shown in the adjacent Scheme. The second step is reduction with diisobutylaluminum which gives access to (±) pumiliotoxin (see box).

Pumiliotoxin C, an alkaloid secreted by the skin of small frogs of the genus *Dendrobates* (occurring in Central America) is much more toxic than curare: a single frog supplies the Indians with enough poison for fifty deadly arrows. The mode of action of this toxin is known: it blocks transport of sodium and potassium ions through complexes formed between membrane channels and acetylcholine receptors. This prevents acetylcholine from triggering off muscular contraction.

References: K. Hattori, Y. Matsumura, T. Miyazaki, K. Maruoka, and H. Yamamoto, *J. Am. Chem. Soc.*, **103**, 7368 (1981).
K. Marvoka, T. Miyazaki, M. Ando, Y. Matsumura, S. Sakane, and H. Yamamoto, *J. Am. Chem. Soc.*, **105**, 2831 (1983).
C. W. Myers and J. W. Daly, *Sci. Amer.*, **1983**, February, p. 96.
J. E. Warnick, P. J. Jessup, L. E. Iverman, M. E. Eldefrwai, Y. Nimit, J. W. Daly, and E. X. Albuquerque, *Mol. Pharm.*, **22**, 565 (1982).

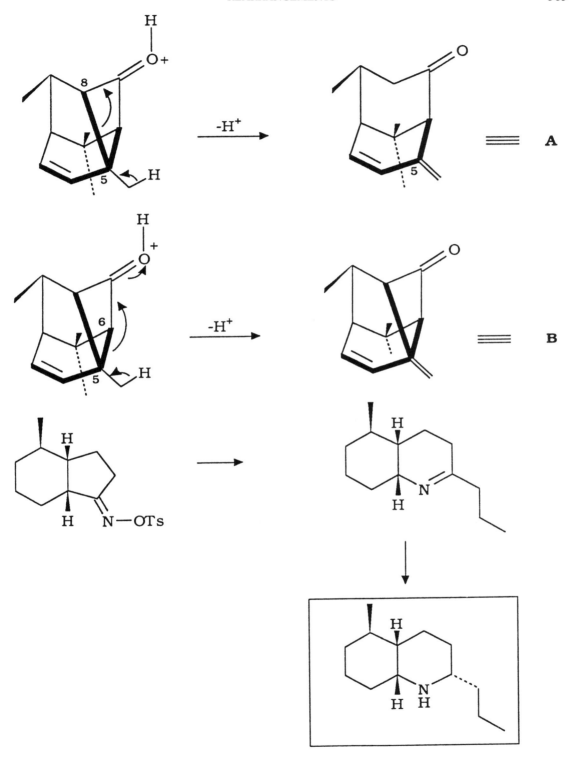

# 6   Some Uses of the Ramberg-Bäcklund Reaction

When the tricyclic α-chlorosulfone is treated with potassium *t*-butoxide in tetrahydrofuran, the propellane shown forms in 54% yield.

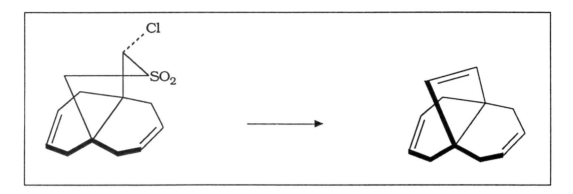

Likewise, the bicyclic α-chlorosulfone, after reaction with aqueous potassium hydroxide, gives Δ$^{1,5}$-bicyclo[3.3.0]octene in 75% yield.

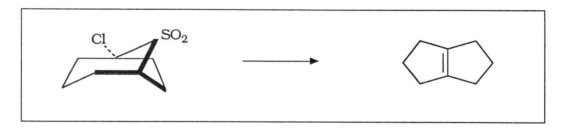

Let us finally describe a general method for the synthesis of 1,3-dienes, useful for instance toward Diels–Alder cycloadditions. Bromomethanesulfonyl bromide BrCH$_2$SO$_2$Br is added photochemically to a mono-olefin; the resulting dibromide is subjected to a strong base which effects a double elimination, resulting in formation of a conjugated diene.

References: L. A. Paquette and J. C. Phillips, *Tetrahedron Lett.*, **1967**, 4645.
       L. A. Paquette and R. A. Hauser, *J. Am. Chem. Soc.*, **91**, 3870 (1969).
       L. A. Paquette, *Accounts Chem. Res.*, **1**, 209 (1968).
       E. Block and M. Aslam, *J. Am. Chem. Soc.*, **105**, 6164–6165 (1983); *Org. Synth.*, **65**, 90–97 (1987).
       E. Block, M. Aslam, V. Eswarakrishnan, and A. Wall, *J. Am. Chem. Soc.*, **105**, 6165–6166 (1983).

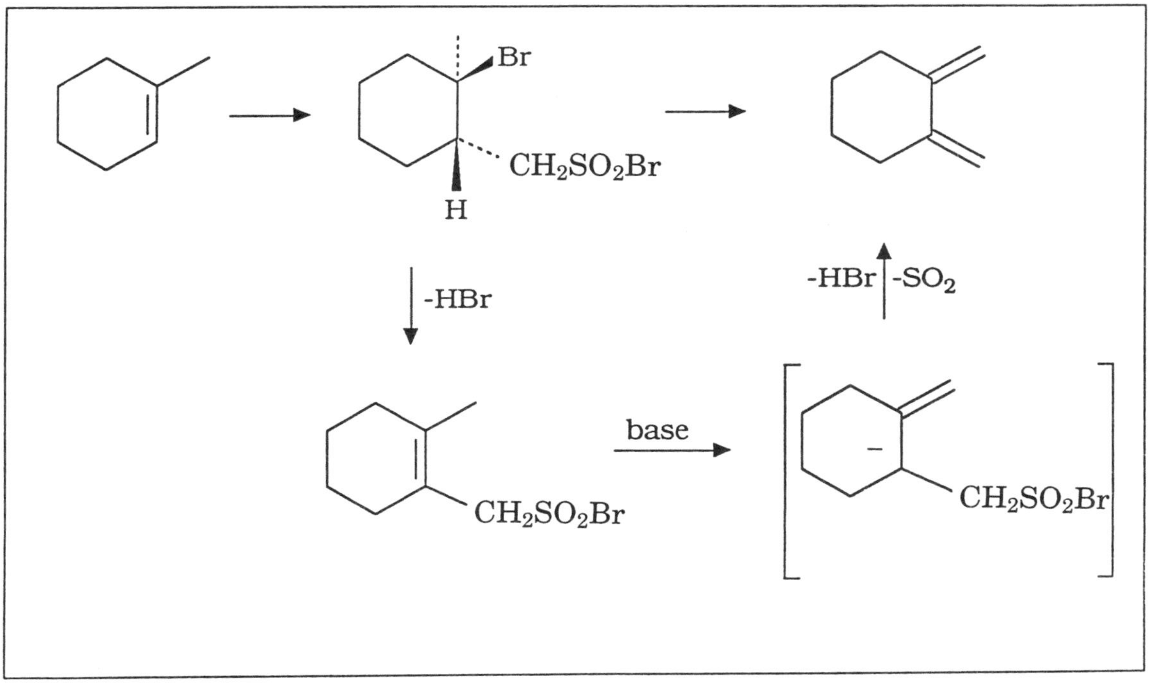
Br
CH2SO2Br
H
-HBr
-HBr
-SO2
base
CH2SO2Br
CH2SO2Br

This is a tool for homologation of mono-olefins into good yields of 1,3-dienes.

The last example is the synthesis of a pheromone (of a mite, the red bollworm moth) in 85% yield and with excellent stereoselectivity: the preference for the Z-diene is 5:1. The high regioselectivity in the formation of these 1,3-dienes is noteworthy.

The mechanism probably involves a first elimination from the dibromide which affords an entity vinylogous to the Ramberg–Bäcklund sulfone. A second equivalent of base deprotonates (due to the steric constraint) one of the allylic protons stereoselectively; this triggers off the Ramberg-Bäcklund reaction: attack of the carbanion at the carbon bearing the bromide, formation of the episulfone and final expulsion of $SO_2$.

## 7  Formation of Small Rings by Rearrangements

When the cyclohexanone is treated with sodium methoxide, formation of two bicycles is observed.

Formation of the bicyclo[3.2.0]heptane has the following explanation: ring expansion with concomitant departure of the tosylate. Under the basic conditions this creates a zwitterion. Formation of a carbon—carbon bond occurs between a donor (−) and an acceptor (+) position on the ring. This forms a cyclobutane ring.

Formation of the bicyclo[3.1.1]heptane is explained by direct attack of the enolate at the carbon bearing the tosylate leaving group ($S_N2$) without prior ring expansion.

In both cases, a cyclobutanone is formed, in spite of its angular strain, because the system has no other intramolecular option to answer the call for electrons, upon cleavage of the leaving group.

# 8 The Giants in the World of Chemistry

**Table 18** Sales and gross profits of major world chemical companies

| | Sales in 1990 $ billion | Gross Profit in 1990 $ billion |
|---|---|---|
| BASF | 18.52 | ? |
| Hoechst | 17.80 | 1.17 |
| ICI | 17.51 | 0.44 |
| Bayer | 16.31 | 0.99 |
| Du Pont | 15.57 | 1.97 |
| Dow | 14.69 | 1.92 |
| Enichem | 12.57 | 0.62 |
| Shell | 12.19 | 0.93 |
| Exxon | 11.15 | 0.74 |
| Rhône–Poulenc | 10.12 | 0.58 |
| Atochem | 9.69 | 0.90 |
| Ciba–Geigy | 8.15 | 0.44 |
| Union Carbide | 7.62 | 1.03 |
| Asahi Chemical | 6.70 | 0.26 |
| Hüls | 6.30 | 0.31 |
| Akzo | 5.83 | 0.33 |
| Monsanto | 5.71 | 0.62 |
| DSM | 5.58 | 0.49 |

Reference: *Chem. Eng. News*, 5 August, 1991, p. 10.

# Synthons: Donor Synthons

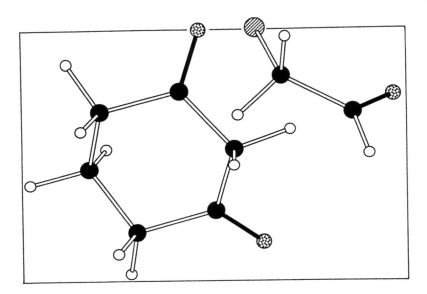

The last part of this text will recapitulate the main topics, and introduce retrosynthetic analysis. The first of these last three chapters defines a synthon. It introduces a convenient notation, and it presents some of the most important electron-donor synthons.

| Synthon | Example | Corresponding reagent |
| --- | --- | --- |
| $d^0$ | $CH_3S^{(-)}$ | $CH_3SH$ |
| $d^1$ | $N\equiv C^{(-)}$ | KCN |
| $d^2$ | $^{(-)}H_2C{-}CHO$ | $CH_3{-}CHO$ |
| $d^3$ | $^{(-)}C\equiv C{-}C{-}CR^1R^2{-}NH_2$ | $Li^{(+)(-)}C\equiv C{-}CR^1R^2{-}NH_2$ |
| Alkyl donor | $CH_3{}^{(-)}$ | $LiCH_3$ |
| $a^0$ | $^{(+)}P(CH_3)_2$ | $(H_3C)_2P{-}Cl$ |
| $a^1$ | $(H_3C)_2C^{(+)}{-}OH$ | $(H_3C)_2 C{=}O$ |
| $a^2$ | $H_2C^{(+)}{-}CO{-}CH_3$ | $Br{-}CH_2{-}CO{-}CH_3$ |
| $a^3$ | $H_2C^{(+)}{-}CH{=}C{\overset{O^{(-)}}{\underset{OR}{<}}}$ | $H_2C{=}CH{-}COOR$ |
| Alkyl acceptor | $CH_3{}^{(+)}$ | $(CH_3)_3S^{(+)}Br^{(-)}$ |

# 1  Donor and Acceptor Synthons

In organic synthesis, a large number of reactions form new carbon−carbon bonds in a polar way, by reaction of a negatively charged center (an electron donor) with a positively charged one (an electron acceptor). This creation of covalent bonds can occur without formation of a *free* carbanionic or carbocationic intermediate. Nevertheless, it is useful to consider the molecules as result of the formal union of a donor *synthon* with an acceptor *synthon*. Thus synthons are *formal* entities, associated with the actual reagents used. For instance, methyl iodide $CH_3I$ is equivalent to a synthon $CH_3{}^+$, whereas methyllithium $CH_3Li$ is equivalent to a synthon $CH_3{}^-$. Another example: ethyl acetate $CH_3CO.OCH_2CH_3$ is equivalent to an acceptor synthon $CH_3.CO^{(+)}$: in fact an addition–elimination reaction (the Claisen condensation) allows for the addition of any nucleophile :Nu to the acceptor synthon $CH_3.CO^{(+)}$.

In organic synthesis, most frequently functionalization is linked to heteroatoms. Accordingly, a useful notation characterizes a synthon by its donor (d) or acceptor (a) character and by the number (shown as an exponent) of bonds that separate the heteroatom from the center considered. Thus, the above example $CH_3.CO^{(+)}$ is an $a^1$ acceptor synthon. The adjacent Table summarizes some of the most common acceptor synthons. Synthons which do not bear any heteroatoms such as carbanions $R^-$ and carbocations $R^+$ are exempted from this notation.

Reference: J. Furhop and G. Penzlin, *Organic Synthesis*, Verlag Chemie, Weinheim, 1983, p. 2.

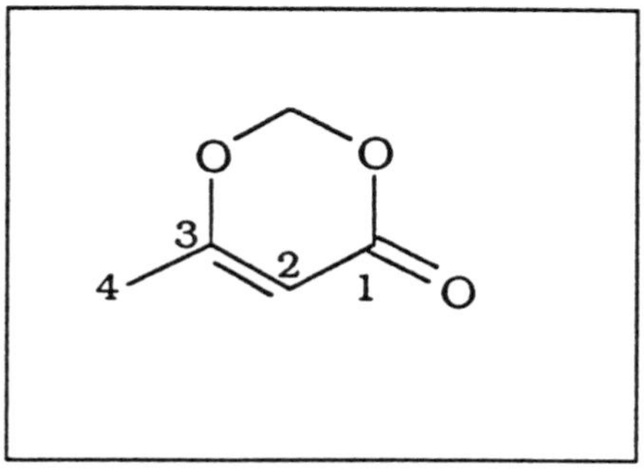
O
O
3
2
4
1
O

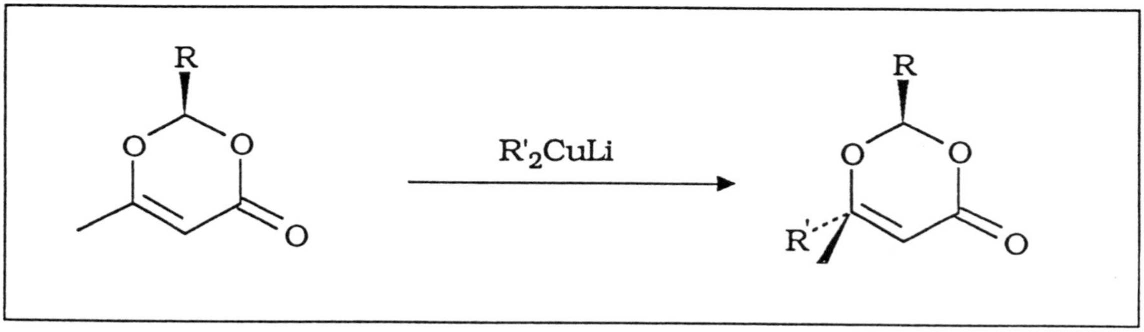
R
O
O
O
R'2CuLi
R
O
O
R'
O

The dioxinone depicted in the adjacent scheme illustrates the identification of the different donor or acceptor centers in a molecule. The electrophilic carbon C-1 is obviously an acceptor of type $a^1$, susceptible to adding nucleophiles. Its vinylogous equivalent C-3 is also an acceptor ($a^3$) as in Michael additions. Dioxinones can be deprotonated in position C-4, a donor center of type $d^4$. C-2 is also a donor center ($d^2$), susceptible to addition of an electrophile.

It is possible to invert donor–acceptor properties. In fact, if the allylic methyl group C-4 is halogenated, the former donor center is transformed into an $a^4$ acceptor since it now bears a leaving group $X^-$.

One of the practical advantages of the dioxinone system is the possibility of strong differentiation between the two faces of the conjugated system by a bulky substituent at the acetal carbon (the carbon atom in-between the two oxygen atoms). In this case (with R = $t$-butyl), the Michael addition occurs almost exclusively in the half-space from which R is absent; the stereoselectivity is better than 95%.

Reference: D. Seebach, S. Roggo, and J. Zimmermann, in *Stereochemistry of Organic and Bioorganic Transformations*, W. Bartmann and K. Barry Sharpless, Eds., VCH, Weinheim, 1987, 85–126.

| RH | $pK_a$ |
|---|---|
| $H_3C\!-\!CH_3$ | 50 |
| $H_2C = CH_2$ | 49 |
| $HC \equiv CH$ | 25 |

$$R\!-\!C \equiv C\!-\!H \; + \; ^{(-)}NH_2 \longrightarrow R\!-\!C \equiv C^{(-)} \; + \; NH_3$$

$$R\!-\!C \equiv C\!-\!H \; + \; EtMgBr \longrightarrow R\!-\!C \equiv C\!-\!MgBr \; + \; EtH\uparrow$$

$$2\,R\!-\!C \equiv C\!-\!H \; + \; M(OAc)_2 \longrightarrow M(\!-\!C \equiv C\!-\!R)_2 \; + \; 2\,AcOH$$

$$M = Cu,\ Hg$$

# 2 Alkyl Donor Synthons

These are carbanions R⁻. Let us recall that the acidity of a C—H bond increases with increasing s character: acetylenes > ethylenes > saturated hydrocarbons. Thus the corresponding carbanions are very reactive entities, as the electronic charge is localized to a large extent on carbon.

In the case of carbanions derived from saturated hydrocarbons, the stability sequence is: tertiary < secondary < primary (i.e. opposite to that for carbocations) for reactions in solution.

The main formation of these alkyl donors R⁻ is in practice metallation of halides RX → RM, being polarized to R⁻ M⁺. It is possible, after the initial metallation, to effect metal–metal exchange (or *transmetallation*).

Acetylides are formed by deprotonation of the acidic proton of a terminal acetylene with a strong base such as sodium amide. Acetylenic Grignard reagents are frequently obtained by proton–metal exchange with an alkyl Grignard reagent. It is also possible to form acetylides under neutral conditions by allowing them to react with a divalent metal such as copper(II) or mercury(II).

$$Me_3Si-CH_2Cl + Mg \longrightarrow Me_3Si-CH_2MgCl$$

$$Me_3Si-CH_2MgCl + HgCl_2 \longrightarrow (Me_3Si-CH_2-)_2\ Hg$$

$$(Me_3Si-CH_2)_2Hg + 2K \longrightarrow 2\ Me_3Si-CH_2{}^-K^+$$

$$Me_3Si-CH_2-SMe + n\text{-BuLi} \longrightarrow (Me_3Si-CH-SMe)^-Li^+$$

$$Me_3Si-CH_2-Ph + n\text{-BuLi} \longrightarrow (Me_3Si-CH-Ph)^-Li^+$$

# 3 d¹ Donor Synthons: Heteroatoms with d-Orbitals

The most important $d^1$ donor synthons are those with heteroatom having d-orbitals of sufficiently low energy to stabilize the charge of the carbanion with a $p\pi$–$d\pi$ type interaction: typically **silicon, sulfur and phosphorus.**

Silylated carbanions can be formed from Grignard reagents. The silylated carbanion resulting from this metallation or from subsequent transmetallations, possesses a labile carbon—metal bond, due to the Coulomb repulsion of (partial) positive charges on the metal and silicon atom. Thus the silylated carbanion is activated.

If furthermore the carbon atom about to become a carbanionic center is flanked by a second activating group (S, Si, P, Ph etc.), then deprotonation with a strong base such as *n*-butyllithium is possible. In this manner, one can form a silylated carbanion, stabilized (see lower box) by delocalization of the lone pair from carbon to silicon on one side, and to sulfur on the other side (in the example shown).

One can also deprotonate allylic silanes in a similar way with delocalization to an allyl anion: the prevailing limiting form is that with the negative charge placed on the carbon adjacent to the electropositive silicon atom, this accounts for the high regioselectivity.

---

Reference: I. Fleming, 'Organic Silicon Chemistry', in *Comprehensive Organic Chemistry*, D. Barton and W. D. Ollis, eds., **3**, 539 (1979).

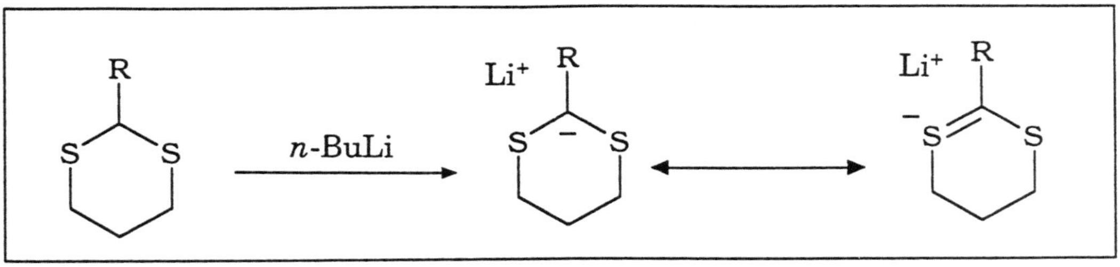
R
S
S
n-BuLi
Li+
R
S
S
Li+
R
S
S

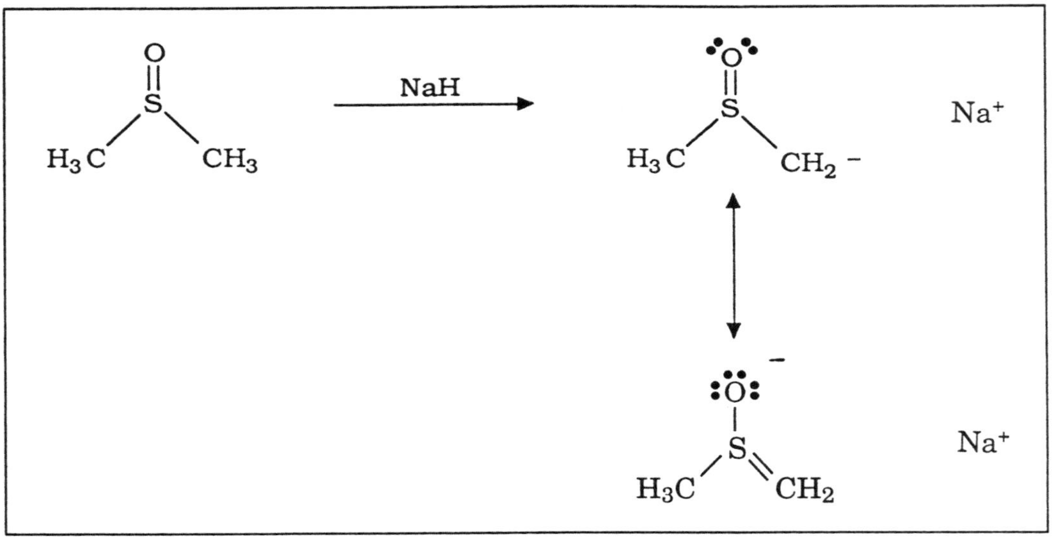
O
S
H3C
CH3
NaH
O
S
H3C
CH2
Na+
O
S
H3C
CH2
Na+

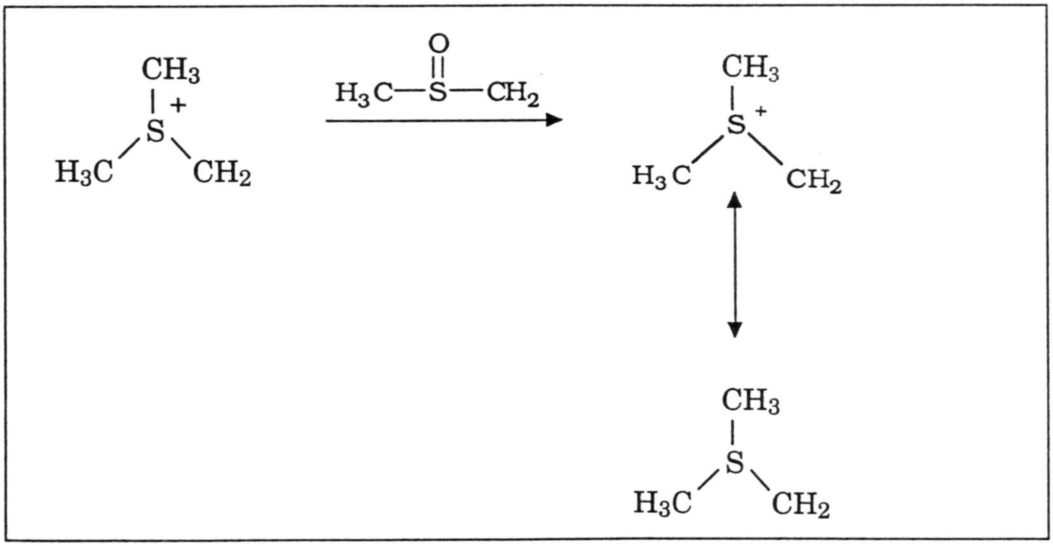
CH3
S
H3C
CH2
O
H3C—S—CH2
CH3
S
H3C
CH2
CH3
S
H3C
CH2

We proceed now to *d¹ donor synthons with sulfur*. We are already acquainted with the first example: that of the dithianes (= bis-thioacetals of aldehydes $RCHO + HS-(CH_2)_3-SH$) which allow for inversion of the polarity at the carbon atom, an $a^1$ acceptor in the original aldehyde.

Another useful reagent is the dimsyl anion, the conjugate base of dimethyl sulfoxide, a common organic solvent. It is formed by dissolving a strong base such as sodium hydride in DMSO. The carbanion is stabilized by the resonance indicated.

One of the applications of the dimsyl anion is formation of sulfur ylides. The dimsyl anion deprotonates a sulfonium (or sulfoxonium) cation. Thus an ylide with a negative charge on carbon and a positive charge on sulfur is formed; the ylide–ylene resonance (see Chapter 12) stabilizes this entity. Ylides derived from sulfonium cations are both generated and used at low temperature. They are stable at room temperature.

Reference: B. M. Trost, *Sulfur Ylides*, Academic Press, New York, 1975.

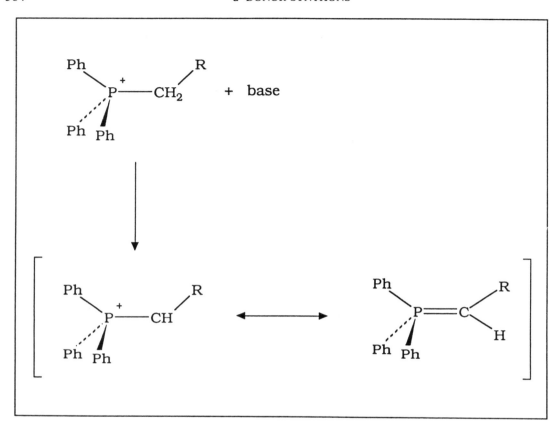

The third important class of $d^1$ donor synthons stabilized by delocalization into d-orbitals of an heteroatom are phosphorus ylides. You remember the obvious analogy between a quaternary ammonium and a phosphonium salt. The latter, generated most frequently by quaternization of triphenylphosphine with an alkyl halide $XCH_2R$, can be deprotonated with a strong base, thus affording ylides. In fact, this entity is a hybrid between a ylide and an ylene form.

The major use of these phosphorus ylides is the Wittig reaction (see Chapter 12).

‘Although this may seem a paradox, all exact science is dominated by the idea of approximation’—Bertrand Russell.

*Note*: The synthons $d^n$ ($n>1$) are of less interest than the $d^1$ synthons discussed here.

# Summary: 'Donor Synthons'

The reader is first reminded of the Wittig reaction, on the example of a natural hydrocarbon, the achiral triene *bisabolene*. Formally it is a linkage of three isoprene fragments (Ruzicka's isoprene rule). Its biosynthesis occurs by acid-catalyzed cyclization of *farnesol*. Retrosynthetic analysis defines the same hydrocarbon as the union of a cyclohexene carboxaldehyde and the adequate phosphorus ylide. The former is a Diels–Alder adduct of isoprene and methylvinylketone. The second results from a chain extension of trimethylethylene (**cyanide method**). The synthesis of Delay and Ohloff (1979) is a good example of a **convergent synthesis**.

A new carbon–carbon bond is frequently the result of **reaction of a donor synthon with an acceptor synthon**. A convenient notation comprises counting the number of bonds from the reactive center to the adjacent heteroatom. $d^1$ donor synthons (e.g. the cyanide anion) are used most frequently. The use of the $d^1$ synthon cyanide for chain extension is illustrated by the synthesis of *juvabione*.

The dioxinone ring is taken as an example of the procedure of identification of the donor or acceptor properties of the various carbons in a molecule. Such a characteristic character of a center is not frozen for eternity; an **inversion of polarity** is possible. The advantage of this particular chiral synthon is the incipient high stereoselectivity. Its alkylation with mixed lithium cuprates provides diastereomeric excesses superior to 95%.

Amongst other important $d^1$ synthons, carbanions with silicon, sulfur or phosphorus atoms, are stabilized by $d\pi$–$p\pi$ interactions. The last two cases are those of ylides, useful in the Corey–Chayakovsky and the Wittig reaction, respectively.

The synthesis of *chrysanthemic acid*, a natural insecticide, applies theses concepts. This useful natural product, found in composites (hence its name) is a good example of the isoprene rule. Its synthesis comprises a Michael addition, followed by a Corey–Chavakovsky elimination.

The synthesis of *cafestol* (Corey) in the first step involves formation of the furan linked to ring A; the second step involves the formation of part of ring B (trimethylsilylated acetylenic carbanion).

## 1  Büchi's Synthesis of α-Sinensal

The starting dienol is transformed into a bromide to improve the leaving group ability. Condensation with the enolate of 4-methyl-4-pentene-2-one affords a triene methylketone. A vinyl Grignard reagent turns it into a tertiary alcohol, bromination of which results in migration of the terminal C=C double bond into a trisubstitued double bond.

This second bromide is used for quaternization of cyanomethyldimethylamine. Use of a strong base ($K^{(+)}$$t$-$BuO^{(-)}$) forms an ammonium ylide by deprotonation of the methylene group, activated by the cyano and the quaternary ammonium groups. This ylide undergoes a sigmatropic [2,3] shift, affording the rearranged aminonitrile, hydrolysis of which leads to the title tetraene aldehyde.

The latter, upon heating in xylene to reflux, undergoes a Cope [3,3] sigmatropic rearrangement; this affords α-sinensal (as well as its 2 $Z$-diastereoisomer which can be isomerized by heating with potassium carbonate). α-Sinensal and its isomer β-sinensal are important constituents of the Chinese orange essential oil.

Reference: G. Büchi and H. Wuest, *J. Am. Chem. Soc.*, **96**, 7573 (1974).

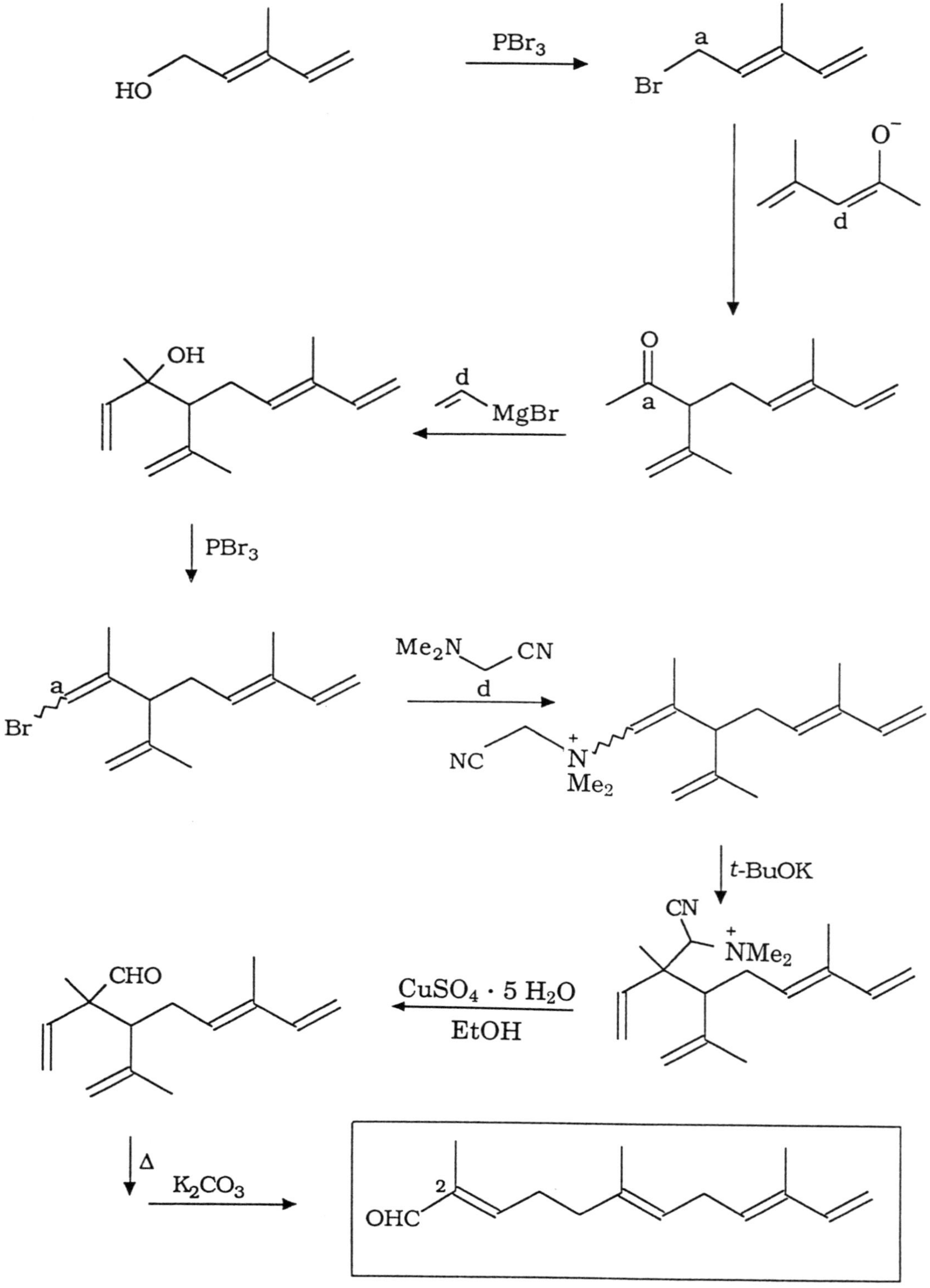
PBr3
HO
Br
a
O⁻
d
O
a
OH
d
MgBr
PBr3
Br
a
Me2N
CN
d
NC
N
Me2
t-BuOK
CN
NMe2
CHO
CuSO4 · 5 H2O
EtOH
Δ
K2CO3
2
OHC

## 2   The First Steps of Corey's Cafestol Synthesis

Cafestol is a natural product from coffee, with anti-inflammatory activity. Other related substances are also found in coffee, such as the 1,2,-dehydro analog named 'kahweol'.

The first step in Corey's synthesis forms the two rings including the furan moiety with a methyl group α to the carbonyl: 1,3-cyclohexadione adds to chloroacetaldehyde, the enolate formed adds to oxygen rather than to carbon, and formation of a five-membered ring occurs more readily than that of a strained three-membered ring.

The synthesis continues by reaction with lithiotrimethylsilylacetylene, and dehydration of the acetylenic alcohol with $MgSO_4$ and pyridinium tosylate; desilylation (KF.2 $H_2O$) affords the free acetylene to be used for making ring B. Resorting to the silylated acetylene carbanion, which stabilizes the carbanion by delocalization towards silicon ($p_C$–$d_{Si}$ interaction), affords a softer nucleophile; furthermore the subsequent purifications (GLC) are simplified. Desilylation makes use of the high thermodynamic stability of the silicon–fluorine bond.

Reference: E. J. Corey, G. Wess, Y. B. Xiang, and A. K. Singh, *J. Am. Chem. Soc.*, **109**, 4717 (1987).

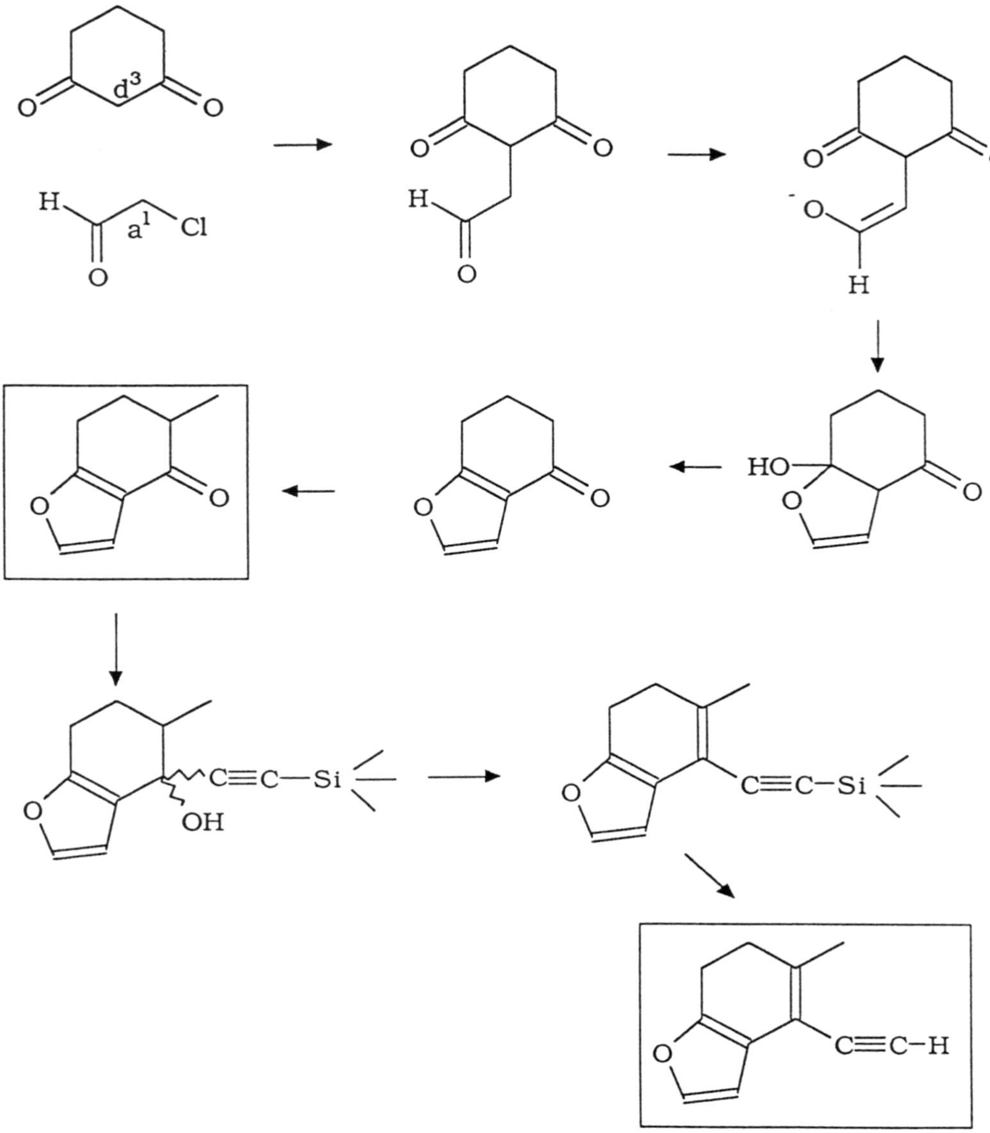

## 3   Synthesis of a Natural Insecticide

Insecticides of the DDT family have caused an ecological disaster worldwide (Rachel Carson, *Silent Spring*). Alternatives had to be found. Gardeners have known for many years that plant-lice do not attack *Compositae*, whose chemical defensive weapon is chrysanthemic acid.

The synthesis of this target molecule shown here uses the Corey–Chaykovsky reaction (see Chapter 12) starting from a sulfur ylide. It provides the methyl ester of chrysanthemic acid.

Reference: W. Carruthers, *Some Modern Methods of Organic Synthesis*, Cambridge University Press, 1971, p. 94.

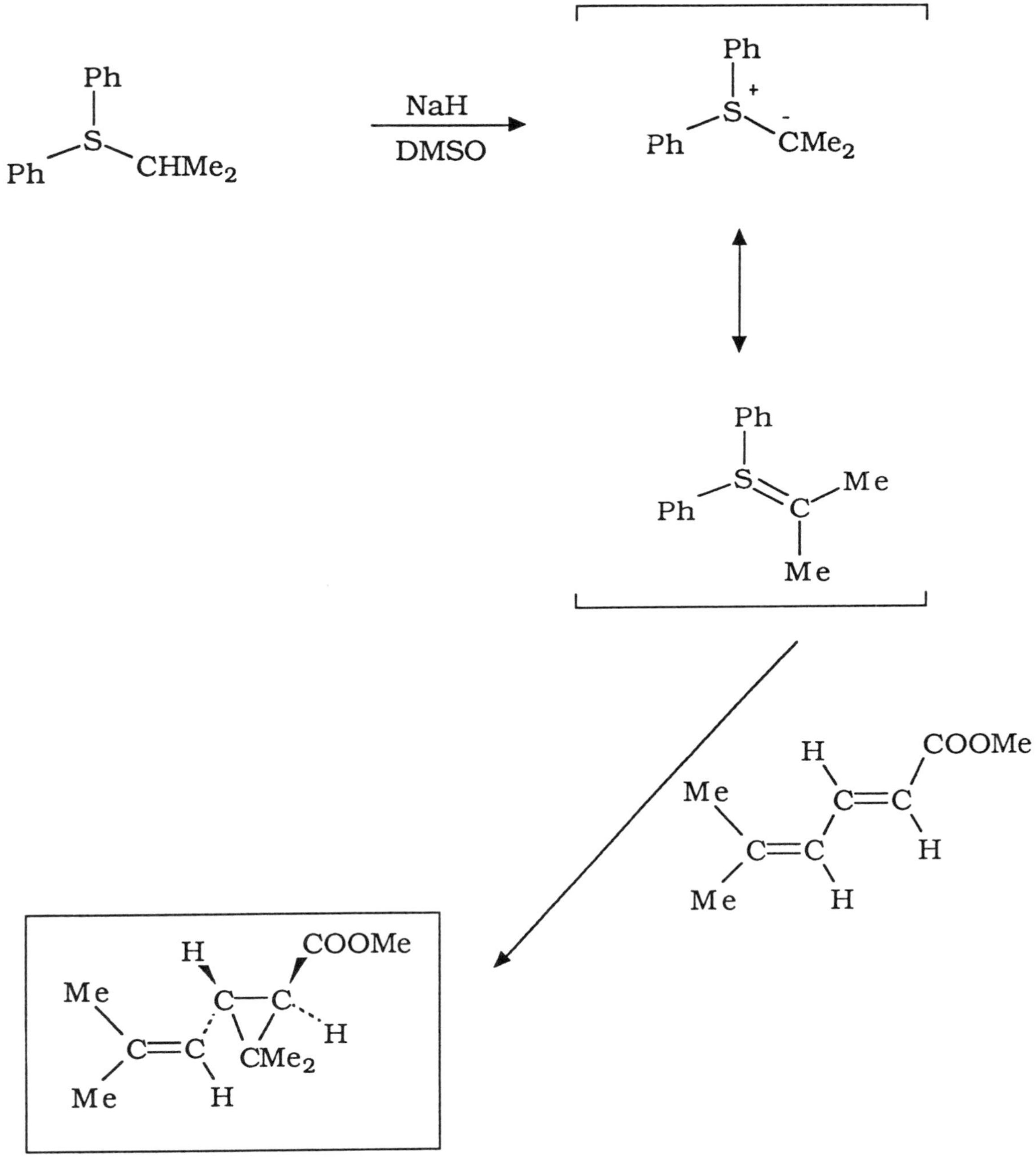
Ph
S
Ph
CHMe2
NaH
DMSO
+
-
CMe2
Me
Me
H
COOMe
C=C
H
H
C—C
CMe2

# 4  Enantioselective Michael Additions

The problem is how to attack the trigonal carbon $\beta$ to the carbonyl group with a reagent equivalent to the donor synthon alkyl $R^-$, from a single face? Regioselectivity, reaction at the $\beta$ carbon, is achieved—as we know already—by use of the Gilman reagent $R_2CuLi$.

Corey's idea is to complex the metal cations with a *chiral* agent, the indicated aminoalcohol, obtained from ephedrine, a natural product available at low cost as either the ($+$) or the ($-$)-enantiomer.

Results, via selective interaction with the *re* face of the cyclohexenone, are gratifying: yields between 75 and 90%, and enantiomeric excesses of 85 to 92%. Furthermore, the aminoalcohol used as an asymmetric inducer can be recovered.

Reference: E. J. Corey, R. Naef, and F. J. Hannon, *J. Am. Chem. Soc;*, **108**, 7114–7116 (1986).

A plausible model of the transition state is depicted in the box: *nucleophilic* copper forms a d,π* complex with the acceptor carbon of the enone; a first lithium bridges the oxygen of the substrate and that of the chiral auxiliary; the second lithium chelates the two nitrogen atoms of the diamine and helps to hold in place the RLi alkyllithium reagent. Such a degree of organization, encompassing the metal atoms, the two organic molecules, the solvent molecules and the counter ions, reminds one of the active site of a metalloenzyme: handsome supramolecular chemistry!

## 5 Fines Guêpes*

Some scientists of the American Department of Agriculture were carrying out a field test of a synthetic pheromone that induced aggregation of the hemipter *Podisus maculiventris*. To their surprise, with several mixtures of volatile constituents they trapped also wasps of the species *Vespula maculifrons*. Which constituents were attracting the wasps? In the first place, the researchers were able to show

---

*Literally 'fine wasp' in French, means 'sharp cookie'.

that the molecules linalool, α-terpineol and (E)-2-hexanal were acting in synergy. These are shown in the following Scheme: from the bottom, the acyclic diene alcohol linalool, the unsaturated cyclic alcohol α-terpineol, and the linear conjugated aldehyde (E)-2-hexanal. Why are *Vespula* attracted by the pheromone of *Podisus*? This is an enigma at first sight, the two types of insects occupy totally different ecological niches. *V. maculifrons* are not predators of *P. maculiventris*: they show no interest in them. The *V. maculifrons* wasps are predators of bees that nest in the soil. The other living preys they fancy are insects living on leaves.

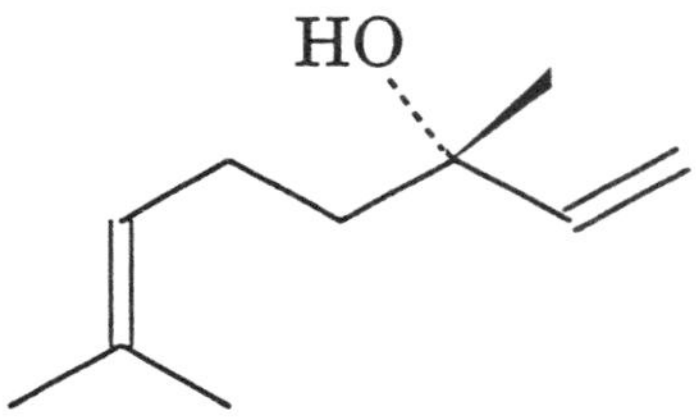

**A**

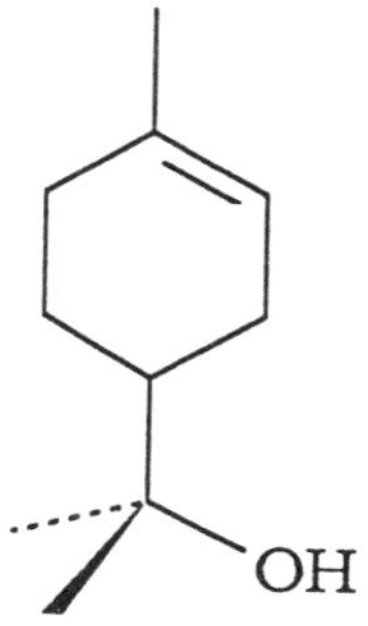

**B**

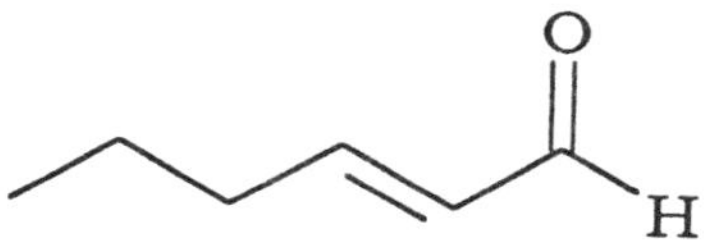

**C**

Now, linalool **A** is a molecule omnipresent in plants; it is one of the important precursors of the $C_{10}$ monoterpenes. Plants contain glucosides of linalool and other monoterpene alcohols. Maceration of leaves hydrolyzes these glucosides, setting free volatile alcohols such as linalool and α-terpineol **B**. (E)-2-hexanal **C** too is known as the *aldehyde of leaves*. It is present in more than 30 plant families: whenever a leaf is torn, enzymes transform linoleic acid **D** into **C**. This molecule is thus the signature for an attacked leaf. Therefore the mixture (**A**, **B**, **C**) has the characteristic odor of crushed and damaged leaves. The wasps *V. maculifrons* use it as a general *kairomone* drawing them to foliovourous insects, i.e. to their future food supply. As a last note, we shall remark that there is a striking similarity between many pheromones, especially those of insects, and the perfumes we are also sensitive to. Linalool and α-terpineol are among the monoterpene alcohols responsible for the aroma of several grapes, especially that of Muscat, and hence of several wines. (E)-2-hexanal is found in large quantities (together with hexanol) in grape juice of the varieties Grenache and Sultana. Is there a universal dictionary of odors? This is one of the important questions raised by chemical communication.

References: J. R. Aldrich, J. P. Kochansky, and J. D. Sexton, *Experientia*, **41**, 420–421 (1985).
A. Rapp and H. Mandery, *Experientia*, **42**, 873–884 (1988).

# 6  C$_1$ Chemistry

A large national reserch and development program—undertaken in Japan and supported by MITI—is the chemistry of C$_1$ precursors. Fourteen large companies were given the following assignments: production of ethanol, acetic acid, ethylene glycol and olefins. The oil crisis in 1973 resulted in this intense interest in C$_1$ chemistry.

Starting from carbon monoxide, the addition reaction $2 CO + H_2$ affords ethylene glycol. This high pressure reduction is carried out at Union Carbide, in the presence of a rhodium catalyst.

Carbon monoxide can also be hydrogenated directly ($4 CO + 8 H_2$) to $t$-butanol with a selectivity of 25% on a laboratory scale and of 11% on an industrial scale, respectively. Dehydration of $t$-butanol can afford $i$-butylene (optimized by BASF). The latter on addition of methanol produces methyl-$t$-butylether (or MTBE), an additive used in lead-free fuel to raise the octane number. The addition reaction $CO + 2H_2$ leads to methanol. Institut Français du Pétrole, in cooperation with Idemitsu (Japan) has optimized several catalysts for this conversion. A typical distribution of alcohols formed by this process is (percent):

| | |
|---|---|
| methanol | 50–70 |
| ethanol and propanol | 16–23 |
| $n$-butanol | 4–7 |
| pentanol | 2–3 |
| hexanol and higher homologs | 1–3 |

The chemistry of methanol is of special importance. Its yearly production is about 16 MT. A pilot plant in New Zealand uses the Mobil process (catalyst: the molecular sieve (or zeolite) H-ZSM-5) to transform methanol into fuel. In the first step, natural gas is converted to methanol.

Previously, the well-known C$_1$ process for gasoline production was the Fischer–Tropsch process, used by Germany in World War II. Using starting mixtures of carbon monoxide and molecular hydrogen, it affords mixtures of saturated hydrocarbons and olefins, mainly linear. It is a reductive oligomerization of CO, following a geometric progression (a distribution known as the Schulz–Flory distribution): the distribution of products can be calculated from the molar ratio of the reactants.

The Monsanto process (carbonylation of methanol with CO) is the best way of producing acetic acid.

Reductive carbonylation of methanol affords either ethanol or acetaldehyde: $CH_3OH + CO + H_2 \rightarrow CH_3CHO + H_2O$. The conversion rate is 97%, and the selectivity for acetaldehyde is 80%.

Carbon dioxide is the third important raw material for C$_1$ chemistry. Its natural abundance ($10^{14}$ tonnes) is tempered by an unfavorable thermodynamic balance, due to its great stability. A major use is production of methanol: $CO_2 + 2H_2 \rightarrow CH_3OH$.

Another use of $CO_2$ is the Sabatier reaction $CO_2 + 4H_2 \rightarrow CH_4 + H_2O$, characterized by a $\Delta G^0$ (298 K) = $-27$ kcal mol$^{-1}$ (favorable due to the formation of *two* molecules of water). The latter reaction is a chemical Mount Everest: high energy intermediates, high barriers (30–171 kJ mol$^{-1}$), formation of by-products. It normally requires relatively high pressures (above atmospheric) and high temperatures ($>300°C$).

Recently, a $Ru/RuO_x$ catalyst system, dispersed on rutile $(TiO_2)$ particles with a diameter of about 20 Å has been discovered. It allows the Sabatier reaction to be run at room temperature.

Furthermore, improvements to the process are possible: rutile—a semi-conductor—upon irradiation with sunlight in a conduction band (3 eV) increases the rate of the methanation reaction (by a factor 4 or 5).

The last use of $CO_2$ quoted here is formation of $(2\pi + 2\pi + 2\pi)$ cycloadducts with two (or four) molecules of 1,3-butadiene, in the presence of several catalysts.

The chemistry of $C_1$ is very exciting, especially on paper! It is a chemistry for times of shortage, starting from cheap raw materials such as carbon, shale oil, biomass, etc. As long as the price of a barrel of crude oil is low (below about US $50), interest in this chemistry remains strictly academic—except for countries located very far from the oil production centers, such as New Zealand and Japan.

References: W. Keim, M. Berger and J. Schlupp, *J. Catal.*, **61**, 359 (1980).

D. L. King, J. A. Cusamano, and R. L. Gorten, *Catal. Rev. -Sci. Eng.*, **23**, 233–263 (1981).

P. C. Ford, *Catalytic Activation of Carbon Monoxide*, ACS Symp. Series 1981, 152, American Chemical Society, Washington.

M. Röper, *Habilitationschrift*, RWTH Aachen, 1985.

P. Courty, D. Durand, E. Freund, and A. Sugier, *J. Mol. Catal.*, **17**, 241–254 (1982).

S. Csicsery and P. Laszlo, in *Preparative Chemistry Using Supported Reagents*, P. Laszlo, Ed., Academic Press, San Diego, CA., 1987.

M. T. Gillies, *C₁ Based Chemicals from Hydrogen and Carbon Monoxide*, Noyes Data Corporation, Park Ridge, N.J., 1982.

J. Falbe, *New Syntheses With Carbon Monoxide*, Springer-Verlag, Berlin, 1980.

W. Keim, *Pure Appl. Chem,*. **58**, 825–832 (1986).

K. R. Thampi, J. Kiwi, and M Grätzel, *Nature,* **327**, 506–508 (1987).

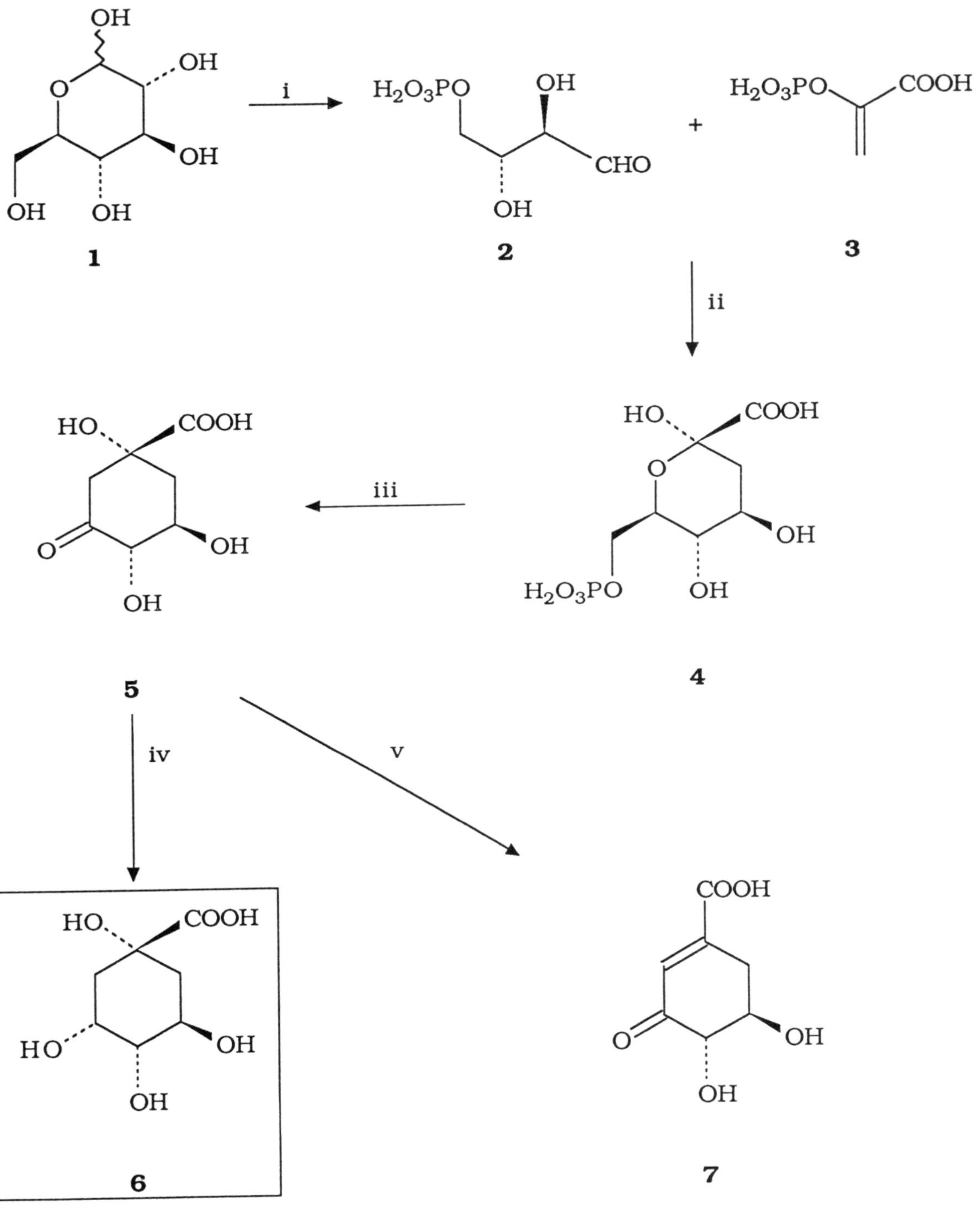
OH
OH
O
OH
OH
OH
1
i
H2O3PO
OH
CHO
OH
2
+
H2O3PO
COOH
3
ii
HO
COOH
O
OH
H2O3PO
OH
4
iii
HO
COOH
O
OH
OH
5
iv
v
HO
COOH
HO
OH
OH
6
COOH
O
OH
OH
7

## 7  Service and Grandeur in Biotechnology

*Biochemistry*: Several enzymes punctuate the pathway from D-glucose **1** to quinic acid **6**. Transketolase (i) converts **1** to D-erythrose-4-phosphate **2**. DAHP synthetase (ii) adds phosphoenol pyruvate **3** affording 3-desoxyheptulosonic acid **4** (or DAHP). DHQ synthetase (iii) transforms **4** into 3-dehydroquinic acid **5** (or DHQ). And quinic acid dehydrogenase (iv) converts **5** to quinic acid **6**.

*Microbiology*: Certain bacteria—for instance *Klebsiella pneumoniae*—are able to live on quinic acid as their only carbon source. The first metabolic step is **6 → 5** oxidation, catalyzed by (iv): let us remind the reader that, by definition, any catalyst increases the rate of *both* the forward and the inverse reaction.

*Conjecture*: Insertion of a gene coding for synthesis of the enzyme (iv) in a bacterium unable to metabolize (i.e. to degrade) quinic acid should result in production of quinic acid.

*Realization: Escherichia coli* modified by genetic engineering produces quinic acid. Furthermore, a mutation suppresses the expression of the dehydratase (v) responsible for the reaction **5 → 7**.

### POTENTIAL USES

*Organic synthesis*: quinic acid with its four chiral centers is a promising starting material for multistep syntheses.

*Chemical industry*: Its oxidation ($H_2SO_4 + MnO_2$) affords benzoquinone in a yield of 70%. It is not forbidden to dream of an industrial use, competing with the transformations phenol → hydroquinone → quinone and benzene → *p*-di-*i*-propylbenzene → hydroquinone → quinone.

*Flavors*: Quinic acid is an acidulant valued in the food industry.

References: K. M. Draths, T. L. Ward, and J. W. Frost, *J. Am. Chem. Soc.*, **114**, 9725–9726 (1992).
J.-L. Montchamp, L. T. Piehler, and J. W. Frost. *J. Am. Chem. Soc.*, **114**, 4453–4459 (1992).
K. M. Draths *et al.*, *J. Am. Chem. Soc.*, **114**, 3956–3962 (1992).

## 8  The French and Their Heart

Be assured that we won't be concerned here either with steroid hormone or with yet another study on the sexual behavior of the French. We the French enjoy also another worldwide reputation: that of being able to gorge on rich foodstuffs such as butter or *foie gras* although this massive intake of cholesterol does not cause too many heart attacks in the population. This absence of correlation between the abundance of cholesterol in the diet and the low incidence of circulatory diseases has caused much discussion. The solution to the paradox may be that one excess cures the other! The French still remain,

Saint Martin drinks the good wine, and lets water run to the mill—old French proverb, quoted by Romain Rolland, *Colas Breugnon*. (The original French phrase plays on the similar word endings in S*ain*t/M*ar*t*in*/v*in*(moul*in*.)

Reference: E. H. Siemann and L. L. Creasy, *Am. J. Enol. Vitic.*, **43**, 49–52 (1992).

statistically-speaking, great wine-drinkers. Some wines—especially red Bordeaux but also Pinot noir from Alsace and its Californian cousin red Zinfandel—contain about a micromole per liter of *resveratrol*. This phenolic compound seems to reduce the concentration of lipids in the liver (evidence from experiments on rats) as well as in the blood serum.

1.

2. $R_3P—Cab + cd=O \longrightarrow cdC=Cab$

3. $ArH + RCl \longrightarrow Ar—R + HCl$

4. $ArH + R•CO•Cl \longrightarrow Ar•CO•R + HCl$

5. $RX + {}^{(-)}:C\equiv N: \longrightarrow R–C\equiv N: + X^{(-)}$

6. $RX + {}^{(-)}:C\equiv C–R' \longrightarrow R–C\equiv C–R' + X^{(-)}$

7. $RX + R'–C(O^{(-)})=CH_2 \longrightarrow R'•CO•CH_2R + X^{(-)}$

8. $R•CO•X + R'–C(O^{(-)})=CH_2 \longrightarrow R•CO•CH_2•CO•R' + X^{(-)}$

9. $R^1•CO•R^2 + R^3M \longrightarrow R^1R^2C(OH)–R^3$

10. $R^2CuLi + R'X \longrightarrow RR'$

$R^2CuLi + H_2C=CH–CO•R' \longrightarrow RCH_2CH_2–CO•R'$

# Synthons: Acceptor Synthons

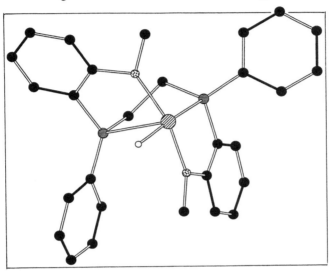

## 1    The Standard C—C Bond Forming Reactions

Let us recapitulate the 10 most important reactions:

*Diels–Alder*: access to cyclohexenes.
*Wittig*: formation of an ethylenic bond by addition of a phosphorus ylide to a carbonyl compound.
*Friedel–Crafts alkylation*: of aromatic substrates.
*Friedel–Crafts acylation*: likewise.
*Use of cyanides*: chain extension by one carbon unit ($S_N2$ substitution).
*Acetylide method*: introduction of a two carbon atom fragment.
*Alkylation of enolates*: introduction of a hydrocarbon residue at a carbon adjacent to a carbonyl group.
*Condensation of enolates*: access to 1,3-diketones.
*Organometallic substitutions*: formation of alcohols.
*Organocuprates*: essential to 1,4-addition (Michael).

The goal of this chapter, after discussing some of the most important acceptor synthons, is to demonstrate the ease of conceiving a synthesis. Every disconnection specifies two possibilities (d + a) or (a + d) for the formation of the corresponding bond. The repertory of reactions provides a concrete solution. If this is not the case, there still remains the possibility

**a**

R$^+$

R—X          X = Cl, Br, I, OTs, OMs, ...

Me$_3$S$^+$ X$^-$

Me$_3$O$^+$ BF$_4^-$

R$^+$ AlCl$_4^-$

---

**a$^1$**

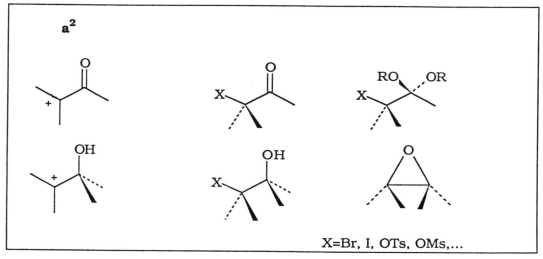

X = Cl, OR', OAc, ...

R—$\overset{+}{C}$=O   AlCl$_4^-$

---

**a$^2$**

X=Br, I, OTs, OMs,...

of using either a *synthetic equivalent* or *inverting the polarity* of a synthon. Furthermore, in this chapter the following two strategies are presented: symmetrization, a way to simplify the overall synthesis; and its converse, dissymmetrization, a method which can render a step stereoselective. These two operations are of primordial importance.

## 2   Acceptor Synthons

Alkyl carbocation synthons R$^+$ form the class **a**. In practice they are obtained from the corresponding halogen derivatives R—X, and in particular from sulfonyl ester (tosylates, mesylates etc.). The sulfonium and oxonium salts (see adjacent scheme)—the latter are the so-called Meerwein salts—are sources of R$^+$, by formation of a neutral molecule of dimethyl sulfide or dimethyl ether. Friedel–Crafts alkylating reagents, for instance RCl plus a Lewis acid such as aluminum chloride, AlCl$_3$, are also sources of carbocations.

The major **a**$^1$ synthons are:

(i) the conjugate acids of carbonyl compounds or their synthetic equivalents, e.g. the O-silylated hemithioacetal which has the important advantage of not being able to enolize;

(ii) the acylium synthon R—$^+$CO, resulting from an acyl halide, an ester, an anhydride, a Friedel–Crafts acylating reagent, or a synthetic equivalent such as an orthoformate RC(OR')$_3$.

The synthons **a**$^2$ are typically obtained by halogenation or oxidation of carbonyl derivatives or olefins. Another type of **a**$^2$ synthon is represented by alcohols with a leaving group at the adjacent carbon, or better yet epoxides which, after opening by a nucleophile, behave like equivalents of **a**$^2$ synthons.

**a³**

$$X = Cl, Br, OTs, OMs, \ldots$$

$a^3$ synthons are typically Michael acceptors such as $\alpha,\beta$-unsaturated carbonyl compounds (R=H, C). Another less frequently used type of $a^3$ synthon is ethylene derivatives with a leaving group in the allylic position: such compounds tend to isomerize with a shift of the double bond at the cost of the structural integrity of the synthon.

Acceptor synthons $a^n$ with $n$ greater than 3 are used very rarely.

Reference: J. Furhop and G. Penzlin, *Organic Synthesis, Concepts, Methods, Starting Materials*, Verlag Chemie, Weinheim, 1983, p. 16.

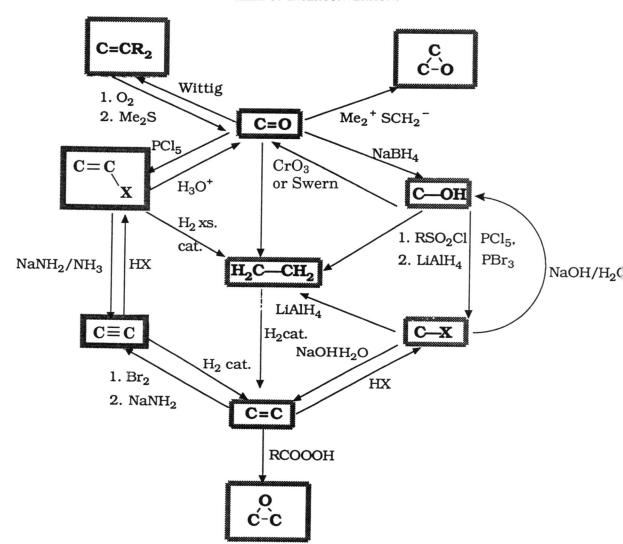

# 3  The Wheel of Interconversions of Functional Groups

In the adjacent Scheme, a summary of the principal functionalizations—those used most frequently in practice—is presented. The carbonyl group is a central function: it points to alkenes via the Wittig reaction; to epoxides via the Corey–Chayakovsky reaction; to alcohols via partial reduction; to hydrocarbons via complete reduction; to vinylic halides via use of phosphorus pentachloride.

To devise any multi-step synthesis one chooses a route compatible with the other functional groups present in the molecule.

This wheel $C{=}O \rightarrow C{-}OH \rightarrow C{-}X \rightarrow C{=}C \rightarrow C{\equiv}C \rightarrow C{=}C{-}X \rightarrow C{=}O$ provides the student also with a useful summary of this text.

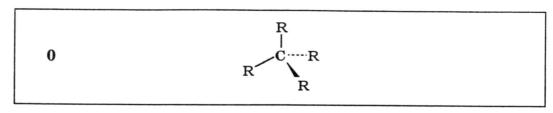

**0**

$$\begin{array}{c} R \\ | \\ R-C\cdots R \\ \diagdown \\ R \end{array}$$

**1**

$$\begin{array}{c} R \\ \diagdown \\ C=C \\ \diagup \\ R \end{array} \qquad \begin{array}{c} A \\ | \\ R-C\cdots R \\ \diagdown \\ R \end{array}$$

**2**

$$R-C\equiv C \qquad C=C=C \qquad \begin{array}{c} R \\ \diagdown \\ C=A \\ \diagup \\ R \end{array} \qquad \begin{array}{c} A \\ | \\ A-C\cdots R \\ \diagdown \\ R \end{array}$$

**3**

$$C=C=A \qquad \begin{array}{c} A \\ \diagup \\ C=C \\ \diagdown \\ A \end{array} \qquad \begin{array}{c} A \\ \diagup \\ R-C=A \end{array} \qquad \begin{array}{c} A \\ | \\ A-C\cdots A \\ \diagdown \\ R \end{array}$$

**4**

$$A=C=A \qquad \begin{array}{c} A \\ \diagup \\ R-C=A \end{array} \qquad \begin{array}{c} A \\ | \\ A-C\cdots A \\ \diagdown \\ A \end{array}$$

$$A = O, \ N, \ S, \ F, \ Cl, \ Br, \ I$$

# 4  Synthetic Equivalents

The (partial) classification opposite of organic molecules is convenient as an introduction to this concept. We start by defining the class of molecules having the same degree of oxidation (formal oxidation, of course) at the central carbon (**bold face**). This shows a first, *horizontal*, type of synthetic equivalents: the functions sharing the same degree of oxidation. For instance, a carbonyl compound, the corresponding acetal, imine etc. all belong to class 2. A slightly more complicated example, a 4,5-dihydro-1,3-oxazoline is equivalent to a carboxylic acid.

Synthetic equivalence also arises from a *vertical* correspondence. Hence in practice an oxidation or reduction is necessary. Thus acetylenes (class 2) are synthetic equivalents of alkenes via a hydrogenation: $Na/NH_3$ (*E*-alkenes) or catalytic hydrogenation $H_2$/cat. (*Z*-alkenes). Similarly, hydrogenation of carbonyl compounds transforms them to primary or secondary alcohols (and vice versa by oxidation). Anisole is an equivalent of cyclohexenone via a Birch reduction.

Let us finally consider the case of olefins $RCH=CH_2$. They are equivalents of (i) primary alcohols (1. $BH_3$ 2. $H_2O_2$), (ii) of secondary schools ($H_3O^+$), (iii) of methylketones (catalytic oxidation with palladium(II)) and (iv) of aldehydes ($O_3$).

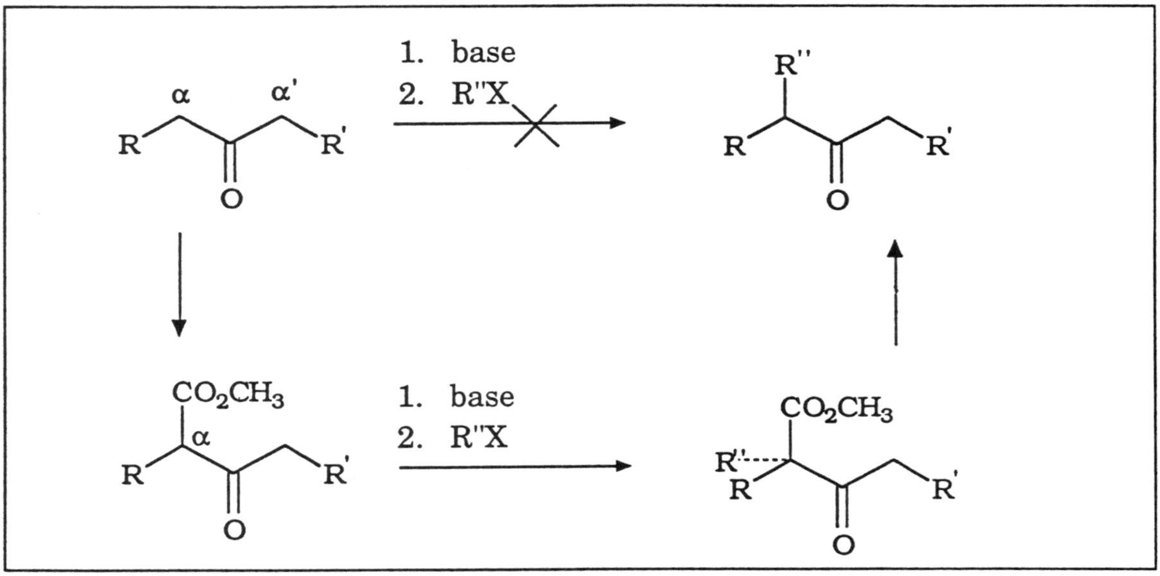
α
α'
R
R'
O
1. base
2. R''X
R''
CO2CH3


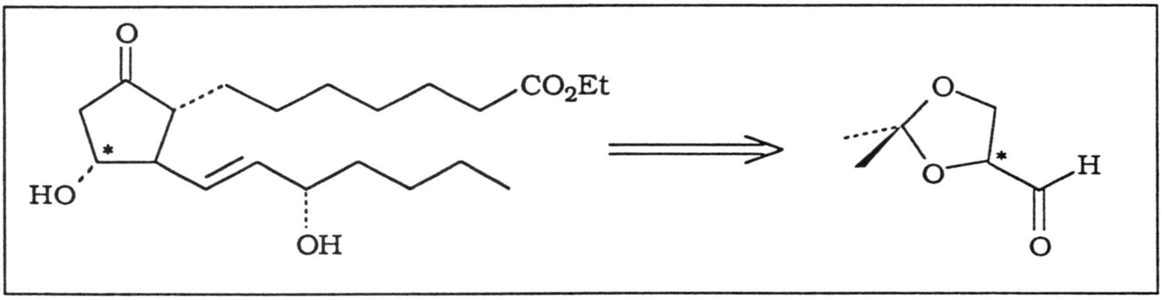
CO2Et
HO
OH
H


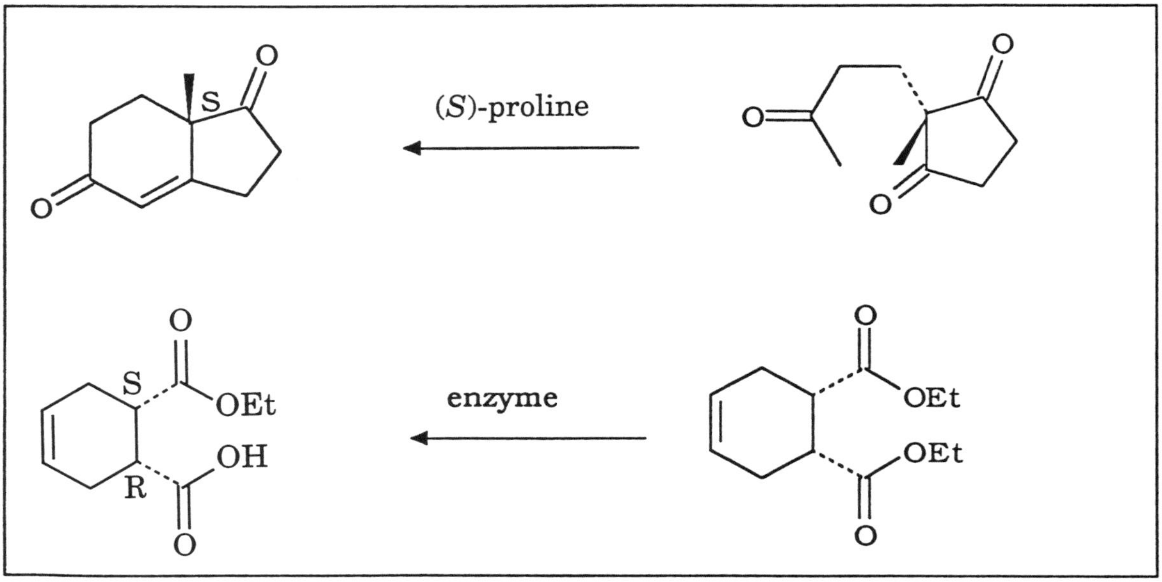
S
(S)-proline
enzyme
OEt
OH
R

# 5   Controlled Connections

Joining a donor center *d* and an acceptor center *a* gives rise to a new bond—in most cases a carbon—carbon bond. Going the other way, disconnecting by heterolytic cleavage a carbon—carbon bond, affords a donor center *d* and an acceptor center *a*. The problem is to make the actual connections occur selectively and to rid them of ambiguities.

Let us recall at this point one example of control of regioselectivity: one should not attempt to directly alkylate a ketone since this would lead to reaction at both the $\alpha$ and $\alpha'$ carbons. Better selectivity arises by way of a carboxylate auxiliary group which limits the alkylation to the $\alpha$ carbon; such an auxiliary is then removed by hydrolysis and thermal decarboxylation.

*Four principal strategies* exist to control stereoselectivity.

(1)  The first is to start from an asymmetric center and to maintain it as such during the whole synthesis; or to change it in a way dictated by its configuration: an $S_N2$ reaction with inversion of configuration; or a transfer of chirality through sigmatropic shifts (pericyclic reactions). The example given is that of Stork's synthesis of prostaglandin $E_2$ starting from (+) glyceraldehyde, with a chiral atom bearing the future OH group of the five-membered ring of the target molecule.

(2)  A second strategy is to elaborate chirality by selection between enantiotopic faces or groups. For instance, the Robinson annulation indicated is carried out in the presence of proline, an asymmetric inducing agent, with an enantiomeric excess of 93%. Pork liver esterase hydrolyzes only one of the two ester functions (ee = 99%).

References: G. Stork and T. Takahashi, *J. Am. Chem. Soc.*, **99**, 1275–1276 (1977).
             Z. G. Hajos and D. R. Parrish, *J. Org. Chem.*, **39**, 1615–1621 (1977).
             H. G. Gais and K. L. Lukas, *Angew. Chem. Int. Ed. Eng.*, **7**, 142 (1984).

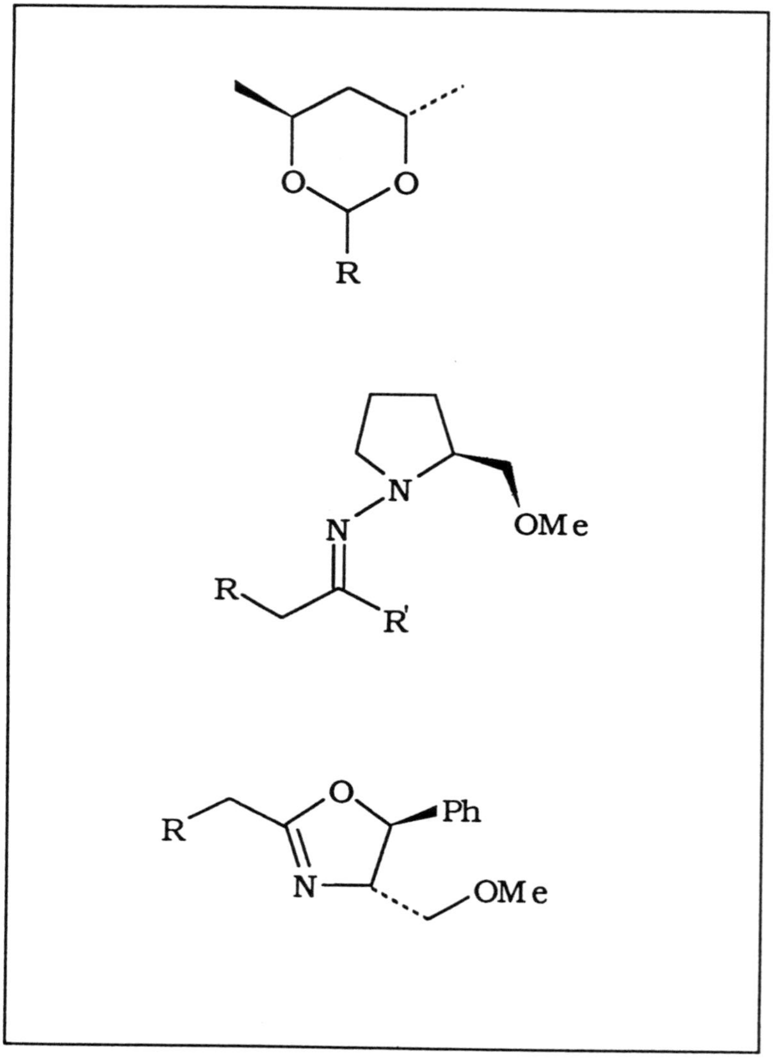
O
O
R
N
N
OMe
R
R'
R
O
Ph
N
OMe

(3) A third strategy uses a chiral auxiliary group to turn the enantiotopic faces (or groups) into diastereotopic faces (or groups). Thus one of the numerous diastereoselective reactions can be used. At the time of writing (1993), enantioselective reactions are by far less numerous. But the intense activity in the field may change the situation rather quickly.

Once the chiral auxiliary group has served its purpose, it is removed, and preferably recycled. Three examples are mentioned again (see the adjacent scheme):
   (i) a chiral acetal, the synthetic equivalent to an aldehyde;
   (ii) a chiral hydrazone, the synthetic equivalent to a ketone, allowing for enantioselective alkylation (Enders method with (S)-AMP or (R)-AMP described in detail earlier in this book);
   (iii) a 4,5-dihydro-1,3-oxazoline derived from a nitrile $R-C\equiv N$ (and consequently a synthetic equivalent of a carboxylic acid) allows for alkylation $\alpha$ to a carbonyl group in an enantioselective manner (method of A. I. Meyers).

(4) The fourth strategy could fall more and more into oblivion: resolution of a racemic mixture. Since this operation eliminates 50% of the population, it is disastrous to the yield—except at the inception of a multi-step synthesis.

References: D. Seebach, R. Imwinkelried and T. Weber, *Modern Synthetic Methods*, **4**, 128 (1986).
          D. Enders, in *Current Trends in Organic Synthesis*, H. Nozaki, Ed., Pergamon, New York, 1983, p. 151.
          K. A. Lutomski and A. I. Meyers, in *Asymmetric Synthesis*, J. D. Morrison, Ed., Academic Press, **3**, 213 (1984).

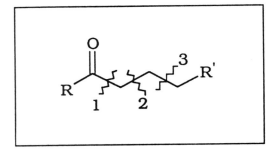

# 6 Routes to Simple Disconnections

Let us consider the carbonyl compound in the upper box (see adjacent scheme) as the target molecule (R=H, alkyl, alkoxy etc.). Let us limit ourselves to the first three disconnections indicated.

The first corresponds—in a retrosynthetic sense—to an acylation of an organometallic compound $R'(CH_2)_3M$. It allows also for an alkylation of a $d^1$ synthon, the acyl anion $R-C^-=O$ with $R'(CH_2)_3X$.

The second disconnection suggests alkylation of an enolate anion $R.CO.CH_2^-$ or a synthetic equivalent with $R'(CH_2)_2X$. Another possibility would be alkylation of an organometallic reagent $R'(CH_2)_2M$ with the $a^2$ synthon $R.CO.CH_2.X$.

The third disconnection allows for two types of answers depending on whether the left-hand side fragment is assumed to be a donor or acceptor synthon: either by alkylation of $d^3$ synthons or by 1,4-addition of an organocuprate to a Michael acceptor.

The second example is that of a target molecule with a multiple bond (an olefin or acetylene). The first disconnection is either a Wittig reaction (or Peterson olefination) or a Ramberg–Bäcklund reaction. The second disconnection suggests an alkylation of a vinylcuprate. As to the third disconnection, the appropriate answers would be as follows:

(i) alkylation of an allyl (or propargyl) anion with $R'X$;
(ii) or allylation (or propargylation) of an organometallic reagent $R'M$.

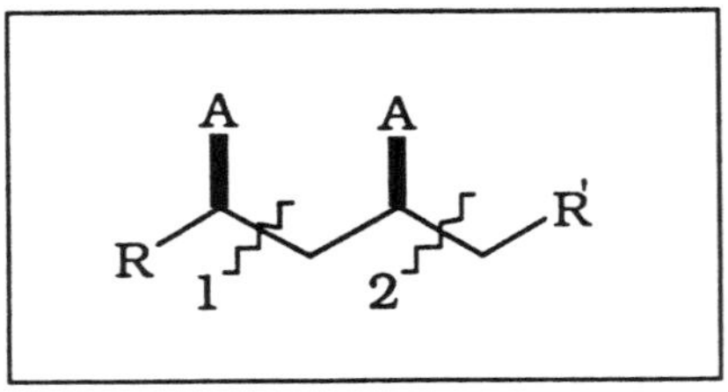
A
A
R
R'
1
2

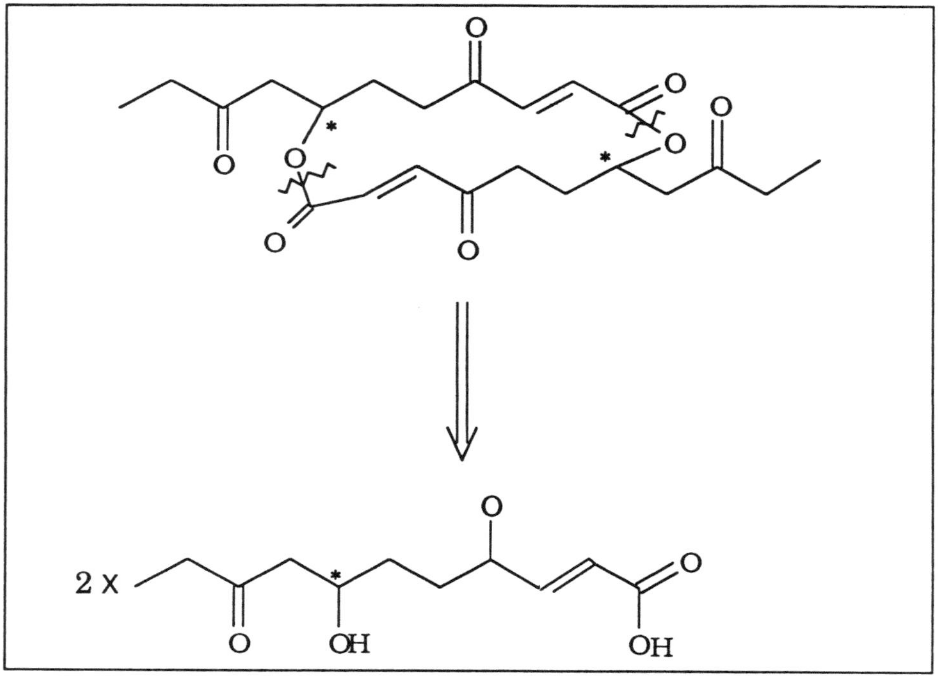
O
O
O
O
O
O
O
O
2 X
O
O
OH
OH

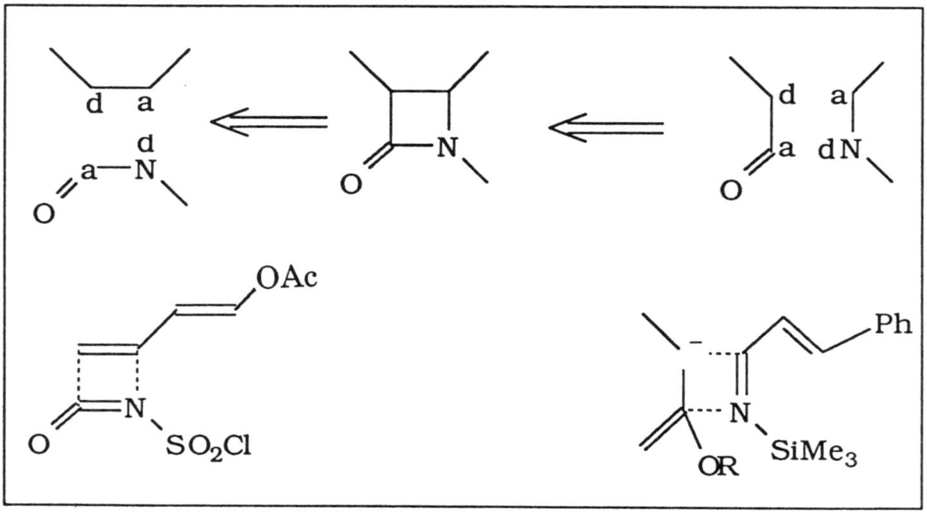
d a
d
a—N
O
N
O
d a
a dN
O
OAc
N
O
SO2Cl
Ph
N
SiMe3
OR

# 7  Routes to Double Disconnections

Let us take as an example the target molecule in the box (see adjacent Scheme): the bold face lines indicate a single or double bond to a heteroatom A. Disconnection 1 suggests a construction via hydroxyalkylation (aldol condensation or related reaction) or a coupling of a $d^1$ synthon with an $a^2$ synthon with inverted polarity. Disconnection 2 calls for an alkylation of a $\beta$-diketone-dianion $R.CO.CH^-.CO^-$ with $R'CH_2X$.

It is often very useful to consider the disconnection not in an isolated and independent way but in pairs. This is obvious when the target molecule is composed of identical or very similar fragments. The example indicated is that of *(−)-vermiculine*: its synthesis is equivalent to that of the halves since the target molecule possesses a $C_2$ symmetry axis; therefore a double disconnection is made.

An example of particular importance in the area of fine chemicals is that of antibiotics with a four-membered ring of the $\beta$-lactam type. It is useful to analyze its synthetic routes with a double disconnection, horizontal or vertical. The first can be effected, for instance, via $[2+2]$ cycloaddition of an alkene and chlorosulfonyl isocyanate; the second is possible via condensation of an imine with an enolate of an ester.

References: D. Seebach, H. O. Kalinowsksi, W. Lubosch, B. Renger, and B. Seuring, *Angew. Chem. Int. Ed. Engl.*, **16**, 264–265 (1977).
F. A. Bouffard, D. B. R. Johnston and B. G. Christensen, *J. Org. Chem.*, **45**, 1130–1135 (1980).
G. Cainelli, M. Contento, D. Giacomini, and M. Panunzio, *Tetrahedron Lett.*, **26**, 937–940 (1985).

# 8   Interconversion with Inverted Polarity

It is possible, starting from a given structural fragment, to transform it from a donor (acceptor) synthon into an acceptor (donor) synthon. An aldehyde, for instance, is a typical $a^1$ synthon. Each of the following equivalent groups is a $d^1$ synthon with inversion of the initial polarity:

$$-CH(SR_2)  -CH(CN)OR  -CH(OR)=PPh_3  -CH_2NO_2$$

Let us consider a fragment $^-CH_2.CO.R$ (i.e. a type $d^2$ synthon, an enolate anion); bromination which affords $BrCH_2.CO.R$ being sufficient to transform it into an $a^2$ synthon. In a similar way, an olefinic bond $abC=Ccd$ is a donor synthon due to its $sp^2$ carbons. After epoxidation with a peracid it becomes an $a^2$ synthon; the epoxide can be attacked by a nucleophile :Nu with ring opening to a compound $NuCab-CcdO^-$.

Other reactions with inversions of polarity are, for instance, the Nef reaction, starting from a nitro-olefin. The carbon atom 2 is therefore, due to the resulting polarization, an acceptor: the fragment $C=C. NO_2$ is an $a^2$ synthon. However in a molecule $R^1R^2C-NO_2^-$ the $NO_2^-$ group can be easily cleaved: with two equivalent of $H_3O^+$ the corresponding ketone $R^1R^2C=O$ (an $a^1$ acceptor) and nitrous acid $HNO_2$ are obtained.

# Summary: 'Acceptor Synthons'

The 10 most important carbon—carbon bond forming reactions are summarized as follows:

Diels–Alder
Wittig
Friedel–Crafts alkylation
Friedel–Crafts acylation
Substitution with cyanide
Substitution with acetylides
Alkylation of enolates
Aldol condensation
Substitution with organometallic reagents
Organocuprates (including Michael addition)

They are all characterized by coupling of a donor center ($d$) with an acceptor center ($a$).

The notation of acceptor synthons corresponds to that of donor synthons. Alkylating agents RX furnish $R^+$ synthons referred to as $a$. The main $a^1$ synthons are the conjugate acids of aldehydes and ketones as well as acylium cations $RCO^+$ derived from acyl chlorides or orthoformates. The most important $a^2$ synton is the epoxide group. Corey's *forskoline* synthesis uses the latter for three purposes: protection of a C=C bond, activation and opening to an alcohol function present in the target molecule. Michael acceptors are $a^3$ synthons.

Then the degree of oxidation of carbon can be defined: It increases by one unit either by introduction of an attracting atom or a unit of unsaturation. This formal term is useful for recognizing horizontally synthetic equivalents and vertically functional group interchange reactions via redox reactions.

*Examples*: carbonyl and acetal groups have the same degree of oxidation (2); carbonyl groups of a ketone (2) and those of an ester or acyl chloride differ by one degree of oxidation.

Let us recall here the wheel of interconversions of the most common functional groups: ketone → alcohol → halide → ethylene → etc.

Systematic exhaustive disconnections have to be evaluated (by hand or by an expert system-type software) in order to identify the bonds to be broken in a retrosynthetic sense. This identification is easier if the starting molecule is fixed.

Once this first disconnection is chosen, the synthetic scheme can be elaborated easily; example: 2-methyl-6-phenyl-3-heptanone prepared from 3-bromopropanol.

The interest of double disconnections consists in either exploiting the pseudosymmetries or similarities between the fragments of the target molecule or constructing the molecular skeleton more rapidly. As shown in the example of a cyclic diene, sometimes a prior functional change is the prerequisite to the unavoidable recourse to the Diels–Alder reaction.

The chapter concludes by recapitulating auxiliary groups for regioselectivity (ester groups) and stereoselectivity (oxaline, Meyers; hydrazone, Enders; acetals, Seeback; chemzymes, Corey) which allow for control of connections.

# 15.II
# Synthons: Acceptor Synthons

## 1   Structural Refinements and Biological Activity: Methotrexate

Uracil is not among the bases in DNA. Thymine, its methylated analog, is one of the four DNA bases. The conversion of deoxyuridylate (dUMP) to deoxythymidylate (dTMP) is the final step in biosynthesis of nucleotides. The methylating agent is $N^5$, $N^{10}$-methylenetetrahydrofolate: the constituents of the methyl group are circled in the following Scheme. Transfer of this group makes $N^5$, $N^{10}$-methylenetetrahydrofolate go to dihydrofolate. The enzyme responsible for the methylation of dUMP to dTMP is thymidylate synthetase. Another enzyme, dihydrofolate reductase, reduces dihydrofolate to tetrahydrofolate.

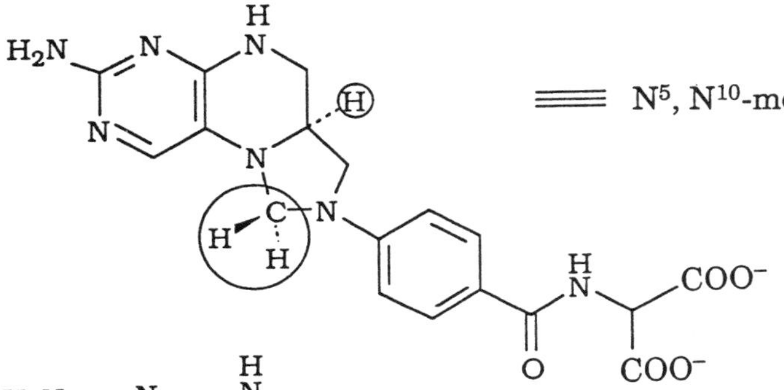

$N^5$, $N^{10}$-methylenetetrahydrofolate

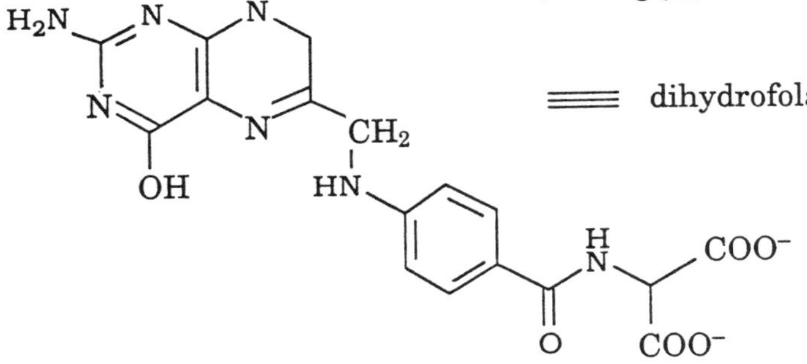

dihydrofolate

References: L. Stryer, *Biochemistry*, 2nd Ed., W. H. Freeman, San Francisco, 1981.
T. H. Cromartie, *J. Chem. Ed.*, **63**, 765–768 (1986).
*Opportunities in Chemistry* (Pimentel Report), National Academy Press, Washington, 1985, pp. 166–167.

Dividing and hence rapidly multiplying cells are large consumers of dTMP for synthesis of their DNA. Dihydrofolate reductase is a universal enzyme, from bacteria to mammals. It has been long established that administration of folic acid to test animals increased the growth rate of tumors. Conversely, one can only hope for a chemotherapy, applicable to several cancers, based on inhibition of dihydrofolate reductase.

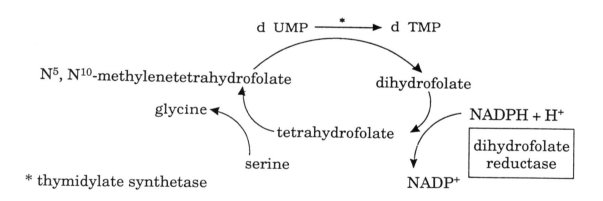

This idea was the basis for a 'systematic' synthesis (the adjective is deceptive, its real meaning is 'trial and error') of numerous analogs to folic acid. In spite of the meager conceptual investment, the enormous synthetic work that was undertaken paid off. Several molecules related to folic acid were thus discovered to be inhibitors of dihydrofolate reductase.

Folic acid is shown at the top of the following scheme, methotrexate at the bottom. The difference is very small, minute one may think: replacement of an OH by an isoelectronic $NH_2$ group of comparable size. Nevertheless, the enzyme detects this minor structural difference. Methotrexate is 50 000 times tighter bound to it than the normal substrate, folate. Since initially the cells contain only small quantities of folic acid, the drug (methotrexate) thus completely inhibits its reduction to tetrahydrofolate, which kills the cells. This is also the drawback: methotrexate unfortunately kills all cells without much discrimination, and its toxicity precludes prolonged therapies. Nevertheless, it is used for drastic treatment of leukemia, bone marrow cancer and Hodgkin's disease.

Other analogs of folate, that also inhibit dihydrofolate reductase, are known. For instance trimethoprim, an antibacterial drug, has an affinity $10^5$ times larger for the dihydrofolate reductase of *Escherichia coli* than for that in vertebrates: this selectivity makes it a highly efficient antibiotic.

Using X-ray crystallography, the structures of several dihydrofolate reductases have been determined: those of the bacteria *Escherichia coli* and *Lactobacillus casei* and of the chicken were the first. Some structures were solved for the drug (methotrexate or trimethoprim) in the active site. Molecular modelling of this interaction is thus providing the explanation for the strength of the binding interaction.

## 2  Functional Variations on a Structural Theme

Steroidal hormones show a great diversity in their biological activity which encompasses sugar metabolism (glucocorticoids), sexual behavior (androgens and estrogens) and maintenance of the electrolytic balance (mineralocorticoids).

All these molecules with highly diverse biological uses share the same parent, with a tetracyclic skeleton. For instance, the apparent structural difference between **A** and **B** is minor: yet, progesterone *A* controls the embedding of the ovule and the continuation of pregnancy whereas testosterone *B* is an androgen controlling libido and hair growth. Progesterone is the precursor of testosterone.

Reference: R. E. Oakey, *Chem. in Britain*, 342–345, April 1987.

The metabolic pathways for the biosynthesis of steroids are also rather complex. In non-pregnant women, androsteronedione **C** is transformed in the ovaries to estrone **E** and estradiol **F**, and the third estrogen estriol **G** is produced in the liver (the physiological role of estradiol **F** is the swelling of the breasts).

The process **C–E–F–G** stops during the last ten weeks of pregnancy; the androgen **D** then serves as starting material (**D–E–F–G**) for the reactions taking place in the placenta (**D–E–F**) and in the liver (**F–G**).

But the major production of estriol is determined by the presence of the foetus: starting from the sulfate **H** via the **H–I–G** conversion in the placenta. The ratio **G/E + F** is 13 times larger for a pregnant woman close to delivery than for a non-pregnant one!

## 3    The Industry of Fragrances and Flavors

Let us give an idea of the size of this industry: sales in 1985 amounted to US$ 5.5 billion. They break down into: flavors (18%); essential oils (17%); components of flavors (31%); perfume compositions (34%). The market is dominated by Europe (37%), followed by the United States (28%), Japan (9%), Latin America (8%), Eastern Europe (5%) and the rest of the world (13%). This market has a slow annual growth rate of 3 to 5% (in dollars). It is a heterogeneous industry in which 15 multinational companies control half of the market share; the other half is shared by hundreds of small companies.

Among flavors, vanilla remains the most important, as already mentioned in this book.

The largest four companies are:

(i) International Flavors and Fragrances (US): 9.5% of the market; benefit in percentage of sales: 14–15% during the last ten years;
(ii) Quest International (= Unilever; UK and Netherlands): 7.2% of the world market;
(iii) Givaudan (= Hofmann–La Roche; Switzerland): 5.8% of the world market;
(iv) Firmenich (Switzerland): 4.3% of the world market; leader in perfumes.

The perfumers in Grasse, on the French Riviera, misjudged the industrial evolution, some twenty years ago. They specialized in natural perfumes. They thought their position unique and secure and saw no threat from the synthetic perfume industry. In fact, the old manufacturing business, based on essential oils obtained by extraction from plants, is on the brink of collapse.

Reference: P. L. Layman, *Chem. Eng. News.*, July 20, 1987, pp. 35–38.

## 4    Industrial Synthesis of L-DOPA

L-DOPA (3,4-dihydroxyphenylalanine) of S configuration is one of the rare drugs active against Parkinson's disease. Its commercial production (Monsanto) follows the Scheme shown: a Knoevenagel condensation—where acetic anhydride serves as base and dehydrating agent (as well as to esterify the phenol)—affords the azlactone. The latter is hydrolyzed to a styrene derivative. The key step in the synthesis is enantioselective hydrogenation of the C=C double bond, affording with high preference the *S* isomer (in 95% enantiomeric excess). Final treatment with HBr gives L-DOPA. The starting material is familiar to the reader (**V**.6), vanillin, abundant and cheap.

References: W. S. Knowles, M. J. Sabacky, B. D. Vineyard, and D. J. Weinkauff, *J. Am. Chem. Soc.*, **97**, 2567 (1975).
W. S. Knowles, *J. Chem. Ed.*, **63**, 222–225 (1986).

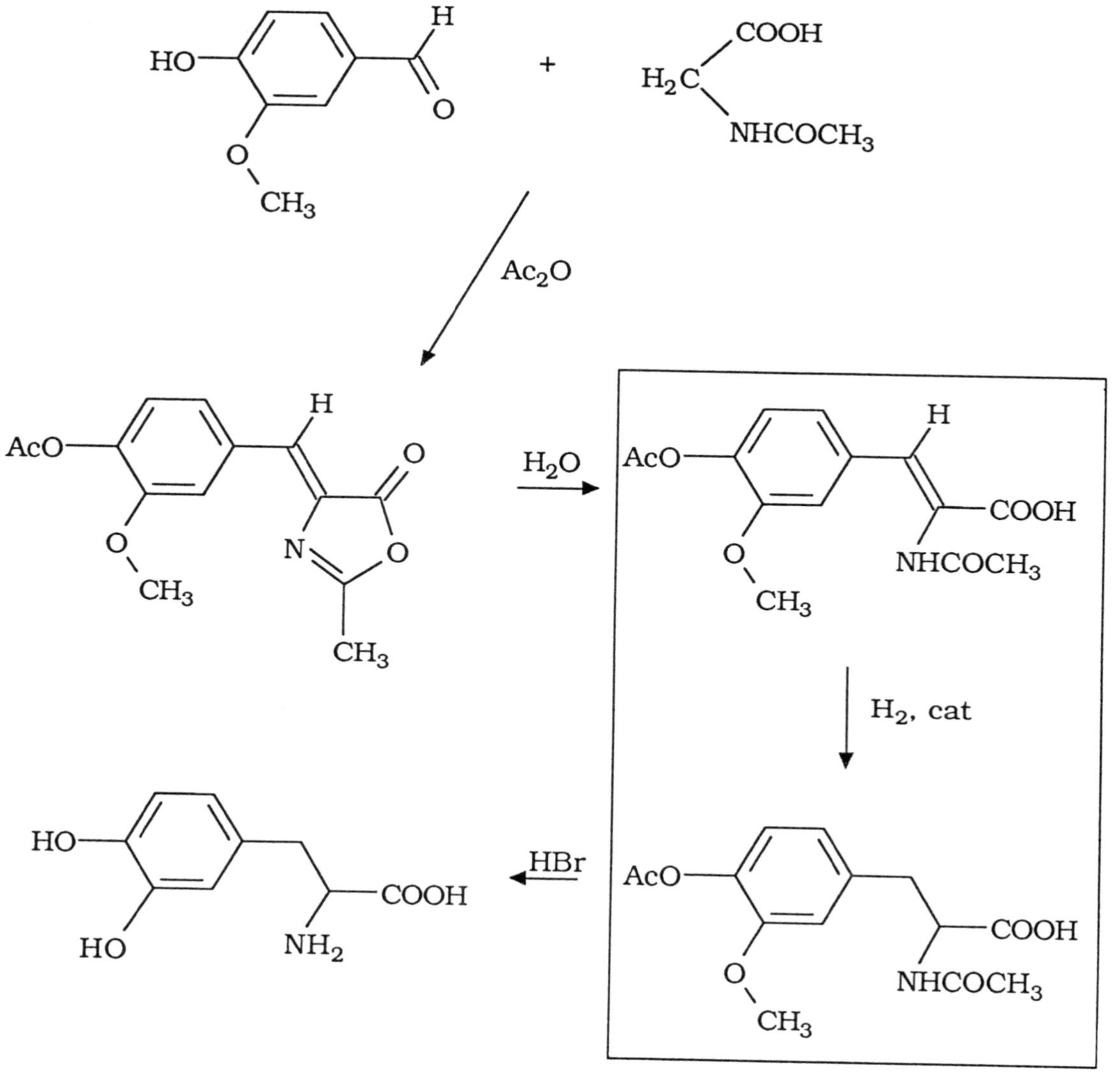
HO
O
CH3
H
O
+
COOH
H2C
NHCOCH3
Ac2O
AcO
O
CH3
H
O
N
O
CH3
H2O
AcO
O
CH3
H
COOH
NHCOCH3
H2, cat
AcO
O
CH3
COOH
NHCOCH3
HBr
HO
HO
COOH
NH2

The enantiotopic hydrogenation is effected by a soluble chiral catalyst of the type invented by Geoffrey Wilkinson (Nobel prize, 1973): Rhodium(I) is at the center of a trigonal bipyramid with a chiral diphosphine ($C_2$ symmetry axis, but no symmetry plane) as ligand. The metal bears also a hydrogen atom. This catalyst is obtained from rhodium(1,4-cyclooctadiene) (Cl) and the diphosphine of (R,R) configuration by hydrogenation under a pressure of 4 bar.

This industrial synthesis illustrates some important principles:

(i) Since few reactions build aromatic rings *de novo*, it is preferable to start with a preformed benzene ring;

(ii) The azlactone method for the synthesis of amino acids;

(iii) Use of a soluble (homogeneous) organometallic catalyst for asymmetric hydrogenation.

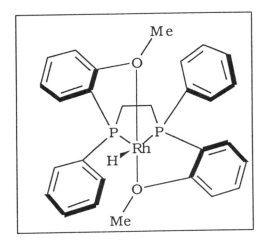

## 5 Biomimetic Synthesis of Steroids

Professor William S. Johnson of Stanford University, who in his late seventies gallantly continues his activity full-time with a team of about a dozen post-doctoral fellows, wrote this splendid new chapter of organic chemistry during the last thirty years. The idea is to synthesize steroid hormones, in like manner as Mother Nature—we extend our gender—categorizing to such abstract entities as God, Nature, Time: Mother Nature and Father Time, this is weird!—i.e. by acid catalyzed cyclization of polyenes (see in the following chapter, Section 16.8).

For instance, under the action of tin tetrachloride as Lewis acid, the allylic alcohol indicated cyclizes to a tetracyclic diene. Note the *trans* stereochemistry at the ring junctions, resulting from *anti* addition to the preceding C=C double bond: the two groups participating in the addition are a carbocation as an electrophile and the ensuing C=C double bond as a nucleophile. A double ozonolysis follows. The final step is a double aldol condensation followed by a two-fold intramolecular cyclodehydration. In this way, (±)-16,17-dehydroprogesterone is made.

($\pm$)-Progesterone itself can be obtained in a similar manner. Ethylene carbonate serves as trapping agent for the particularly unstable, and hence very reactive vinyl cation.

The next step, conceptual and practical, is to control the absolute stereochemistry of the product. An excellent example is that shown: the asymmetric center C-11, in spite of being quite remote from the initiating and terminating groups in the cyclization, controls the absolute stereochemistry, probably because it masks one of the diastereotopic faces of the central C=C bond.

A = H           30%
A = CH=CMe₂   77%

A = H           1 h   : 1-2%
A = CH=CMe₂   1 mn : 83%

The last step is to refine the synthesis so as to increase the yields. Johnson was successful in doing so, by auxiliary groups A that stabilize the carbocation. Yields are increased and the reaction times considerably reduced. Whereas the conversion, in the example shown, is only 1–2% after one hour in the absence of the auxiliary group, the reaction becomes quantitative after only one minute when an auxiliary group which stabilizes carbocations is introduced.

The acceleration is also a conceptual find: Johnson is convinced that he has thus located one of the key roles of the enzyme in the biological cyclization (in the microsomes of the liver) of 2,3-epoxysqualene to protolanosterol. He believes that the enzyme bears negative charges in the active site, at roughly the points shown by the asterisks in the Scheme, so that upon attachment of the epoxysqualene substrate its positive charges can each be stabilized locally—making for fast and quantitative cyclization.

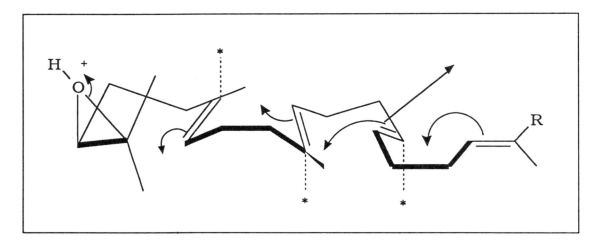

What is remarkable here is the interplay between synthesis of artificial constructs and the information that they provide indirectly on biosynthesis. The last word belongs to Berthelot who foresaw more than a century ago that

'to launch into the reproduction of a natural product (molecule), it is often necessary to devise a whole construction, founded upon formation of artificial entities. Examination of the latter leads to identification of the general laws for the composition of natural products and of the way by which their study, whether analytic or synthetic, can proceed with any hope of success. But conversely the success of any given synthesis bears amazing fecundity whenever it results from discovery of a general law.'

References: W. S. Johnson, R. S. Brinkmeyer, V. M. Kapoor, and T. M. Yarnell, *J. Am. Chem. Soc.*, **99**, 8341 (1977).
W. S. Johnson, S. J. Telfer, S. Cheng, and U. Schubert, *J. Am. Chem. Soc.*, **109**, 2517–2518 (1987).
W. S. Johnson, S. D. Lindell, and J. Steele, *J. Am. Chem. Soc.*, **109**, 5852–5853 (1987).
M. Berthelot, *La synthèse chimique*, Félix Alcan, Paris, 6th Ed, 1887, p. 185.

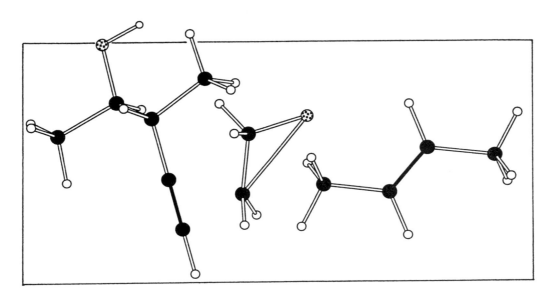

This closing chapter introduces retrosynthetic analysis. Design of an organic synthesis is henceforth helped by the computer. A succession of logical and ordered regressions make the target molecule stem from all its conceivable precursors. The chemist then chooses the preferable route. This becomes a rational decision, based on the costs of the starting materials, on those of the reagents, and on the yields to be expected for the reactions called upon. This chapter introduces retrosynthetic thinking, on elementary examples, and in so doing goes through some of the most important earlier notions.

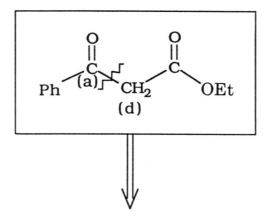

# 1 Definitions and Terminology

Retrosynthetic analysis lists all the very diverse access paths to a target molecule. It proceeds in a systematic way by logical regression. Thus an ordered list of *transforms* is defined. These, in the direction opposite to the usual progression from reactants to products and thus known as the *retro*synthetic sense, link the target molecule to starting materials. These are chosen for their simplicity, low cost, and commercial availability. Synthesis is akin to the building of a house: normally, as a synthesis proceeds, more and more complex systems are built by addition of new atoms and of new carbon—carbon bonds. Conversely, retrosynthesis proceeds most frequently by substraction. It reviews in a systematic manner all the possible breaks of each carbon—carbon bond. Retrosynthetic *transforms* are denoted in flowsheets by the *double arrow* of logical implication to distinguish them from ordinary reactions, which follow Time's Arrow.

Another definition is in order. A *synthon* is a formal entity (as opposed to an actual chemical species), called for by a retrosynthetic transform. Let us consider the example of the framed molecule in the adjacent Scheme. One of the possible disconnections produces an acceptor synthon $Ph-CO^+$ and a donor synthon $^-H_2C-CO-OEt$. In practice, a reagent with a good leaving group such as $Ph-CO-OEt$ can serve this actual acceptor role (in the synthetic sense). The actual donor can be generated from ethyl acetate with a base (e.g. $EtO^-$). The reaction of choice is a Claisen condensation.

A synthon is thus an intermediate in retrosynthetic transforms.

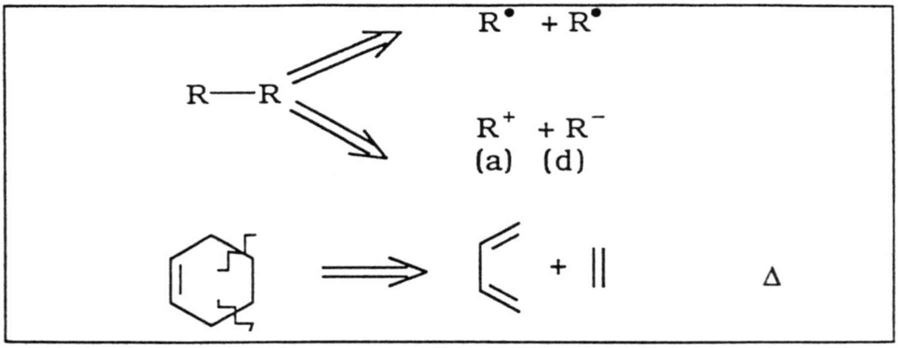

$R-R \Rightarrow R^{\bullet} + R^{\bullet}$

$R-R \Rightarrow R^{+} + R^{-}$

(a) (d)

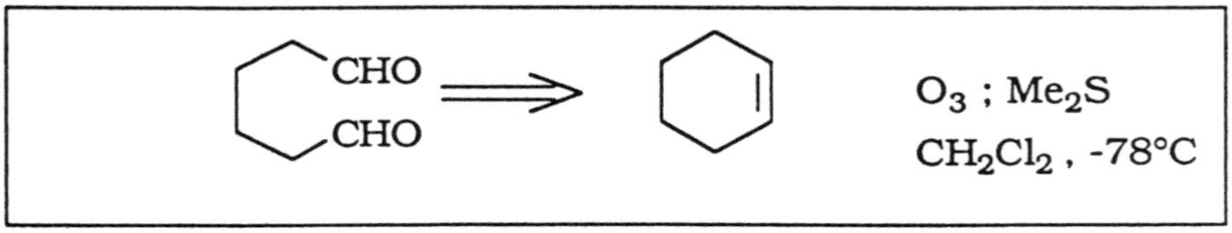

$\Rightarrow$ + ||       $\Delta$

CHO / CHO $\Rightarrow$      $O_3$ ; $Me_2S$

$CH_2Cl_2$ , -78°C

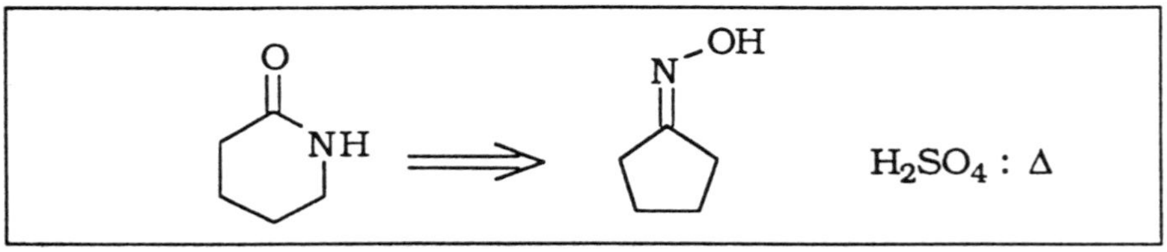

$\Rightarrow$ N$-$OH      $H_2SO_4 : \Delta$

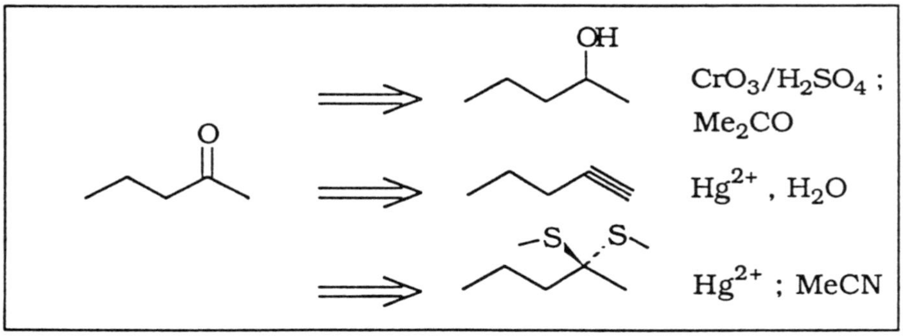

$\Rightarrow$ OH      $CrO_3/H_2SO_4$ ; $Me_2CO$

$\Rightarrow$      $Hg^{2+}$ , $H_2O$

$\Rightarrow$ $-S$ $S-$      $Hg^{2+}$ ; MeCN

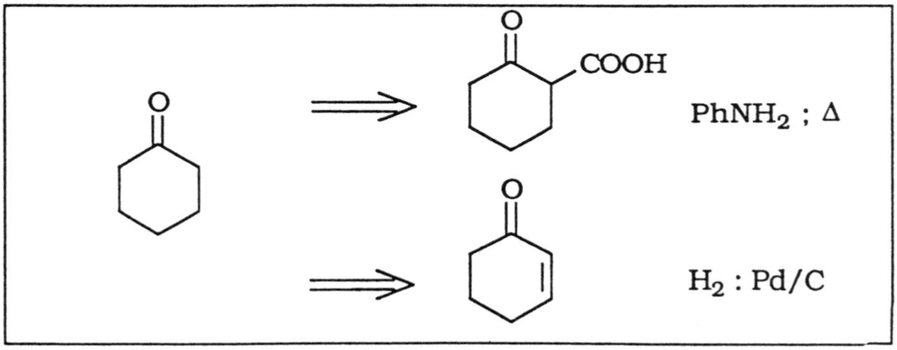

$\Rightarrow$ COOH      $PhNH_2$ ; $\Delta$

$\Rightarrow$      $H_2 : Pd/C$

## 2 List of Transforms

The disconnection of C—C bonds can occur in a *homolytic* or *heterolytic* manner. The whole emphasis of this book has been on the latter, i.e. on creating new carbon—carbon bonds by the union of donor and acceptor centers. Disconnection can affect several C—C bonds simultaneously, in concerted transforms: the case that springs to mind is that of electrocyclic reactions.

Nevertheless, one has to be wary not to omit *connecting* transforms: these correspond, in the opposite, synthetic sense, to degradative reactions. Examples include oxidation of a 1,2-diol with periodic acid; ozonolysis of a C=C bond, etc.

A third group of transforms corresponds to reactions which in the synthetic sense are *rearrangements* and, consequently, modify the carbon skeleton. Example: the transform inverse to the Beckmann rearrangement goes from a lactam to an oxime.

*Functional group interchanges*—earlier referred to as functional adaptations—are transforms affecting only substituents with retention of the carbon skeleton. They consist mainly of substitutions, oxidations, reductions, and protections/regenerations.

Finally, less frequent transforms relate to introduction or removal of an auxiliary function absent from the final product of the synthesis.

## EXAMPLE OF ISOVALERALDEHYDE

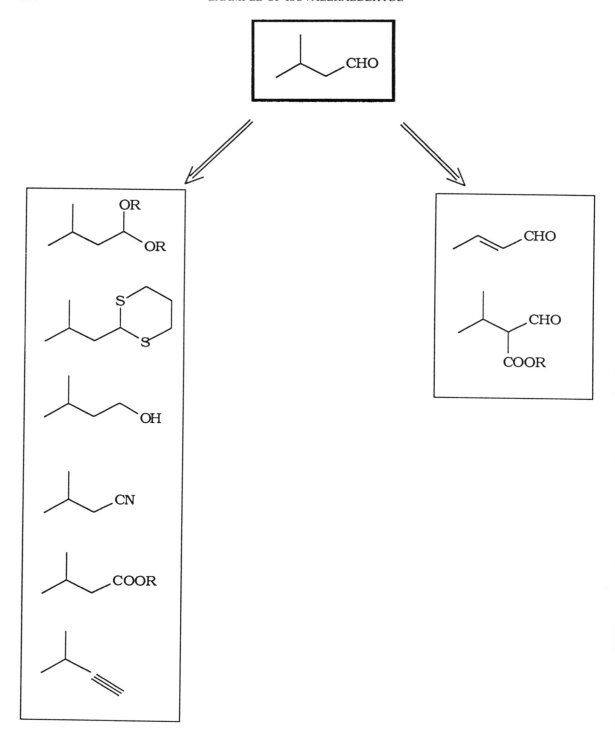

## 3 Example of Isovaleraldehyde

This target molecule can be obtained by functional group interchange reactions (left column): regeneration from an acetal by acidic hydrolysis; regeneration from a bisthioketal (mercuric salts $Hg^{2+}$; $H_2O$); hydrolysis after hydrogenation of the nitrile (di-$i$-butylaluminum hydride $i$-Bu$_2$AlH); hydrolysis of the ester to the carboxylic acid followed by reduction of the latter to the aldehyde (DIBAL, $i$-Bu$_2$AlH again); hydroboration of the corresponding acetylenic bond (diborane $B_2H_6$, followed by hydrogen peroxide $H_2O_2$ in an alkaline medium).

Other transforms (right column) involve auxiliary groups. The double bond is removed by catalytic hydrogenation ($H_2$/Pd; risk of low chemoselectivity). The ester group $-$COOR is removed by thermal decarboxylation after hydrolysis of the $\beta$-ketoester to the corresponding acid.

Accordingly, all the opposite molecules are equivalent to the target molecule. Therefore, they can be considered themselves in turn, and set as targets for retrosynthetic analysis.

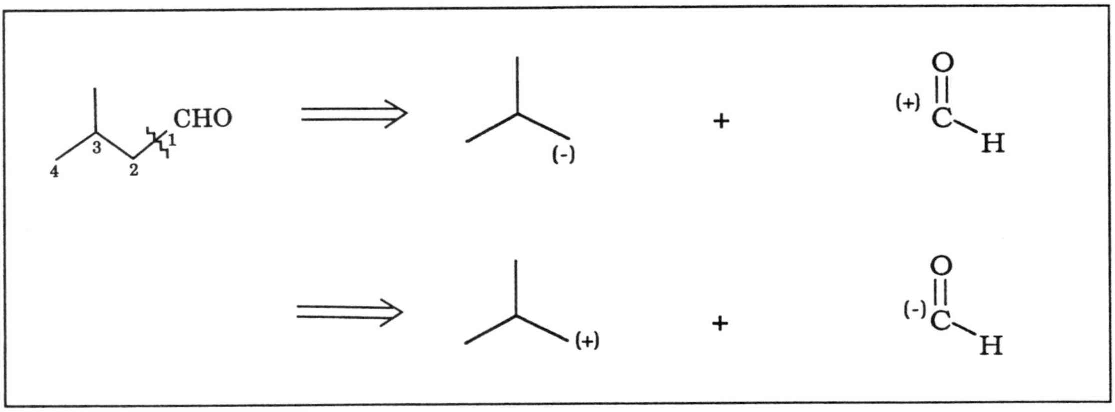
CHO
1
2
3
4
(-)
+
(+)
O
C
H
(+)
+
(-)
O
C
H

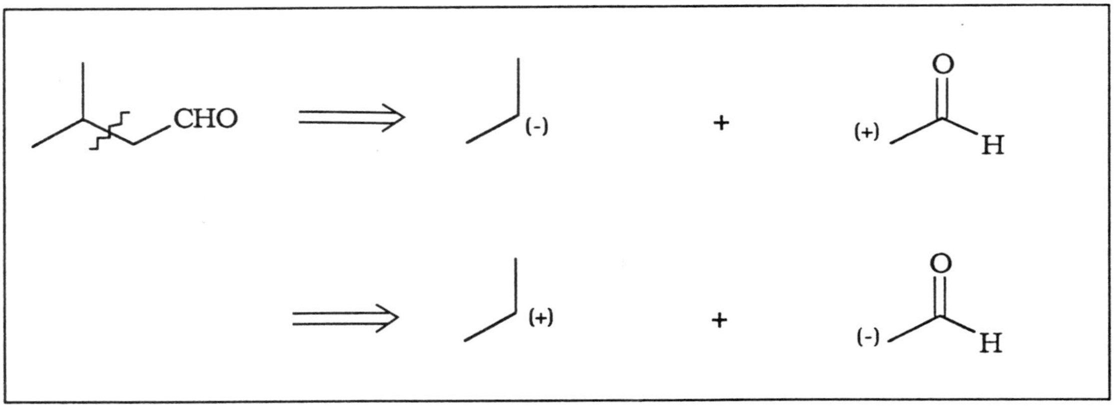
CHO
(-)
+
(+)
O
H
(+)
+
(-)
O
H

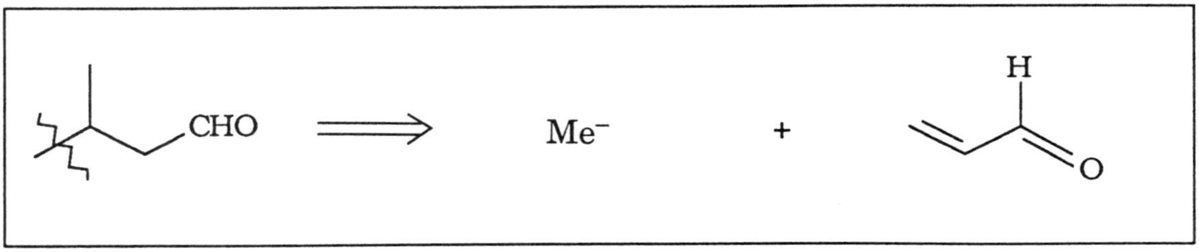
CHO
Me⁻
+
H
O

Let us now consider various disconnections, applied to the same target molecule. The disconnection between C(1) and C(2) defines the cation $(H-C=O)^+$ and the anion $Me_2CH-CH_2^-$: the corresponding reaction would be condensation between $HCONR_2$ and an organometallic such as the $Me_2CH-CH_2MgBr$ Grignard reagent (THF, low temperature) followed by acid hydrolysis to cleave off a molecule of the amine $HNR_2$. On the other hand the cationic synthon $Me_2CH-CH_2^+$ can be condensed with the anion $(H-C=O)^-$ by means of an organometallic reagent $Fe(CO)_4^2$.

There are two possible disconnections, likewise, between C(2) and C(3). The first suggests an isopropyl anion synthon ($iPr_2CuLi$) to be combined with the cation synthon $^+H_2C-CHO$ ($BrCH_2-CHO$). The second suggests addition of the enolate of acetaldehyde, as the anionic synthon, to isopropyl bromide as the cationic synthon.

Note that in the synthetic sense these two disconnections suffer from the necessary protection of the carbonyl group. Indeed, the two nucleophiles involved, namely the isopropyl anion and the enolate from acetaldehyde could also be trapped by the carbonyl groups of the starting aldehyde or of the final product.

Finally, one can disconnect C(3) from C(4); the reverse can occur in practice by a 1,4-addition (Michael addition) of methyl cuprate ($Me_2CuLi$) to $H_2C=CH-CH=O$ (acrolein) in THF, at low temperature, followed by hydrolysis with ammonium chloride in water $NH_4Cl/H_2O$.

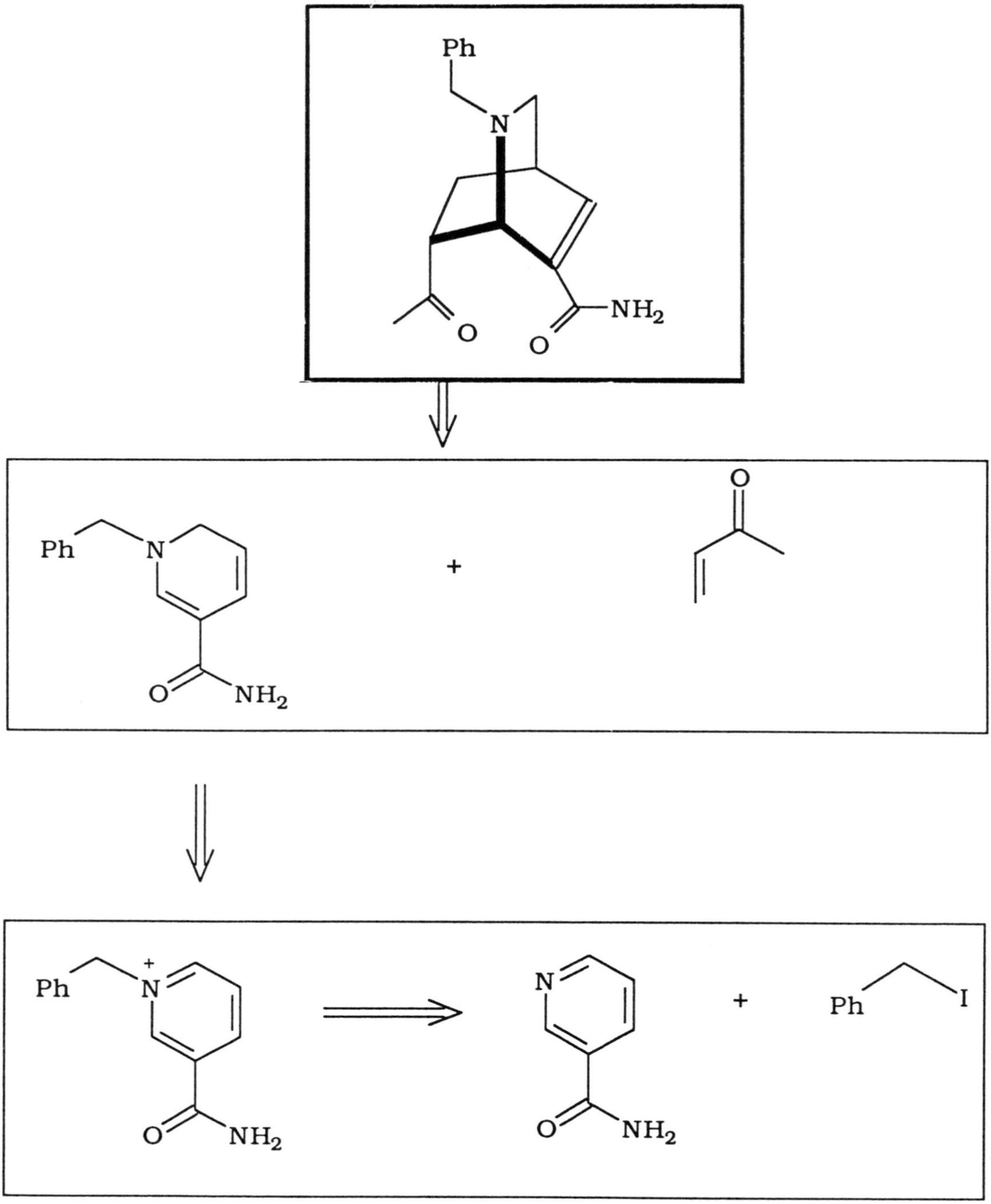

Ph
N
O
O
NH2
Ph
N
+
O
O
NH2
Ph
N+
O
NH2
N
O
NH2
+
Ph
I

# 4  Six-Membered Rings

Presence in the target molecule of a six-membered carbocycle, in particular of a cyclohexene ring, should immediately bring to mind the Diels–Alder reaction as the choice method of construction. In the example in the adjacent scheme, the double disconnection—by cycloreversion—defines methylvinylketone as a dienophile; the diene is a 1,3-disubstituted cyclohexadiene-like hetero ring system with a nitrogen atom bearing a —$CH_2Ph$ benzyl group.

One cannot resist writing-in another double bond. Such an aromatization transform leads back to a pyridine, quaternized at nitrogen by the —$CH_2Ph$ benzyl group. It is obtained in practice by reaction of the non-quaternized pyridine with a benzyl halide (Mentschukin reaction).

This is a very common way of constructing six-membered rings. We have already covered the regioselectivity and stereoselectivity of the Diels–Alder reaction.

Reference: G. Büchi, D. L. Coffen, K. Kocsis, P. E. Sonnet, and F. E. Ziegler, *J. Am. Chem. Soc.*, **87**, 2073 (1965); **88**, 3099–3109 (1966).

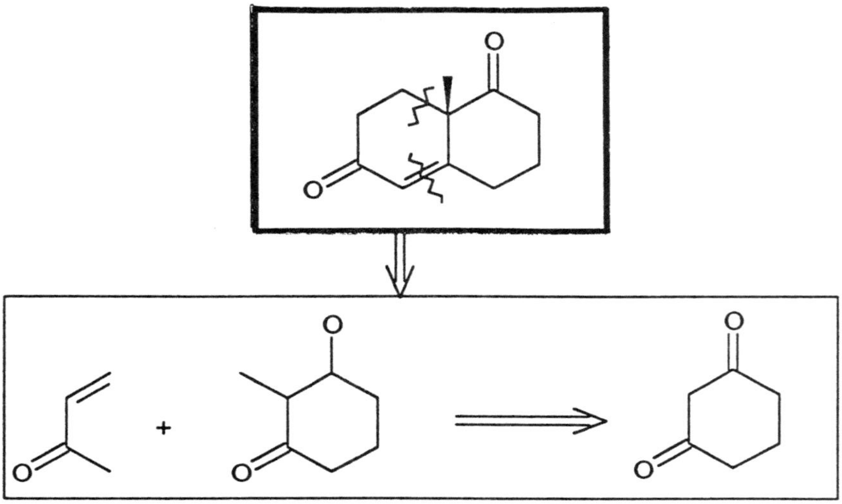

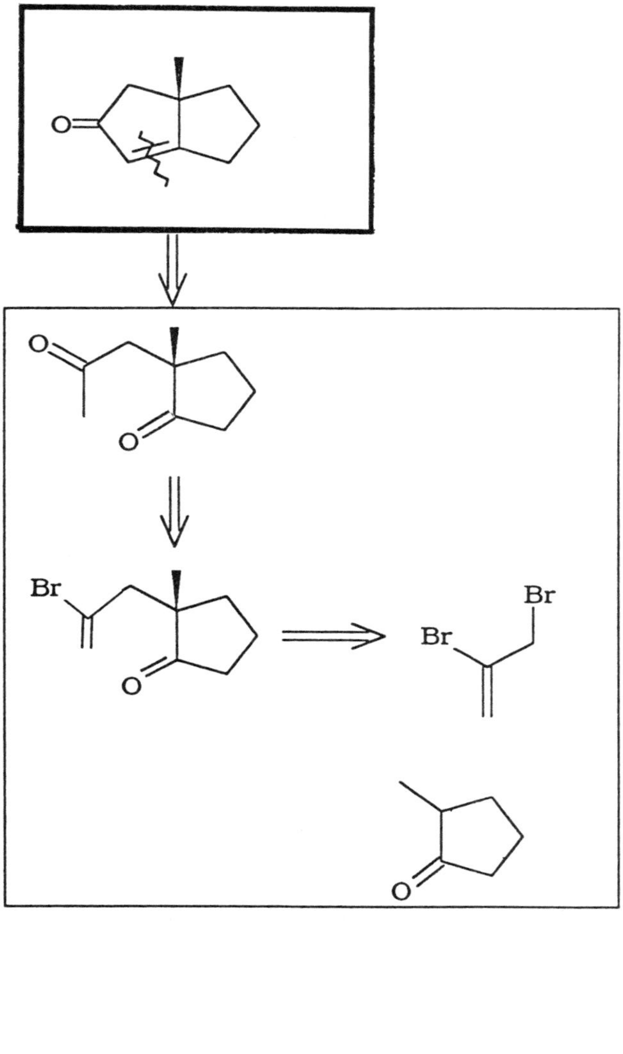

# 5  Rings from Robinson Annulation

The target molecule (known as the 'Wieland–Miescher ketone') includes a cyclohexenone ring. This structural element suggests the double disconnection corresponding to the Robinson annulation between methylvinylketone and 2-methyl-1,3-cycloexadione.

Let us recall here that the Robinson annulation is initiated by a Michael addition, here of the enolate, the conjugate base of the dione, onto methylvinylketone. In the second step, the adduct is again enolized under basic conditions and the terminal methylene donor group adds to one of the carbonyl groups of the dione. The reaction sequence is completed by crotonization, i.e. dehydration of the tertiary alcohol (the system phosphorus oxytrichloride $POCl_3$/pyridine can be used for cleavage of $H_2O$).

The second target molecule shows the consequences of a minor structural modification ('nothing is simple'). For sure, the first transform remains the same, a retro-aldol reaction. However, unlike the previous case, it is not possible to continue with a retro-Michael reaction; a carbon is missing. Thus the methylketone is now transformed into a vinyl bromide (the inverse reaction is catalyzed by electrophiles such as $H^+$ or $Hg^{2+}$). The latter can be obtained by condensation of the thermodynamic enolate (conjugate base of 2-methylcyclopentanone) with the dibromide indicated: 2,3-dibromopropene is commercial and cheap.

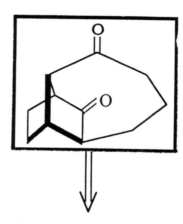

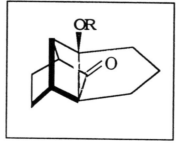

# 6 Retroaldol Transform

The target molecule invites use of the aldol condensation for its construction: it includes the characteristic pattern of two oxygens (a trigonal and a tetrahedral one), separated by three carbon atoms. The appropriate retrosynthetic transform will thus be the indicated disconnection. The donor will be the enolate anion of a $C_5$ aldehyde and the acceptor will be formaldehyde.

In the same way, a variety of 1,3-difunctional molecules can be traced back to an aldol-type condensation. The aldol condensation or the inverse reaction (retro-aldol) is the reaction of choice for the construction of polycyclic systems. We quote here—from a retrosynthetic point of view—the key step of Oppolzer's *longifolene* synthesis (see the second part of Chapter 3).

---

'There is no science without fancy, and no art without facts'—Vladimir Nabokov.

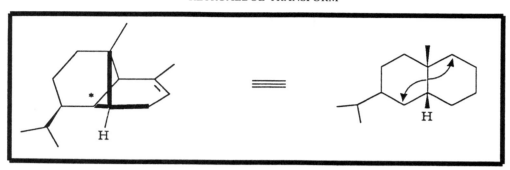

A related example is found in the synthesis of *copaene*, a sesquiterpene having a four-membered ring. The disconnection of one of these four bonds, marked by an asterisk in the framed structure, considerably simplifies the synthesis: it leads to a precursor with two fused six-membered rings with a *cis* ring junction.

This would suggest starting from the *cis*-decalin having a donor and an acceptor center (with a leaving group X), as required by this disconnection. Examination of a molecular model shows the spatial proximity of these two centers. The lefthand ring now needs an auxiliary function to form the donor center: a carbonyl group, the enolization of which, were it to be regioselective, would fill the bill.

Indeed, if this *cis*-decalone is allowed to react with the dimsyl anion $H_3C-S(O)-CH_2^-$ in dimethyl sulfoxide at 75°C, the new cyclization takes place affording the tricycle indicated. This is the key step in this copaene synthesis. Please note that the *cis*-decalone is available in nine steps from the Wieland–Miescher ketone (see a previous paragraph).

References: C. H. Heathcock, *J. Am. Chem. Soc.*, **88**, 4110–4112 (1966).
C. H. Heathcock, R. A. Badger, and J. W. Patterson, Jr., *J. Am. Chem. Soc.*, **89**, 4133–4145 (1967).

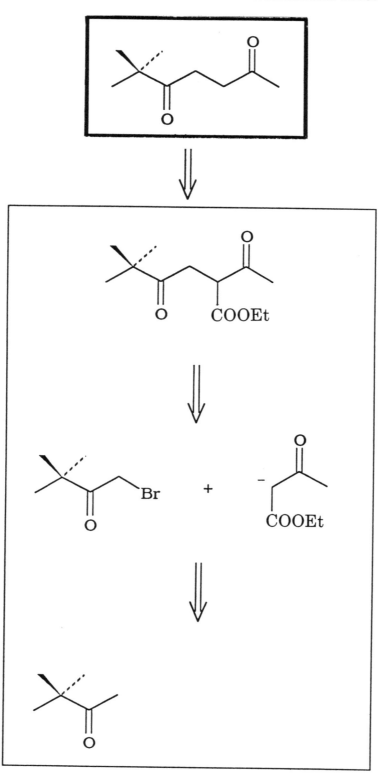

# 7 Regioselectivity by Assistance of a Carboxylate Group

The trick in obtaining this 1,4-diketone consists in starting the retrosynthetic analysis by adding the —COOEt auxiliary group. In fact, the key step in the proposed scheme is condensation of the conjugate base of ethyl acetylacetate (a commercially available product, of low cost) with the bromide from bromination of *t*-butylmethylketone, also commercial.

This simplifies everything, since the proportion of side-products is vastly reduced, and only a weak base is required to form the enolate. Were the enolate of acetone to be used, a strong base would be necessary, it would also enolize the bromoketone; clearly other products would be formed, such as Claisen condensation products.

The auxiliary group can be removed at the end of the sequence by hydrolysis followed by thermal decarboxylation.

'Science moves, but slowly, slowly, creeping on from point to point'—Alfred Lord Tennyson, *Locksley Hall* (1842).

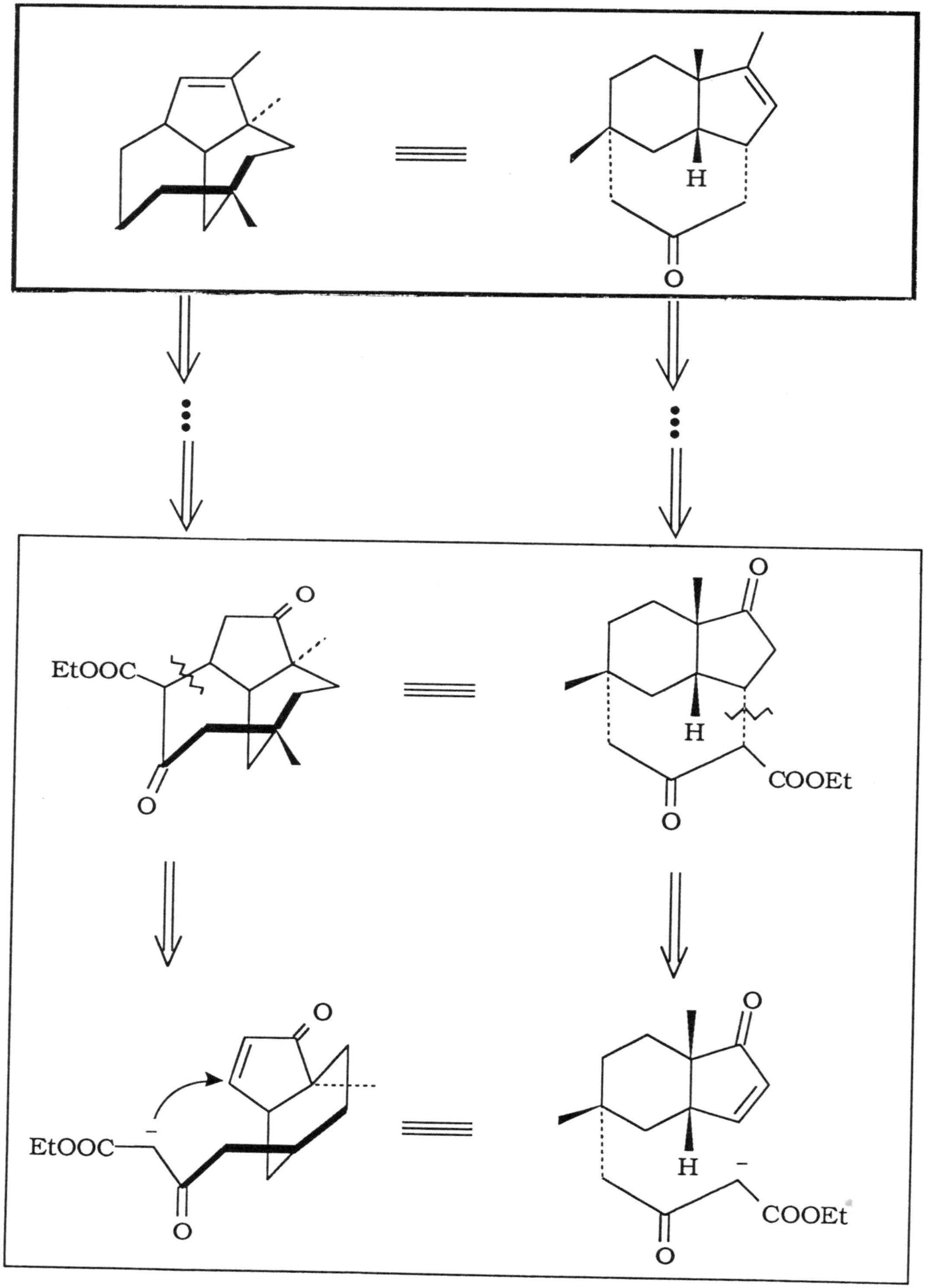
H
O
EtOOC
O
O
O
H
COOEt
O
O
EtOOC
O
O
H
COOEt
O

The target molecule *isoclovene* with its nine-membered ring is depicted in two ways for better visualization.

Methylcyclopentene is equivalent to a cyclopentanone: through addition of methyllithium followed by dehydration of the resulting tertiary alcohol.

The intermediate cyclopentanone itself is accessible (see the indicated disconnection) through 1,4-addition (Michael). This implies therefore the involvement of an enolate as the donor center, i.e. an adjacent carbonyl group. In order to make the reaction regionselective, i.e. to avoid formation of both enolates both of which can attack the cyclopentenone, the auxiliary group $-COOEt$ is used. All the auxiliary groups can be removed once this key cyclization step is achieved; firstly the carboxylate ($-COOEt \rightarrow -COOH \rightarrow CO_2$) and secondly the ketone (for example: $C=O \rightarrow C=N-NH_2 \rightarrow CH_2$).

Reference: P.G. Baraldi, A. Barco, S. Benetti, G. P. Pollini, And D. Simoni, *Tetrahedron Lett.*, **24**, 5669 (1963).

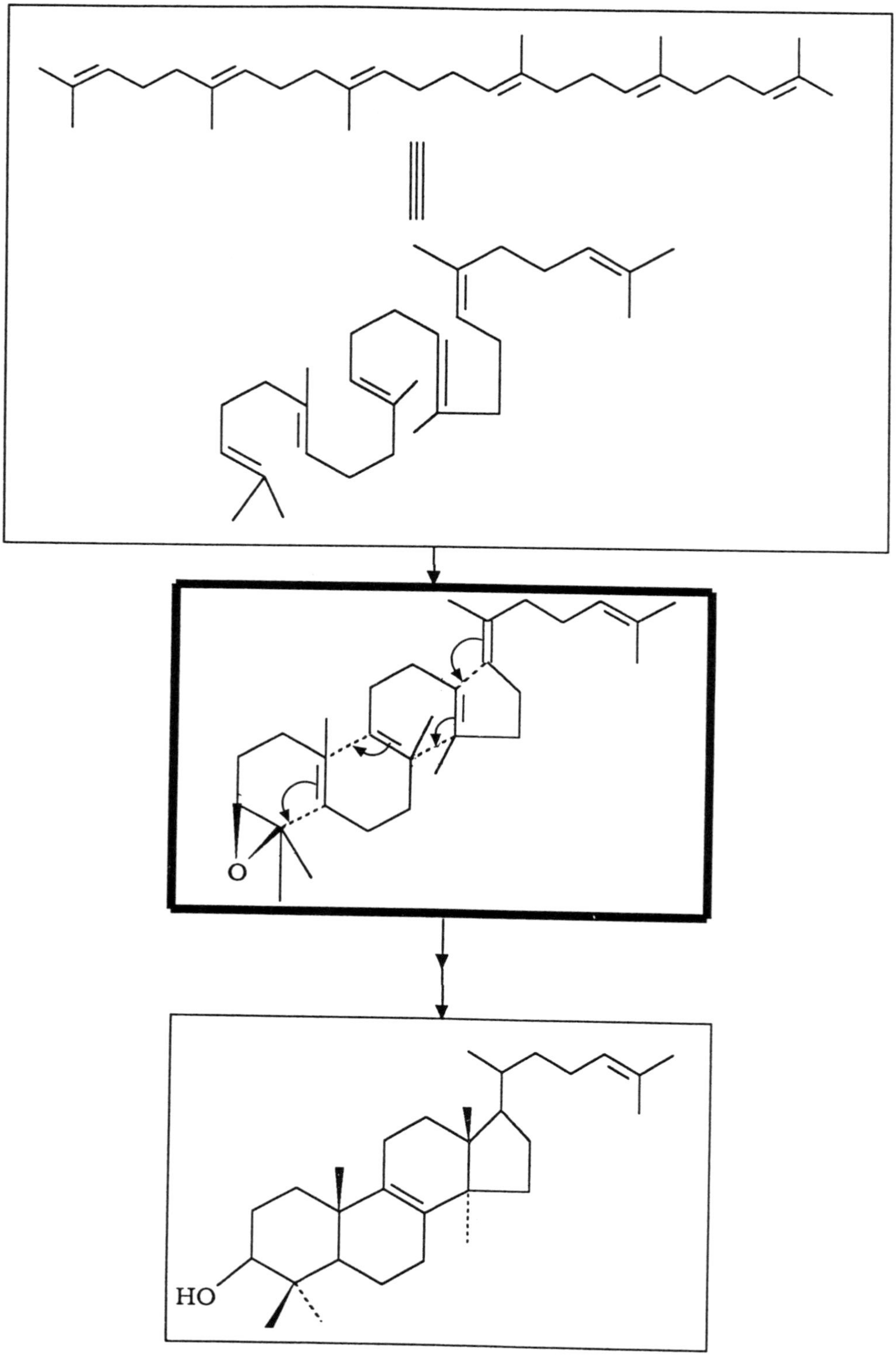
O
HO

# 8   Activation by Opening of Small Rings

Retrosynthetic thinking prepares a choice, that of the pathway to be followed, amongst all those that are conceivable. The chemist can rely on the computer which lists all possible precursors of a target molecule after systematically checking connections, disconnections, functionalizations etc. It has stored the ranges of typical yields for the various steps as well as a catalog of commercially available products together with their prices. Thus the chemist can evaluate the approximate cost of a product that he plans to synthesize. His final decision will also take other criteria into account such as the originality of, say, a particular rearrangement and the elegance of the overall concept.

Nevertheless, it is not uncommon that one or several steps create(s) difficulties. Often the chemist resorts to an activation procedure (see Chapter 6). Let us recall here—because of its simplicity and frequent use—a concept, the opening of strained small rings. It provides abundant driving force for reactions otherwise difficult to carry out, ring contractions for example.

Nature provides us with an example. In the biosynthesis of steriods, in the microsomes of the liver, multiple cyclization of epoxysqualene to a precursor of lanosterol (afterwards transformed into cholesterol) is triggered by the opening of an epoxide, catalyzed by (enzymatic) fixation of an electrophilic group to oxygen.

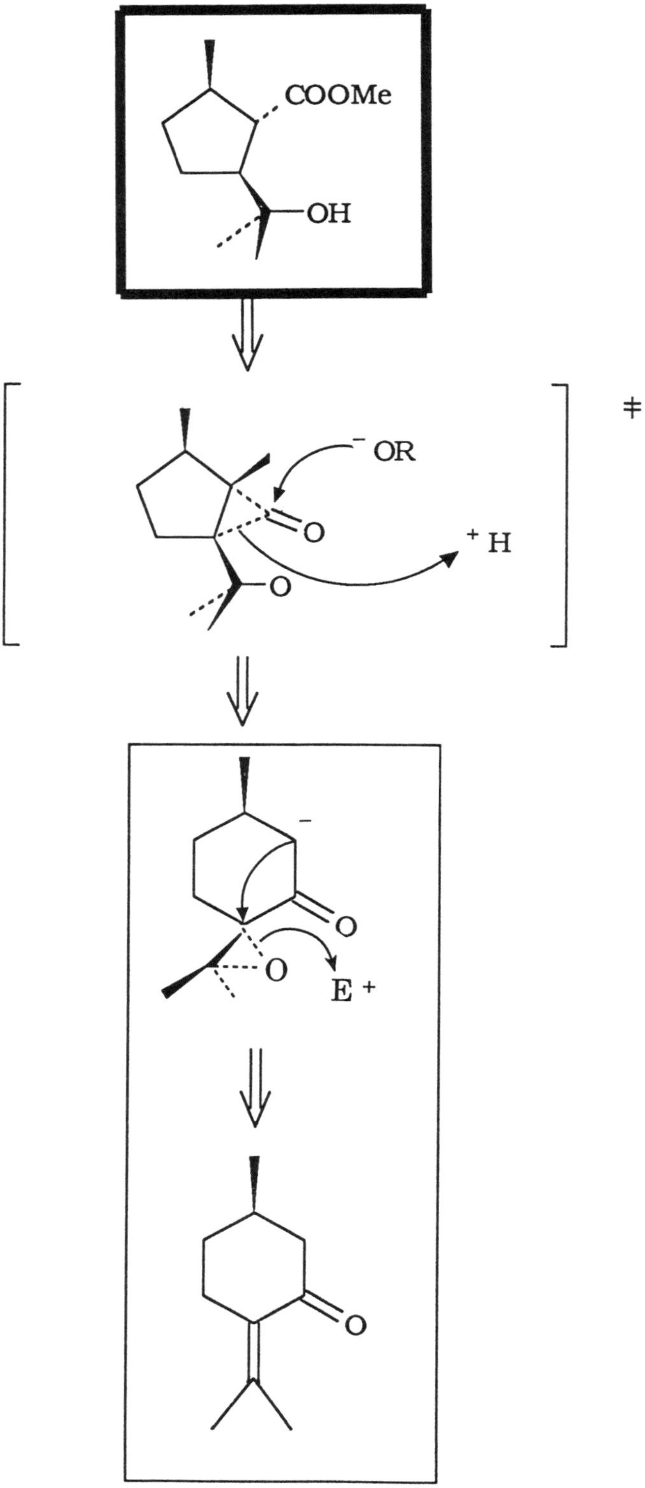
COOMe
OH
OR
O
H
O
‡
O
O
E +
O

The indicated target molecule combines a relative simplicity of constitution with a great sterochemical sophistication, due to three adjacent asymmetric centers with the *a priori* favorable *trans, trans* orientation. Regarding the tertiary alcohol, the idea is to generate it as in biosynthesis of steroids, from an epoxide ring as precursor. The next idea is to couple the opening of an epoxide with the presence of a pre-existing ester group, in one form or another, since opening of an epoxide with a nucleophile results in a *trans* relationship of the attacking nucleophile and the oxygen atom after the ring opening; as in the target. This carries into the more sophisticated idea to use another small ring opening reaction following that of the epoxide.

The unstable intermediate cyclopropanone indicated, formed by a Favorskii rearrangement, can be attacked by the methoxide anion ($R = CH_3$), leading to the target molecule. The driving force for the ring contraction is the opening of a strained epoxide. This epoxide is readily obtainable from the inexpensive natural product *pulegone* by epoxidation, for example, with *m*-chloroperbenzoic acid. As a ketone with only one enolizable carbon, it undergoes a regioselective Favorskii rearrangement. The stereoselectivity of the epoxidation is that required. Even though the methyl group responsible for it is at a distance from the C=C bond, it directs the epoxidation to the less hindered face. Noteworthy is the inversion of polarity resulting from epoxidation of the C=C bond in pulegone.

Reference: G. W. K. Cavill and C. D. Hall, *Tetrahedron*, **23**, 1119–1128 (1967).

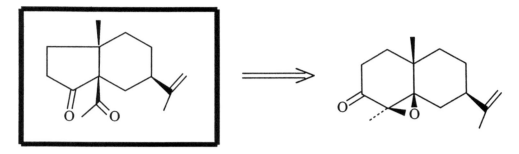

The top target molecule in the adjacent Scheme is readily available through ring contraction, supplied with energy by ring opening of the epoxide, with catalysis by Lewis acids such as boron trifluoride. This carbocationic rearrangement defines this particular synthesis of *cyperelone*.

Let us now mention the (big) problem of the regioselectivity of the opening of an epoxide by a nucleophile. The reason for selecting such a reaction is that the vast majority of ring opening reactions of epoxides under basic conditions are of $S_N2$ type, i.e. are stereochemically controlled. Two solutions will be presented here for controlling also their regiochemistry. The first solution uses an attracting group A. Thus the epoxide oxygen will become an improved leaving group if it remains attached to the carbon bearing the attracting group, which assists in stabilizing the negative charge on oxygen.

The second solution is geometric, by means of an intramolecular nucleophilic attack, which provides an excellent control of the regiochemistry of the ring opening of the epoxide.

---

'He applied on the boss a daily control, made extraordinarily acute and lucid by a combination of spite and envy'—Alexandre Arnoux, *Crimes innocents*, 1952.

References: H. Hikino, N. Suzuki, and T. Takemoto, *Chem. Pharm. Bull.*, **14**, 1441–1443 (1966).
C. H. Behrens and K. B. Sharpless, *Aldrichimica Acta*, **16**, 67 (1983).
N. Minami, S. S. Ko, and Y. Kishi, *J. Am. Chem. Soc.*, **104**, 1109–1111 (1982).
E. J. Corey, P. B. Hopkins, J. E. Munrose, A. Marfat, and S. Hashimoto, *J. Am. Chem. Soc.*, **102**, 7986–7987 (1980).
W. R. Roush and R. J. Brown, *J. Org. Chem.*, **48**, 5093–5101 (1983).

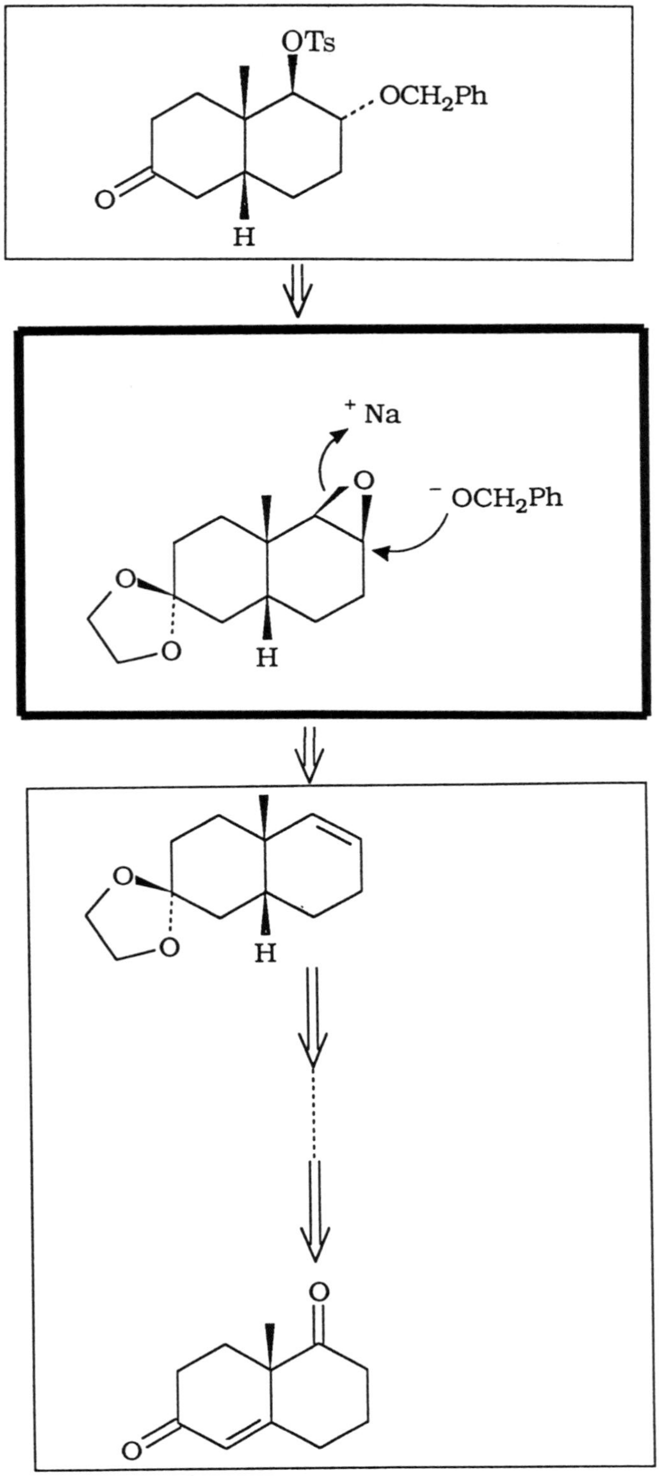

OTs
OCH₂Ph
O
H
⁺ Na
O
⁻ OCH₂Ph
O
O
H
O
O
H
O
O

The same synthesis of *copaene* (Professor Heathcock, Berkeley), already quoted in this chapter, provides us with another example of retrosynthetic analysis using an epoxide as intermediate. Examining again the *cis*-decalone, required for the key step (aldol cyclization), we see two adjacent oxygen groups *trans* to one another. Hence the idea is to form them by opening of an epoxide. The expected regioselectivity arises from attack at the least hindered carbon. To avoid attack by the nucleophile of the carbonyl group of the cyclohexanone, the latter is protected. This is in fact how the synthesis is carried out. The epoxide is formed with *m*-chloroperbenzoic acid from the Wieland–Miescher ketone in four steps.

Language consists of propositions (excluding for the moment so-called mathematical propositions). A proposition is a picture of reality, and we compare proposition with reality. We give prescriptions for action in propositions, and these prescriptions must have some picture–pictured relation with reality. The prescriptions we give (the signals, symbols which we use) must have a general, arranged significance and must be interpreted in particular instances (. . . )

Reference: L. Wittgenstein, *Cambridge Lectures 1930–1932*, D. Lee, ed., Lecture AI.

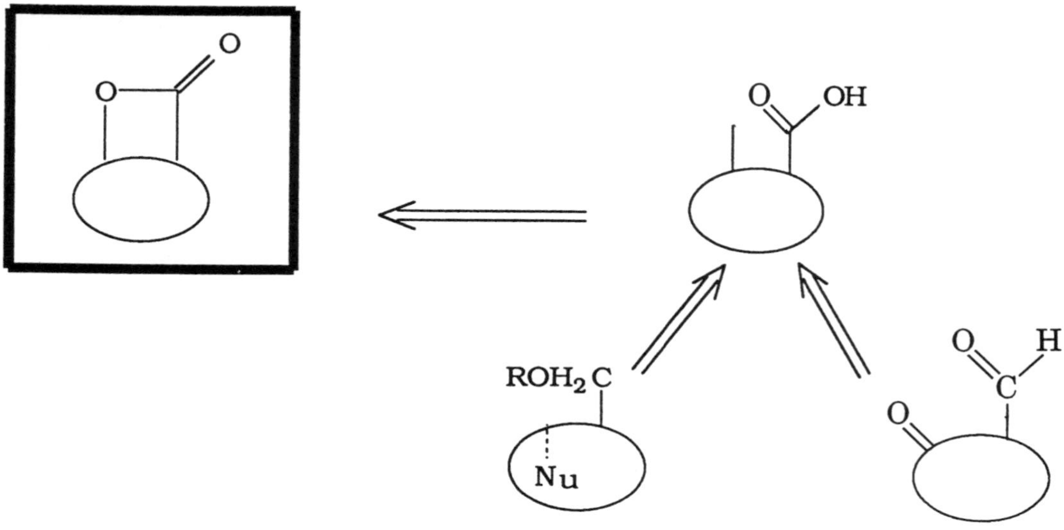
O
O
O
OH
ROH2C
Nu
O
H
C
O

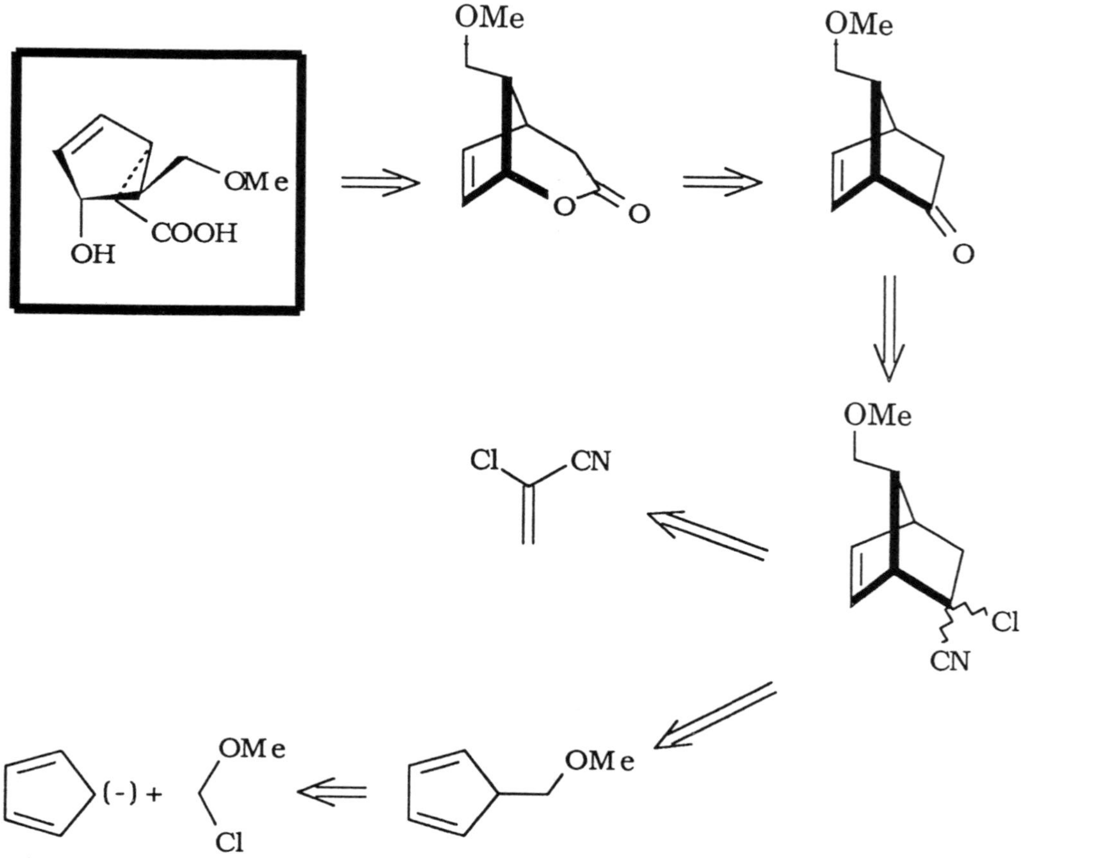
OMe
COOH
OH
OMe
O
O
OMe
O
OMe
Cl
CN
Cl
CN
OMe
(-) +
OMe
Cl
OMe

# 9 Resorting to Lactones

Lactones are ambivalent functional groups of many virtues. A lactone combines two groups with very different degrees of oxidation, an alcohol and a carboxylic acid. The alcohol group can be substituted (via an ester) with a number of groups and can also be oxidized. Conversely, the carboxylic acid can be reduced to the corresponding aldehyde, or even to the primary alcohol. Another major asset of lactones is that they are stereochemical bridges, maintaining the two OH and COOH groups on the same face of a molecule.

The first example illustrates the stereochemical control that the lactone group provides. The target molecule possesses three asymmetric tetrahedral carbons in a cyclopentene that can be formed by hydrolysis of a lactone. The latter is produced in the key step by way of a Baeyer–Villiger rearrangement: the oxygen ends up (see Chapter 13) next to the most substituted carbon (here carbon-1 of the norbornenone) since the latter migrates preferentially. The bicyclic ketone results from a Diels–Alder reaction between cyclopentadiene bearing the ether side chain, and α-chloroacrylonitrile, followed by hydrolysis of the chlorocyanohydrin. This elegant sequence of reactions is part of the now classical prostaglandin synthesis by Corey.

References: E. J. Corey, N. M. Weinshenker, T. K. Schaaf, and W. Huber, *J. Am. Chem. Soc.*, **91**, 5675–5677 (1969).
M. P. L. Caton, *Tetrahedron*, **35**, 2705–2742 (1979).

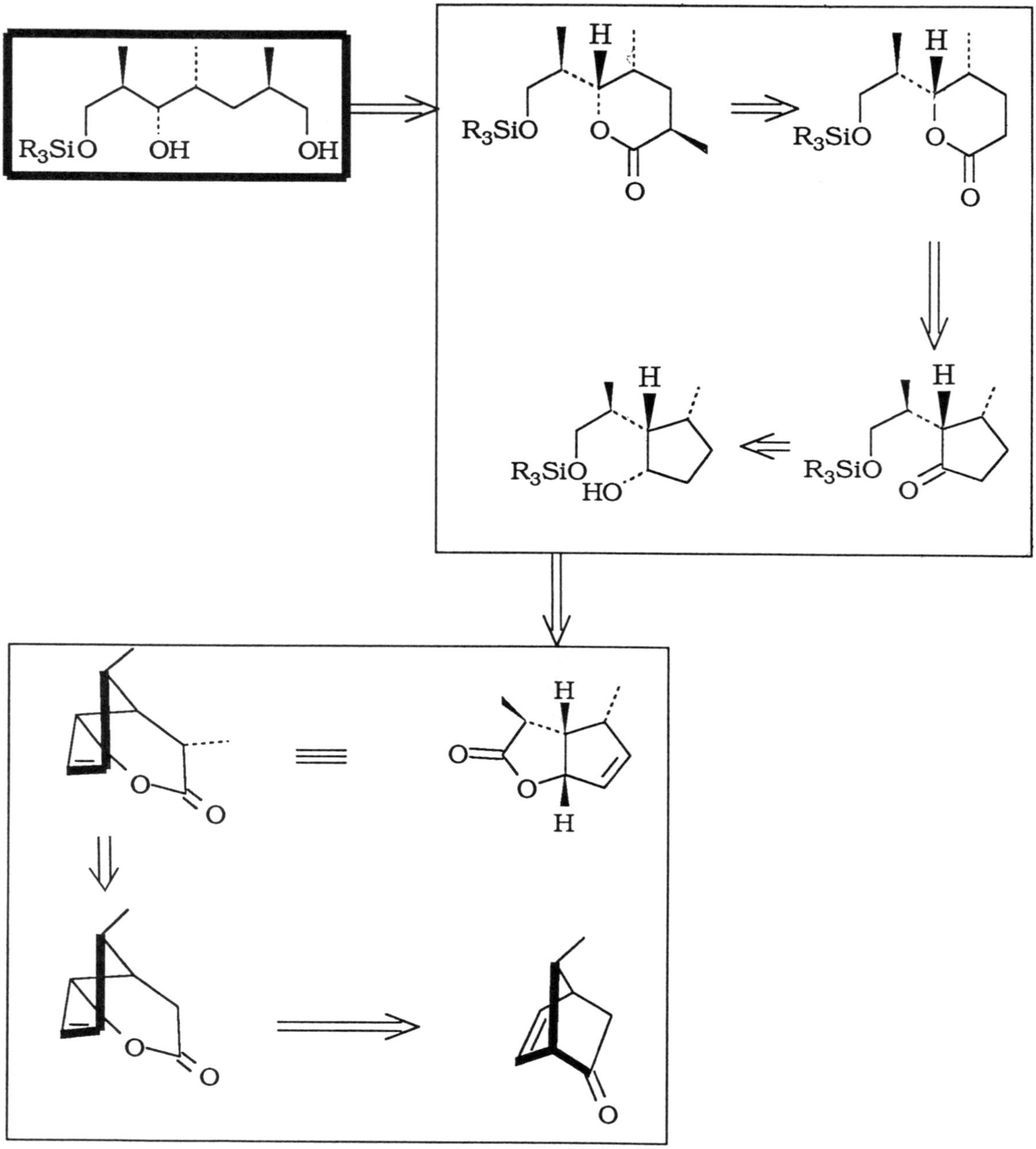
R3SiO
OH
OH
H
R3SiO
O
O
H
R3SiO
O
O
H
R3SiO
O
H
R3SiO
HO
O
O
H
O
H
O
O
O
O

The box highlighted in the adjacent scheme represents the basic aliphatic synthon for the construction of the important natural product *calcimycin*. It contains no less than four asymmetric carbons. The goal is to form selectively one among 64 possible stereoisomers. The striking advantage of the Baeyer–Villiger oxidation for this problem is that it occurs with retention of configuration.

From the target molecule, oxidation of the primary alcohol (in the retrosynthetic sense) leads to the carboxylic acid which forms the lactone upon reaction with the secondary alcohol. This lactone, in another anthetical regression, could be alkylated, in the position α to the carbonyl group and from the last hindered face. This six-membered lactone can be obtained from regioselective Baeyer-Villiger oxidation (using *m*-chloroperbenzoic acid) of the cyclopentanone. The latter in turn results from oxidation (pyridine disolvate of chromic anhydride $CrO_3.py_2$) of a cyclopentanol. This alcohol is obtained by reduction of a five-membered ring lactone, also produced by alkylation of the corresponding precursor α to the carbonyl group. This first five-membered ring lactone arises from Baeyer–Villiger oxidation of *syn*-7-methyl-5-norbornenone followed by a boron trifluoride ($BF_3. Et_2O$)-catalyzed rearrangement.

It is remarkable how the Baeyer–Villiger oxidation assists in this chemical elaboration (with so few steps), at the same time maintaining the stereochemical integrity of the previously introduced chiral centers.

Reference: G. R. Martinez, P. A. Grieco, E. Williams, K. Kanai, and C. V. Srinivasan, *J. Am. Chem. Soc.*, **104**, 1436–1438 (1982).

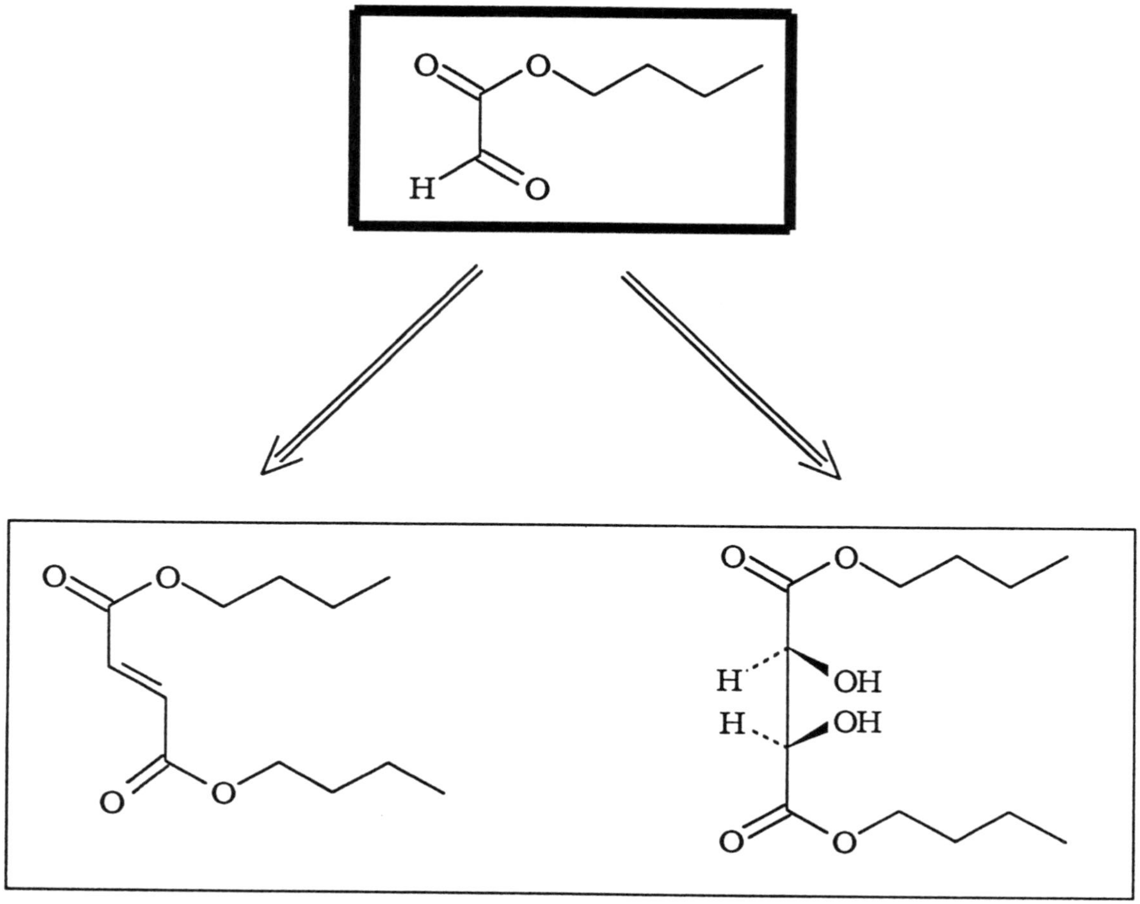

# 10  Increasing the Symmetry

To conclude this chapter, the paragraph title precept may be useful. The example given is that of *n*-butyl glyoxylate. Symmetrization, as a first step of retrosynthetic analysis suggests starting either from *n*-butyl fumarate or from *n*-butyl maleate (not represented here). The target molecule is obtained by ozonolysis. Another possibility is an oxidation with periodate which results in cleavage of the 1,2-glycol.

---

Mathematicians always rely initially on something that is assumed but unproved, such as *postulates* or *axioms*. However, these things are adequate, as they say. They appear to us as clear and evident. We readily accept as certain the deductions that follow logically. This is a science based on deductions. Mathematics, in a way, is the science of the mind. It relies on laws of the mind. It consists of logic, very simply.

In the natural sciences, we wish to put the logic that we have in our minds and we believe that such a logic is indeed present. However, the difficulty is to find a starting point for a deduction: because deduction is the only method that the mind understands well—Claude Bernard, *Notebook (1850–1860)* (M. D. Grmek ed.), Gallimard, Paris, 1965, p. 139.

# Summary: 'Strategy of Synthesis'

Systematic application of **disconnections** to the carbon—carbon bonds in a target molecule allows classification of the access pathways. This is called **retrosynthetic analysis**.

Some frequently used reactions are emphasized. Cycloaddition of a diene and a dienophile (**Diels–Alder**) leads to cyclohexenes. The **Robinson annulation** is another general ring-forming procedure. The **aldol addition** forms fragments with two oxygens, one trigonal and one tetrahedral separated by three carbon atoms.

**Squalene**, a $C_{30}$ hydrocarbon, incorporates six isoprene fragments with a $C_2$ symmetry. Biosynthesis of steroids occurs by opening of the epoxide ring of epoxysqualene. Following this model, biomimetic syntheses (W. S. Johnson) stabilize a carbocation by a favorable orientation of C=C double bonds to form the four rings A—D.

**Copaene** is another hydrocarbon subject to the isoprene rule. Its synthesis by Heathcock from the Wieland–Miescher ketone is presented; the latter in turn results from a Robinson annulation.

**Resorting to lactones**—accessible by Baeyer–Villiger oxidation of ketones—allows for the simultaneous introduction of an alcohol and a carboxylic acid function, and for control of the stereochemistry.

The advantages of epoxides are recalled: protection of a C=C double bond; inversion of polarity ($d \rightarrow a^2$); activation (driving force: release of ring strain); incorporation of a leaving group. These assets are demonstrated in a transformation of **pulegone** (another natural product embodying the isoprene rule).

**Longifolene** is the last example given of a sesquiterpene, also obeying the isoprene rule. This chapter concludes with presentation of the third synthesis (after the classics by Corey and by Oppolzer) of this hydrocarbon from the slopes of the Himalaya (*Pinus longifolia*). This is the elegant and courageous recent synthesis we owe to Alex Fallis.

# 16.II
# Retrosynthetic Analysis

## 1 Symmetrization in Retrosynthesis

It often pays to start from a precursor with higher symmetry than the target molecule; the starting material is thus more readily available or made.

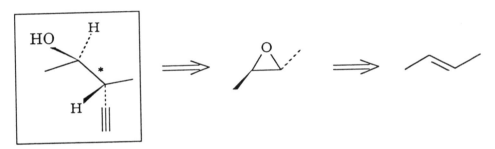

An example is provided by the acetylenic alcohol as the target molecule. Let us focus on the bond marked with an asterisk. If seen as resulting from a dissymmetric addition to a C=C double bond it is pseudo-symmetrical. Hence the idea to start from butene, *trans* 2-butene, to be more precise, since the *erythro* alcohol is required. The next conceptual step is to posit the epoxide intermediate for its inherent activation. Attack by acetylide is sufficient to form the required alcohol.

The second example comes from a Danishefsky synthesis. The first retrosynthetic step, and it is essential, is symmetrization by introduction of a second carboxylic group. This second COOH will be readily removed, in the synthetic sense, since it occupies a vinylogous position (i.e. there is an interposed C=C bond) in a β-ketoacid. This makes for easy thermal decarboxylation. The diacid is also a cyclo-hexenone. A cyclohexene implies in turn formation by a Diels–Alder reaction. To continue the retrosynthetic planning, a functional group interconversion will relocate the double bond, with an adequate oxygen substituent, to the position where it would result from such a cycloaddition. This defines the diene and dienophile types for the Diels–Alder reaction: with *normal* electron demand. Such reasoning thus concludes by use of the commercial methylmethoxyvinylketone as precursor of the diene. As the for dienophile, it originates from—via a 1,2-glycol intermediate obtained in a regioselective manner—to the cycloadduct of a first Diels–Alder addition between cyclopentadiene and methyl acetylene dicarboxylate.

Reference; S. Danishefsky, M. Hirama, K. Gombatz, T. Harayama, E. Berman and P. F. Schada, *J. Am. Chem. Soc.*, **101**, 7020 (1979).

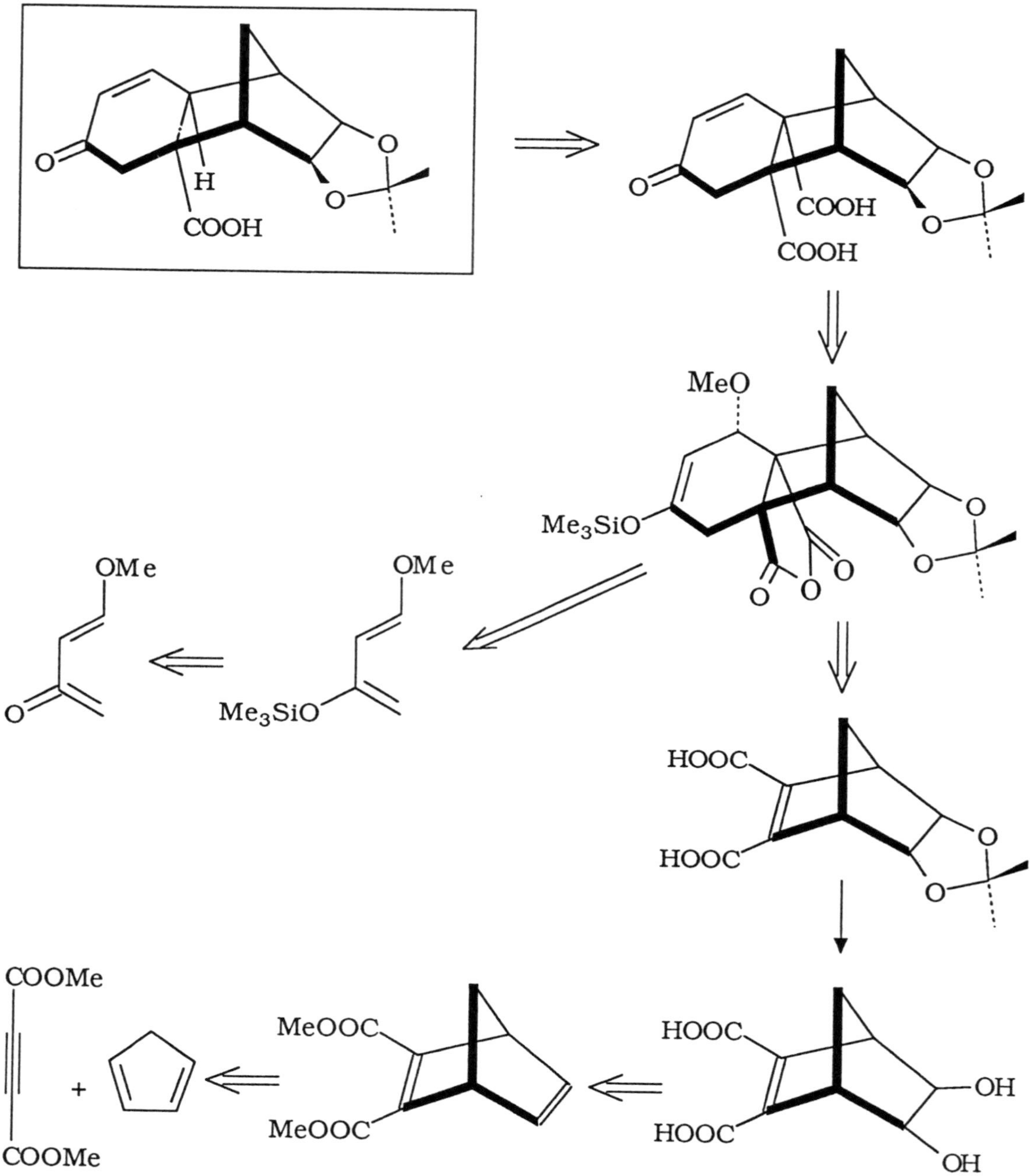
H
COOH
COOH
COOH
MeO
$Me_3SiO$
OMe
OMe
$Me_3SiO$
HOOC
HOOC
HOOC
HOOC
OH
OH
COOMe
COOMe
MeOOC
MeOOC
+

## 2 Utility of Connections in Retrosynthetic Reasoning

The target molecule shows alcohol and ester functions separated by five carbons. The structural feature immediately suggests a ring-opening reaction starting from a six-membered ring, i.e. from a benzene or cyclohexane ring. In whichever case, going through a cyclohexene is advisable since one of the best ways to cleave a carbon—carbon bond is ozonolysis of an olefin.

A last access route involves 3-methylanisole as the starting material, to submit to Birch reduction into 1,4-cyclohexadiene, as indicated. Ozonolysis is regioselective, the reactive electrophile adding to the electron-rich double bond bearing the methoxy substituent. Finally, hydrogenation of the C=C double bond and reduction of the aldehyde afford the required product.

A second route starts from 3-methylcyclohexanone: enolization and trapping of the enol as methyl ether. After ozonolysis, the aldehyde has to be reduced to the alcohol.

The third route is suggested by the structural similarity of the target molecule to the abundant and cheap naturally occurring product citronellal. The aldehyde is oxidized to the acid, and, after esterification, the double bond is cleaved with ozone.

The first two routes illustrate the importance of connections in the retrosynthetic sense: in this case the joining of carbons 1 and 6.

# 3  Stereochemical Control by an Auxiliary Group

The target molecule is a lactone, thus the equivalent to the sum of an acid plus an alcohol. The peculiarity to be noted here is the *trans* arrangement of the OH group of the alcohol and the isopropylidene chain on the adjacent carbon.

The alcohol could be obtained by reduction of the corresponding ketone; however this reduction with metal hydrides would occur from the least hindered face, affording the *cis* diastereoisomer.

This unfavorable event can be avoided by introduction of an auxiliary group, which will force reduction of the ketone from the rear, i.e. from the most hindered face. Hence the idea of using a tertiary alcohol with the dual function of an auxiliary for the stereoselective reduction, and of an incipient leaving group (via its mandatory improvement as a leaving group) toward formation of the ethylenic double bond in the side chain.

This tertiary alcohol—an obvious disconnection—results from an aldol condensation between acetone and the enolate, that can but form in a regioselective way.

The carboxylic function has to be protected (ArCOOH → ArCOOMe) during the key step of the ketone reduction.

The corresponding acid is accessible from oxidation at the benzylic position affording a derivative of a carboxylic acid such as the amide (X—NH$_2$); a regioselective reaction due to the spatial proximity of this group.

One could start from anthracene-1-carboxylic acid, followed by di-hydrogenation and functional group interconversion (—COOH → —CONH$_2$).

Reference: F. Bohlmann and E. Eickeler, *Chem. Ber.*, **112**, 2811 (1979).

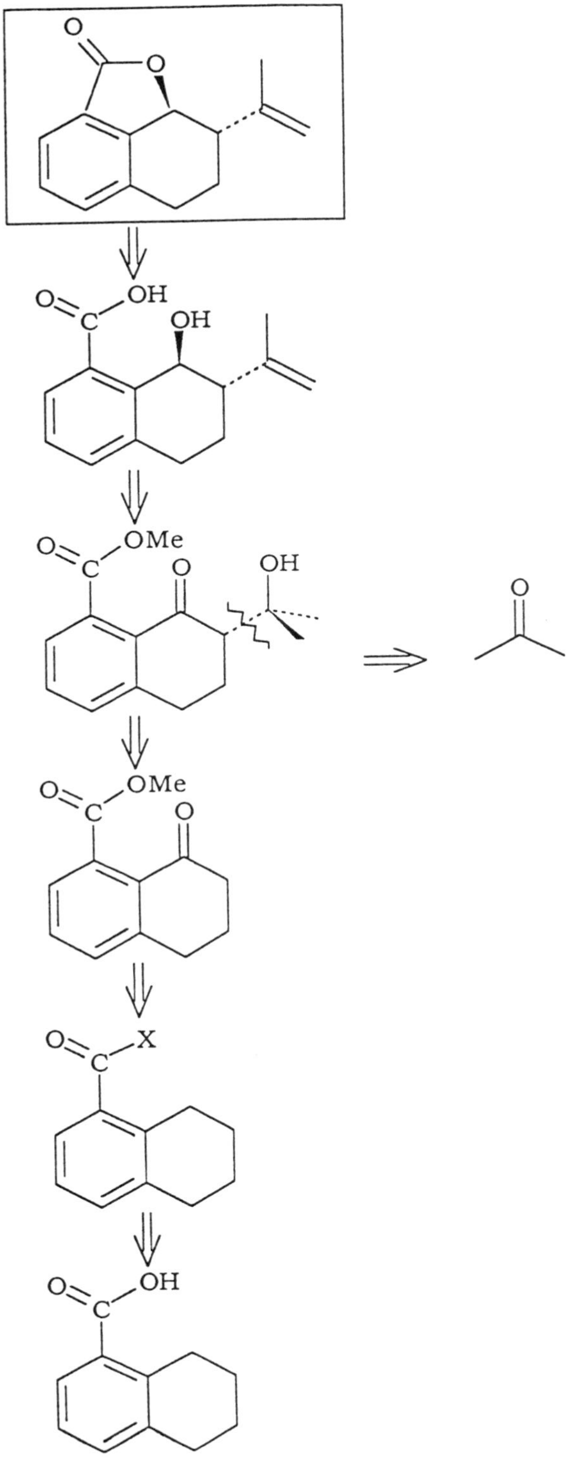
O
O
OH
OH
OMe
O
OH
O
OMe
O
X
OH

# 4  Oxidizing Retrosynthesis

As unsaturated hydrocarbons, these target molecules, the sesquiterpenes bazzanene and trichodiene, of plant origin, offer little hold to retrosynthetic reasoning. Heeding the French revolutionary Danton's precept: 'let's be bold, and bold again, and always bold', it might be worthwhile to replace one of the methyl groups on the quaternary carbon by an aldehyde function —CHO.

Which? That giving rise to two unsaturated groups C=C and C=O, separated by two tetrahedral carbons: indeed, such a structural module would result from a sigmatropic [3,3] rearrangement; hence the notion of a Claisen rearrangement.

This defines the starting allyl vinyl ether as the substrate for the Claisen rearrangement. This ether could arise from the disconnection shown, i.e. by a Wittig reaction. The Wittig partners are formed, on the one hand, by a Diels–Alder cycloaddition (the phosphorus ylide is made from the corresponding bromide,) and, on the other hand, by reduction ($LiAlH_4$) of 2-methyl-cyclopentene-1-carboxylic acid to the primary alcohol, the latter then esterified with formic acid.

Finally, the aldehyde RCHO has to be reduced to the methyl derivative $RCH_3$. One way to effect this reduction is to convert the aldehyde to the corresponding hydrazone (hydrazine $H_2N-NH_2$, as the reagent) and subsequent treatment with potassium $t$-butoxide.

Reference: M. Suda, *Tetrahedron Lett*, **23**, 427–428 (1982).

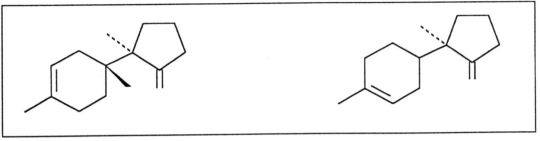

# 5  Protecting Our Forests. Synthesis of Racemic Frontalin

Several insect pheromones share a 6,8-dioxabicyclo[3.2.1]octaine skeleton: at the top of the facing Scheme are depicted, besides the aggregation pheromone for the parasite of pine trees in the southern United States *Dendroctonus frontalis*—our target molecule—the pheromones responsible for sexual attraction in *Dendroctonus brevicomis*, his cousin in the western USA, and in *Scolytus multistriatus*, a bark parasite of the European elm tree.

The idea is to use an epoxide to make easier formation of a strained bicyclic system. The disconnection indicated suggests the acid-catalyzed ring opening of an epoxide as a key step, with participation of the neighboring keto group. This 'bicyclization' is catalyzed with 0.1 MHClO$_4$.

The epoxyketone can be prepared from the corresponding olefin (with *m*-chloroperbenzoic acid). In practice, it is not necessary to protect the carbonyl group against side reactions from this reagent, such as Bayer–Villiger oxidation.

The unsaturated ketone is available from condensation of ethyl acetylacetate and the tosylate, derived from the corresponding commercial alcohol, as shown.

The β-ketoester is decarboxylated by the hydrolysis–thermolysis standard sequence.

*Note*: A recent synthesis of both enantiomers of frontalin was discussed in an earlier chapter (Monneret, XII).

References: K. Mori, S. Kobayshi, and M. Matuni, *Agric. Biol. Chem.*, **39**, 1889 (1975).
         T. E. Bellas, R. G. Brownlee, and R. M. Silverstein, *Tetrahedron*, **25**, 5149 (1969).
         P. A. Bartlett and J. Myerson, *J. Org. Chem.*, **44**, 1625 (1979).
         P. A. Bartlett *et al.*, *J. Chem. Ed.*, **61**, 816–817 (1984).

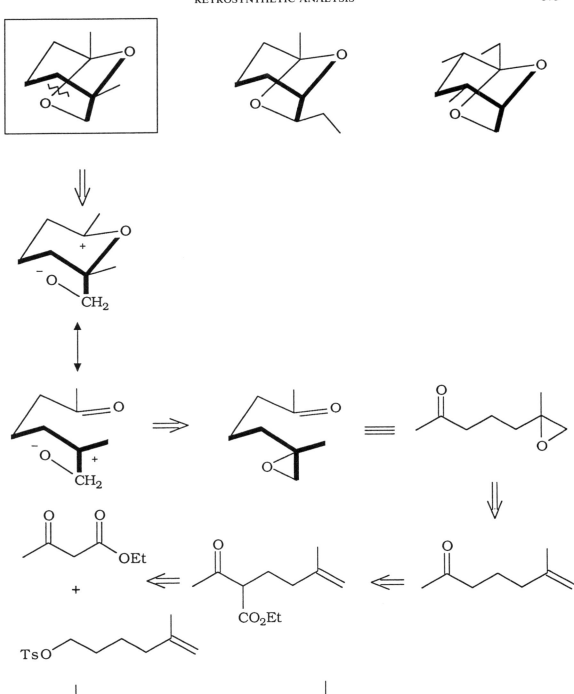

# A View to the Future . . . Fullerenes

A whole new chapter in organic chemistry has opened with the discovery and structure determination of $C_{60}$, the prototype for a new class of molecules known as *fullerenes*. That these, furthermore, constitute a third allotropic form of the element carbon, besides diamond and graphite, makes the serendipitous discovery even more remarkable. Even though fullerenes have not yet entered the realm of organic synthesis except on tiptoe, so to say, it is easy to foresee that they will: after only a few years of existence, they have won for themselves a prime niche in the thinking and in the imagination of chemists; and their future is certain to be very bright. Our concern here is only to document briefly some of their known properties, focusing near-exclusively on the first member of this family, also the most famous, $C_{60}$, aka 'buckyball' or 'footballene'.

## Prehistory

One of the biggest surprises to anyone, whether casual observer or professional historian, monitoring the advancement of learning is how major discoveries have been heralded by earlier work. The finding of $C_{60}$ by Kroto and Smalley makes no exception. Davidson in 1981 calculated the Hückel MOs for this, at the time hypothetical, molecule. Back in 1966, a witty piece in *New Scientist* by David Jones, writing under the pseudonym 'Daedalus', had proposed such a form of carbon. In 1970, Eiji Osawa had already published a similar conjecture.

In 1972, Bochvar and Gal'pern presented the first Hückel calculations. As for preparation of $C_{60}$, it was achieved by several groups in the 1980s. Nevertheless, credit for the discovery of $C_{60}$ with its football (soccerball) structure must go to Smalley, Kroto and their co-workers at Rice University in 1985.

References: R. A. Davidson, *Theor. Chim. Acta*, **58**, 193 (1981).
D. E. H. Jones, *New Scientist*, 3rd November 1966, 245.
E. Osawa, *Kagaku* (Kyoto), **25**, 854 (1970).
D. A. Bochvar and E. G. Gal'pern, *Dokl. Akad. Nauk SSSR*, **209**, 610 (1973).
E. A. Rohlfing, D. M. Cox, and A. Kaldor, *J. Chem. Phys.*, **81**, 3322 (1984).
Bloomfield *et al.*, *Chem. Phys. Lett.*, **121**, 33 (1985).
H. W. Kroto, J. R. Heath, S. C. O'Brian, R. F. Curl, and R. E. Smalley, *Nature*, **318**, 162 (1985).

# SELF-ASSEMBLING

Smalley's great merit is to have postulated the structure of a football, or of a geodesic dome (hence the name 'buckminsterfullerene') for particularly stable clusters of 60 carbon atoms. The history of the discovery at Rice University is by now well documented. See for instance:

The infinitely rich domain of this new allotrope of carbon opened up with its large-scale preparation by Krätschmer in 1990.

The mechanism of fullerene formation has yet to be fully elucidated. One school of thought hypothesizes that self-assembly occurs by accretion of small carbon fragments to a growing network with a pseudo-graphite structure, made of hexagonal and pentagonal rings, with the largest possible number of the latter while avoiding mutual adjacency of pentagons. If this were the case, such a structure would have to snap close very fast, since a genuine benzenoid graphite layer enjoys greater thermodynamic stability.

Another school of thought stresses rearrangement into fullerenes of coalesced polyyne rings.

References: H. W. Kroto, *Angew. Chem. Int. Ed. Engl.*, **31**, 111 (1992).
W. Krätschmer, K. Fostiropoulos, and D. R. Huffman, *Chem. Phys. Lett.*, **170**, 167 (1990); see also the review by W. Krätschmer, *Il Nuovo Cimento*, **107A**, 1077 (1994).
R. E. Smalley, *Accounts Chem. Res.*, **25**, 98 (1992).
S. W. McElvany, F. N. Diederich, *et al.*, *Science*, **259**, 1594 (1993).

# HIGHER HOMOLOGS

Numerous other parent carbon *n*-mers have been prepared, isolated and characterized. A first class includes molecules similar to $C_{60}$, with a globular shape such that a single shell of carbon atoms makes up its surface, usually ellipsoidal. Oligomers such as $C_{70}$, $C_{76}$ or $C_{78}$ belong to this class. In the latter case, two isomers are known, one with $C_{2v}$, the other with $D_3$ symmetry.

A second class is that of the onion-like giant fullerenes, whose formulas include thousands of carbon atoms. They are made of concentric shells of more or less closed and more or less spheroid graphite-like networks. Denoting by the @ sign inclusion within a fullerene, one such megafullerene, with a quasi-icosahedral shape, conforms to the formula $C_{960}@C_{1500}@C_{2160}@C_{2940}$, Russian-doll like.

There are other classes of fullerenes yet. Some are microtubules or nanotubes, others are fibers. They can be grafted onto polymers: the only limit is the imagination, the potential morphologic diversity is truly immense.

References: F. Diederich and R. L. Whetter, *Accounts Chem. Res.*, **25**, 119 (1992).
H. W. Kroto, J. P. Hare, A. Sarkar, K. Hsu, M. Terrones, and J. R. Abeysinghe, *MRS Bulletin*, **19**, 51 (1994).
D. T. Colbert, J. Zhang, S. M. McClure, P. Nikolaev, Z. Chen, J. H. Hafner, D. W. Owens, P. G. Kotula, C. B. Carter, J. H. Weaver, A. G. Rinzler, and R. E. Smalley, *Science*, **266**, 1218 (1994).

# SYMMETRY AND STRUCTURE

Smalley's hypothesis (1985) of a truncated icosahedron (the football), i.e. $I_h$ point group symmetry, implies strict equivalence of the 60 carbon atoms. It is proven by the $^{13}C$ NMR spectrum of $C_{60}$ with only a single resonance.

As for full structural elucidation, using X-ray diffraction, it was performed after the buckyball had been immobilized on its lattice sites with a neat little trick, attachment of the $OsO_4$ (4-*t*-butylpyridine)$_2$ complex to one double bond per molecule. Indeed, $C_{60}$ shows a truncated icosahedral structure in which every carbon atom is equivalent, sharing a double bond and two single bonds with its nearest neighbours. A truncated icosahedron has (only) two types of edges indeed. The 30 double bonds, 1.40 Å, are shared by adjacent hexagons. Each pentagon has single bond-sides, 1.45 Å long. This determination by X-ray on a derivative of $C_{60}$ was later confirmed on the parent molecule by NMR.

References: R. D. Johnson, G. Meijer, and D. S. Bethune, *J. Am. Chem. Soc.*, **112**, 8983–8984 (1990); R. Taylor, J. P. Hare, A. K. Abdul-Sada, and H. W. Kroto, *J. Chem. Soc. Chem. Comm.*, **1990**, 1423.
J. M. Hawkins, A. Meyer, T. A. Lewis, S. Loren, and F. J. Hollander, *Science*, **252**, 312–313 (1991).
C. S. Yannoni, P. P. Bernier, D. S. Bethune, G. Meijer, and J. R. Salem, *J. Am. Chem. Soc.*, **113**, 3190–3192 (1991).

# ORGANOMETALLIC COMPLEXES

Besides oxygen (by epoxidation), numerous metals can be inserted into the short $C=C$ bond of $C_{60}$.

References: F. Wudl, *Accounts Chem. Res.*, **25**, 157 (1992).
P. J. Fagan, J. C. Calabrese, and B. Malone, *Accounts Chem. Res.*, **25**, 134 (1992).

$X = O,\ CAr^1Ar^2, ML_2$

$(M = Pd,\ Pt,$

# STRUCTURE–REACTIVITY RELATIONSHIP

$C_{60}$ is an easy molecule to characterize with a succinct statement: due to strain, it is an activated (and quite reactive) aromatic system. As pointed out originally by Haddon, the aromaticity stems from the graphite-like shell made of trigonal carbons, with $\pi$ electronic delocalization encompassing the whole three-dimensional structure. Very many Kekulé structures can be written for the $C_{60}$ molecule! However, each of the 60 trigonal carbons is non-planar, due to the requirement of closing the fullerene cage on itself—to the tune of *ca* 6 kcal.mol$^{-1}$ of strain per carbon.

    The aromatic character of fullerenes was gauged by NMR. The theoretical prediction is that pentagons are associated with diamagnetic and hexagons with paramagnetic ring currents. A superb NMR experiment was performed, measurement of the $^3$He chemical shift for both $^3$He@C$_{60}$ and $^3$He@C$_{70}$. There is a downfield shift for the encapsulated $^3$He nucleus of 6.3 and 29 ppm relative to the free $^3$He. These inclusion complexes were made on samples produced if the carbon-arc discharge, used for making fullerenes, is operated in a helium atmosphere.

References: R. C. Haddon, *Science*, **261**, 1545–1550 (1993).
        M. Saunders, H. A. Jiménez-Vásquez, R. J. Cross, S. Mcrozkowski, D. I. Freedberg, and F. A. L. Anet, *Nature*, **367**, 256 (1994).

# INCLUSION COMPOUNDS

Footballene $C_{60}$ is a hollow sphere, with a 7.1 Å cage diameter. More generally, fullerenes $C_n$ are ellipsoids capable of encapsulating other atoms or ions in their inner cavity. As we have just seen, one preparative route consists in preparing the fullerenes in the presence of the atoms to be enclosed. There will be, statistically, a proportion of the latter ending up inside the forming fullerene cage. Another strategy, with the same end result, is to shoot guest atoms at the fullerene, hoping that they will thus get in. Both strategies work. In the above case, when monoatomic inert gases are inserted, pressurizing a fullerene sample up to 2500 atm with the inert gas for a few hours at temperature up to 610 °C makes for incorporation of helium, neon, argon, krypton, and even xenon, as shown by Saunders *et al.*, in the above reference.

Among other $M@C_n$ inclusion complexes already characterized are $La@C_{60}$; $La_2@C_{80}$; $K@C_{44}$; $Cs@C_{48}$; $Zr@C_{28}$; $Hf@C_{28}$; $Ti@C_{28}$; $La@C_{82}$; $La_2@C_{80}$; $Y@C_{82}$; $Y_2@C_{82}$; and $Fe@C_{60}$.

Let us give an idea of the fascinating structural detail with which these endohedral fullerenes (as they are named) start being known. The three scandium atoms in $Sc_3@C_{82}$ form an equilateral triangle—as had been previously suggested for $Sc_3$ molecules isolated in a cryogenic rare-gas matrix:

Let us consider a simpler, textbook case: assume that we include a neutral sodium atom inside a $C_{60}$ cage. The sodium atom, it can be expected, will donate its valence electron to the fullerene cage (into a $\pi^*$ level). Thus a complex best described as $Na^+@C_{60}^-$ should (and does) form. The sodium cation will reside in an equilibrium position away from the center of the cage, closer to the walls. The ensuing electrostatic dipole moment interacts favorably with the polarizable electronic distribution associated with the $C_{60}$ cage. The vibrational motion of the encaged ion has a characteristic frequency around $100 \, cm^{-1}$. The rotation–vibration spectrum of such an endohedral fullerene as $Na@C_{60}$ or $Fe@C_{60}$ thus displays a wealth of IR-active absorptions in the $1–500 \, cm^{-1}$ part of the spectrum: it surely is 'the world's smallest rattle', in a felicitous phrase by one of the authors.

This raises once again the question of the status of science fiction, or at least that of the problem and the solution, or equivalently that of the proposal and of its implementation: is it more important to conceptualize or to realize? There was a visionary scientist, Serratosa, who had the first intuition of these inclusion complexes of 'footballene', about ten years in anticipation of their actual preparation.

References: Y. Chai, T. Guo, C. Jin, R. E. Haufler, L. P. F. Chibante, J. Fure, L. Wang, M. J. Alford, and R. E. Smalley, *J. Phys. Chem.*, **95**, 7564 (1991).
R. M. Baum, *Chem. Eng. News*, June 1, 1992, 25.
C. S. Yannoni, M. Hoinkis, M. S. de Vries, D. S. Bethune, J. R. Salem, M. S. Crowder, and R. D. Johnson, *Science*, **256**, 1191 (1992).
I. Holleman, M. G. H. Boogaarts, and G. Meijer, *Rec. Trav. Chim. Pays-Bas*, **113**, 543 (1994).
J. Castells and F. Serratosa, *J. Chem. Ed.*, **60**, 941 (1983).

# CONDUCTIVITY AND SUPERCONDUCTIVITY

Count on the chemists for a jump start! Had superconductivity become a closed chapter of physics? Superconducting ceramics (Raveau of Caen; taken-up with Nobel Prize-winning success by Bednorz and Müller) have brought renewed vitality to this area. And here we find doped $C_{60}$ (and other analogs) displaying remarkable electrical conductivities.

Let us consider again an endohedral fullerene such as $Na@C_{60}$. Combination of the low ionization potential of alkali metals and of the high electron affinity of the $C_{60}$ cage (2.65 eV) leads one to expect full electron transfer, as expressed in the $Na^+@C_{60}^-$ formula. Furthermore, the face-centered cubic crystals of $C_{60}$ leave ample (26%) volume for incorporation of other atoms next to (outside) the $C_{60}$ spheres. In the fcc structure, the most spacious sites are the octahedral sites (2.06 Å radius), with six nearest-neighbor $C_{60}$ molecules, and the tetrahedral sites (1.12 Å radius), with four nearest-neighbor $C_{60}$ molecules. Thus, it is possible to fill three such sites, one octahedral and two tetrahedral, without changing the original fcc structure. Indeed, some time after the first large-scale production of $C_{60}$, scientists of AT&T Bell, in Murray Hill, New Jersey showed that fullerenes of the formulas $K_3C_{60}$ and $Rb_3C_{60}$ become superconductive at 18 K and 28 K, respectively.

Other scientists of the NEC company at Tsukuba, after having formed a compound $Cs_2RbC_{60}$ demonstrated that it had a critical temperature of 33 K.

At Harvard, Lieber tested the BSC (Bardeen–Cooper–Schrieffer) theory by measuring the depression of the critical temperature $T_c$ (the sample shows superconductivity below this temperature) between $K_3\,^{12}C_{60}$ and $K_3\,^{13}C_{60}$. This isotope effect $\Delta T_c = 0.45 \pm 0.1\,K$, inconsistent with the BSC model, suggests an important role for phonons.

The current understanding is a conventional electron-phonon mechanism for superconductivity in these fullerene compounds.

The pressure dependence of $T_c$ for $K_3C_{60}$ ($\Delta T_c = 11\,K$ between 1 bar and 21 kbar) suggests also an electron-phonon coupling for electron pairing as in the BSC model.

Fullerenes show many other strange, wonderful and useful properties: $C_{60}$ doped with tetrakis (dimethylaminoethylene) becomes ferromagnetic at 16.1 K.

Films of the polyvinylcarbazole polymer, doped with a mixture of $C_{60}$ and $C_{70}$, become photoconductors of high quality.

References: R. C. Haddon, A. F. Hebard, M. J. Rosseinsky, D. W. Murphy, S. J. Duclos, *et al.*, *Nature*, **350**, 320 (1991).
A. F. Hebard, *Ann. Rev. Mater. Sci.*, **23**, 159 (1993).
K. Tanigawa, T. W. Ebbesen, S. Saito, J. Mizuki, J. S. Tsai, *et al.*, *Nature*, **352**, 222 (1991).
C. M. Lieber and C. C. Chen, *J. Am. Chem. Soc.*, **114**, 3141 (1992).
S.-M. Huang, R. R. Kanier, F. N. Diederich, R. L. Whetten, G. Grüner, K. Holczer, G. Sparn, and J. D. Thompson, *Science*, **152**, 1829 (1991).
F. Wudl, P.-M. Allemand, *et al.*, *Science*, **253**, 301 (1991).
Y. Wang, *Nature*, **356**, 585 (1992).

# FROM TERRESTRIAL ROCKS TO INTERSTELLAR SPACE

The $C_{60}$ and $C_{70}$ fullerenes were found in trace amounts in rocks dating from the Precambrian, i.e. 600 million to a billion years old. The exact origin of this shungite is not known. It was named for the Russian city Shunga, 400 km to the NE of Saint Petersburg. Is this shungite of eruptive, metamorphic, or biogenic origin? At least two aspects are puzzling: the fullerenes are obviously separated from other carbon-containing parts of the rock; they seem to have been formed in the condensed phase.

Shungite is thought to have formed from carbonaceous material creeping into fissures of a Precambrian rock that metamorphized under extreme pressures. The same $C_{60}$ and $C_{70}$ fullerenes have also been found in fulgurite, a mineral formed when lightning strikes rocks or soils. The $C_{60}$ footballene molecule is present, at concentrations of 0.1 to 0.2 ppm of the associated soot in the cretaceous–tertiary boundary layer, from two sites in New Zealand. It has been surmised that it is due to the extensive wildfires associated with the cataclysmic impact event that terminated the Mesozoic era about 65 million years ago, when (as is conjectured) a meteorite impacted the Yucatán Peninsula near Chicxulub, an event responsible in the attractive Alvarez hypothesis for the demise of dinosaurs (among other large animals of the secondary era). $C_{60}$ is also a component of soot. Thus, it was already a molecule in the environment. It has also been searched for, and perhaps even been found in outer space: the 10 m-long Long Duration Exposure Facility spacecraft orbited Earth from 1984 to 1990. When the space shuttle *Columbia* retrieved it from space and brought it back, investigators found fullerenes, $C_{60}$ and $C_{70}$ in particular, in a tiny crater less than 0.2 mm in diameter, on the aluminum skin of the spacecraft.

It is conjectured that this crater was caused by the impact of a meteorite that hit the spacecraft with a velocity between 6 km s$^{-1}$ and 13 km s$^{-1}$, and that the fullerenes were present in this meteorite.

Although the dust clouds omnipresent in the cosmos contain carbon-containing molecules, and no doubt also carbon in several forms, fullerenes have not yet been identified for sure.

References: P. R. Buseck, S. J. Tsipursky, and R. Hettich, *Science*, **257**, 215 (1992).
T. K. Daly, P. R. Buseck, P. Williams, and C. F. Lewis, *Science*, **259**, 1599 (1993).
D. Heymann, L. P. Felipe Chibante, R. R. Brooks, W. S. Wolbach, and R. E. Smalley, *Science*, **265**, 645 (1994).
F. Radicati di Brozolo, R. H. Fleming, T. E. Bunch, and J. Macklin, *Nature*, **369**, 37 (1994).
J. P. Hare and H. W. Kroto, *Accounts Chem. Res.*, **25**, 106 (1992).

# AND BEYOND THAT?

I am concluding this text with $C_{60}$: this novel dynamic and fertile chapter of organic chemistry—more than 700 publications had already appeared before 1992, they number now several thousands—in spite of its apparent heterogeneity and fragmentation, due to the myriad of fascinating properties of the fullerenes, bears witness to the superb vitality of the discipline and to the creativity of its practitioners.

I can find no more eloquent witness to those than the prophetic and oniric words penned, at the beginning of the century, by Georges Darzens, a graduate and later a professor (1913–1937) in the institution where I teach, the Ecole polytechnique:

> 'The chemist appears to me like a traveller climbing a boundless mountain, clouds mark his perspective, he fancies that he sees his goal afar in these trees, in these grandiose landscapes, beyond which nothing yet takes shape. However, as soon as he reaches these, as soon as he overcomes the fog, beyond this first horizon others appear, also wrapped in the same deceptive mist. Then, he is seized with the crazy desire to go yet further, and to get where he will be alone to contemplate the wonders that he has left behind. Indeed his solitary and fearless climb provides him with so many wonders to look at that he becomes captivated, enthralled. He starts dreaming of what might exist further on and his intuition does not lead him astray.

G. Darzens, *Initiation chimique*, Hachette, Paris, 1912, pp. 128–129.

# Index

Note: Figures and Tables are indicated by *italic page numbers*, Footnotes by suffix 'n'

*Index compiled by P. Nash*